T0336872

Fundamentals of Wireless Communication

The past decade has seen many advances in physical-layer wireless communication theory and their implementation in wireless systems. This textbook takes a unified view of the fundamentals of wireless communication and explains the web of concepts underpinning these advances at a level accessible to an audience with a basic background in probability and digital communication. Topics covered include MIMO (multiple input multiple output) communication, space-time coding, opportunistic communication, OFDM and CDMA. The concepts are illustrated using many examples from wireless systems such as GSM, IS-95 (CDMA), IS-856 ($1\times$ EV-DO), Flash OFDM and ArrayComm SDMA systems. Particular emphasis is placed on the interplay between concepts and their implementation in systems. An abundant supply of exercises and figures reinforce the material in the text. This book is intended for use on graduate courses in electrical and computer engineering and will also be of great interest to practicing engineers.

David Tse is a professor at the Department of Electrical Engineering and Computer Sciences, University of California at Berkeley.

Pramod Viswanath is an assistant professor at the Department of Electrical and Computer Engineering, University of Illinois at Urbana-Champaign.

Fundamentals of Wireless Communication

David Tse
University of California, Berkeley

and

Pramod Viswanath
University of Illinois, Urbana-Champaign

CAMBRIDGE
UNIVERSITY PRESS

CAMBRIDGE
UNIVERSITY PRESS

University Printing House, Cambridge CB2 8BS, United Kingdom

One Liberty Plaza, 20th Floor, New York, NY 10006, USA

477 Williamstown Road, Port Melbourne, VIC 3207, Australia

4843/24, 2nd Floor, Ansari Road, Daryaganj, Delhi - 110002, India

79 Anson Road, #06-04/06, Singapore 079906

Cambridge University Press is part of the University of Cambridge.

It furthers the University's mission by disseminating knowledge in the pursuit of education, learning and research at the highest international levels of excellence.

www.cambridge.org
Information on this title: www.cambridge.org/9780521845274

© Cambridge University Press 2005

First published 2005
11th printing 2016

A catalogue record for this publication is available from the British Library

ISBN 978-0-521-84527-4 Hardback

To my family and Lavínia
DT

To my parents and to Suma
PV

Contents

Preface

Why we wrote this book

The writing of this book was prompted by two main developments in wireless communication in the past decade. First is the huge surge of research activities in physical-layer wireless communication theory. While this has been a subject of study since the sixties, recent developments such as opportunistic and multiple input multiple output (MIMO) communication techniques have brought completely new perspectives on how to communicate over wireless channels. Second is the rapid evolution of wireless systems, particularly cellular networks, which embody communication concepts of increasing sophistication. This evolution started with second-generation digital standards, particularly the IS-95 Code Division Multiple Access standard, continuing to more recent third-generation systems focusing on data applications. This book aims to present modern wireless communication concepts in a coherent and unified manner and to illustrate the concepts in the broader context of the wireless systems on which they have been applied.

Structure of the book

This book is a web of interlocking concepts. The concepts can be structured roughly into three levels:

1. channel characteristics and modeling;
2. communication concepts and techniques;
3. application of these concepts in a system context.

A wireless communication engineer should have an understanding of the concepts at all three levels as well as the tight interplay between the levels. We emphasize this interplay in the book by interlacing the chapters across these levels rather than presenting the topics sequentially from one level to the next.

- Chapter 2: basic properties of multipath wireless channels and their modeling (level 1).
- Chapter 3: point-to-point communication techniques that increase reliability by exploiting time, frequency and spatial diversity (2).
- Chapter 4: cellular system design via a case study of three systems, focusing on multiple access and interference management issues (3).
- Chapter 5: point-to-point communication revisited from a more fundamental capacity point of view, culminating in the modern concept of opportunistic communication (2).
- Chapter 6: multiuser capacity and opportunistic communication, and its application in a third-generation wireless data system (3).
- Chapter 7: MIMO channel modeling (1).
- Chapter 8: MIMO capacity and architectures (2).
- Chapter 9: diversity–multiplexing tradeoff and space-time code design (2).
- Chapter 10: MIMO in multiuser channels and cellular systems (3).

How to use this book

This book is written as a textbook for a first-year graduate course in wireless communication. The expected background is solid undergraduate/beginning graduate courses in signals and systems, probability and digital communication. This background is supplemented by the two appendices in the book. Appendix A summarizes some basic facts in vector detection and estimation in Gaussian noise which are used repeatedly throughout the book. Appendix B covers the underlying information theory behind the channel capacity results used in this book. Even though information theory has played a significant role in many of the recent developments in wireless communication, in the main text we only introduce capacity results in a heuristic manner and use them mainly to motivate communication concepts and techniques. No background in information theory is assumed. The appendix is intended for the reader who wants to have a more in-depth and unified understanding of the capacity results.

At Berkeley and Urbana-Champaign, we have used earlier versions of this book to teach one-semester (15 weeks) wireless communication courses. We have been able to cover most of the materials in Chapters 1 through 8 and parts of 9 and 10. Depending on the background of the students and the time available, one can envision several other ways to structure a course around this book. Examples:

- A senior level advanced undergraduate course in wireless communication: Chapters 2, 3, 4.
- An advanced graduate course for students with background in wireless channels and systems: Chapters 3, 5, 6, 7, 8, 9, 10.

- A short (quarter) course focusing on MIMO and space-time coding: Chapters 3, 5, 7, 8, 9.

The more than 230 exercises form an integral part of the book. Working on at least some of them is essential in understanding the material. Most of them elaborate on concepts discussed in the main text. The exercises range from relatively straightforward derivations of results in the main text, to "back-of-envelope" calculations for actual wireless systems, to "get-your-hands-dirty" MATLAB types, and to reading exercises that point to current research literature. The small bibliographical notes at the end of each chapter provide pointers to literature that is very closely related to the material discussed in the book; we do not aim to exhaust the immense research literature related to the material covered here.

Acknowledgements

We would like first to thank the students in our research groups for the selfless help they provided. In particular, many thanks to: Sanket Dusad, Raúl Etkin and Lenny Grokop, who between them painstakingly produced most of the figures in the book; Aleksandar Jovičić, who drew quite a few figures and proofread some chapters; Ada Poon whose research shaped significantly the material in Chapter 7 and who drew several figures in that chapter as well as in Chapter 2; Saurabha Tavildar and Lizhong Zheng whose research led to Chapter 9; Tie Liu and Vinod Prabhakaran for their help in clarifying and improving the presentation of Costa precoding in Chapter 10.

Several researchers read drafts of the book carefully and provided us with very useful comments on various chapters of the book: thanks to Stark Draper, Atilla Eryilmaz, Irem Koprulu, Dana Porrat and Pascal Vontobel. This book has also benefited immensely from critical comments from students who have taken our wireless communication courses at Berkeley and Urbana-Champaign. In particular, sincere thanks to Amir Salman Avestimehr, Alex Dimakis, Krishnan Eswaran, Jana van Greunen, Nils Hoven, Shridhar Mubaraq Mishra, Jonathan Tsao, Aaron Wagner, Hua Wang, Xinzhou Wu and Xue Yang.

Earlier drafts of this book have been used in teaching courses at several universities: Cornell, ETHZ, MIT, Northwestern and University of Colorado at Boulder. We would like to thank the instructors for their feedback: Helmut Bölcskei, Anna Scaglione, Mahesh Varanasi, Gregory Wornell and Lizhong Zheng. We would like to thank Ateet Kapur, Christian Peel and Ulrich Schuster from Helmut's group for their very useful feedback. Thanks are also due to Mitchell Trott for explaining to us how the ArrayComm systems work.

This book contains the results of many researchers, but it owes an intellectual debt to two individuals in particular. Bob Gallager's research and teaching style have greatly inspired our writing of this book. He has taught us that good theory, by providing a unified and conceptually simple understanding of a morass of results, should *shrink* rather than *grow* the knowledge tree. This book is an attempt to implement this dictum. Our many discussions with

Rajiv Laroia have significantly influenced our view of the system aspects of wireless communication. Several of his ideas have found their way into the "system view" discussions in the book.

Finally we would like to thank the National Science Foundation, whose continual support of our research led to this book.

Notation

Some specific sets

$\mathcal{R}$ Real numbers

$\mathcal{C}$ Complex numbers

$\mathcal{S}$ A subset of the users in the uplink of a cell

Scalars

m	Non-negative integer representing discrete-time
L	Number of diversity branches
ℓ	Scalar, indexing the diversity branches
K	Number of users
N	Block length
N_c	Number of tones in an OFDM system
T_c	Coherence time
T_d	Delay spread
W	Bandwidth
n_t	Number of transmit antennas
n_r	Number of receive antennas
$n_{\min}$	Minimum of number of transmit and receive antennas
$h[m]$	Scalar channel, complex valued, at time m
h^*	Complex conjugate of the complex valued scalar h
$x[m]$	Channel input, complex valued, at time m
$y[m]$	Channel output, complex valued, at time m
$\mathcal{N}(\mu, \sigma^2)$	Real Gaussian random variable with mean μ and variance σ^2
$\mathcal{CN}(0, \sigma^2)$	Circularly symmetric complex Gaussian random variable: the real and imaginary parts are i.i.d. $\mathcal{N}(0, \sigma^2/2)$
N_0	Power spectral density of white Gaussian noise
$\{w[m]\}$	Additive Gaussian noise process, i.i.d. $\mathcal{CN}(0, N_0)$ with time m
$z[m]$	Additive colored Gaussian noise, at time m
P	Average power constraint measured in joules/symbol
$\bar{P}$	Average power constraint measured in watts
SNR	Signal-to-noise ratio
SINR	Signal-to-interference-plus-noise ratio

$\mathcal{E}_{b}$ Energy per received bit
P_{e} Error probability

Capacities

C_{awgn} Capacity of the additive white Gaussian noise channel
C_{ϵ} ϵ-Outage capacity of the slow fading channel
C_{sum} Sum capacity of the uplink or the downlink
C_{sym} Symmetric capacity of the uplink or the downlink
C_{ϵ}^{sym} ϵ-Outage symmetric capacity of the slow fading uplink channel
p_{out} Outage probability of a scalar fading channel
p_{out}^{Ala} Outage probability when employing the Alamouti scheme
p_{out}^{rep} Outage probability with the repetition scheme
p_{out}^{ul} Outage probability of the uplink
p_{out}^{mimo} Outage probability of the MIMO fading channel
$p_{out}^{ul-mimo}$ Outage probability of the uplink with multiple antennas at the
base-station

Vectors and matrices

h   Vector, complex valued, channel
x   Vector channel input
y   Vector channel output
$\mathcal{CN}(0, \mathbf{K})$ Circularly symmetric Gaussian random vector with
mean zero and covariance matrix **K**
w   Additive Gaussian noise vector $\mathcal{CN}(0, N_0\mathbf{I})$
h* Complex conjugate-transpose of **h**
d   Data vector
$\tilde{\mathbf{d}}$ Discrete Fourier transform of **d**
H   Matrix, complex valued, channel
$\mathbf{K}_x$ Covariance matrix of the random complex vector **x**
H* Complex conjugate-transpose of **H**
$\mathbf{H}^t$ Transpose of matrix **H**
Q, U, V   Unitary matrices
$\mathbf{I}_n$ Identity $n \times n$ matrix
Λ, Ψ Diagonal matrices
$\text{diag}\{p_1, \ldots, p_n\}$ Diagonal matrix with the diagonal entries equal
to $p_1, \ldots, p_n$
C   Circulant matrix
D   Normalized codeword difference matrix

Operations

$\mathbb{E}[x]$ Mean of the random variable x
$\mathbb{P}\{A\}$ Probability of an event A
$\text{Tr}[\mathbf{K}]$ Trace of the square matrix **K**
$\text{sinc}(t)$ Defined to be the ratio of $\sin(\pi t)$ to πt
$Q(a)$ $\int_a^\infty (1/\sqrt{2\pi}) \exp^{-x^2/2} \, dx$
$\mathcal{L}(\cdot, \cdot)$ Lagrangian function

1 Introduction

1.1 Book objective

Wireless communication is one of the most vibrant areas in the communication field today. While it has been a topic of study since the 1960s, the past decade has seen a surge of research activities in the area. This is due to a confluence of several factors. First, there has been an explosive increase in demand for tetherless connectivity, driven so far mainly by cellular telephony but expected to be soon eclipsed by wireless data applications. Second, the dramatic progress in VLSI technology has enabled small-area and low-power implementation of sophisticated signal processing algorithms and coding techniques. Third, the success of second-generation (2G) digital wireless standards, in particular, the IS-95 Code Division Multiple Access (CDMA) standard, provides a concrete demonstration that good ideas from communication theory can have a significant impact in practice. The research thrust in the past decade has led to a much richer set of perspectives and tools on how to communicate over wireless channels, and the picture is still very much evolving.

There are two fundamental aspects of wireless communication that make the problem challenging and interesting. These aspects are by and large not as significant in wireline communication. First is the phenomenon of *fading*: the time variation of the channel strengths due to the small-scale effect of multipath fading, as well as larger-scale effects such as path loss via distance attenuation and shadowing by obstacles. Second, unlike in the wired world where each transmitter–receiver pair can often be thought of as an isolated point-to-point link, wireless users communicate over the air and there is significant *interference* between them. The interference can be between transmitters communicating with a common receiver (e.g., uplink of a cellular system), between signals from a single transmitter to multiple receivers (e.g., downlink of a cellular system), or between different transmitter–receiver pairs (e.g., interference between users in different cells). How to deal with fading and with interference is central to the design of wireless communication systems and will be the central theme of this book. Although this book takes a physical-layer perspective, it will be seen that in fact the management of fading and interference has ramifications across multiple layers.

Traditionally the design of wireless systems has focused on increasing the *reliability* of the air interface; in this context, fading and interference are viewed as *nuisances* that are to be countered. Recent focus has shifted more towards increasing the *spectral efficiency*; associated with this shift is a new point of view that fading can be viewed as an *opportunity* to be exploited. The main objective of the book is to provide a unified treatment of wireless communication from both these points of view. In addition to traditional topics such as diversity and interference averaging, a substantial portion of the book will be devoted to more modern topics such as opportunistic and multiple input multiple output (MIMO) communication.

An important component of this book is the *system view* emphasis: the successful implementation of a theoretical concept or a technique requires an understanding of how it interacts with the wireless system as a whole. Unlike the derivation of a concept or a technique, this system view is less malleable to mathematical formulations and is primarily acquired through experience with designing actual wireless systems. We try to help the reader develop some of this intuition by giving numerous examples of how the concepts are applied in actual wireless systems. Five examples of wireless systems are used. The next section gives some sense of the scope of the wireless systems considered in this book.

1.2 Wireless systems

Wireless communication, despite the hype of the popular press, is a field that has been around for over a hundred years, starting around 1897 with Marconi's successful demonstrations of wireless telegraphy. By 1901, radio reception across the Atlantic Ocean had been established; thus, rapid progress in technology has also been around for quite a while. In the intervening hundred years, many types of wireless systems have flourished, and often later disappeared. For example, television transmission, in its early days, was broadcast by wireless radio transmitters, which are increasingly being replaced by cable transmission. Similarly, the point-to-point microwave circuits that formed the backbone of the telephone network are being replaced by optical fiber. In the first example, wireless technology became outdated when a wired distribution network was installed; in the second, a new wired technology (optical fiber) replaced the older technology. The opposite type of example is occurring today in telephony, where wireless (cellular) technology is partially replacing the use of the wired telephone network (particularly in parts of the world where the wired network is not well developed). The point of these examples is that there are many situations in which there is a choice between wireless and wire technologies, and the choice often changes when new technologies become available.

In this book, we will concentrate on cellular networks, both because they are of great current interest and also because the features of many other wireless systems can be easily understood as special cases or simple generalizations of the features of cellular networks. A cellular network consists of a large number of wireless subscribers who have cellular telephones (users), that can be used in cars, in buildings, on the street, or almost anywhere. There are also a number of fixed base-stations, arranged to provide coverage of the subscribers.

The area covered by a base-station, i.e., the area from which incoming calls reach that base-station, is called a cell. One often pictures a cell as a hexagonal region with the base-station in the middle. One then pictures a city or region as being broken up into a hexagonal lattice of cells (see Figure 1.1a). In reality, the base-stations are placed somewhat irregularly, depending on the location of places such as building tops or hill tops that have good communication coverage and that can be leased or bought (see Figure 1.1b). Similarly, mobile users connected to a base-station are chosen by good communication paths rather than geographic distance.

When a user makes a call, it is connected to the base-station to which it appears to have the best path (often but not always the closest base-station). The base-stations in a given area are then connected to a *mobile telephone switching office* (MTSO, also called a *mobile switching center* MSC) by high-speed wire connections or microwave links. The MTSO is connected to the public wired telephone network. Thus an incoming call from a mobile user is first connected to a base-station and from there to the MTSO and then to the wired network. From there the call goes to its destination, which might be an ordinary wire line telephone, or might be another mobile subscriber. Thus, we see that a cellular network is not an independent network, but rather an appendage to the wired network. The MTSO also plays a major role in coordinating which base-station will handle a call to or from a user and when to handoff a user from one base-station to another.

When another user (either wired or wireless) places a call to a given user, the reverse process takes place. First the MTSO for the called subscriber is found,

Figure 1.1 Cells and base-stations for a cellular network. (a) An oversimplified view in which each cell is hexagonal. (b) A more realistic case where base-stations are irregularly placed and cell phones choose the best base-station.

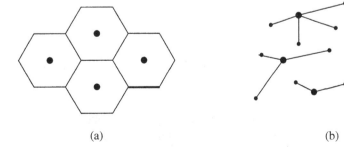

(a)   (b)

then the closest base-station is found, and finally the call is set up through the MTSO and the base-station. The wireless link from a base-station to the mobile users is interchangeably called the *downlink* or the *forward channel*, and the link from the users to a base-station is called the *uplink* or a *reverse channel*. There are usually many users connected to a single base-station, and thus, for the downlink channel, the base-station must multiplex together the signals to the various connected users and then broadcast one waveform from which each user can extract its own signal. For the uplink channel, each user connected to a given base-station transmits its own waveform, and the base-station receives the sum of the waveforms from the various users plus noise. The base-station must then separate out the signals from each user and forward these signals to the MTSO.

Older cellular systems, such as the AMPS (advanced mobile phone service) system developed in the USA in the eighties, are analog. That is, a voice waveform is modulated on a carrier and transmitted without being transformed into a digital stream. Different users in the same cell are assigned different modulation frequencies, and adjacent cells use different sets of frequencies. Cells sufficiently far away from each other can reuse the same set of frequencies with little danger of interference.

Second-generation cellular systems are digital. One is the GSM (global system for mobile communication) system, which was standardized in Europe but now used worldwide, another is the TDMA (time-division multiple access) standard developed in the USA (IS-136), and a third is CDMA (code division multiple access) (IS-95). Since these cellular systems, and their standards, were originally developed for telephony, the current data rates and delays in cellular systems are essentially determined by voice requirements. Third-generation cellular systems are designed to handle data and/or voice. While some of the third-generation systems are essentially evolution of second-generation voice systems, others are designed from scratch to cater for the specific characteristics of data. In addition to a requirement for higher rates, data applications have two features that distinguish them from voice:

- Many data applications are extremely bursty; users may remain inactive for long periods of time but have very high demands for short periods of time. Voice applications, in contrast, have a fixed-rate demand over long periods of time.
- Voice has a relatively tight latency requirement of the order of 100 ms. Data applications have a wide range of latency requirements; real-time applications, such as gaming, may have even tighter delay requirements than voice, while many others, such as http file transfers, have a much laxer requirement.

In the book we will see the impact of these features on the appropriate choice of communication techniques.

As mentioned above, there are many kinds of wireless systems other than cellular. First there are the broadcast systems such as AM radio, FM radio, TV and paging systems. All of these are similar to the downlink part of cellular networks, although the data rates, the sizes of the areas covered by each broadcasting node and the frequency ranges are very different. Next, there are wireless LANs (local area networks). These are designed for much higher data rates than cellular systems, but otherwise are similar to a single cell of a cellular system. These are designed to connect laptops and other portable devices in the local area network within an office building or similar environment. There is little mobility expected in such systems and their major function is to allow portability. The major standards for wireless LANs are the IEEE 802.11 family. There are smaller-scale standards like Bluetooth or a more recent one based on ultra-wideband (UWB) communication whose purpose is to reduce cabling in an office and simplify transfers between office and hand-held devices. Finally, there is another type of LAN called an *ad hoc network*. Here, instead of a central node (base-station) through which all traffic flows, the nodes are all alike. The network organizes itself into links between various pairs of nodes and develops routing tables using these links. Here the network layer issues of routing, dissemination of control information, etc. are important concerns, although problems of relaying and distributed cooperation between nodes can be tackled from the physical-layer as well and are active areas of current research.

1.3 Book outline

The central object of interest is the wireless fading channel. Chapter 2 introduces the multipath fading channel model that we use for the rest of the book. Starting from a continuous-time passband channel, we derive a discrete-time complex baseband model more suitable for analysis and design. Key physical parameters such as coherence time, coherence bandwidth, Doppler spread and delay spread are explained and several statistical models for multipath fading are surveyed. There have been many statistical models proposed in the literature; we will be far from exhaustive here. The goal is to have a small set of example models in our repertoire to evaluate the performance of basic communication techniques we will study.

Chapter 3 introduces many of the issues of communicating over fading channels in the simplest point-to-point context. As a baseline, we start by looking at the problem of detection of uncoded transmission over a narrowband fading channel. We find that the performance is very poor, much worse than over the additive white Gaussian noise (AWGN) channel with the same average signal-to-noise ratio (SNR). This is due to a significant probability that the channel is in *deep fade*. Various *diversity techniques* to mitigate this adverse effect of fading are then studied. Diversity techniques increase reliability by sending the same information through multiple independently faded paths so that the probability of successful transmission is higher. Some of the techniques studied include:

- interleaving of coded symbols over time to obtain time diversity;
- inter-symbol equalization, multipath combining in spread-spectrum systems and coding over sub-carriers in orthogonal frequency division multiplexing (OFDM) systems to obtain frequency diversity;
- use of multiple transmit and/or receive antennas, via *space-time* coding, to obtain spatial diversity.

In some scenarios, there is an interesting interplay between channel uncertainty and the diversity gain: as the number of diversity branches increases, the performance of the system first improves due to the diversity gain but then subsequently deteriorates as channel uncertainty makes it more difficult to combine signals from the different branches.

In Chapter 4 the focus is shifted from point-to-point communication to studying cellular systems as a whole. Multiple access and inter-cell interference management are the key issues that come to the forefront. We explain how existing digital wireless systems deal with these issues. The concepts of frequency reuse and cell sectorization are discussed, and we contrast narrowband systems such as GSM and IS-136, where users within the same cell are kept orthogonal and frequency is reused only in cells far away, and CDMA systems, such as IS-95, where the signals of users both within the same cell and across different cells are spread across the same spectrum, i.e., frequency reuse factor of 1. Due to the full reuse, CDMA systems have to manage intra-cell and inter-cell interference more efficiently: in addition to the diversity techniques of time-interleaving, multipath combining and soft handoff, *power control* and *interference averaging* are the key interference management mechanisms. All the five techniques strive toward the same system goal: to maintain the channel quality of each user, as measured by the signal-to-interference-and-noise ratio (SINR), as constant as possible. This chapter is concluded with the discussion of a wideband OFDM system, which combines the advantages of both the CDMA and the narrowband systems.

Chapter 5 studies the capacity of wireless channels. This provides a higher level view of the tradeoffs involved in the earlier chapters and also lays the foundation for understanding the more modern developments in the subsequent chapters. The performance over the (non-faded) AWGN channel is used as a baseline for comparison. We introduce the concept of *channel capacity* as the basic performance measure. The capacity of a channel provides the fundamental limit of communication achievable by any scheme. For the fading channel, there are several capacity measures, relevant for different scenarios. Two distinct scenarios provide particular insight: (1) the *slow* fading channel, where the channel stays the same (random value) over the entire time-scale of communication, and (2) the *fast* fading channel, where the channel varies significantly over the time-scale of communication.

In the slow fading channel, the key event of interest is *outage*: this is the situation when the channel is so poor that no scheme can communicate reliably at a certain target data rate. The largest rate of reliable communication at a certain outage probability is called the outage capacity. In the fast fading channel, in contrast, outage can be avoided due to the ability to average over the time variation of the channel, and one can define a positive capacity at which arbitrarily reliable communication is possible. Using these capacity measures, several resources associated with a fading channel are defined: (1) diversity; (2) number of degrees of freedom; (3) received power. These three resources form a basis for assessing the nature of performance gain by the various communication schemes studied in the rest of the book.

Chapters 6 to 10 cover the more recent developments in the field. In Chapter 6 we revisit the problem of multiple access over fading channels from a more fundamental point of view. Information theory suggests that if both the transmitters and the receiver can track the fading channel, the optimal strategy to maximize the total system throughput is to allow only the user with the best channel to transmit at any time. A similar strategy is also optimal for the downlink. Opportunistic strategies of this type yield a system-wide *multiuser diversity* gain: the more users in the system, the larger the gain, as there is more likely to be a user with a very strong channel. To implement this concept in a real system, three important considerations are: *fairness* of the resource allocation across users; *delay* experienced by the individual user waiting for its channel to become good; and *measurement inaccuracy* and *delay* in feeding back the channel state to the transmitters. We discuss how these issues are addressed in the context of IS-865 (also called HDR or CDMA 2000 1× EV-DO), a third-generation wireless data system.

A wireless system consists of multiple dimensions: time, frequency, space and users. Opportunistic communication maximizes the spectral efficiency by measuring when and where the channel is good and only transmits in those degrees of freedom. In this context, channel fading is *beneficial* in the sense that the fluctuation of the channel across the degrees of freedom ensures that there will be some degrees of freedom in which the channel is very good. This is in sharp contrast to the diversity-based approach in Chapter 3, where channel fluctuation is always detrimental and the design goal is to average out the fading to make the overall channel as constant as possible. Taking this philosophy one step further, we discuss a technique, called *opportunistic beamforming*, in which channel fluctuation can be *induced* in situations when the natural fading has small dynamic range and/or is slow. From the cellular system point of view, this technique also increases the fluctuations of the *interference* imparted on adjacent cells, and presents an opposing philosophy to the notion of interference averaging in CDMA systems.

Chapters 7, 8, 9 and 10 discuss multiple input multiple output (MIMO) communication. It has been known for a while that the uplink with multiple receive antennas at the base-station allow several users to simultaneously communicate to the receiver. The multiple antennas in effect increase the number of degrees of freedom in the system and allow spatial separation of the signals from the different users. It has recently been shown that a similar effect occurs for point-to-point channels with multiple transmit *and* receive antennas, i.e., even when the antennas of the multiple users are co-located. This holds provided that the scattering environment is rich enough to allow the receive antennas to separate out the signal from the different transmit antennas, allowing the *spatial multiplexing* of information. This is yet another example where channel fading is beneficial to communication. Chapter 7 studies the properties of the multipath environment that determine the amount of spatial multiplexing possible and defines an *angular domain* in which such properties are seen most explicitly. We conclude with a class of statistical MIMO channel models, based in the angular domain, which will be used in later chapters to analyze the performance of communication techniques.

Chapter 8 discusses the capacity and capacity-achieving transceiver architectures for MIMO channels, focusing on the fast fading scenario. It is demonstrated that the fast fading capacity increases linearly with the minimum of the number of transmit and receive antennas at all values of SNR. At high SNR, the linear increase is due to the increase in degrees of freedom from spatial multiplexing. At low SNR, the linear increase is due to a power gain from receive beamforming. At intermediate SNR ranges, the linear increase is due to a combination of both these gains. Next, we study the transceiver architectures that achieve the capacity of the fast fading channel. The focus is on the V-BLAST architecture, which multiplexes independent data streams, one onto each of the transmit antennas. A variety of receiver structures are considered: these include the decorrelator and the linear minimum mean square-error (MMSE) receiver. The performance of these receivers can be enhanced by successively canceling the streams as they are decoded; this is known as successive interference cancellation (SIC). It is shown that the MMSE–SIC receiver achieves the capacity of the fast fading MIMO channel.

The V-BLAST architecture is very suboptimal for the slow fading MIMO channel: it does not code across the transmit antennas and thus the diversity gain is limited by that obtained with the receive antenna array. A modification, called D-BLAST, where the data streams are *interleaved* across the transmit antenna array, achieves the outage capacity of the slow fading MIMO channel. The boost of the outage capacity of a MIMO channel as compared to a single antenna channel is due to a combination of both diversity and spatial multiplexing gains. In Chapter 9, we study a fundamental *tradeoff* between the diversity and multiplexing gains that can be simultaneously harnessed over a slow fading MIMO channel. This formulation is then used as a unified framework to assess both the diversity and multiplexing performance of several schemes that have appeared earlier in the book. This framework is also used to motivate the construction of new tradeoff-optimal space-time codes. In particular, we discuss an approach to design *universal* space-time codes that are tradeoff-optimal.

Finally, Chapter 10 studies the use of multiple transmit and receive antennas in multiuser and cellular systems; this is also called *space-division multiple access* (SDMA). Here, in addition to providing spatial multiplexing and diversity, multiple antennas can also be used to mitigate interference between different users. In the uplink, interference mitigation is done at the base-station via the SIC receiver. In the downlink, interference mitigation is also done at the base-station and this requires *precoding*: we study a precoding scheme, called Costa or dirty-paper precoding, that is the natural analog of the SIC receiver in the uplink. This study allows us to relate the performance of an SIC receiver in the uplink with a corresponding precoding scheme in a *reciprocal* downlink. The ArrayComm system is used as an example of an SDMA cellular system.

2 The wireless channel

A good understanding of the wireless channel, its key physical parameters and the modeling issues, lays the foundation for the rest of the book. This is the goal of this chapter.

A defining characteristic of the mobile wireless channel is the variations of the channel strength over time and over frequency. The variations can be roughly divided into two types (Figure 2.1):

- *Large-scale fading*, due to path loss of signal as a function of distance and shadowing by large objects such as buildings and hills. This occurs as the mobile moves through a distance of the order of the cell size, and is typically frequency independent.
- *Small-scale fading*, due to the constructive and destructive interference of the multiple signal paths between the transmitter and receiver. This occurs at the spatial scale of the order of the carrier wavelength, and is frequency dependent.

We will talk about both types of fading in this chapter, but with more emphasis on the latter. Large-scale fading is more relevant to issues such as cell-site planning. Small-scale multipath fading is more relevant to the design of reliable and efficient communication systems – the focus of this book.

We start with the physical modeling of the wireless channel in terms of electromagnetic waves. We then derive an input/output linear time-varying model for the channel, and define some important physical parameters. Finally, we introduce a few statistical models of the channel variation over time and over frequency.

2.1 Physical modeling for wireless channels

Wireless channels operate through electromagnetic radiation from the transmitter to the receiver. In principle, one could solve the electromagnetic field equations, in conjunction with the transmitted signal, to find the

Figure 2.1 Channel quality varies over multiple time-scales. At a slow scale, channel varies due to large-scale fading effects. At a fast scale, channel varies due to multipath effects.

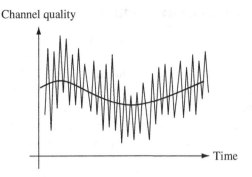

electromagnetic field impinging on the receiver antenna. This would have to be done taking into account the obstructions caused by ground, buildings, vehicles, etc. in the vicinity of this electromagnetic wave.[1]

Cellular communication in the USA is limited by the Federal Communication Commission (FCC), and by similar authorities in other countries, to one of three frequency bands, one around 0.9 GHz, one around 1.9 GHz, and one around 5.8 GHz. The wavelength λ of electromagnetic radiation at any given frequency f is given by $\lambda = c/f$, where $c = 3 \times 10^8$ m/s is the speed of light. The wavelength in these cellular bands is thus a fraction of a meter, so to calculate the electromagnetic field at a receiver, the locations of the receiver and the obstructions would have to be known within sub-meter accuracies. The electromagnetic field equations are therefore too complex to solve, especially on the fly for mobile users. Thus, we have to ask what we really need to know about these channels, and what approximations might be reasonable.

One of the important questions is where to choose to place the base-stations, and what range of power levels are then necessary on the downlink and uplink channels. To some extent this question must be answered experimentally, but it certainly helps to have a sense of what types of phenomena to expect. Another major question is what types of modulation and detection techniques look promising. Here again, we need a sense of what types of phenomena to expect. To address this, we will construct stochastic models of the channel, assuming that different channel behaviors appear with different probabilities, and change over time (with specific stochastic properties). We will return to the question of why such stochastic models are appropriate, but for now we simply want to explore the gross characteristics of these channels. Let us start by looking at several over-idealized examples.

---

[1] By obstructions, we mean not only objects in the line-of-sight between transmitter and receiver, but also objects in locations that cause non-negligible changes in the electromagnetic field at the receiver; we shall see examples of such obstructions later.

2.1.1 Free space, fixed transmit and receive antennas

First consider a fixed antenna radiating into free space. In the far field,[2] the electric field and magnetic field at any given location are perpendicular both to each other and to the direction of propagation from the antenna. They are also proportional to each other, so it is sufficient to know only one of them (just as in wired communication, where we view a signal as simply a voltage waveform or a current waveform). In response to a transmitted sinusoid $\cos 2\pi ft$, we can express the electric far field at time t as

$$E(f, t, (r, \theta, \psi)) = \frac{\alpha_s(\theta, \psi, f) \cos 2\pi f(t - r/c)}{r}. \tag{2.1}$$

Here, (r, θ, ψ) represents the point $\mathbf{u}$ in space at which the electric field is being measured, where r is the distance from the transmit antenna to $\mathbf{u}$ and where (θ, ψ) represents the vertical and horizontal angles from the antenna to $\mathbf{u}$ respectively. The constant c is the speed of light, and $\alpha_s(\theta, \psi, f)$ is the radiation pattern of the sending antenna at frequency f in the direction (θ, ψ); it also contains a scaling factor to account for antenna losses. Note that the phase of the field varies with fr/c, corresponding to the delay caused by the radiation traveling at the speed of light.

We are not concerned here with actually finding the radiation pattern for any given antenna, but only with recognizing that antennas have radiation patterns, and that the free space far field behaves as above.

It is important to observe that, as the distance r increases, the electric field decreases as r^{-1} and thus the power per square meter in the free space wave decreases as r^{-2}. This is expected, since if we look at concentric spheres of increasing radius r around the antenna, the total power radiated through the sphere remains constant, but the surface area increases as r^2. Thus, the power per unit area must decrease as r^{-2}. We will see shortly that this r^{-2} reduction of power with distance is often not valid when there are obstructions to free space propagation.

Next, suppose there is a fixed receive antenna at the location $\mathbf{u} = (r, \theta, \psi)$. The received waveform (in the absence of noise) in response to the above transmitted sinusoid is then

$$E_r(f, t, \mathbf{u}) = \frac{\alpha(\theta, \psi, f) \cos 2\pi f(t - r/c)}{r}, \tag{2.2}$$

where $\alpha(\theta, \psi, f)$ is the product of the antenna patterns of transmit and receive antennas in the given direction. Our approach to (2.2) is a bit odd since we started with the free space field at $\mathbf{u}$ in the absence of an antenna. Placing a

---

[2] The far field is the field sufficiently far away from the antenna so that (2.1) is valid. For cellular systems, it is a safe assumption that the receiver is in the far field.

receive antenna there changes the electric field in the vicinity of **u**, but this is taken into account by the antenna pattern of the receive antenna.

Now suppose, for the given **u**, that we define

$$H(f) := \frac{\alpha(\theta, \psi, f)e^{-j2\pi fr/c}}{r}. \tag{2.3}$$

We then have $E_r(f, t, \mathbf{u}) = \Re\left[H(f)e^{j2\pi ft}\right]$. We have not mentioned it yet, but (2.1) and (2.2) are both linear in the input. That is, the received field (waveform) at **u** in response to a weighted sum of transmitted waveforms is simply the weighted sum of responses to those individual waveforms. Thus, $H(f)$ is the system function for an LTI (linear time-invariant) channel, and its inverse Fourier transform is the impulse response. The need for understanding electromagnetism is to determine what this system function is. We will find in what follows that linearity is a good assumption for all the wireless channels we consider, but that the time invariance does not hold when either the antennas or obstructions are in relative motion.

2.1.2 Free space, moving antenna

Next consider the fixed antenna and free space model above with a receive antenna that is moving with speed v in the direction of increasing distance from the transmit antenna. That is, we assume that the receive antenna is at a moving location described as $\mathbf{u}(t) = (r(t), \theta, \psi)$ with $r(t) = r_0 + vt$. Using (2.1) to describe the free space electric field at the moving point $\mathbf{u}(t)$ (for the moment with no receive antenna), we have

$$E(f, t, (r_0 + vt, \theta, \psi)) = \frac{\alpha_s(\theta, \psi, f)\cos 2\pi f(t - r_0/c - vt/c)}{r_0 + vt}. \tag{2.4}$$

Note that we can rewrite $f(t - r_0/c - vt/c)$ as $f(1 - v/c)t - fr_0/c$. Thus, the sinusoid at frequency f has been converted to a sinusoid of frequency $f(1 - v/c)$; there has been a *Doppler shift* of $-fv/c$ due to the motion of the observation point.[3] Intuitively, each successive crest in the transmitted sinusoid has to travel a little further before it gets observed at the moving observation point. If the antenna is now placed at $\mathbf{u}(t)$, and the change of field due to the antenna presence is again represented by the receive antenna pattern, the received waveform, in analogy to (2.2), is

$$E_r(f, t, (r_0 + vt, \theta, \psi)) = \frac{\alpha(\theta, \psi, f)\cos 2\pi f[(1 - v/c)t - r_0/c]}{r_0 + vt}. \tag{2.5}$$

---

[3] The reader should be familiar with the Doppler shift associated with moving cars. When an ambulance is rapidly moving toward us we hear a higher frequency siren. When it passes us we hear a rapid shift toward a lower frequency.

This channel cannot be represented as an LTI channel. If we ignore the time-varying attenuation in the denominator of (2.5), however, we can represent the channel in terms of a system function followed by translating the frequency f by the Doppler shift $-fv/c$. It is important to observe that the amount of shift depends on the frequency f. We will come back to discussing the importance of this Doppler shift and of the time-varying attenuation after considering the next example.

The above analysis does not depend on whether it is the transmitter or the receiver (or both) that are moving. So long as $r(t)$ is interpreted as the distance between the antennas (and the relative orientations of the antennas are constant), (2.4) and (2.5) are valid.

2.1.3 Reflecting wall, fixed antenna

Consider Figure 2.2 in which there is a fixed antenna transmitting the sinusoid $\cos 2\pi ft$, a fixed receive antenna, and a single perfectly reflecting large fixed wall. We assume that in the absence of the receive antenna, the electromagnetic field at the point where the receive antenna will be placed is the sum of the free space field coming from the transmit antenna plus a reflected wave coming from the wall. As before, in the presence of the receive antenna, the perturbation of the field due to the antenna is represented by the antenna pattern. An additional assumption here is that the presence of the receive antenna does not appreciably affect the plane wave impinging on the wall. In essence, what we have done here is to approximate the solution of Maxwell's equations by a method called *ray tracing*. The assumption here is that the received waveform can be approximated by the sum of the free space wave from the transmitter plus the reflected free space waves from each of the reflecting obstacles.

In the present situation, if we assume that the wall is very large, the reflected wave at a given point is the same (except for a sign change[4]) as the free space wave that would exist on the opposite side of the wall if the wall were not present (see Figure 2.3). This means that the reflected wave from the wall has the intensity of a free space wave at a distance equal to the distance to the wall and then

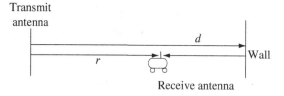

Figure 2.2 Illustration of a direct path and a reflected path.

Transmit antenna

d

Wall

r

Receive antenna

---

[4] By basic electromagnetics, this sign change is a consequence of the fact that the electric field is parallel to the plane of the wall for this example.

Figure 2.3 Relation of reflected wave to wave without wall.

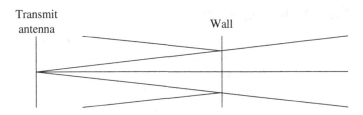

Transmit antenna

Wall

back to the receive antenna, i.e., $2d - r$. Using (2.2) for both the direct and the reflected wave, and assuming the same antenna gain α for both waves, we get

$$E_r(f, t) = \frac{\alpha \cos 2\pi f(t - r/c)}{r} - \frac{\alpha \cos 2\pi f(t - (2d - r)/c)}{2d - r}. \tag{2.6}$$

The received signal is a superposition of two waves, both of frequency f. The phase difference between the two waves is

$$\Delta\theta = \left(\frac{2\pi f(2d - r)}{c} + \pi\right) - \left(\frac{2\pi f r}{c}\right) = \frac{4\pi f}{c}(d - r) + \pi. \tag{2.7}$$

When the phase difference is an integer multiple of 2π, the two waves add *constructively*, and the received signal is strong. When the phase difference is an odd integer multiple of π, the two waves add *destructively*, and the received signal is weak. As a function of r, this translates into a spatial pattern of constructive and destructive interference of the waves. The distance from a peak to a valley is called the *coherence distance*:

$$\boxed{\Delta x_c := \frac{\lambda}{4},} \tag{2.8}$$

where $\lambda := c/f$ is the wavelength of the transmitted sinusoid. At distances much smaller than Δx_c, the received signal at a particular time does not change appreciably.

The constructive and destructive interference pattern also depends on the frequency f: for a fixed r, if f changes by

$$\frac{1}{2}\left(\frac{2d - r}{c} - \frac{r}{c}\right)^{-1}, \tag{2.9}$$

we move from a peak to a valley. The quantity

$$\boxed{T_d := \frac{2d - r}{c} - \frac{r}{c}} \tag{2.10}$$

is called the *delay spread* of the channel: it is the difference between the propagation delays along the two signal paths. The constructive and destructive interference pattern does not change appreciably if the frequency changes by an amount much smaller than $1/T_d$. This parameter is called the *coherence bandwidth*.

2.1.4 Reflecting wall, moving antenna

Suppose the receive antenna is now moving at a velocity v (Figure 2.4). As it moves through the pattern of constructive and destructive interference created by the two waves, the strength of the received signal increases and decreases. This is the phenomenon of *multipath fading*. The time taken to travel from a peak to a valley is $c/(4fv)$: this is the time-scale at which the fading occurs, and it is called the *coherence time* of the channel.

An equivalent way of seeing this is in terms of the Doppler shifts of the direct and the reflected waves. Suppose the receive antenna is at location r_0 at time 0. Taking $r = r_0 + vt$ in (2.6), we get

$$E_r(f, t) = \frac{\alpha \cos 2\pi f[(1 - v/c)t - r_0/c]}{r_0 + vt}$$
$$- \frac{\alpha \cos 2\pi f[(1 + v/c)t + (r_0 - 2d)/c]}{2d - r_0 - vt}. \tag{2.11}$$

The first term, the direct wave, is a sinusoid at frequency $f(1 - v/c)$, experiencing a Doppler shift $D_1 := -fv/c$. The second is a sinusoid at frequency $f(1 + v/c)$, with a Doppler shift $D_2 := +fv/c$. The parameter

$$\boxed{D_s := D_2 - D_1} \tag{2.12}$$

is called the *Doppler spread*. For example, if the mobile is moving at 60 km/h and $f = 900$ MHz, the Doppler spread is 100 Hz. The role of the Doppler spread can be visualized most easily when the mobile is much closer to the wall than to the transmit antenna. In this case the attenuations are roughly the same for both paths, and we can approximate the denominator of the second term by $r = r_0 + vt$. Then, combining the two sinusoids, we get

$$E_r(f, t) \approx \frac{2\alpha \sin 2\pi f [vt/c + (r_0 - d)/c] \sin 2\pi f[t - d/c]}{r_0 + vt}. \tag{2.13}$$

This is the product of two sinusoids, one at the input frequency f, which is typically of the order of GHz, and the other one at $fv/c = D_s/2$, which might be of the order of 50 Hz. Thus, the response to a sinusoid at f is another sinusoid at f with a time-varying envelope, with peaks going to zeros around every 5 ms (Figure 2.5). The envelope is at its widest when the mobile is at a peak of the

Figure 2.4 Illustration of a direct path and a reflected path.

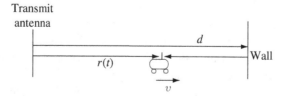

Figure 2.5 The received
waveform oscillating at
frequency *f* with a slowly
varying envelope at frequency
$D_s/2$.

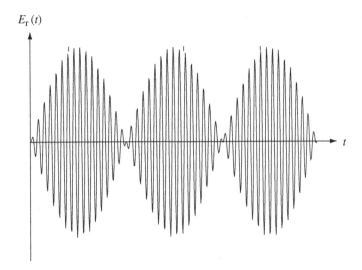

$E_r(t)$

interference pattern and at its narrowest when the mobile is at a valley. Thus, the Doppler spread determines the rate of traversal across the interference pattern and is inversely proportional to the coherence time of the channel.

We now see why we have partially ignored the denominator terms in (2.11) and (2.13). When the difference in the length between two paths changes by a quarter wavelength, the phase difference between the responses on the two paths changes by $\pi/2$, which causes a very significant change in the overall received amplitude. Since the carrier wavelength is very small relative to the path lengths, the time over which this phase effect causes a significant change is far smaller than the time over which the denominator terms cause a significant change. The effect of the phase changes is of the order of milliseconds, whereas the effect of changes in the denominator is of the order of seconds or minutes. In terms of modulation and detection, the time-scales of interest are in the range of milliseconds and less, and the denominators are effectively constant over these periods.

The reader might notice that we are constantly making approximations in trying to understand wireless communication, much more so than for wired communication. This is partly because wired channels are typically time-invariant over a very long time-scale, while wireless channels are typically time-varying, and appropriate models depend very much on the time-scales of interest. For wireless systems, the most important issue is what approximations to make. Thus, it is important to understand these modeling issues thoroughly.

2.1.5 Reflection from a ground plane

Consider a transmit and a receive antenna, both above a plane surface such as a road (Figure 2.6). When the horizontal distance *r* between the antennas becomes very large relative to their vertical displacements from the ground

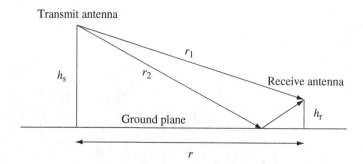

plane (i.e., height), a very surprising thing happens. In particular, the difference between the direct path length and the reflected path length goes to zero
as r^{-1} with increasing r (Exercise 2.5). When r is large enough, this difference
between the path lengths becomes small relative to the wavelength c/f. Since
the sign of the electric field is reversed on the reflected path[5], these two waves
start to cancel each other out. The electric wave at the receiver is then attenuated as r^{-2}, and the received power decreases as r^{-4}. This situation is particularly important in rural areas where base-stations tend to be placed on roads.

2.1.6 Power decay with distance and shadowing

The previous example with reflection from a ground plane suggests that the
received power can decrease with distance faster than r^{-2} in the presence of
disturbances to free space. In practice, there are several obstacles between
the transmitter and the receiver and, further, the obstacles might also absorb
some power while scattering the rest. Thus, one expects the power decay to
be considerably faster than r^{-2}. Indeed, empirical evidence from experimental
field studies suggests that while power decay near the transmitter is like r^{-2},
at large distances the power can even decay *exponentially* with distance.

   The ray tracing approach used so far provides a high degree of numerical
accuracy in determining the electric field at the receiver, but requires a precise
physical model including the location of the obstacles. But here, we are only
looking for the order of decay of power with distance and can consider an
alternative approach. So we look for a model of the physical environment with
the fewest parameters but one that still provides useful global information
about the field properties. A simple probabilistic model with two parameters
of the physical environment, the density of the obstacles and the fraction of
energy each object absorbs, is developed in Exercise 2.6. With each obstacle

---

[5] This is clearly true if the electric field is parallel to the ground plane. It turns out that this is
    also true for arbitrary orientations of the electric field, as long as the ground is not a perfect
    conductor and the angle of incidence is small enough. The underlying electromagnetics is
    analyzed in Chapter 2 of Jakes [62].

absorbing the same fraction of the energy impinging on it, the model allows us to show that the power decays exponentially in distance at a rate that is proportional to the density of the obstacles.

With a limit on the transmit power (either at the base-station or at the mobile), the largest distance between the base-station and a mobile at which communication can reliably take place is called the *coverage* of the cell. For reliable communication, a minimal received power level has to be met and thus the fast decay of power with distance constrains cell coverage. On the other hand, rapid signal attenuation with distance is also helpful; it reduces the *interference* between adjacent cells. As cellular systems become more popular, however, the major determinant of cell size is the number of mobiles in the cell. In engineering jargon, the cell is said to be *capacity* limited instead of coverage limited. The size of cells has been steadily decreasing, and one talks of micro cells and pico cells as a response to this effect. With capacity limited cells, the inter-cell interference may be intolerably high. To alleviate the inter-cell interference, neighboring cells use different parts of the frequency spectrum, and frequency is reused at cells that are far enough. Rapid signal attenuation with distance allows frequencies to be reused at closer distances.

The density of obstacles between the transmit and receive antennas depends very much on the physical environment. For example, outdoor plains have very little by way of obstacles while indoor environments pose many obstacles. This randomness in the environment is captured by modeling the density of obstacles and their absorption behavior as random numbers; the overall phenomenon is called *shadowing*.[6] The effect of shadow fading differs from multipath fading in an important way. The duration of a shadow fade lasts for multiple seconds or minutes, and hence occurs at a much slower time-scale compared to multipath fading.

2.1.7 Moving antenna, multiple reflectors

Dealing with multiple reflectors, using the technique of ray tracing, is in principle simply a matter of modeling the received waveform as the sum of the responses from the different paths rather than just two paths. We have seen enough examples, however, to understand that finding the magnitudes and phases of these responses is no simple task. Even for the very simple large wall example in Figure 2.2, the reflected field calculated in (2.6) is valid only at distances from the wall that are small relative to the dimensions of the wall. At very large distances, the total power reflected from the wall is proportional to both d^{-2} and to the area of the cross section of the wall. The power reaching the receiver is proportional to $(d - r(t))^{-2}$. Thus, the power attenuation from transmitter to receiver (for the large distance case) is proportional to $(d(d - r(t)))^{-2}$ rather

---

[6] This is called shadowing because it is similar to the effect of clouds partly blocking sunlight.

than to $(2d - r(t))^{-2}$. This shows that ray tracing must be used with some caution. Fortunately, however, linearity still holds in these more complex cases.

Another type of reflection is known as *scattering* and can occur in the atmosphere or in reflections from very rough objects. Here there are a very large number of individual paths, and the received waveform is better modeled as an integral over paths with infinitesimally small differences in their lengths, rather than as a sum.

Knowing how to find the amplitude of the reflected field from each type of reflector is helpful in determining the coverage of a base-station (although ultimately experimentation is necessary). This is an important topic if our objective is trying to determine where to place base-stations. Studying this in more depth, however, would take us afield and too far into electromagnetic theory. In addition, we are primarily interested in questions of modulation, detection, multiple access, and network protocols rather than location of base-stations. Thus, we turn our attention to understanding the nature of the aggregate received waveform, given a representation for each reflected wave. This leads to modeling the input/output behavior of a channel rather than the detailed response on each path.

2.2 Input/output model of the wireless channel

We derive an input/output model in this section. We first show that the multipath effects can be modeled as a linear time-varying system. We then obtain a baseband representation of this model. The continuous-time channel is then sampled to obtain a discrete-time model. Finally we incorporate additive noise.

2.2.1 The wireless channel as a linear time-varying system

In the previous section we focused on the response to the sinusoidal input $\phi(t) = \cos 2\pi f t$. The received signal can be written as $\sum_i a_i(f, t)\phi(t - \tau_i(f, t))$, where $a_i(f, t)$ and $\tau_i(f, t)$ are respectively the overall attenuation and propagation delay at time t from the transmitter to the receiver on path i. The overall attenuation is simply the product of the attenuation factors due to the antenna pattern of the transmitter and the receiver, the nature of the reflector, as well as a factor that is a function of the distance from the transmitting antenna to the reflector and from the reflector to the receive antenna. We have described the channel effect at a particular frequency f. If we further assume that the $a_i(f, t)$ and the $\tau_i(f, t)$ do not depend on the frequency f, then we can use the principle of superposition to generalize the above input/output relation to an arbitrary input $x(t)$ with non-zero bandwidth:

$$y(t) = \sum_i a_i(t)x(t - \tau_i(t)). \tag{2.14}$$

In practice the attenuations and the propagation delays are usually slowly varying functions of frequency. These variations follow from the time-varying path lengths and also from frequency-dependent antenna gains. However, we are primarily interested in transmitting over bands that are narrow relative to the carrier frequency, and over such ranges we can omit this frequency dependence. It should however be noted that although the *individual* attenuations and delays are assumed to be independent of the frequency, the *overall* channel response can still vary with frequency due to the fact that different paths have different delays.

For the example of a perfectly reflecting wall in Figure 2.4, then,

$$a_1(t) = \frac{|\alpha|}{r_0 + vt}, \qquad\qquad a_2(t) = \frac{|\alpha|}{2d - r_0 - vt}, \qquad (2.15)$$

$$\tau_1(t) = \frac{r_0 + vt}{c} - \frac{\angle\phi_1}{2\pi f}, \qquad \tau_2(t) = \frac{2d - r_0 - vt}{c} - \frac{\angle\phi_2}{2\pi f}, \qquad (2.16)$$

where the first expression is for the direct path and the second for the reflected path. The term $\angle\phi_j$ here is to account for possible phase changes at the transmitter, reflector, and receiver. For the example here, there is a phase reversal at the reflector so we take $\phi_1 = 0$ and $\phi_2 = \pi$.

Since the channel (2.14) is linear, it can be described by the response $h(\tau, t)$ at time t to an impulse transmitted at time $t - \tau$. In terms of $h(\tau, t)$, the input/output relationship is given by

$$y(t) = \int_{-\infty}^{\infty} h(\tau, t)x(t - \tau)\,d\tau. \qquad (2.17)$$

Comparing (2.17) and (2.14), we see that the impulse response for the fading multipath channel is

$$h(\tau, t) = \sum_i a_i(t)\delta(\tau - \tau_i(t)). \qquad (2.18)$$

This expression is really quite nice. It says that the effect of mobile users, arbitrarily moving reflectors and absorbers, and all of the complexities of solving Maxwell's equations, finally reduce to an input/output relation between transmit and receive antennas which is simply represented as the impulse response of a linear time-varying channel filter.

The effect of the Doppler shift is not immediately evident in this representation. From (2.16) for the single reflecting wall example, $\tau_i'(t) = v_i/c$ where v_i is the velocity with which the ith path length is increasing. Thus, the Doppler shift on the ith path is $-f\tau_i'(t)$.

In the special case when the transmitter, receiver and the environment are all stationary, the attenuations $a_i(t)$ and propagation delays $\tau_i(t)$ do not depend on time t, and we have the usual linear time-invariant channel with an impulse response

$$h(\tau) = \sum_i a_i \delta(\tau - \tau_i). \tag{2.19}$$

For the time-varying impulse response $h(\tau, t)$, we can define a time-varying frequency response

$$H(f; t) := \int_{-\infty}^{\infty} h(\tau, t) e^{-j2\pi f\tau} \, d\tau = \sum_i a_i(t) e^{-j2\pi f\tau_i(t)}. \tag{2.20}$$

In the special case when the channel is time-invariant, this reduces to the usual frequency response. One way of interpreting $H(f; t)$ is to think of the system as a slowly varying function of t with a frequency response $H(f; t)$ at each fixed time t. Corresponding, $h(\tau, t)$ can be thought of as the impulse response of the system at a fixed time t. This is a legitimate and useful way of thinking about many multipath fading channels, as the time-scale at which the channel varies is typically much longer than the delay spread (i.e., the amount of memory) of the impulse response at a fixed time. In the reflecting wall example in Section 2.1.4, the time taken for the channel to change significantly is of the order of milliseconds while the delay spread is of the order of microseconds. Fading channels which have this characteristic are sometimes called *underspread* channels.

2.2.2 Baseband equivalent model

In typical wireless applications, communication occurs in a passband $[f_c - W/2, f_c + W/2]$ of bandwidth W around a center frequency f_c, the spectrum having been specified by regulatory authorities. However, most of the processing, such as coding/decoding, modulation/demodulation, synchronization, etc., is actually done at the baseband. At the transmitter, the last stage of the operation is to "up-convert" the signal to the carrier frequency and transmit it via the antenna. Similarly, the first step at the receiver is to "down-convert" the RF (radio-frequency) signal to the baseband before further processing. Therefore from a communication system design point of view, it is most useful to have a baseband equivalent representation of the system. We first start with defining the baseband equivalent representation of signals.

Consider a real signal $s(t)$ with Fourier transform $S(f)$, band-limited in $[f_c - W/2, f_c + W/2]$ with $W < 2f_c$. Define its *complex baseband equivalent* $s_b(t)$ as the signal having Fourier transform:

$$S_b(f) = \begin{cases} \sqrt{2}S(f + f_c) & f + f_c > 0, \\ 0 & f + f_c \le 0. \end{cases} \tag{2.21}$$

Figure 2.7 Illustration of the relationship between a passband spectrum *S(f)* and its baseband equivalent S$_b$*(f).*

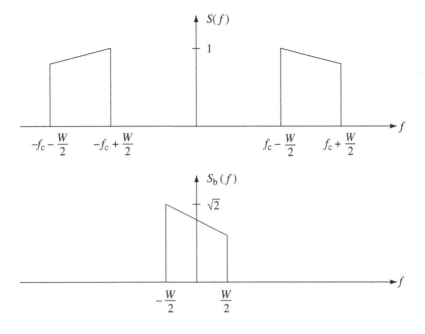

Since $s(t)$ is real, its Fourier transform satisfies $S(f) = S^*(-f)$, which means that $s_b(t)$ contains exactly the same information as $s(t)$. The factor of $\sqrt{2}$ is quite arbitrary but chosen to normalize the energies of $s_b(t)$ and $s(t)$ to be the same. Note that $s_b(t)$ is band-limited in $[-W/2, W/2]$. See Figure 2.7.

To reconstruct $s(t)$ from $s_b(t)$, we observe that

$$\sqrt{2}S(f) = S_b(f - f_c) + S_b^*(-f - f_c). \tag{2.22}$$

Taking inverse Fourier transforms, we get

$$s(t) = \frac{1}{\sqrt{2}}\left[s_b(t)e^{j2\pi f_c t} + s_b^*(t)e^{-j2\pi f_c t}\right] = \sqrt{2}\Re\left[s_b(t)e^{j2\pi f_c t}\right]. \tag{2.23}$$

In terms of real signals, the relationship between $s(t)$ and $s_b(t)$ is shown in Figure 2.8. The passband signal $s(t)$ is obtained by modulating $\Re[s_b(t)]$ by $\sqrt{2}\cos 2\pi f_c t$ and $\Im[s_b(t)]$ by $-\sqrt{2}\sin 2\pi f_c t$ and summing, to get $\sqrt{2}\Re\left[s_b(t)e^{j2\pi f_c t}\right]$ (up-conversion). The baseband signal $\Re[s_b(t)]$ (respectively $\Im[s_b(t)]$) is obtained by modulating $s(t)$ by $\sqrt{2}\cos 2\pi f_c t$ (respectively $-\sqrt{2}\sin 2\pi f_c t$) followed by ideal low-pass filtering at the baseband $[-W/2, W/2]$ (down-conversion).

Let us now go back to the multipath fading channel (2.14) with impulse response given by (2.18). Let $x_b(t)$ and $y_b(t)$ be the complex baseband equivalents of the transmitted signal $x(t)$ and the received signal $y(t)$, respectively. Figure 2.9 shows the system diagram from $x_b(t)$ to $y_b(t)$. This implementation of a passband communication system is known as *quadrature amplitude modulation* (QAM). The signal $\Re[x_b(t)]$ is sometimes called the

Figure 2.8 Illustration of
upconversion from $s_b(t)$ to
$s(t)$, followed by
downconversion from $s(t)$
back to $s_b(t)$.

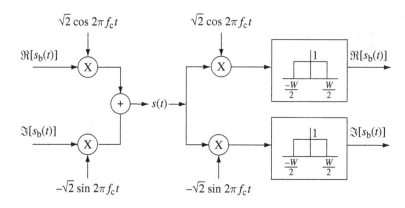

Figure 2.9 System diagram
from the baseband transmitted
signal $x_b(t)$ to the baseband
received signal $y_b(t)$.

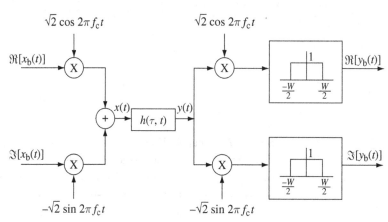

in-phase component I and $\Im[x_b(t)]$ the quadrature component Q (rotated
by $\pi/2$). We now calculate the baseband equivalent channel. Substituting
$x(t) = \sqrt{2}\Re[x_b(t)e^{j2\pi f_c t}]$ and $y(t) = \sqrt{2}\Re[y_b(t)e^{j2\pi f_c t}]$ into (2.14) we get

$$
\begin{aligned}
\Re[y_b(t)e^{j2\pi f_c t}] &= \sum_i a_i(t)\Re[x_b(t-\tau_i(t))e^{j2\pi f_c(t-\tau_i(t))}] \\
&= \Re\left[\left\{\sum_i a_i(t)x_b(t-\tau_i(t))e^{-j2\pi f_c\tau_i(t)}\right\}e^{j2\pi f_c t}\right].
\end{aligned}
\tag{2.24}
$$

Similarly, one can obtain (Exercise 2.13)

$$
\Im[y_b(t)e^{j2\pi f_c t}] = \Im\left[\left\{\sum_i a_i(t)x_b(t-\tau_i(t))e^{-j2\pi f_c\tau_i(t)}\right\}e^{j2\pi f_c t}\right].
\tag{2.25}
$$

Hence, the baseband equivalent channel is

$$
y_b(t) = \sum_i a_i^b(t)x_b(t-\tau_i(t)),
\tag{2.26}
$$

where

$$a_i^b(t) := a_i(t)e^{-j2\pi f_c \tau_i(t)}.\tag{2.27}$$

The input/output relationship in (2.26) is also that of a linear time-varying system, and the baseband equivalent impulse response is

$$h_b(\tau, t) = \sum_i a_i^b(t)\delta(\tau - \tau_i(t)).\tag{2.28}$$

This representation is easy to interpret in the time domain, where the effect of the carrier frequency can be seen explicitly. The baseband output is the sum, over each path, of the delayed replicas of the baseband input. The magnitude of the ith such term is the magnitude of the response on the given path; this changes slowly, with significant changes occurring on the order of seconds or more. The phase is changed by $\pi/2$ (i.e., is changed significantly) when the delay on the path changes by $1/(4f_c)$, or equivalently, when the path length changes by a quarter wavelength, i.e., by $c/(4f_c)$. If the path length is changing at velocity v, the time required for such a phase change is $c/(4f_c v)$. Recalling that the Doppler shift D at frequency f is fv/c, and noting that $f \approx f_c$ for narrowband communication, the time required for a $\pi/2$ phase change is $1/(4D)$. For the single reflecting wall example, this is about 5 ms (assuming $f_c = 900\,\text{MHz}$ and $v = 60\,\text{km/h}$). The phases of both paths are rotating at this rate but in opposite directions.

Note that the Fourier transform $H_b(f; t)$ of $h_b(\tau, t)$ for a fixed t is simply $H(f + f_c; t)$, i.e., the frequency response of the original system (at a fixed t) shifted by the carrier frequency. This provides another way of thinking about the baseband equivalent channel.

2.2.3 A discrete-time baseband model

The next step in creating a useful channel model is to convert the continuous-time channel to a discrete-time channel. We take the usual approach of the sampling theorem. Assume that the input waveform is band-limited to W. The baseband equivalent is then limited to $W/2$ and can be represented as

$$x_b(t) = \sum_n x[n]\text{sinc}(Wt - n),\tag{2.29}$$

where $x[n]$ is given by $x_b(n/W)$ and $\text{sinc}(t)$ is defined as

$$\text{sinc}(t) := \frac{\sin(\pi t)}{\pi t}.\tag{2.30}$$

This representation follows from the sampling theorem, which says that any waveform band-limited to $W/2$ can be expanded in terms of the orthogonal basis $\{\text{sinc}(Wt-n)\}_n$, with coefficients given by the samples (taken uniformly at integer multiples of $1/W$).

Using (2.26), the baseband output is given by

$$y_b(t) = \sum_n x[n] \sum_i a_i^b(t)\text{sinc}(Wt - W\tau_i(t) - n). \qquad (2.31)$$

The sampled outputs at multiples of $1/W$, $y[m] := y_b(m/W)$, are then given by

$$y[m] = \sum_n x[n] \sum_i a_i^b(m/W)\text{sinc}[m - n - \tau_i(m/W)W]. \qquad (2.32)$$

The sampled output $y[m]$ can equivalently be thought of as the projection of the waveform $y_b(t)$ onto the waveform $W\text{sinc}(Wt - m)$. Let $\ell := m - n$. Then

$$y[m] = \sum_\ell x[m - \ell] \sum_i a_i^b(m/W)\text{sinc}[\ell - \tau_i(m/W)W]. \qquad (2.33)$$

By defining

$$\boxed{h_\ell[m] := \sum_i a_i^b(m/W)\text{sinc}[\ell - \tau_i(m/W)W],} \qquad (2.34)$$

(2.33) can be written in the simple form

$$y[m] = \sum_\ell h_\ell[m]\, x[m - \ell]. \qquad (2.35)$$

We denote $h_\ell[m]$ as the ℓth (complex) channel filter tap at time m. Its value is a function of mainly the gains $a_i^b(t)$ of the paths, whose delays $\tau_i(t)$ are close to ℓ/W (Figure 2.10). In the special case where the gains $a_i^b(t)$ and the delays $\tau_i(t)$ of the paths are time-invariant, (2.34) simplifies to

$$h_\ell = \sum_i a_i^b \,\text{sinc}[\ell - \tau_i W], \qquad (2.36)$$

and the channel is linear time-invariant. The ℓth tap can be interpreted as the sample (ℓ/W)th of the low-pass filtered baseband channel response $h_b(\tau)$ (cf. (2.19)) convolved with $\text{sinc}(W\tau)$.

We can interpret the sampling operation as modulation and demodulation in a communication system. At time n, we are modulating the complex symbol $x[m]$ (in-phase plus quadrature components) by the sinc pulse before the up-conversion. At the receiver, the received signal is sampled at times m/W

Figure 2.10 Due to the decay
of the sinc function, the ith
path contributes most
significantly to the ℓth tap if
its delay falls in the window
$[\ell/W - 1/(2W), \ell/W + 1/(2W)]$.

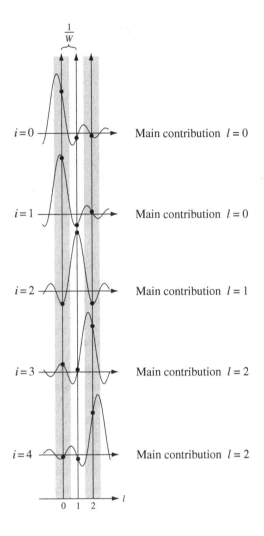

at the output of the low-pass filter. Figure 2.11 shows the complete system. In practice, other transmit pulses, such as the raised cosine pulse, are often used in place of the sinc pulse, which has rather poor time-decay property and tends to be more susceptible to timing errors. This necessitates sampling above the Nyquist sampling rate, but does not alter the essential nature of the model. Hence we will confine to Nyquist sampling.

Due to the Doppler spread, the bandwidth of the output $y_b(t)$ is generally slightly larger than the bandwidth $W/2$ of the input $x_b(t)$, and thus the output samples $\{y[m]\}$ do not fully represent the output waveform. This problem is usually ignored in practice, since the Doppler spread is small (of the order of tens to hundreds of Hz) compared to the bandwidth W. Also, it is very convenient for the sampling rate of the input and output to be the same. Alternatively, it would be possible to sample the output at twice the rate of the input. This would recapture all the information in the received waveform.

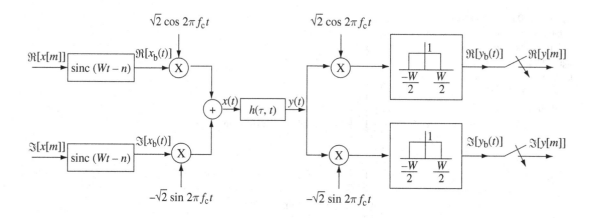

Figure 2.11 System diagram from the baseband transmitted symbol $x[m]$ to the baseband sampled received signal $y[m]$.

The number of taps would be almost doubled because of the reduced sample interval, but it would typically be somewhat less than doubled since the representation would not spread the path delays so much.

Discussion 2.1 Degrees of freedom

The symbol $x[m]$ is the mth sample of the transmitted signal; there are W samples per second. Each symbol is a complex number; we say that it represents one (complex) *dimension* or *degree of freedom*. The continuous-time signal $x(t)$ of duration one second corresponds to W discrete symbols; thus we could say that the band-limited, continuous-time signal has W degrees of freedom, per second.

The mathematical justification for this interpretation comes from the following important result in communication theory: the signal space of complex continuous-time signals of duration T which have most of their energy within the frequency band $[-W/2, W/2]$ has dimension approximately WT. (A precise statement of this result is in standard communication theory text/books; see Section 5.3 of [148] for example.) This result reinforces our interpretation that a continuous-time signal with bandwidth W can be represented by W complex dimensions per second.

The received signal $y(t)$ is also band-limited to approximately W (due to the Doppler spread, the bandwidth is slightly larger than W) and has W complex dimensions per second. From the point of view of communication over the channel, the *received* signal space is what matters because it dictates the number of different signals which can be reliably distinguished at the receiver. Thus, we define the *degrees of freedom of the channel* to be the dimension of the received signal space, and whenever we refer to the signal space, we implicitly mean the received signal space unless stated otherwise.

2.2.4 Additive white noise

As a last step, we include additive noise in our input/output model. We make the standard assumption that $w(t)$ is zero-mean additive white Gaussian noise (AWGN) with power spectral density $N_0/2$ (i.e., $E[w(0)w(t)] = (N_0/2)\delta(t)$). The model (2.14) is now modified to be

$$y(t) = \sum_i a_i(t)x(t - \tau_i(t)) + w(t). \tag{2.37}$$

See Figure 2.12. The discrete-time baseband-equivalent model (2.35) now becomes

$$y[m] = \sum_\ell h_\ell[m]x[m - \ell] + w[m], \tag{2.38}$$

where $w[m]$ is the low-pass filtered noise at the sampling instant m/W. Just like the signal, the white noise $w(t)$ is down-converted, filtered at the baseband and ideally sampled. Thus, it can be verified (Exercise 2.11) that

$$\Re(w[m]) = \int_{-\infty}^{\infty} w(t)\psi_{m,1}(t)\mathrm{d}t, \tag{2.39}$$

$$\Im(w[m]) = \int_{-\infty}^{\infty} w(t)\psi_{m,2}(t)\mathrm{d}t, \tag{2.40}$$

where

$$\psi_{m,1}(t) := \sqrt{2W}\cos(2\pi f_c t)\mathrm{sinc}(Wt - m),$$
$$\psi_{m,2}(t) := -\sqrt{2W}\sin(2\pi f_c t)\mathrm{sinc}(Wt - m). \tag{2.41}$$

It can further be shown that $\{\psi_{m,1}(t), \psi_{m,2}(t)\}_m$ forms an *orthonormal set* of waveforms, i.e., the waveforms are orthogonal to each other (Exercise 2.12). In Appendix A we review the definition and basic properties of white Gaussian random *vectors* (i.e., vectors whose components are independent and identically distributed (i.i.d.) Gaussian random variables). A key property is that the projections of a white Gaussian random vector onto any orthonormal vectors are independent and identically distributed Gaussian random variables. Heuristically, one can think of continuous-time Gaussian white noise as an infinite-dimensional white random vector and the above property carries through: the projections onto orthogonal waveforms are uncorrelated and hence independent. Hence the discrete-time noise process $\{w[m]\}$ is white, i.e., independent over time; moreover, the real and imaginary components are i.i.d. Gaussians with variances $N_0/2$. A complex Gaussian random variable X whose real and imaginary components are i.i.d. satisfies a *circular symmetry* property: $e^{j\phi}X$ has the same distribution as X for any ϕ. We shall call such a random variable *circular symmetric complex*

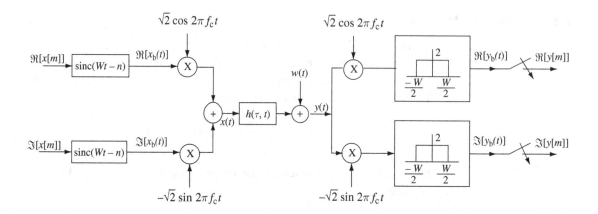

Figure 2.12 A complete system diagram.

Gaussian, denoted by $\mathcal{CN}(0, \sigma^2)$, where $\sigma^2 = E[|X|^2]$. The concept of circular symmetry is discussed further in Section A.1.3 of Appendix A.

The assumption of AWGN essentially means that we are assuming that the primary source of the noise is at the receiver or is radiation impinging on the receiver that is independent of the paths over which the signal is being received. This is normally a very good assumption for most communication situations.

2.3 Time and frequency coherence

2.3.1 Doppler spread and coherence time

An important channel parameter is the time-scale of the variation of the channel. How fast do the taps $h_\ell[m]$ vary as a function of time m? Recall that

$$h_\ell[m] = \sum_i a_i^b(m/W)\mathrm{sinc}[\ell - \tau_i(m/W)W]$$

$$= \sum_i a_i(m/W)e^{-j2\pi f_c\tau_i(m/W)}\mathrm{sinc}[\ell - \tau_i(m/W)W]. \quad (2.42)$$

Let us look at this expression term by term. From Section 2.2.2 we gather that significant changes in a_i occur over periods of seconds or more. Significant changes in the phase of the ith path occur at intervals of $1/(4D_i)$, where $D_i = f_c\tau_i'(t)$ is the Doppler shift for that path. When the different paths contributing to the ℓth tap have different Doppler shifts, the magnitude of $h_\ell[m]$ changes significantly. This is happening at the time-scale inversely proportional to the largest difference between the Doppler shifts, the *Doppler spread* D_s:

$$\boxed{D_s := \max_{i, j} f_c|\tau_i'(t) - \tau_j'(t)|,} \quad (2.43)$$

where the maximum is taken over all the paths that contribute significantly to a tap.[7] Typical intervals for such changes are on the order of 10 ms. Finally, changes in the sinc term of (2.42) due to the time variation of each $\tau_i(t)$ are proportional to the bandwidth, whereas those in the phase are proportional to the carrier frequency, which is typically much larger. Essentially, it takes much longer for a path to move from one tap to the next than for its phase to change significantly. Thus, the fastest changes in the filter taps occur because of the phase changes, and these are significant over delay changes of $1/(4D_s)$.

The coherence time T_c of a wireless channel is defined (in an order of magnitude sense) as the interval over which $h_\ell[m]$ changes significantly as a function of m. What we have found, then, is the important relation

$$T_c = \frac{1}{4D_s}. \qquad (2.44)$$

This is a somewhat imprecise relation, since the largest Doppler shifts may belong to paths that are too weak to make a difference. We could also view a phase change of $\pi/4$ to be significant, and thus replace the factor of 4 above by 8. Many people instead replace the factor of 4 by 1. The important thing is to recognize that the major effect in determining time coherence is the Doppler spread, and that the relationship is reciprocal; the larger the Doppler spread, the smaller the time coherence.

In the wireless communication literature, channels are often categorized as *fast fading* and *slow fading*, but there is little consensus on what these terms mean. In this book, we will call a channel fast fading if the coherence time T_c is much shorter than the delay requirement of the application, and slow fading if T_c is longer. The operational significance of this definition is that, in a fast fading channel, one can transmit the coded symbols over multiple fades of the channel, while in a slow fading channel, one cannot. Thus, whether a channel is fast or slow fading depends not only on the environment but also on the application; voice, for example, typically has a short delay requirement of less than 100 ms, while some types of data applications can have a laxer delay requirement.

2.3.2 Delay spread and coherence bandwidth

Another important general parameter of a wireless system is the multipath delay spread, T_d, defined as the difference in propagation time between the

---

[7] The Doppler spread can in principle be different for different taps. Exercise 2.10 explores this possibility.

longest and shortest path, counting only the paths with significant energy. Thus,

$$T_{\mathrm{d}} := \max_{i,j} |\tau_i(t) - \tau_j(t)|. \qquad (2.45)$$

This is defined as a function of t, but we regard it as an order of magnitude quantity, like the time coherence and Doppler spread. If a cell or LAN has a linear extent of a few kilometers or less, it is very unlikely to have path lengths that differ by more than 300 to 600 meters. This corresponds to path delays of one or two microseconds. As cells become smaller due to increased cellular usage, T_{d} also shrinks. As was already mentioned, typical wireless channels are underspread, which means that the delay spread T_{d} is much smaller than the coherence time T_{c}.

The bandwidths of cellular systems range between several hundred kilohertz and several megahertz, and thus, for the above multipath delay spread values, all the path delays in (2.34) lie within the peaks of two or three sinc functions; more often, they lie within a single peak. Adding a few extra taps to each channel filter because of the slow decay of the sinc function, we see that cellular channels can be represented with at most four or five channel filter taps. On the other hand, there is a recent interest in *ultra-wideband* (UWB) communication, operating from 3.1 to 10.6 GHz. These channels can have up to a few hundred taps.

When we study modulation and detection for cellular systems, we shall see that the receiver must estimate the values of these channel filter taps. The taps are estimated via transmitted and received waveforms, and thus the receiver makes no explicit use of (and usually does not have) any information about individual path delays and path strengths. This is why we have not studied the details of propagation over multiple paths with complicated types of reflection mechanisms. All we really need is the aggregate values of gross physical mechanisms such as Doppler spread, coherence time, and multipath spread.

The delay spread of the channel dictates its *frequency coherence*. Wireless channels change both in time and frequency. The time coherence shows us how quickly the channel changes in time, and similarly, the frequency coherence shows how quickly it changes in frequency. We first understood about channels changing in time, and correspondingly about the duration of fades, by studying the simple example of a direct path and a single reflected path. That same example also showed us how channels change with frequency. We can see this in terms of the frequency response as well.

Recall that the frequency response at time t is

$$H(f; t) = \sum_i a_i(t) \mathrm{e}^{-\mathrm{j}2\pi f \tau_i(t)}. \qquad (2.46)$$

The contribution due to a particular path has a phase linear in f. For multiple paths, there is a differential phase, $2\pi f(\tau_i(t) - \tau_k(t))$. This differential

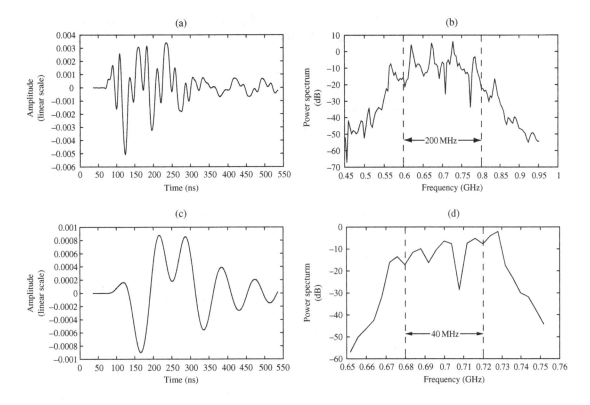

Figure 2.13 (a) A channel over 200 MHz is frequency-selective, and the impulse response has many taps. (b) The spectral content of the same channel. (c) The same channel over 40 MHz is flatter, and has for fewer taps. (d) The spectral contents of the same channel, limited to 40 MHz bandwidth. At larger bandwidths, the same physical paths are resolved into a finer resolution.

phase causes selective fading in frequency. This says that $E_r(f, t)$ changes significantly, not only when t changes by $1/(4D_s)$, but also when f changes by $1/(2T_d)$. This argument extends to an arbitrary number of paths, so the *coherence bandwidth*, W_c, is given by

$$W_c = \frac{1}{2T_d}. \tag{2.47}$$

This relationship, like (2.44), is intended as an order of magnitude relation, essentially pointing out that the coherence bandwidth is reciprocal to the multipath spread. When the bandwidth of the input is considerably less than W_c, the channel is usually referred to as *flat fading*. In this case, the delay spread T_d is much less than the symbol time $1/W$, and a single channel filter tap is sufficient to represent the channel. When the bandwidth is much larger than W_c, the channel is said to be *frequency-selective*, and it has to be represented by multiple taps. Note that flat or frequency-selective fading is not a property of the channel alone, but of the relationship between the bandwidth W and the coherence bandwidth T_d (Figure 2.13).

The physical parameters and the time-scale of change of key parameters of the discrete-time baseband channel model are summarized in Table 2.1. The different types of channels are summarized in Table 2.2.

Table 2.1 A summary of the physical parameters of the channel and the time-scale of change of the key parameters in its discrete-time baseband model.

Key channel parameters and time-scales	Symbol	Representative values
Carrier frequency	f_c	1 GHz
Communication bandwidth	W	1 MHz
Distance between transmitter and receiver	d	1 km
Velocity of mobile	v	64 km/h
Doppler shift for a path	$D = f_c v/c$	50 Hz
Doppler spread of paths corresponding to a tap	D_s	100 Hz
Time-scale for change of path amplitude	d/v	1 minute
Time-scale for change of path phase	$1/(4D)$	5 ms
Time-scale for a path to move over a tap	$c/(vW)$	20 s
Coherence time	$T_c = 1/(4D_s)$	2.5 ms
Delay spread	T_d	1 µs
Coherence bandwidth	$W_c = 1/(2T_d)$	500 kHz

Table 2.2 A summary of the types of wireless channels and their defining characteristics.

Types of channel	Defining characteristic
Fast fading	$T_c \ll$ delay requirement
Slow fading	$T_c \gg$ delay requirement
Flat fading	$W \ll W_c$
Frequency-selective fading	$W \gg W_c$
Underspread	$T_d \ll T_c$

2.4 Statistical channel models

2.4.1 Modeling philosophy

We defined Doppler spread and multipath spread in the previous section as quantities associated with a given receiver at a given location, velocity, and time. However, we are interested in a characterization that is valid over some range of conditions. That is, we recognize that the channel filter taps $\{h_\ell[m]\}$ must be measured, but we want a statistical characterization of how many taps are necessary, how quickly they change and how much they vary.

Such a characterization requires a probabilistic model of the channel tap values, perhaps gathered by statistical measurements of the channel. We are familiar with describing additive noise by such a probabilistic model (as a Gaussian random variable). We are also familiar with evaluating error probability while communicating over a channel using such models. These error probability evaluations, however, depend critically on the independence and Gaussian distribution of the noise variables.

It should be clear from the description of the physical mechanisms generating Doppler spread and multipath spread that probabilistic models for the channel filter taps are going to be far less believable than the models for additive noise. On the other hand, we need such models, even if they are quite inaccurate. Without models, systems are designed using experience and experimentation, and creativity becomes somewhat stifled. Even with highly over-simplified models, we can compare different system approaches and get a sense of what types of approaches are worth pursuing.

To a certain extent, all analytical work is done with simplified models. For example, white Gaussian noise (WGN) is often assumed in communication models, although we know the model is valid only over sufficiently small frequency bands. With WGN, however, we expect the model to be quite good when used properly. For wireless channel models, however, probabilistic models are quite poor and only provide order-of-magnitude guides to system design and performance. We will see that we can define Doppler spread, multipath spread, etc. much more cleanly with probabilistic models, but the underlying problem remains that these channels are very different from each other and cannot really be characterized by probabilistic models. At the same time, there is a large literature based on probabilistic models for wireless channels, and it has been highly useful for providing insight into wireless systems. However, it is important to understand the robustness of results based on these models.

There is another question in deciding what to model. Recall the continuous-time multipath fading channel

$$y(t) = \sum_i a_i(t)x(t - \tau_i(t)) + w(t). \tag{2.48}$$

This contains an exact specification of the delay and magnitude of each path. From this, we derived a discrete-time baseband model in terms of channel filter taps as

$$y[m] = \sum_\ell h_\ell[m]x[m - \ell] + w[m], \tag{2.49}$$

where

$$h_\ell[m] = \sum_i a_i(m/W)e^{-j2\pi f_c\tau_i(m/W)}\text{sinc}[\ell - \tau_i(m/W)W]. \tag{2.50}$$

We used the sampling theorem expansion in which $x[m] = x_b(m/W)$ and $y[m] = y_b(m/W)$. Each channel tap $h_\ell[m]$ contains an aggregate of paths, with the delays smoothed out by the baseband signal bandwidth.

Fortunately, it is the filter taps that must be modeled for input/output descriptions, and also fortunately, the filter taps often contain a sufficient path aggregation so that a statistical model might have a chance of success.

2.4.2 Rayleigh and Rician fading

The simplest probabilistic model for the channel filter taps is based on the assumption that there are a large number of statistically independent reflected and scattered paths with random amplitudes in the delay window corresponding to a single tap. The phase of the ith path is $2\pi f_c \tau_i$ modulo 2π. Now, $f_c \tau_i = d_i/\lambda$, where d_i is the distance travelled by the ith path and λ is the carrier wavelength. Since the reflectors and scatterers are far away relative to the carrier wavelength, i.e., $d_i \gg \lambda$, it is reasonable to assume that the phase for each path is uniformly distributed between 0 and 2π and that the phases of different paths are independent. The contribution of each path in the tap gain $h_\ell[m]$ is

$$a_i(m/W)e^{-j2\pi f_c \tau_i(m/W)}\mathrm{sinc}[\ell - \tau_i(m/W)W] \tag{2.51}$$

and this can be modeled as a circular symmetric complex random variable.[8] Each tap $h_\ell[m]$ is the sum of a large number of such small independent circular symmetric random variables. It follows that $\Re(h_\ell[m])$ is the sum of many small independent real random variables, and so by the Central Limit Theorem, it can reasonably be modeled as a zero-mean Gaussian random variable. Similarly, because of the uniform phase, $\Re(h_\ell[m]e^{j\phi})$ is Gaussian with the same variance for any fixed ϕ. This assures us that $h_\ell[m]$ is in fact circular symmetric $\mathcal{CN}(0, \sigma_\ell^2)$ (see Section A.1.3 in Appendix A for an elaboration). It is assumed here that the variance of $h_\ell[m]$ is a function of the tap ℓ, but independent of time m (there is little point in creating a probabilistic model that depends on time). With this assumed Gaussian probability density, we know that the magnitude $|h_\ell[m]|$ of the ℓth tap is a *Rayleigh* random variable with density (cf. (A.20) in Appendix A and Exercise 2.14)

$$\boxed{\frac{2x}{\sigma^2}\exp\left\{\frac{-x^2}{\sigma^2}\right\}, \quad x \geq 0,} \tag{2.52}$$

and the squared magnitude $|h_\ell[m]|^2$ is exponentially distributed with density

$$\boxed{\frac{1}{\sigma_\ell^2}\exp\left\{\frac{-x}{\sigma_\ell^2}\right\}, \quad x \geq 0.} \tag{2.53}$$

This model, which is called *Rayleigh fading*, is quite reasonable for scattering mechanisms where there are many small reflectors, but is adopted primarily for its simplicity in typical cellular situations with a relatively small number of reflectors. The word *Rayleigh* is almost universally used for this

---

[8] See Section A.1.3 in Appendix A for a more in-depth discussion of circular symmetric random variables and vectors.

model, but the assumption is that the tap gains are circularly symmetric complex Gaussian random variables.

There is a frequently used alternative model in which the line-of-sight path (often called a *specular* path) is large and has a known magnitude, and that there are also a large number of independent paths. In this case, $h_\ell[m]$, at least for one value of ℓ, can be modeled as

$$h_\ell[m] = \sqrt{\frac{\kappa}{\kappa+1}}\sigma_\ell e^{j\theta} + \sqrt{\frac{1}{\kappa+1}}\mathcal{CN}(0, \sigma_\ell^2) \qquad (2.54)$$

with the first term corresponding to the specular path arriving with uniform phase θ and the second term corresponding to the aggregation of the large number of reflected and scattered paths, independent of θ. The parameter κ (so-called K-factor) is the ratio of the energy in the specular path to the energy in the scattered paths; the larger κ is, the more deterministic is the channel. The magnitude of such a random variable is said to have a *Rician* distribution. Its density has quite a complicated form; it is often a better model of fading than the Rayleigh model.

2.4.3 Tap gain auto-correlation function

Modeling each $h_\ell[m]$ as a complex random variable provides part of the statistical description that we need, but this is not the most important part. The more important issue is how these quantities vary with time. As we will see in the rest of the book, the rate of channel variation has significant impact on several aspects of the communication problem. A statistical quantity that models this relationship is known as the *tap gain auto-correlation function*, $R_\ell[n]$. It is defined as

$$R_\ell[n] := \mathbb{E}\left\{h_\ell^*[m]h_\ell[m+n]\right\}. \qquad (2.55)$$

For each tap ℓ, this gives the auto-correlation function of the sequence of random variables modeling that tap as it evolves in time. We are tacitly assuming that this is not a function of time m. Since the sequence of random variables $\{h_\ell[m]\}$ for any given ℓ has both a mean and covariance function that does not depend on m, this sequence is wide-sense stationary. We also assume that, as a random variable, $h_\ell[m]$ is independent of $h_{\ell'}[m']$ for all $\ell \neq \ell'$ and all m, m'. This final assumption is intuitively plausible since paths in different ranges of delay contribute to $h_\ell[m]$ for different values of ℓ.[9]

The coefficient $R_\ell[0]$ is proportional to the energy received in the ℓth tap. The multipath spread T_d can be defined as the product of $1/W$ times the range of ℓ which contains most of the total energy $\sum_{\ell=0}^{\infty} R_\ell[0]$. This is

---

[9]   One could argue that a moving reflector would gradually travel from the range of one tap to another, but as we have seen, this typically happens over a very large time-scale.

somewhat preferable to our previous "definition" in that the statistical nature of T_d becomes explicit and the reliance on some sort of stationarity becomes explicit. Now, we can also define the coherence time T_c more explicitly as the smallest value of $n > 0$ for which $R_\ell[n]$ is significantly different from $R_\ell[0]$. With both of these definitions, we still have the ambiguity of what "significant" means, but we are now facing the reality that these quantities must be viewed as statistics rather than as instantaneous values.

The tap gain auto-correlation function is useful as a way of expressing the statistics for how tap gains change given a particular bandwidth W, but gives little insight into questions related to choice of a bandwidth for communication. If we visualize increasing the bandwidth, we can see several things happening. First, the ranges of delay that are separated into different taps ℓ become narrower ($1/W$ seconds), so there are fewer paths corresponding to each tap, and thus the Rayleigh approximation becomes poorer. Second, the sinc functions of (2.50) become narrower, and $R_\ell[0]$ gives a finer grained picture of the amount of power being received in the ℓth delay window of width $1/W$. In summary, as we try to apply this model to larger W, we get more detailed information about delay and correlation at that delay, but the information becomes more questionable.

Example 2.2 Clarke's model

This is a popular statistical model for flat fading. The transmitter is fixed, the mobile receiver is moving at speed v, and the transmitted signal is scattered by stationary objects around the mobile. There are K paths, the ith path arriving at an angle $\theta_i := 2\pi i/K$, $i = 0, \ldots, K - 1$, with respect to the direction of motion. K is assumed to be large. The scattered path arriving at the mobile at the angle θ has a delay of $\tau_\theta(t)$ and a time-invariant gain a_θ, and the input/output relationship is given by

$$y(t) = \sum_{i=0}^{K-1} a_{\theta_i} x(t - \tau_{\theta_i}(t)) \qquad (2.56)$$

The most general version of the model allows the received power distribution $p(\theta)$ and the antenna gain pattern $\alpha(\theta)$ to be arbitrary functions of the angle θ, but the most common scenario assumes uniform power distribution and isotropic antenna gain pattern, i.e., the amplitudes $a_\theta = a/\sqrt{K}$ for all angles θ. This models the situation when the scatterers are located in a ring around the mobile (Figure 2.14). We scale the amplitude of each path by $\sqrt{K}$ so that the total received energy along all paths is a^2; for large K, the received energy along each path is a small fraction of the total energy.

Suppose the communication bandwidth W is much smaller than the reciprocal of the delay spread. The complex baseband channel can be represented by a single tap at each time:

$$y[m] = h_0[m]x[m] + w[m]. \qquad (2.57)$$

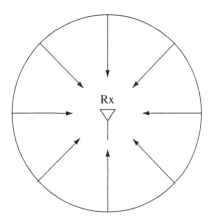

Figure 2.14 The one-ring model.

The phase of the signal arriving at time 0 from an angle θ is $2\pi f_c \tau_\theta(0)$ mod 2π, where f_c is the carrier frequency. Making the assumption that this phase is uniformly distributed in $[0, 2\pi]$ and independently distributed across all angles θ, the tap gain process $\{h_0[m]\}$ is a sum of many small independent contributions, one from each angle. By the Central Limit Theorem, it is reasonable to model the process as Gaussian. Exercise 2.17 shows further that the process is in fact stationary with an autocorrelation function $R_0[n]$ given by:

$$R_0[n] = 2a^2 \pi J_0 \left(n\pi D_s/W \right) \tag{2.58}$$

where $J_0(\cdot)$ is the zeroth-order Bessel function of the first kind:

$$J_0(x) := \frac{1}{\pi} \int_0^\pi e^{jx\cos\theta} d\theta. \tag{2.59}$$

and $D_s = 2f_c v/c$ is the Doppler spread. The power spectral density $S(f)$, defined on $[-1/2, +1/2]$, is given by

$$S(f) = \begin{cases} \frac{4a^2 W}{D_s \sqrt{1 - (2fW/D_s)^2}} & -D_s/(2W) \leqslant f \leqslant +D_s/(2W) \\ 0 & \text{else.} \end{cases} \tag{2.60}$$

This can be verified by computing the inverse Fourier transform of (2.60) to be (2.58). Plots of the autocorrelation function and the spectrum for are shown in Figure 2.15. If we define the coherence time T_c to be the value of n/W such that $R_0[n] = 0.05R_0[0]$, then

$$T_c = \frac{J_0^{-1}(0.05)}{\pi D_s}, \tag{2.61}$$

i.e., the coherence time is inversely proportional to D_s.

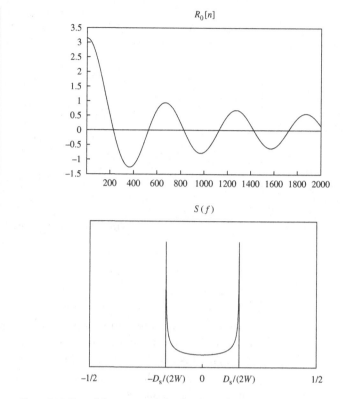

Figure 2.15 Plots of the auto-correlation function and Doppler spectrum in Clarke's model.

In Exercise 2.17, you will also verify that $S(f)\mathrm{d}f$ has the physical interpretation of the received power along paths that have Doppler shifts in the range $[f, f + \mathrm{d}f]$. Thus, $S(f)$ is also called the *Doppler spectrum*. Note that $S(f)$ is zero beyond the maximum Doppler shift.

Chapter 2 The main plot

Large-scale fading

Variation of signal strength over distances of the order of cell sizes. Received power decreases with distance r like:

$$\frac{1}{r^2} \quad \text{(free space)}$$

$$\frac{1}{r^4} \quad \text{(reflection from ground plane)}.$$

Decay can be even faster due to shadowing and scattering effects.

Small-scale fading

Variation of signal strength over distances of the order of the carrier wavelength, due to constructive and destructive interference of multipaths. Key parameters:

$$\text{Doppler spread } D_s \longleftrightarrow \text{coherence time } T_c \sim 1/D_s$$

Doppler spread is proportional to the velocity of the mobile and to the angular spread of the arriving paths.

$$\text{delay spread } T_d \longleftrightarrow \text{coherence bandwidth } W_c \sim 1/T_d$$

Delay spread is proportional to the difference between the lengths of the shortest and the longest paths.

Input/output channel models

- Continuous-time passband (2.14):

$$y(t) = \sum_i a_i(t)x(t - \tau_i(t)).$$

- Continuous-time complex baseband (2.26):

$$y_b(t) = \sum_i a_i(t)e^{-j2\pi f_c \tau_i(t)} x_b(t - \tau_i(t)).$$

- Discrete-time complex baseband with AWGN (2.38):

$$y[m] = \sum_\ell h_\ell[m]x[m - \ell] + w[m].$$

The ℓth tap is the aggregation of the physical paths with delays in $[\ell/W - 1/(2W), \ell/W + 1/(2W)]$.

Statistical channel models

- $\{h_\ell[m]\}_m$ is modeled as circular symmetric processes independent across the taps.
- If for all taps,

$$h_\ell[m] \sim \mathcal{CN}(0, \sigma_\ell^2),$$

the model is called *Rayleigh*.
- If for one tap,

$$h_\ell[m] = \sqrt{\frac{\kappa}{\kappa + 1}}\sigma_\ell e^{j\theta} + \sqrt{\frac{1}{\kappa + 1}}\mathcal{CN}(0, \sigma_\ell^2),$$

the model is called *Rician* with K-factor κ.

> • The tap gain auto-correlation function $R_\ell[n] := \mathbb{E}[h_\ell^*[0]h_\ell[n]]$ models the dependency over time.
> • The delay spread is $1/W$ times the range of taps ℓ which contains most of the total gain $\sum_{\ell=0}^{\infty} R_\ell[0]$. The coherence time is $1/W$ times the range of n for which $R_\ell[n]$ is significantly different from $R_\ell[0]$.

2.5 Bibliographical notes

This chapter was modified from R. G. Gallager's MIT 6.450 course notes on digital communication. The focus is on small-scale multipath fading. Large-scale fading models are discussed in many texts; see for example Rappaport [98]. Clarke's model was introduced in [22] and elaborated further in [62]. Our derivation here of the Clarke power spectrum follows the approach of [111].

2.6 Exercises

Exercise 2.1 (Gallager) Consider the electric field in (2.4).

1. It has been derived under the assumption that the motion is in the direction of the line-of-sight from sending antenna to receive antenna. Find the electric field assuming that ϕ is the angle between the line-of-sight and the direction of motion of the receiver. Assume that the range of time of interest is small enough so that changes in (θ, ψ) can be ignored.
2. Explain why, and under what conditions, it is a reasonable approximation to ignore the change in (θ, ψ) over small intervals of time.

Exercise 2.2 (Gallager) Equation (2.13) was derived under the assumption that $r(t) \approx d$. Derive an expression for the received waveform for general $r(t)$. Break the first term in (2.11) into two terms, one with the same numerator but the denominator $2d - r_0 - vt$ and the other with the remainder. Interpret your result.

Exercise 2.3 In the two-path example in Sections 2.1.3 and 2.1.4, the wall is on the right side of the receiver so that the reflected wave and the direct wave travel in opposite directions. Suppose now that the reflecting wall is on the left side of transmitter. Redo the analysis. What is the nature of the multipath fading, both over time and over frequency? Explain any similarity or difference with the case considered in Sections 2.1.3 and 2.1.4.

Exercise 2.4 A mobile receiver is moving at a speed v and is receiving signals arriving along two reflected paths which make angles θ_1 and θ_2 with the direction of motion. The transmitted signal is a sinusoid at frequency f.

1. Is the above information enough for estimating (i) the coherence time T_c; (ii) the coherence bandwidth W_c? If so, express them in terms of the given parameters. If not, specify what additional information would be needed.
2. Consider an environment in which there are reflectors and scatterers in all directions from the receiver and an environment in which they are clustered within a small

angular range. Using part (1), explain how the channel would differ in these two environments.

Exercise 2.5 Consider the propagation model in Section 2.1.5 where there is a reflected path from the ground plane.

1. Let r_1 be the length of the direct path in Figure 2.6. Let r_2 be the length of the reflected path (summing the path length from the transmitter to the ground plane and the path length from the ground plane to the receiver). Show that $r_2 - r_1$ is asymptotically equal to b/r and find the value of the constant b. *Hint*: Recall that for x small, $\sqrt{1+x} \approx 1 + x/2$ in the sense that $(\sqrt{1+x} - 1)/x \to 1/2$ as $x \to 0$.

2. Assume that the received waveform at the receive antenna is given by

$$E_r(f, t) = \frac{\alpha \cos 2\pi[ft - fr_1/c]}{r_1} - \frac{\alpha \cos 2\pi[ft - fr_2/c]}{r_2}. \qquad (2.62)$$

Approximate the denominator r_2 by r_1 in (2.62) and show that $E_r \approx \beta/r^2$ for r^{-1} much smaller than c/f. Find the value of β.

3. Explain why this asymptotic expression remains valid without first approximating the denominator r_2 in (2.62) by r_1.

Exercise 2.6 Consider the following simple physical model in just a *single* dimension. The source is at the origin and transmits an isotropic wave of angular frequency ω. The physical environment is filled with uniformly randomly located obstacles. We will model the inter-obstacle distance as an exponential random variable, i.e., it has the density[10]

$$\eta e^{-\eta r}, \qquad r \geq 0. \qquad (2.63)$$

Here $1/\eta$ is the mean distance between obstacles and captures the *density* of the obstacles. Viewing the source as a stream of photons, suppose each obstacle independently (from one photon to the other and independent of the behavior of the other obstacles) either absorbs the photon with probability γ or scatters it either to the left or to the right (both with equal probability $(1 - \gamma)/2$).

Now consider the path of a photon transmitted either to the left or to the right with equal probability from some fixed point on the line. The probability density function of the distance (denoted by r) to the first obstacle (the distance can be on either side of the starting point, so r takes values on the entire line) is equal to

$$q(r) := \frac{\eta e^{-\eta|r|}}{2}, \qquad r \in \mathcal{R}. \qquad (2.64)$$

So the probability density function of the distance at which the photon is absorbed upon hitting the first obstacle is equal to

$$f_1(r) := \gamma q(r), \qquad r \in \mathcal{R}. \qquad (2.65)$$

---

[10] This random arrangement of points on a line is called a *Poisson point process*.

1. Show that the probability density function of the distance from the origin at which the second obstacle is absorbed is

$$f_2(r) := \int_{-\infty}^{\infty} (1-\gamma)q(x)f_1(r-x)dx, \qquad r \in \mathcal{R}. \tag{2.66}$$

2. Denote by $f_k(r)$ the probability density function of the distance from the origin at which the photon is absorbed by exactly the kth obstacle it hits and show the recursive relation

$$f_{k+1}(r) = \int_{-\infty}^{\infty} (1-\gamma)q(x)f_k(r-x)dx, \qquad r \in \mathcal{R}. \tag{2.67}$$

3. Conclude from the previous step that the probability density function of the distance from the source at which the photon is absorbed (by some obstacle), denoted by $f(r)$, satisfies the recursive relation

$$f(r) = \gamma q(r) + (1-\gamma) \int_{-\infty}^{\infty} q(x)f(r-x)dx, \qquad r \in \mathcal{R}. \tag{2.68}$$

Hint: Observe that $f(r) = \sum_{k=1}^{\infty} f_k(r)$.
4. Show that

$$f(r) = \frac{\sqrt{\gamma}\eta}{2}e^{-\eta\sqrt{\gamma}|r|} \tag{2.69}$$

is a solution to the recursive relation in (2.68). *Hint*: Observe that the convolution between the probability densities $q(\cdot)$ and $f(\cdot)$ in (2.68) is more easily represented using Fourier transforms.

5. Now consider the photons that are absorbed at a distance of more than r from the source. This is the radiated power density at a distance r and is found by integrating $f(x)$ over the range (r, ∞) if $r > 0$ and $(-\infty, r)$ if $r < 0$. Calculate the radiated power density to be

$$\frac{e^{-\eta\sqrt{\gamma}|r|}}{2}, \tag{2.70}$$

and conclude that the power decreases exponentially with distance r. Also observe that with very low absorption ($\gamma \to 0$) or very few obstacles ($\eta \to 0$), the power density converges to 0.5; this is expected since the power splits equally on either side of the line.

Exercise 2.7 In Exercise 2.6, we considered a single-dimensional physical model of a scattering and absorption environment and concluded that power decays exponentially with distance. A reading exercise is to study [42], which considers a natural extension of this simple model to two- and three-dimensional spaces. Further, it extends the analysis to two- and three-dimensional physical models. While the analysis is more complicated, we arrive at the same conclusion: the radiated power decays exponentially with distance.

Exercise 2.8 (Gallager) Assume that a communication channel first filters the transmitted passband signal before adding WGN. Suppose the channel is known and the channel filter has an impulse response $h(t)$. Suppose that a QAM scheme with symbol duration T is developed without knowledge of the channel filtering. A baseband filter $\theta(t)$ is developed satisfying the Nyquist property that $\{\theta(t - kT)\}_k$ is an orthonormal set. The matched filter $\theta(-t)$ is used at the receiver before sampling and detection.

If one is aware of the channel filter $h(t)$, one may want to redesign either the baseband filter at the transmitter or the baseband filter at the receiver so that there is no intersymbol interference between receiver samples and so that the noise on the samples is i.i.d.

1. Which filter should one redesign?
2. Give an expression for the impulse response of the redesigned filter (assume a carrier frequency f_c).
3. Draw a figure of the various filters at passband to show why your solution is correct. (We suggest you do this before answering the first two parts.)

Exercise 2.9 Consider the two-path example in Section 2.1.4 with $d = 2\,\text{km}$ and the receiver at $1.5\,\text{km}$ from the transmitter moving at velocity $60\,\text{km/h}$ away from the transmitter. The carrier frequency is $900\,\text{MHz}$.

1. Plot in MATLAB the magnitudes of the taps of the discrete-time baseband channel at a fixed time t. Give a few plots for several bandwidths W so as to exhibit both flat and frequency-selective fading.
2. Plot the time variation of the phase and magnitude of a typical tap of the discrete-time baseband channel for a bandwidth where the channel is (approximately) flat and for a bandwidth where the channel is frequency-selective. How do the time-variations depend on the bandwidth? Explain.

Exercise 2.10 For each tap of the discrete-time channel response, the Doppler spread is the range of Doppler shifts of the paths contributing to that tap. Give an example of an environment (i.e. location of reflectors/scatterers with respect to the location of the transmitter and the receiver) in which the Doppler spread is the same for different taps and an environment in which they are different.

Exercise 2.11 Verify (2.39) and (2.40).

Exercise 2.12 In this problem we consider generating passband orthogonal waveforms from baseband ones.

1. Show that if the waveforms $\{\theta(t - nT)\}_n$ form an orthogonal set, then the waveforms $\{\psi_{n,1}, \psi_{n,2}\}_n$ also form an orthogonal set, provided that $\theta(t)$ is band-limited to $[-f_c, f_c]$. Here,

$$\psi_{n,1}(t) = \theta(t - nT)\cos 2\pi f_c t,$$
$$\psi_{n,2}(t) = \theta(t - nT)\sin 2\pi f_c t.$$

How should we normalize the energy of $\theta(t)$ to make the $\psi(t)$ *orthonormal*?

2. For a given f_c, find an example where the result in part (1) is false when the condition that $\theta(t)$ is band-limited to $[-f_c, f_c]$ is violated.

Exercise 2.13 Verify (2.25). Does this equation contain any more information about the communication system in Figure 2.9 beyond what is in (2.24)? Explain.

Exercise 2.14 Compute the probability density function of the magnitude $|X|$ of a complex circular symmetric Gaussian random variable X with variance σ^2.

Exercise 2.15 In the text we have discussed the various reasons why the channel tap gains, $h_\ell[m]$, vary in time (as a function of m) and how the various dynamics operate at different time-scales. The analysis is based on the assumption that communication takes place on a bandwidth W around a carrier frequency f_c with $f_c \gg W$. This assumption is not valid for *ultra-wideband* (UWB) communication systems, where the transmission bandwidth is from 3.1 GHz to 10.6 GHz, as regulated by the FCC. Redo the analysis for this system. What is the main mechanism that causes the tap gains to vary at the fastest time-scale, and what is this fastest time-scale determined by?

Exercise 2.16 In Section 2.4.2, we argue that the channel gain $h_\ell[m]$ at a particular time m can be assumed to be circular symmetric. Extend the argument to show that it is also reasonable to assume that the complex random vector

$$\mathbf{h} := \begin{pmatrix} h_\ell[m] \\ h_\ell[m+1] \\ \vdots \\ h_\ell[m+n] \end{pmatrix}$$

is circular symmetric for any n.

Exercise 2.17 In this question, we will analyze in detail Clarke's one-ring model discussed at the end of the chapter. Recall that the scatterers are assumed to be located in a ring around the receiver moving at speed v. There are K paths coming in at angles $\theta_i = 2\pi i/K$ with respect to the direction of motion of the mobile, $i = 0, \ldots, K-1$. The path coming at angle θ has a delay of $\tau_\theta(t)$ and a time-invariant gain $a/\sqrt{K}$ (not dependent on the angle), and the input/output relationship is given by

$$y(t) = \frac{a}{\sqrt{K}} \sum_{i=0}^{K-1} x(t - \tau_{\theta_i}(t)). \tag{2.71}$$

1. Give an expression for the impulse response $h(\tau, t)$ for this channel, and give an expression for $\tau_\theta(t)$ in terms of $\tau_\theta(0)$. (You can assume that the distance the mobile travelled in $[0, t]$ is small compared to the radius of the ring.)
2. Suppose communication takes place at carrier frequency f_c and over a narrowband of bandwidth W such that the delay spread of the channel T_d satisfies $T_d \ll 1/W$. Argue that the discrete-time baseband model can be approximately represented by a single tap

$$y[m] = h_0[m]x[m] + w[m], \tag{2.72}$$

and give an approximate expression for that tap in terms of the a_θ's and $\tau_\theta(t)$'s. *Hint*: Your answer should contain no sinc functions.
3. Argue that it is reasonable to assume that the phase of the path from an angle θ at time 0,

$$2\pi f_c \tau_\theta(0) \quad \mathrm{mod}\ 2\pi$$

is uniformly distributed in $[0, 2\pi]$ and that it is i.i.d. across θ.

4. Based on the assumptions in part (3), for large K one can use the Central Limit Theorem to approximate $\{h_0[m]\}$ as a Gaussian process. Verify that the limiting process is stationary and the autocorrelation function $R_0[n]$ is given by (2.58).

5. Verify that the Doppler spectrum $S(f)$ is given by (2.60). *Hint*: It is easier to show that the inverse Fourier transform of (2.60) is (2.58).

6. Verify that $S(f)df$ is indeed the received power from the paths that have Doppler shifts in $[f, f + df]$. Is this surprising?

Exercise 2.18 Consider a one-ring model where there are K scatterers located at angles $\theta_i = 2\pi i/K$, $i = 0, \ldots, K-1$, on a circle of radius 1 km around the receiver and the transmitter is 2 km away. (The angles are with respect to the line joining the transmitter and the receiver.) The transmit power is P. The power attenuation along a path from the transmitter to a scatterer to the receiver is

$$\frac{G}{K} \cdot \frac{1}{s^2} \cdot \frac{1}{r^2}, \tag{2.73}$$

where G is a constant and r and s are the distance from the transmitter to the scatterer and the distance from the scatterer to the receiver respectively. Communication takes place at a carrier frequency $f_c = 1.9$ GHz and the bandwidth is W Hz. You can assume that, at any time, the phases of each arriving path in the baseband representation of the channel are independent and uniformly distributed between 0 and 2π.

1. What are the key differences and the similarities between this model and the Clarke's model in the text?

2. Find approximate conditions on the bandwidth W for which one gets a flat fading channel.

3. Suppose the bandwidth is such that the channel is frequency selective. For large K, find approximately the amount of power in tap ℓ of the discrete-time baseband impulse response of the channel (i.e., compute the power-delay profile.). Make any simplifying assumptions but state them. (You can leave your answers in terms of integrals if you cannot evaluate them.)

4. Compute and sketch the power-delay profile as the bandwidth becomes very large (and K is large).

5. Suppose now the receiver is moving at speed v towards the (fixed) transmitter. What is the Doppler spread of tap ℓ? Argue heuristically from physical considerations what the Doppler spectrum (i.e., power spectral density) of tap ℓ is, for large K.

6. We have made the assumptions that the scatterers are all on a circle of radius 1km around the receiver and the paths arrive with independent and uniform distributed phases at the receiver. Mathematically, are the two assumptions consistent? If not, do you think it matters, in terms of the validity of your answers to the earlier parts of this question?

Exercise 2.19 Often in modeling multiple input multiple output (MIMO) fading channels the fading coefficients between different transmit and receive antennas are assumed to be independent random variables. This problem explores whether this is a reasonable assumption based on Clarke's one-ring scattering model and the antenna separation.

1. (Antenna separation at the mobile) Assume a mobile with velocity v moving away from the base-station, with uniform scattering from the ring around it.

(a) Compute the Doppler spread D_s for a carrier frequency f_c, and the corresponding coherence time T_c.

(b) Assuming that fading states separated by T_c are approximately uncorrelated, at what distance should we place a second antenna at the mobile to get an independently faded signal? *Hint*: How much distance does the mobile travel in T_c?

2. (Antenna separation at the base-station) Assume that the scattering ring has radius R and that the distance between the base-station and the mobile is d. Further assume for the time being that the base-station is moving away from the mobile with velocity v'. Repeat the previous part to find the minimum antenna spacing at the base-station for uncorrelated fading. *Hint*: Is the scattering still uniform around the base-station?

3. Typically, the scatterers are local around the mobile (near the ground) and far away from the base-station (high on a tower). What is the implication of your result in part (2) for this scenario?

3

Point-to-point communication: detection, diversity, and channel uncertainty

In this chapter we look at various basic issues that arise in communication over fading channels. We start by analyzing uncoded transmission in a narrowband fading channel. We study both coherent and non-coherent detection. In both cases the error probability is much higher than in a non-faded AWGN channel. The reason is that there is a significant probability that the channel is in a deep fade. This motivates us to investigate various *diversity* techniques that improve the performance. The diversity techniques operate over time, frequency or space, but the basic idea is the same. By sending signals that carry the same information through different paths, multiple independently faded replicas of data symbols are obtained at the receiver end and more reliable detection can be achieved. The simplest diversity schemes use *repetition coding*. More sophisticated schemes exploit channel diversity and, at the same time, efficiently use the degrees of freedom in the channel. Compared to repetition coding, they provide *coding gains* in addition to *diversity gains*. In space diversity, we look at both transmit and receive diversity schemes. In frequency diversity, we look at three approaches:

- single-carrier with inter-symbol interference equalization,
- direct-sequence spread-spectrum,
- orthogonal frequency division multiplexing.

Finally, we study the impact of channel uncertainty on the performance of diversity combining schemes. We will see that, in some cases, having too many diversity paths can have an adverse effect due to channel uncertainty.

To familiarize ourselves with the basic issues, the emphasis of this chapter is on concrete techniques for communication over fading channels. In Chapter 5 we take a more fundamental and systematic look and use information theory to derive the *best* performance one can achieve. At that fundamental level, we will see many of the issues discussed here recur.

The derivations in this chapter make repeated use of a few key results in vector detection under Gaussian noise. We develop and summarize the basic results in Appendix A, emphasizing the underlying geometry. The reader is encouraged to take a look at the appendix before proceeding with this chapter and to refer back to it often. In particular, a thorough understanding of the canonical detection problem in Summary A.2 will be very useful.

3.1 Detection in a Rayleigh fading channel

3.1.1 Non-coherent detection

We start with a very simple detection problem in a fading channel. For simplicity, let us assume a flat fading model where the channel can be represented by a single discrete-time complex filter tap $h_0[m]$, which we abbreviate as $h[m]$:

$$y[m] = h[m]x[m] + w[m], \tag{3.1}$$

where $w[m] \sim \mathcal{CN}(0, N_0)$. We suppose Rayleigh fading, i.e., $h[m] \sim \mathcal{CN}(0, 1)$, where we normalize the variance to be 1. For the time being, however, we do not specify the dependence between the fading coefficients $h[m]$ at different times m nor do we make any assumption on the prior knowledge the receiver might have of $h[m]$. (This latter assumption is sometimes called *non-coherent* communication.)

First consider uncoded binary antipodal signaling (or binary phase-shift-keying, BPSK) with amplitude a, i.e., $x[m] = \pm a$, and the symbols $x[m]$ are independent over time. This signaling scheme fails completely, even in the absence of noise, since the phase of the received signal $y[m]$ is uniformly distributed between 0 and 2π regardless of whether $x[m] = a$ or $x[m] = -a$ is transmitted. Further, the received amplitude is independent of the transmitted symbol. Binary antipodal signaling is binary phase modulation and it is easy to see that phase modulation in general is similarly flawed. Thus, signal structures are required in which either different signals have different magnitudes, or coding between symbols is used. Next we look at orthogonal signaling, a special type of coding between symbols.

Consider the following simple orthogonal modulation scheme: a form of binary pulse-position modulation. For a pair of time samples, transmit either

$$\mathbf{x}_A := \begin{pmatrix} x[0] \\ x[1] \end{pmatrix} = \begin{pmatrix} a \\ 0 \end{pmatrix}, \tag{3.2}$$

or

$$\mathbf{x}_B := \begin{pmatrix} 0 \\ a \end{pmatrix}. \tag{3.3}$$

We would like to perform detection based on

$$\mathbf{y} := \begin{pmatrix} y[0] \\ y[1] \end{pmatrix}. \tag{3.4}$$

This is a simple hypothesis testing problem, and it is straightforward to derive the maximum likelihood (ML) rule:

$$\Lambda(\mathbf{y}) \overset{\mathbf{x}_A}{\underset{\mathbf{x}_B}{\gtrless}} 0, \qquad (3.5)$$

where $\Lambda(\mathbf{y})$ is the log-likelihood ratio

$$\Lambda(\mathbf{y}) := \ln\left\{\frac{f(\mathbf{y}|\mathbf{x}_A)}{f(\mathbf{y}|\mathbf{x}_B)}\right\}. \qquad (3.6)$$

It can be seen that, if $\mathbf{x}_A$ is transmitted, $y[0] \sim \mathcal{CN}(0, a^2 + N_0)$ and $y[1] \sim \mathcal{CN}(0, N_0)$ and $y[0], y[1]$ are independent. Similarly, if $\mathbf{x}_B$ is transmitted, $y[0] \sim \mathcal{CN}(0, N_0)$ and $y[1] \sim \mathcal{CN}(0, a^2 + N_0)$. Further, $y[0]$ and $y[1]$ are independent. Hence the log-likelihood ratio can be computed to be

$$\Lambda(\mathbf{y}) = \frac{\left\{|y[0]|^2 - |y[1]|^2\right\} a^2}{(a^2 + N_0)N_0}. \qquad (3.7)$$

The optimal rule is simply to decide $\mathbf{x}_A$ is transmitted if $|y[0]|^2 > |y[1]|^2$ and decide $\mathbf{x}_B$ otherwise. Note that the rule does not make use of the phases of the received signal, since the random unknown phases of the channel gains $h[0], h[1]$ render them useless for detection. Geometrically, we can interpret the detector as projecting the received vector $\mathbf{y}$ onto each of the two possible transmit vectors $\mathbf{x}_A$ and $\mathbf{x}_B$ and comparing the energies of the projections (Figure 3.1). Thus, this detector is also called an *energy* or a *square-law* detector. It is somewhat surprising that the optimal detector does not depend on how $h[0]$ and $h[1]$ are correlated.

We can analyze the error probability of this detector. By symmetry, we can assume that $\mathbf{x}_A$ is transmitted. Under this hypothesis, $y[0]$ and $y[1]$ are

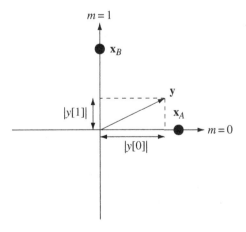

Figure 3.1 The non-coherent detector projects the received vector **y** onto each of the two orthogonal transmitted vectors $\mathbf{x}_A$ and $\mathbf{x}_B$ and compares the lengths of the projections.

independent circular symmetric complex Gaussian random variables with variances $a^2 + N_0$ and N_0 respectively. (See Section A.1.3 in the appendices for a discussion on circular symmetric Gaussian random variables and vectors.) As shown there, $|y[0]|^2$, $|y[1]|^2$ are exponentially distributed with mean $a^2 + N_0$ and N_0 respectively.[1] The probability of error can now be computed by direct integration:

$$p_e = \mathbb{P}\left\{|y[1]|^2 > |y[0]|^2 | \mathbf{x}_A\right\} = \left[2 + \frac{a^2}{N_0}\right]^{-1}. \tag{3.8}$$

We make the general definition

$$\boxed{\mathrm{SNR} := \frac{\text{average received signal energy per (complex) symbol time}}{\text{noise energy per (complex) symbol time}}} \tag{3.9}$$

which we use consistently throughout the book *for any modulation scheme.* The noise energy per complex symbol time is N_0.[2] For the orthogonal modulation scheme here, the average received energy per symbol time is $a^2/2$ and so

$$\mathrm{SNR} := \frac{a^2}{2N_0}. \tag{3.10}$$

Substituting into (3.8), we can express the error probability of the orthogonal scheme in terms of SNR:

$$p_e = \frac{1}{2(1 + \mathrm{SNR})}. \tag{3.11}$$

This is a very discouraging result. To get an error probability $p_e = 10^{-3}$ one would require $\mathrm{SNR} \approx 500$ (27 dB). Stupendous amounts of power would be required for more reliable communication.

3.1.2 Coherent detection

Why is the performance of the non-coherent maximum likelihood (ML) receiver on a fading channel so bad? It is instructive to compare its performance with detection in an AWGN channel without fading:

$$y[m] = x[m] + w[m]. \tag{3.12}$$

---

[1] Recall that a random variable U is exponentially distributed with mean μ if its pdf is $f_U(u) = \frac{1}{\mu} e^{-u/\mu}$.

[2] The orthogonal modulation scheme considered here uses only real symbols and hence transmits only on the I channel. Hence it may seem more natural to define the SNR in terms of noise energy per *real* symbol, i.e., $N_0/2$. However, later we will consider modulation schemes that use complex symbols and hence transmit on both the I and Q channels. In order to be consistent throughout, we choose to define SNR this way.

For antipodal signaling (BPSK), $x[m] = \pm a$, a sufficient statistic is $\Re\{y[m]\}$ and the error probability is

$$p_e = Q\left(\frac{a}{\sqrt{N_0/2}}\right) = Q\left(\sqrt{2\mathsf{SNR}}\right), \qquad (3.13)$$

where $\mathsf{SNR} = a^2/N_0$ is the received signal-to-noise ratio per symbol time, and $Q(\cdot)$ is the complementary cumulative distribution function of an $N(0, 1)$ random variable. This function decays exponentially with x^2; more specifically,

$$Q(x) < e^{-x^2/2}, \qquad x > 0 \qquad (3.14)$$

and

$$Q(x) > \frac{1}{\sqrt{2\pi}x}\left(1 - \frac{1}{x^2}\right)e^{-x^2/2}, \qquad x > 1. \qquad (3.15)$$

Thus, *the detection error probability decays exponentially in SNR in the AWGN channel while it decays only inversely with the SNR in the fading channel*. To get an error probability of 10^{-3}, an SNR of only about 7 dB is needed in an AWGN channel (as compared to 27 dB in the non-coherent fading channel). Note that $2\sqrt{\mathsf{SNR}}$ is the separation between the two constellation points as a multiple of the standard deviation of the Gaussian noise; the above observation says that when this separation is much larger than 1, the error probability is very small.

Compared to detection in the AWGN channel, the detection problem considered in the previous section has two differences: the channel gains $h[m]$ are random, and the receiver is assumed not to know them. Suppose now that the channel gains are tracked at the receiver so that they are known at the receiver (but still random). In practice, this is done either by sending a known sequence (called a *pilot* or training sequence) or in a decision directed manner, estimating the channel using symbols detected earlier. The accuracy of the tracking depends, of course, on how fast the channel varies. For example, in a narrowband 30-kHz channel (such as that used in the North American TDMA cellular standard IS-136) with a Doppler spread of 100 Hz, the coherence time T_c is roughly 80 symbols and in this case the channel can be estimated with minimal overhead expended in the pilot.[3] For our current purpose, let us suppose that the channel estimates are perfect.

Knowing the channel gains, *coherent* detection of BPSK can now be performed on a symbol by symbol basis. We can focus on one symbol time and drop the time index

$$y = hx + w \qquad (3.16)$$

---

[3] The channel estimation problem for a broadband channel with many taps in the impulse response is more difficult; we will get to this in Section 3.5.

Detection of x from y can be done in a way similar to that in the AWGN case; the decision is now based on the sign of the real sufficient statistic

$$r := \Re\{(h/|h|)^* y\} = |h|x + z, \tag{3.17}$$

where $z \sim N(0, N_0/2)$. If the transmitted symbol is $x = \pm a$, then, for a given value of h, the error probability of detecting x is

$$Q\left(\frac{a|h|}{\sqrt{N_0/2}}\right) = Q\left(\sqrt{2|h|^2 \mathsf{SNR}}\right) \tag{3.18}$$

where $\mathsf{SNR} = a^2/N_0$ is the average received signal-to-noise ratio per symbol time. (Recall that we normalized the channel gain such that $\mathbb{E}[|h|^2] = 1$.) We average over the random gain h to find the overall error probability. For Rayleigh fading when $h \sim \mathcal{CN}(0, 1)$, direct integration yields

$$p_e = \mathbb{E}\left[Q\left(\sqrt{2|h|^2 \mathsf{SNR}}\right)\right] = \frac{1}{2}\left(1 - \sqrt{\frac{\mathsf{SNR}}{1 + \mathsf{SNR}}}\right). \tag{3.19}$$

(See Exercise 3.1.) Figure 3.2 compares the error probabilities of coherent BPSK and non-coherent orthogonal signaling over the Rayleigh fading channel, as well as BPSK over the AWGN channel. We see that while the error probability for BPSK over the AWGN channel decays very fast with the SNR, the error probabilities for the Rayleigh fading channel are much worse,

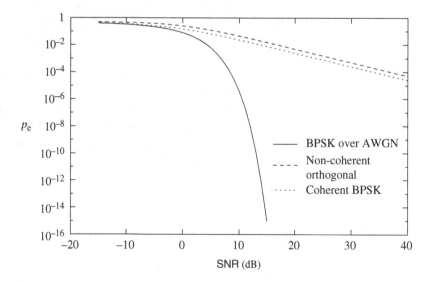

Figure 3.2 Performance of coherent BPSK vs. non-coherent orthogonal signaling over Rayleigh fading channel vs. BPSK over AWGN schannel.

whether the detection is coherent or non-coherent. At high SNR, Taylor series expansion yields

$$\sqrt{\frac{\mathsf{SNR}}{1+\mathsf{SNR}}} = 1 - \frac{1}{2\mathsf{SNR}} + O\left(\frac{1}{\mathsf{SNR}^2}\right). \tag{3.20}$$

Substituting into (3.19), we get the approximation

$$p_{\mathrm{e}} \approx \frac{1}{4\mathsf{SNR}}, \tag{3.21}$$

which decays inversely proportional to the SNR, just as in the non-coherent orthogonal signaling scheme (cf. (3.11)). There is only a 3 dB difference in the required SNR between the coherent and non-coherent schemes; in contrast, at an error probability of 10^{-3}, there is a 17 dB difference between the performance on the AWGN channel and coherent detection on the Rayleigh fading channel.[4]

We see that the main reason why detection in the fading channel has poor performance is not because of the lack of knowledge of the channel at the receiver. It is due to the fact that the channel gain is random and there is a significant probability that the channel is in a "deep fade". At high SNR, we can in fact be more precise about what a "deep fade" means by inspecting (3.18). The quantity $|h|^2\mathsf{SNR}$ is the instantaneous received SNR. Under typical channel conditions, i.e., $|h|^2\mathsf{SNR} \gg 1$, the conditional error probability is very small, since the tail of the Q-function decays very rapidly. In this regime, the separation between the constellation points is much larger than the standard deviation of the Gaussian noise. On the other hand, when $|h|^2\mathsf{SNR}$ is of the order of 1 or less, the separation is of the same order as the standard deviation of the noise and the error probability becomes significant. The probability of this event is

$$\mathbb{P}\{|h|^2\mathsf{SNR} < 1\} = \int_0^{1/\mathsf{SNR}} \mathrm{e}^{-x}\mathrm{d}x \tag{3.22}$$

$$= \frac{1}{\mathsf{SNR}} + O\left(\frac{1}{\mathsf{SNR}^2}\right). \tag{3.23}$$

This probability has the same order of magnitude as the error probability itself (cf. (3.21)). Thus, we can define a "deep fade" via an order-of-magnitude approximation:

$$\boxed{\begin{array}{l} \text{Deep fade event} : |h|^2 < \dfrac{1}{\mathsf{SNR}}. \\[2mm] \qquad\mathbb{P}\{\text{deep fade}\} \approx \dfrac{1}{\mathsf{SNR}}. \end{array}}$$

---

[4] Communication engineers often compare schemes based on the difference in the required SNR to attain the same error probability. This corresponds to the horizontal gap between the error probability versus SNR curves of the two schemes.

We conclude that high-SNR error events most often occur because the channel is in deep fade and not as a result of the additive noise being large. In contrast, in the AWGN channel the only possible error mechanism is for the additive noise to be large. Thus, the error probability performance over the AWGN channel is much better.

We have used the explicit error probability expression (3.19) to help identify the typical error event at high SNR. We can in fact turn the table around and use it as a basis for an approximate analysis of the high-SNR performance (Exercises 3.2 and 3.3). Even though the error probability p_e can be directly computed in this case, the approximate analysis provides much insight as to how typical errors occur. Understanding typical error events in a communication system often suggests how to improve it. Moreover, the approximate analysis gives some hints as to how robust the conclusion is to the Rayleigh fading model. In fact, the only aspect of the Rayleigh fading model that is important to the conclusion is the fact that $\mathbb{P}\{|h|^2 < \epsilon\}$ is proportional to ϵ for ϵ small. This holds whenever the pdf of $|h|^2$ is positive and continuous at 0.

3.1.3 From BPSK to QPSK: exploiting the degrees of freedom

In Section 3.1.2, we have considered BPSK modulation, $x[m] = \pm a$. This uses only the real dimension (the I channel), while in practice both the I and Q channels are used simultaneously in coherent communication, increasing spectral efficiency. Indeed, an extra bit can be transmitted by instead using QPSK (quadrature phase-shift-keying) modulation, i.e., the constellation is

$$\{a(1+j), a(1-j), a(-1+j), a(-1-j)\}; \tag{3.24}$$

in effect, a BPSK symbol is transmitted on each of the I and Q channels simultaneously. Since the noise is independent across the I and Q channels, the bits can be detected separately and the bit error probability on the AWGN channel (cf. (3.12)) is

$$Q\left(\sqrt{\frac{2a^2}{N_0}}\right), \tag{3.25}$$

the same as BPSK (cf. (3.13)). For BPSK, the SNR (as defined in (3.9)) is given by

$$\mathrm{SNR} = \frac{a^2}{N_0}, \tag{3.26}$$

while for QPSK,

$$\mathrm{SNR} = \frac{2a^2}{N_0}, \tag{3.27}$$

is twice that of BPSK since both the I and Q channels are used. Equivalently, for a given SNR, the bit error probability of BPSK is $Q(\sqrt{2\mathsf{SNR}})$ (cf. (3.13)) and that of QPSK is $Q(\sqrt{\mathsf{SNR}})$. The error probability of QPSK under Rayleigh fading can be similarly obtained by replacing SNR by $\mathsf{SNR}/2$ in the corresponding expression (3.19) for BPSK to yield

$$p_e = \frac{1}{2}\left(1 - \sqrt{\frac{\mathsf{SNR}}{2 + \mathsf{SNR}}}\right) \approx \frac{1}{2\mathsf{SNR}}. \tag{3.28}$$

at high SNR. For expositional simplicity, we will consider BPSK modulation in many of the discussions in this chapter, but the results can be directly mapped to QPSK modulation.

One important point worth noting is that it is much more energy-efficient to use both the I and Q channels rather than just one of them. For example, if we had to send the two bits carried by the QPSK symbol on the I channel alone, then we would have to transmit a 4-PAM symbol. The constellation is $\{-3b, -b, b, 3b\}$ and the average error probability on the AWGN channel is

$$\frac{3}{2}Q\left(\sqrt{\frac{2b^2}{N_0}}\right). \tag{3.29}$$

To achieve approximately the same error probability as QPSK, the argument inside the Q-function should be the same as that in (3.25) and hence b should be the same as a, i.e., the same minimum separation between points in the two constellations (Figure 3.3). But QPSK requires a transmit energy of $2a^2$ per symbol, while 4-PAM requires a transmit energy of $5b^2$ per symbol. Hence, for the same error probability, approximately 2.5 times more transmit energy is needed: a 4 dB worse performance. Exercise 3.4 shows that this loss is even more significant for larger constellations. The loss is due to the fact that it is more energy efficient to pack, for a desired minimum distance separation, a

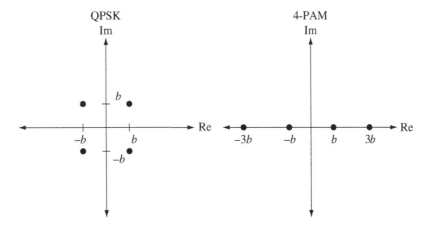

Figure 3.3 QPSK versus 4-PAM: for the same minimum separation between constellation points, the 4-PAM constellation requires higher transmit power.

given number of constellation points in a higher-dimensional space than in a lower-dimensional space. We have thus arrived at a general design principle (cf. Discussion 2.1):

> A good communication scheme exploits all the available degrees of freedom in the channel.

This important principle will recur throughout the book, and in fact will be shown to be of a fundamental nature as we talk about channel capacity in Chapter 5. Here, the choice is between using just the I channel and using both the I and Q channels, but the same principle applies to many other situations. As another example, the non-coherent orthogonal signaling scheme discussed in Section 3.1.1 conveys one bit of information and uses one real dimension per two symbol times (Figure 3.4). This scheme does not assume any relationship between consecutive channel gains, but if we assume that they do not change much from symbol to symbol, an alternative scheme is *differential* BPSK, which conveys information in the relative phases of consecutive transmitted symbols. That is, if the BPSK information symbol is $u[m]$ at time m ($u[m] = \pm 1$), the transmitted symbol at time m is given by

$$x[m] = u[m]x[m-1].\qquad(3.30)$$

Exercise 3.5 shows that differential BPSK can be demodulated non-coherently at the expense of a 3-dB loss in performance compared to coherent BPSK (at high SNR). But since non-coherent orthogonal modulation also has a 3-dB worse performance compared to coherent BPSK, this implies that differential BPSK and non-coherent orthogonal modulation have the *same* error probability performance. On the other hand, differential BPSK conveys one

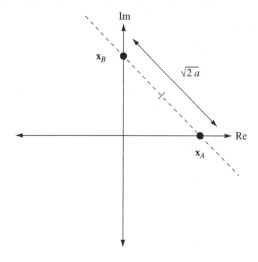

Figure 3.4 Geometry of orthogonal modulation. Signaling is performed over one real dimension, but two (complex) symbol times are used.

bit of information and uses one real dimension per *single* symbol time, and therefore has twice the spectral efficiency of orthogonal modulation. Better performance is achieved because differential BPSK uses more efficiently the available degrees of freedom.

3.1.4 Diversity

The performance of the various schemes considered so far for fading channels is summarized in Table 3.1. Some schemes are spectrally more efficient than others, but from a practical point of view, they are all bad: the error probabilities all decay very slowly, like 1/SNR. From Section 3.1.2, it can be seen that the root cause of this poor performance is that reliable communication depends on the strength of a single signal path. There is a significant probability that this path will be in a deep fade. When the path is in a deep fade, any communication scheme will likely suffer from errors. A natural solution to improve the performance is to ensure that the information symbols pass through multiple signal paths, each of which fades independently, making sure that reliable communication is possible as long as one of the paths is strong. This technique is called *diversity*, and it can dramatically improve the performance over fading channels.

There are many ways to obtain diversity. Diversity over **time** can be obtained via *coding* and *interleaving*: information is coded and the coded symbols are dispersed over time in different coherence periods so that different parts of the codewords experience independent fades. Analogously, one can also exploit diversity over **frequency** if the channel is frequency-selective. In a channel with multiple transmit or receive antennas spaced sufficiently, diversity can be obtained over **space** as well. In a cellular network, **macrodiversity** can be exploited by the fact that the signal from a mobile can be received at two base-stations. Since diversity is such an important resource, a wireless system typically uses several types of diversity.

In the next few sections, we will discuss diversity techniques in time, frequency and space. In each case, we start with a simple scheme based on *repetition coding*: the same information symbol is transmitted over several signal paths. While repetition coding achieves the maximal diversity gain, it is usually quite wasteful of the degrees of freedom of the channel. More sophisticated schemes can increase the data rate and achieve a *coding gain* along with the diversity gain.

To keep the discussion simple we begin by focusing on the coherent scenario: the receiver has perfect knowledge of the channel gains and can coherently combine the received signals in the diversity paths. As discussed in the previous section, this knowledge is learnt via training (pilot) symbols and the accuracy depends on the coherence time of the channel and the received power of the transmitted signal. We discuss the impact of channel measurement error and non-coherent diversity combining in Section 3.5.

Table 3.1 Performance of coherent and non-coherent schemes under Rayleigh fading. The data rates are in bits/s/Hz, which is the same as bits per complex symbol time. The performance of differential QPSK is derived in Exercise 3.5. It is also 3-dB worse than coherent QPSK.

Scheme	Bit error prob. (High SNR)	Data rate (bits/s/Hz)
Coherent BPSK	1/(4SNR)	1
Coherent QPSK	1/(2SNR)	2
Coherent 4-PAM	5/(4SNR)	2
Coherent 16-QAM	5/(2SNR)	4
Non-coherent orth. mod.	1/(2SNR)	1/2
Differential BPSK	1/(2SNR)	1
Differential QPSK	1/SNR	2

3.2 Time diversity

Time diversity is achieved by averaging the fading of the channel over time. Typically, the channel coherence time is of the order of tens to hundreds of symbols, and therefore the channel is highly correlated across consecutive symbols. To ensure that the coded symbols are transmitted through independent or nearly independent fading gains, *interleaving* of codewords is required (Figure 3.5). For simplicity, let us consider a flat fading channel. We transmit a codeword $\mathbf{x} = [x_1, \ldots, x_L]^t$ of length L symbols and the received signal is given by

$$y_\ell = h_\ell x_\ell + w_\ell, \qquad \ell = 1, \ldots, L. \tag{3.31}$$

Assuming ideal interleaving so that consecutive symbols x_ℓ are transmitted sufficiently far apart in time, we can assume that the h_ℓ are independent. The parameter L is commonly called the number of *diversity branches*. The additive noises $w_1, \ldots, w_L$ are i.i.d. $\mathcal{CN}(0, N_0)$ random variables.

3.2.1 Repetition coding

The simplest code is a *repetition code*, in which $x_\ell = x_1$ for $\ell = 1, \ldots, L$. In vector form, the overall channel becomes

$$\mathbf{y} = \mathbf{h}x_1 + \mathbf{w}, \tag{3.32}$$

where $\mathbf{y} = [y_1, \ldots, y_L]^t$, $\mathbf{h} = [h_1, \ldots, h_L]^t$ and $\mathbf{w} = [w_1, \ldots, w_L]^t$.

Figure 3.5 The codewords are transmitted over consecutive symbols (top) and interleaved (bottom). A deep fade will wipe out the entire codeword in the former case but only one coded symbol from each codeword in the latter. In the latter case, each codeword can still be recovered from the other three unfaded symbols.

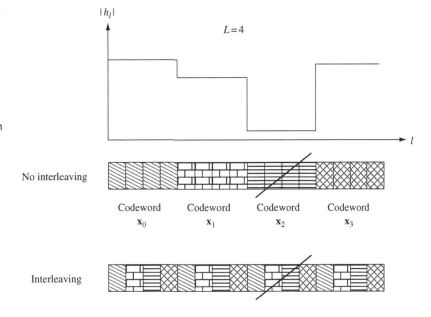

Consider now coherent detection of x_1, i.e., the channel gains are known to the receiver. This is the canonical vector Gaussian detection problem in Summary A.2 of Appendix A. The scalar

$$\frac{\mathbf{h}^*}{\|\mathbf{h}\|}\mathbf{y} = \|\mathbf{h}\|x_1 + \frac{\mathbf{h}^*}{\|\mathbf{h}\|}\mathbf{w} \tag{3.33}$$

is a sufficient statistic. Thus, we have an equivalent scalar detection problem with noise $(\mathbf{h}^*/\|\mathbf{h}\|)\mathbf{w} \sim \mathcal{CN}(0, N_0)$. The receiver structure is a *matched filter* and is also called a *maximal ratio combiner*: it weighs the received signal in each branch in proportion to the signal strength and also aligns the phases of the signals in the summation to maximize the output SNR. This receiver structure is also called *coherent combining*.

Consider BPSK modulation, with $x_1 = \pm a$. The error probability, conditional on $\mathbf{h}$, can be derived exactly as in (3.18):

$$Q\left(\sqrt{2\|\mathbf{h}\|^2\mathsf{SNR}}\right) \tag{3.34}$$

where as before $\mathsf{SNR} = a^2/N_0$ is the average received signal-to-noise ratio per (complex) symbol time, and $\|\mathbf{h}\|^2\mathsf{SNR}$ is the received SNR for a given channel vector $\mathbf{h}$. We average over $\|\mathbf{h}\|^2$ to find the overall error probability. Under Rayleigh fading with each gain h_ℓ i.i.d. $\mathcal{CN}(0, 1)$,

$$\|\mathbf{h}\|^2 = \sum_{\ell=1}^{L} |h_\ell|^2 \tag{3.35}$$

is a sum of the squares of $2L$ independent real Gaussian random variables, each term $|h_\ell|^2$ being the sum of the squares of the real and imaginary parts of h_ℓ. It is Chi-square distributed with $2L$ degrees of freedom, and the density is given by

$$f(x) = \frac{1}{(L-1)!} x^{L-1} e^{-x}, \qquad x \ge 0. \tag{3.36}$$

The average error probability can be explicitly computed to be (cf. Exercise 3.6)

$$
\begin{aligned}
p_e &= \int_0^\infty Q\left(\sqrt{2x\mathsf{SNR}}\right) f(x)\mathrm{d}x \\
&= \left(\frac{1-\mu}{2}\right)^L \sum_{\ell=0}^{L-1} \binom{L-1+\ell}{\ell} \left(\frac{1+\mu}{2}\right)^\ell,
\end{aligned} \tag{3.37}
$$

where

$$\mu := \sqrt{\frac{\mathsf{SNR}}{1+\mathsf{SNR}}}. \tag{3.38}$$

The error probability as a function of the SNR for different numbers of diversity branches L is plotted in Figure 3.6. Increasing L dramatically decreases the error probability.

At high SNR, we can see the role of L analytically: consider the leading term in the Taylor series expansion in 1/SNR to arrive at the approximations

$$\frac{1+\mu}{2} \approx 1, \quad \text{and} \quad \frac{1-\mu}{2} \approx \frac{1}{4\mathsf{SNR}}. \tag{3.39}$$

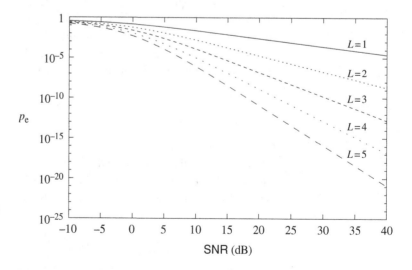

Figure 3.6 Error probability as a function of SNR for different numbers of diversity branches L.

Furthermore,

$$\sum_{\ell=0}^{L-1} \binom{L-1+\ell}{\ell} = \binom{2L-1}{L}. \tag{3.40}$$

Hence,

$$p_e \approx \binom{2L-1}{L} \frac{1}{(4\mathsf{SNR})^L} \tag{3.41}$$

at high SNR. In particular, the error probability decreases as the Lth power of SNR, corresponding to a slope of $-L$ in the error probability curve (in dB/dB scale).

To understand this better, we examine the probability of the deep fade event, as in our analysis in Section 3.1.2. The typical error event at high SNR is when the overall channel gain is small. This happens with probability

$$\mathbb{P}\{\|\mathbf{h}\|^2 < 1/\mathsf{SNR}\}. \tag{3.42}$$

Figure 3.7 plots the distribution of $\|\mathbf{h}\|^2$ for different values of L; clearly the tail of the distribution near zero becomes lighter for larger L. For small x, the probability density function of $\|\mathbf{h}\|^2$ is approximately

$$f(x) \approx \frac{1}{(L-1)!} x^{L-1} \tag{3.43}$$

and so

$$\mathbb{P}\{\|\mathbf{h}\|^2 < 1/\mathsf{SNR}\} \approx \int_0^{\frac{1}{\mathsf{SNR}}} \frac{1}{(L-1)!} x^{L-1} \mathrm{d}x = \frac{1}{L!} \frac{1}{\mathsf{SNR}^L}. \tag{3.44}$$

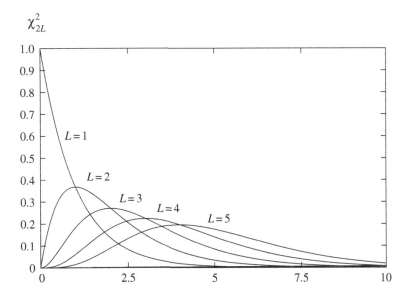

Figure 3.7 The probability density function of $\|\mathbf{h}\|^2$ for different values of L. The larger the L, the faster the probability density function drops off around 0.

This analysis is too crude to get the correct constant before the $1/\text{SNR}^L$ term in (3.41), but does get the correct exponent L. Basically, an error occurs when $\sum_{\ell=1}^{L} |h_\ell|^2$ is of the order of or smaller than $1/\text{SNR}$, and this happens when *all* the magnitudes of the gains $|h_\ell|^2$ are small, of the order of $1/\text{SNR}$. Since the probability that each $|h_\ell|^2$ is less than $1/\text{SNR}$ is approximately $1/\text{SNR}$ and the gains are independent, the probability of the overall gain being small is of the order $1/\text{SNR}^L$. Typically, L is called the *diversity gain* of the system.

3.2.2 Beyond repetition coding

The repetition code is the simplest possible code. Although it achieves a diversity gain, it does not exploit the degrees of freedom available in the channel effectively because it simply repeats the same symbol over the L symbol times. By using more sophisticated codes, a coding gain can also be obtained beyond the diversity gain. There are many possible codes that one can use. We first focus on the example of a *rotation code* to explain some of the issues in code design for fading channels.

Consider the case $L = 2$. A repetition code which repeats a BPSK symbol $u = \pm a$ twice obtains a diversity gain of 2 but would only transmit one bit of information over the two symbol times. Transmitting two independent BPSK symbols u_1, u_2 over the two times would use the available degrees of freedom more efficiently, but of course offers no diversity gain: an error would be made whenever one of the two channel gains h_1, h_2 is in deep fade. To get both benefits, consider instead a scheme that transmits the vector

$$\mathbf{x} = \mathbf{R} \begin{bmatrix} u_1 \\ u_2 \end{bmatrix} \tag{3.45}$$

over the two symbol times, where

$$\mathbf{R} := \begin{bmatrix} \cos\theta & -\sin\theta \\ \sin\theta & \cos\theta \end{bmatrix} \tag{3.46}$$

is a rotation matrix (for some $\theta \in (0, 2\pi)$). This is a code with four codewords:

$$\mathbf{x}_A = \mathbf{R} \begin{bmatrix} a \\ a \end{bmatrix}, \qquad \mathbf{x}_B = \mathbf{R} \begin{bmatrix} -a \\ a \end{bmatrix}, \qquad \mathbf{x}_C = \mathbf{R} \begin{bmatrix} -a \\ -a \end{bmatrix}, \qquad \mathbf{x}_D = \mathbf{R} \begin{bmatrix} a \\ -a \end{bmatrix}; \tag{3.47}$$

they are shown in Figure 3.8(a).[5] The received signal is given by

$$y_\ell = h_\ell x_\ell + w_\ell, \qquad \ell = 1, 2. \tag{3.48}$$

---

[5] Here communication is over the (real) I channel since both x_1 and x_2 are real, but as in Section 3.1.3, the spectral efficiency can be doubled by using both the I and the Q channels. Since the two channels are orthogonal, one can apply the same code separately to the symbols transmitted in the two channels to get the same performance gain.

Figure 3.8 (a) Codewords of rotation code. (b) Codewords of repetition code.

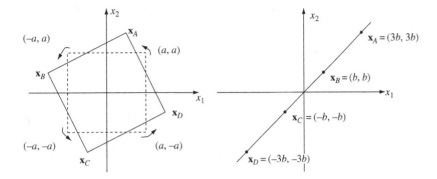

It is difficult to obtain an explicit expression for the exact error probability. So, we will proceed by looking at the union bound. Due to the symmetry of the code, without loss of generality we can assume $\mathbf{x}_A$ is transmitted. The union bound says that

$$p_e \leq \mathbb{P}\{\mathbf{x}_A \to \mathbf{x}_B\} + \mathbb{P}\{\mathbf{x}_A \to \mathbf{x}_C\} + \mathbb{P}\{\mathbf{x}_A \to \mathbf{x}_D\}, \qquad (3.49)$$

where $\mathbb{P}\{\mathbf{x}_A \to \mathbf{x}_B\}$ is the pairwise error probability of confusing $\mathbf{x}_A$ with $\mathbf{x}_B$ when $\mathbf{x}_A$ is transmitted and when these are the only two hypotheses. Conditioned on the channel gains h_1 and h_2, this is just the binary detection problem in Summary A.2 of Appendix A, with

$$\mathbf{u}_A = \begin{bmatrix} h_1 x_{A1} \\ h_2 x_{A2} \end{bmatrix} \quad \text{and} \quad \mathbf{u}_B = \begin{bmatrix} h_1 x_{B1} \\ h_2 x_{B2} \end{bmatrix}. \qquad (3.50)$$

Hence,

$$\mathbb{P}\{\mathbf{x}_A \to \mathbf{x}_B | h_1, h_2\} = Q\left(\frac{\|\mathbf{u}_A - \mathbf{u}_B\|}{2\sqrt{N_0/2}}\right) = Q\left(\sqrt{\frac{\mathsf{SNR}(|h_1|^2|d_1|^2 + |h_2|^2|d_2|^2)}{2}}\right),$$
$$(3.51)$$

where $\mathsf{SNR} = a^2/N_0$ and

$$\mathbf{d} := \frac{1}{a}(\mathbf{x}_A - \mathbf{x}_B) = \begin{bmatrix} 2\cos\theta \\ 2\sin\theta \end{bmatrix} \qquad (3.52)$$

is the normalized difference between the codewords, normalized such that the transmit energy is 1 per symbol time. We use the upper bound $Q(x) \leq e^{-x^2/2}$, for $x > 0$, in (3.51) to get

$$\mathbb{P}\{\mathbf{x}_A \to \mathbf{x}_B | h_1, h_2\} \leq \exp\left(\frac{-\mathsf{SNR}(|h_1|^2|d_1|^2 + |h_2|^2|d_2|^2)}{4}\right). \qquad (3.53)$$

Averaging with respect to h_1 and h_2 under the independent Rayleigh fading assumption, we get

$$
\begin{aligned}
\mathbb{P}\{\mathbf{x}_A \rightarrow \mathbf{x}_B\} &\leq \mathbb{E}_{h_1,h_2}\left[\exp\left(\frac{-\mathsf{SNR}(|h_1|^2|d_1|^2 + |h_2|^2|d_2|^2)}{4}\right)\right] \\
&= \left(\frac{1}{1+\mathsf{SNR}|d_1|^2/4}\right)\left(\frac{1}{1+\mathsf{SNR}|d_2|^2/4}\right).
\end{aligned}
\tag{3.54}
$$

Here we have used the fact that the moment generating function for a unit mean exponential random variable X is $\mathbb{E}[e^{sX}] = 1/(1-s)$ for $s < 1$. While it is possible to get an exact expression for the pairwise error probability, this upper bound is more explicit; moreover, it is asymptotically tight at high SNR (Exercise 3.7).

We first observe that if $d_1 = 0$ or $d_2 = 0$, then the diversity gain of the code is only 1. If they are both non-zero, then at high SNR the above bound on the pairwise error probability becomes

$$
\mathbb{P}\{\mathbf{x}_A \rightarrow \mathbf{x}_B\} \leq \frac{16}{|d_1 d_2|^2}\mathsf{SNR}^{-2},
\tag{3.55}
$$

Call

$$
\boxed{\delta_{AB} := |d_1 d_2|^2,}
\tag{3.56}
$$

the *squared product distance* between $\mathbf{x}_A$ and $\mathbf{x}_B$, when the average energy of the code is normalized to be 1 per symbol time (cf. (3.52)). This determines the pairwise error probability between the two codewords. Similarly, we can define δ_{ij} to be the squared product distance between $\mathbf{x}_i$ and $\mathbf{x}_j$, $i, j = A, B, C, D$. Combining (3.55) with (3.49) yields a bound on the overall error probability:

$$
\begin{aligned}
p_{\mathrm{e}} &\leq 16\left(\frac{1}{\delta_{AB}} + \frac{1}{\delta_{AC}} + \frac{1}{\delta_{AD}}\right)\mathsf{SNR}^{-2} \\
&\leq \frac{48}{\min_{j=B,C,D}\delta_{Aj}}\mathsf{SNR}^{-2}.
\end{aligned}
\tag{3.57}
$$

We see that as long as $\delta_{ij} > 0$ for all i, j, we get a diversity gain of 2. The minimum squared product distance $\min_{j=B,C,D}\delta_{Aj}$ then determines the *coding gain* of the scheme beyond the diversity gain. This parameter depends on θ, and we can optimize over θ to maximize the coding gain. Here

$$
\delta_{AB} = \delta_{AD} = 4\sin^2 2\theta, \qquad \text{and} \qquad \delta_{AC} = 16\cos^2 2\theta.
\tag{3.58}
$$

The angle θ^* that maximizes the minimum squared product distance makes δ_{AB} equal δ_{AC}, yielding $\theta^* = (1/2)\tan^{-1}2$ and $\min \delta_{ij} = 16/5$. The bound in (3.57) now becomes

$$p_e \leq 15\,\mathsf{SNR}^{-2}. \tag{3.59}$$

To get more insight into why the product distance is important, we see from (3.51) that the typical way for $\mathbf{x}_A$ to be confused with $\mathbf{x}_B$ is for the squared Euclidean distance $|h_1|^2|d_1|^2 + |h_2|^2|d_2|^2$ between the *received* codewords to be of the order of $1/\mathsf{SNR}$. This event holds roughly when both $|h_1|^2|d_1|^2$ and $|h_2|^2|d_2|^2$ are of the order of $1/\mathsf{SNR}$, and this happens with probability approximately

$$\left(\frac{1}{|d_1|^2\mathsf{SNR}}\right)\left(\frac{1}{|d_2|^2\mathsf{SNR}}\right) = \frac{1}{|d_1|^2|d_2|^2}\mathsf{SNR}^{-2}. \tag{3.60}$$

Thus, it is important that both $|d_1|^2$ and $|d_2|^2$ are large to ensure diversity against fading in both components.

It is interesting to see how this code compares to the repetition scheme. To keep the bit rate the same (2 bits over 2 real-valued symbols), the repetition scheme would be using 4-PAM modulation $\{-3b, -b, b, 3b\}$. The codewords of the repetition scheme are shown in Figure 3.8(b). From (3.51), the pairwise error probability between two adjacent codewords (say, $\mathbf{x}_A$ and $\mathbf{x}_B$) is

$$\mathbb{P}\{\mathbf{x}_A \rightarrow \mathbf{x}_B\} = \mathbb{E}\left[Q\left(\sqrt{\mathsf{SNR}/2 \cdot (|h_1|^2|d_1|^2 + |h_2|^2|d_2|^2)}\right)\right]. \tag{3.61}$$

But now $\mathsf{SNR} = 5b^2/N_0$ is the average SNR per symbol time for the 4-PAM constellation,[6] and $d_1 = d_2 = 2/\sqrt{5}$ are the normalized component differences between the adjacent codewords. The minimum squared product distance for the repetition code is therefore $16/25$ and we can compare this to the minimum squared product distance of $16/5$ for the previous rotation code. Since the error probability is proportional to SNR^{-2} in both cases, we conclude that the rotation code has an improved *coding gain* over the repetition code in terms of a saving in transmit power by a factor of $\sqrt{5}$ (3.5 dB) for the same product distance. This improvement comes from increasing the overall product distance, and this is in turn due to spreading the codewords in the two-dimensional space rather than packing them on a single-dimensional line as in the repetition code. This is the same reason that QPSK is more efficient than BPSK (as we have discussed in Section 3.1.3).

We summarize and generalize the above development to any time diversity code.

---

[6] As we have seen earlier, the 4-PAM constellation requires five times more energy than BPSK for the same separation between the constellation points.

Summary 3.1 Time diversity code design criterion

Ideal time-interleaved channel

$$y_\ell = h_\ell x_\ell + w_\ell, \qquad \ell = 1, \ldots, L, \tag{3.62}$$

where h_ℓ are i.i.d. $\mathcal{CN}(0, 1)$ Rayleigh faded channel gains.

$\mathbf{x}_1, \ldots, \mathbf{x}_M$ are the codewords of a time diversity code with block length L, normalized such that

$$\frac{1}{ML} \sum_{i=1}^{M} \|\mathbf{x}_i\|^2 = 1. \tag{3.63}$$

Union bound on overall probability of error:

$$p_e \leq \frac{1}{M} \sum_{i \neq j} \mathbb{P}\{\mathbf{x}_i \to \mathbf{x}_j\} \tag{3.64}$$

Bound on pairwise error probability:

$$\mathbb{P}\{\mathbf{x}_i \to \mathbf{x}_j\} \leq \prod_{\ell=1}^{L} \frac{1}{1 + \mathsf{SNR}|x_{i\ell} - x_{j\ell}|^2/4} \tag{3.65}$$

where $x_{i\ell}$ is the ℓth component of codeword $\mathbf{x}_i$, and $\mathsf{SNR} := 1/N_0$.

Let L_{ij} be the number of components on which the codewords $\mathbf{x}_i$ and $\mathbf{x}_j$ differ. Diversity gain of the code is

$$\min_{i \neq j} L_{ij}. \tag{3.66}$$

If $L_{ij} = L$ for all $i \neq j$, then the code achieves the full diversity L of the channel, and

$$p_e \leq \frac{4^L}{M} \sum_{i \neq j} \frac{1}{\delta_{ij}} \mathsf{SNR}^{-L} \leq \frac{4^L(M-1)}{\min_{i \neq j} \delta_{ij}} \mathsf{SNR}^{-L} \tag{3.67}$$

where

$$\delta_{ij} := \prod_{\ell=1}^{L} |x_{i\ell} - x_{j\ell}|^2 \tag{3.68}$$

is the squared product distance between $\mathbf{x}_i$ and $\mathbf{x}_j$.

The rotation code discussed above is specifically designed to exploit time diversity in *fading* channels. In the AWGN channel, however, rotation of the constellation does not affect performance since the i.i.d. Gaussian noise is invariant to rotations. On the other hand, codes that are designed for the AWGN channel, such as linear block codes or convolutional codes, can be used to extract time diversity in fading channels when combined with interleaving. Their performance can be analyzed using the general framework above. For example, the diversity gain of a binary linear block code where the coded symbols are ideally interleaved is simply the *minimum Hamming distance* between the codewords or equivalently the minimum weight of a codeword; the diversity gain of a binary convolutional code is given by the *free distance* of the code, which is the minimum weight of the coded sequence of the convolutional code. The performance analysis of these codes and various decoding techniques is further pursued in Exercise 3.11.

It should also be noted that the above code design criterion is derived assuming i.i.d. Rayleigh fading across the symbols. This can be generalized to the case when the coded symbols pass through *correlated* fades of the channel (see Exercise 3.12). Generalization to the case when the fading is Rician is also possible and is studied in Exercise 3.18. Nevertheless these code design criteria all depend on the specific channel statistics assumed. Motivated by information theoretic considerations, we take a completely different approach in Chapter 9 where we seek a *universal* criterion which works for *all* channel statistics. We will also be able to define what it means for a time-diversity code to be *optimal*.

Example 3.1 Time diversity in GSM

Global System for Mobile (GSM) is a digital cellular standard developed in Europe in the 1980s. GSM is a frequency division duplex (FDD) system and uses two 25-MHz bands, one for the uplink (mobiles to base-station) and one for the downlink (base-station to mobiles). The original bands set aside for GSM are the 890–915 MHz band (uplink) and the 935–960 MHz band (downlink). The bands are further divided into 200-kHz sub-channels and each sub-channel is shared by eight users in a time-division fashion (time-division multiple access (TDMA)). The data of each user are sent over time slots of length 577 microseconds (μs) and the time slots of the eight users together form a frame of length 4.615 ms (Figure 3.9).

Voice is the main application for GSM. Voice is coded by a speech encoder into speech frames each of length 20 ms. The bits in each speech frame are encoded by a convolutional code of rate 1/2, with the two generator polynomials $D^4 + D^3 + 1$ and $D^4 + D^3 + D + 1$. The number of coded bits for each speech frame is 456. To achieve time diversity, these coded bits are interleaved across eight consecutive time slots assigned to that specific user: the 0th, 8th, ..., 448th bits are put into the first time slot, the 1st, 9th, ..., 449th bits are put into the second time slot, etc.

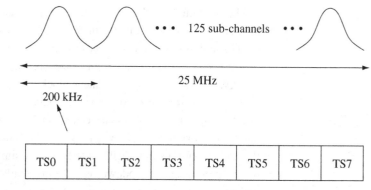

Figure 3.9 The 25-MHz band of a GSM system is divided into 200-kHz sub-channels, which are further divided into time slots for eight different users.

Since one time slot occurs every 4.615 ms for each user, this translates into a delay of roughly 40 ms, a delay judged tolerable for voice. The eight time slots are shared between two 20-ms speech frames. The interleaving structure is summarized in Figure 3.10.

The maximum possible time diversity gain is 8, but the actual gain that can be obtained depends on how fast the channel varies, and that depends primarily on the mobile speed. If the mobile speed is v, then the largest possible Doppler spread (assuming full scattering in the environment) is $D_s = 2f_c v/c$, where f_c is the carrier frequency and c is the speed of light. (Recall the example in Section 2.1.4.) The coherence time is roughly $T_c = 1/(4D_s) = c/(8f_c v)$ (cf. (2.44)). For the channel to fade more or less independently across the different time slots for a user, the coherence time should be less than 5 ms. For $f_c = 900$ MHz, this translates into a mobile speed of at least 30 km/h.

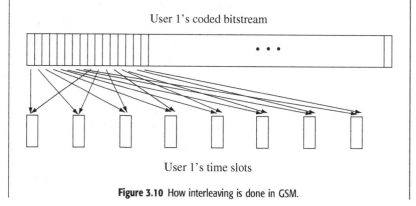

Figure 3.10 How interleaving is done in GSM.

> For a walking speed of say 3 km/h, there may be too little time diversity. In this case, GSM can go into a *frequency hopping* mode, where consecutive frames (each composed of the time slots of the eight users) can hop from one 200-kHz sub-channel to another. With a typical delay spread of about 1 μs, the coherence bandwidth is 500 kHz (cf. Table 2.1). The total bandwidth equal to 25 MHz is thus much larger than the typical coherence bandwidth of the channel and the consecutive frames can be expected to fade independently. This provides the same effect as having time diversity. Section 3.4 discusses other ways to exploit frequency diversity.

3.3 Antenna diversity

To exploit time diversity, interleaving and coding over several coherence time periods is necessary. When there is a strict delay constraint and/or the coherence time is large, this may not be possible. In this case other forms of diversity have to be obtained. Antenna diversity, or spatial diversity, can be obtained by placing multiple antennas at the transmitter and/or the receiver. If the antennas are placed sufficiently far apart, the channel gains between different antenna pairs fade more or less independently, and independent signal paths are created. The required antenna separation depends on the local scattering environment as well as on the carrier frequency. For a mobile which is near the ground with many scatterers around, the channel decorrelates over shorter spatial distances, and typical antenna separation of half to one carrier wavelength is sufficient. For base-stations on high towers, larger antenna separation of several to tens of wavelengths may be required. (A more careful discussion of these issues is found in Chapter 7.)

We will look at both *receive diversity*, using multiple receive antennas (single input multiple output or SIMO channels), and *transmit diversity*, using multiple transmit antennas (multiple input single output or MISO channels). Interesting coding problems arise in the latter and have led to recent excitement in *space-time codes*. Channels with multiple transmit *and* multiple receive antennas (so-called multiple input multiple output or MIMO channels) provide even more potential. In addition to providing diversity, MIMO channels also provide additional *degrees of freedom* for communication. We will touch on some of the issues here using a 2×2 example; the full study of MIMO communication will be the subject of Chapters 7 to 10.

3.3.1 Receive diversity

In a flat fading channel with 1 transmit antenna and L receive antennas (Figure 3.11(a)), the channel model is as follows:

$$y_\ell[m] = h_\ell[m]x[m] + w_\ell[m] \qquad \ell = 1, \ldots, L \qquad (3.69)$$

Figure 3.11 (a) Receive
diversity; (b) transmit diversity;
(c) transmit and receive
diversity.

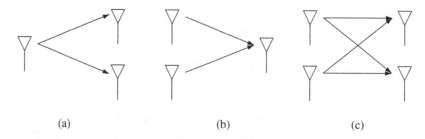

(a)   (b)   (c)

where the noise $w_\ell[m] \sim \mathcal{CN}(0, N_0)$ and is independent across the antennas.
We would like to detect $x[1]$ based on $y_1[1], \ldots, y_L[1]$. This is exactly the
same detection problem as in the use of a repetition code and interleaving
over time, with L diversity branches now over space instead of over time. If
the antennas are spaced sufficiently far apart, we can assume that the gains
$h_\ell[1]$ are independent Rayleigh, and we get a diversity gain of L.

With receive diversity, there are actually two types of gain as we increase L.
This can be seen by looking at the expression (3.34) for the error probability
of BPSK conditional on the channel gains:

$$Q\left(\sqrt{2\|\mathbf{h}\|^2 \mathsf{SNR}}\right). \tag{3.70}$$

We can break up the total received SNR conditioned on the channel gains
into a product of two terms:

$$\|\mathbf{h}\|^2 \mathsf{SNR} = L\mathsf{SNR} \cdot \frac{1}{L}\|\mathbf{h}\|^2. \tag{3.71}$$

The first term corresponds to a *power gain* (also called *array gain*): by having
multiple receive antennas and coherent combining at the receiver, the effective
total received signal power increases linearly with L: doubling L yields a
3-dB power gain.[7] The second term reflects the *diversity gain*: by averaging
over multiple independent signal paths, the probability that the overall gain
is small is decreased. The diversity gain L is reflected in the SNR exponent
in (3.41); the power gain affects the constant before the $1/\mathsf{SNR}^L$. Note that if
the channel gains $h_\ell[1]$ are fully correlated across all branches, then we only
get a power gain but no diversity gain as we increase L. On the other hand,
even when all the h_ℓ are independent there is a diminishing marginal return
as L increases: due to the law of large numbers, the second term in (3.71),

$$\frac{1}{L}\|\mathbf{h}\|^2 = \frac{1}{L}\sum_{\ell=1}^{L}|h_\ell[1]|^2, \tag{3.72}$$

[7] Although mathematically the same situation holds in the time diversity repetition coding
case, the increase in received SNR there comes from increasing the total *transmit* energy
required to send a single bit; it is therefore not appropriate to call that a power gain.

converges to 1 with increasing L (assuming each of the channel gains is normalized to have unit variance). The power gain, on the other hand, suffers from no such limitation: a 3-dB gain is obtained for every doubling of the number of antennas.[8]

3.3.2 Transmit diversity: space-time codes

Now consider the case when there are L transmit antennas and 1 receive antenna, the MISO channel (Figure 3.11(b)). This is common in the downlink of a cellular system since it is often cheaper to have multiple antennas at the base-station than to have multiple antennas at every handset. It is easy to get a diversity gain of L: simply transmit the same symbol over the L different antennas during L symbol times. At any one time, only one antenna is turned on and the rest are silent. This is simply a repetition code, and, as we have seen in the previous section, repetition codes are quite wasteful of degrees of freedom. More generally, *any* time diversity code of block length L can be used on this transmit diversity system: simply use one antenna at a time and transmit the coded symbols of the time diversity code successively over the different antennas. This provides a coding gain over the repetition code. One can also design codes specifically for the transmit diversity system. There have been a lot of research activities in this area under the rubric of *space-time coding* and here we discuss the simplest, and yet one of the most elegant, space-time code: the so-called Alamouti scheme. This is the transmit diversity scheme proposed in several third-generation cellular standards. The Alamouti scheme is designed for two transmit antennas; generalization to more than two antennas is possible, to some extent.

Alamouti scheme

With flat fading, the two transmit, single receive channel is written as

$$y[m] = h_1[m]x_1[m] + h_2[m]x_2[m] + w[m], \qquad (3.73)$$

where h_i is the channel gain from transmit antenna i. The Alamouti scheme transmits two complex symbols u_1 and u_2 over two symbol times: at time 1, $x_1[1] = u_1$, $x_2[1] = u_2$; at time 2, $x_1[2] = -u_2^*$, $x_2[2] = u_1^*$. If we assume that the channel remains constant over the two symbol times and set $h_1 = h_1[1] = h_1[2]$, $h_2 = h_2[1] = h_2[2]$, then we can write in matrix form:

$$\begin{bmatrix} y[1] & y[2] \end{bmatrix} = \begin{bmatrix} h_1 & h_2 \end{bmatrix} \begin{bmatrix} u_1 & -u_2^* \\ u_2 & u_1^* \end{bmatrix} + \begin{bmatrix} w[1] & w[2] \end{bmatrix}. \qquad (3.74)$$

---

[8] This will of course ultimately not hold since the received power cannot be larger than the transmit power, but the number of antennas for our model to break down will have to be humongous.

We are interested in detecting u_1, u_2, so we rewrite this equation as

$$\begin{bmatrix} y[1] \\ y[2]^* \end{bmatrix} = \begin{bmatrix} h_1 & h_2 \\ h_2^* & -h_1^* \end{bmatrix} \begin{bmatrix} u_1 \\ u_2 \end{bmatrix} + \begin{bmatrix} w[1] \\ w[2]^* \end{bmatrix}. \tag{3.75}$$

We observe that the columns of the square matrix are orthogonal. Hence, the detection problem for u_1, u_2 decomposes into two separate, orthogonal, scalar problems. We project $\mathbf{y}$ onto each of the two columns to obtain the sufficient statistics

$$r_i = \|\mathbf{h}\| u_i + w_i, \qquad i = 1, 2, \tag{3.76}$$

where $\mathbf{h} = [h_1, h_2]^t$ and $w_i \sim \mathcal{CN}(0, N_0)$ and w_1, w_2 are independent. Thus, the diversity gain is 2 for the detection of each symbol. Compared to the repetition code, two symbols are now transmitted over two symbol times instead of one symbol, but with half the power in each symbol (assuming that the total transmit power is the same in both cases).

The Alamouti scheme works for any constellation for the symbols u_1, u_2, but suppose now they are BPSK symbols, thus conveying a total of two bits over two symbol times. In the repetition scheme, we need to use 4-PAM symbols to achieve the same data rate. To achieve the same minimum distance as the BPSK symbols in the Alamouti scheme, we need five times the energy per symbol. Taking into account the factor of 2 energy saving since we are only transmitting one symbol at a time in the repetition scheme, we see that the repetition scheme requires a factor of 2.5 (4 dB) more power than the Alamouti scheme. Again, the repetition scheme suffers from an inefficient utilization of the available degrees of freedom in the channel: over the two symbol times, bits are packed into only one dimension of the received signal space, namely along the direction $[h_1, h_2]^t$. In contrast, the Alamouti scheme spreads the information onto two dimensions – along the orthogonal directions $[h_1, h_2^*]^t$ and $[h_2, -h_1^*]^t$.

The determinant criterion for space-time code design

In Section 3.2, we saw that a good code exploiting time diversity should maximize the minimum product distance between codewords. Is there an analogous notion for space-time codes? To answer this question, let us think of a space-time code as a set of complex codewords $\{\mathbf{X}_i\}$, where each $\mathbf{X}_i$ is an L by N matrix. Here, L is the number of transmit antennas and N is the block length of the code. For example, in the Alamouti scheme, each codeword is of the form

$$\begin{bmatrix} u_1 & -u_2^* \\ u_2 & u_1^* \end{bmatrix}, \tag{3.77}$$

with $L = 2$ and $N = 2$. In contrast, each codeword in the repetition scheme is of the form

$$\begin{bmatrix} u & 0 \\ 0 & u \end{bmatrix}. \tag{3.78}$$

More generally, any block length L time diversity code with codewords $\{\mathbf{x}_i\}$ translates into a block length L transmit diversity code with codeword matrices $\{\mathbf{X}_i\}$, where

$$\mathbf{X}_i = \text{diag}\{x_{i1}, \ldots, x_{iL}\}. \tag{3.79}$$

For convenience, we normalize the codewords so that the average energy per symbol time is 1, hence $\mathsf{SNR} = 1/N_0$. Assuming that the channel remains constant for N symbol times, we can write

$$\mathbf{y}^t = \mathbf{h}^* \mathbf{X} + \mathbf{w}^t, \tag{3.80}$$

where

$$\mathbf{y} := \begin{bmatrix} y[1] \\ \vdots \\ y[N] \end{bmatrix}, \qquad \mathbf{h} := \begin{bmatrix} h_1^* \\ \vdots \\ h_L^* \end{bmatrix}, \qquad \mathbf{w} := \begin{bmatrix} w[1] \\ \vdots \\ w[N] \end{bmatrix}. \tag{3.81}$$

To bound the error probability, consider the pairwise error probability of confusing $\mathbf{X}_B$ with $\mathbf{X}_A$, when $\mathbf{X}_A$ is transmitted. Conditioned on the fading gains $\mathbf{h}$, we have the familiar vector Gaussian detection problem (see Summary A.2): here we are deciding between the vectors $\mathbf{h}^* \mathbf{X}_A$ and $\mathbf{h}^* \mathbf{X}_B$ under additive circular symmetric white Gaussian noise. A sufficient statistic is $\Re\{\mathbf{v}^* \mathbf{y}\}$, where $\mathbf{v} := \mathbf{h}^* (\mathbf{X}_A - \mathbf{X}_B)$. The conditional pairwise error probability is

$$\mathbb{P}\{\mathbf{X}_A \to \mathbf{X}_B \mid \mathbf{h}\} = Q\left(\frac{\|\mathbf{h}^* (\mathbf{X}_A - \mathbf{X}_B)\|}{2\sqrt{N_0/2}} \right). \tag{3.82}$$

Hence, the pairwise error probability averaged over the channel statistics is

$$\mathbb{P}\{\mathbf{X}_A \to \mathbf{X}_B\} = \mathbb{E}\left[Q\left(\sqrt{\frac{\mathsf{SNR}\, \mathbf{h}^* (\mathbf{X}_A - \mathbf{X}_B)(\mathbf{X}_A - \mathbf{X}_B)^* \mathbf{h}}{2}} \right) \right]. \tag{3.83}$$

The matrix $(\mathbf{X}_A - \mathbf{X}_B)(\mathbf{X}_A - \mathbf{X}_B)^*$ is Hermitian[9] and is thus diagonalizable by a unitary transformation, i.e., we can write $(\mathbf{X}_A - \mathbf{X}_B)(\mathbf{X}_A - \mathbf{X}_B)^* = \mathbf{U}\Lambda\mathbf{U}^*$,

---

[9] A complex square matrix $\mathbf{X}$ is Hermitian if $\mathbf{X}^* = \mathbf{X}$.

where $\mathbf{U}$ is unitary[10] and $\Lambda = \text{diag}\{\lambda_1^2, \ldots, \lambda_L^2\}$. Here λ_ℓ are the *singular values* of the codeword difference matrix $\mathbf{X}_A - \mathbf{X}_B$. Therefore, we can rewrite the pairwise error probability as

$$\mathbb{P}\{\mathbf{X}_A \rightarrow \mathbf{X}_B\} = \mathbb{E}\left[Q\left(\sqrt{\frac{\text{SNR} \sum_{\ell=1}^{L} |\tilde{h}_\ell|^2 \lambda_\ell^2}{2}} \right) \right], \qquad (3.84)$$

where $\tilde{\mathbf{h}} := \mathbf{U}^* \mathbf{h}$. In the Rayleigh fading model, the fading coefficients h_ℓ are i.i.d. $\mathcal{CN}(0, 1)$ and then $\tilde{\mathbf{h}}$ has the same distribution as $\mathbf{h}$ (cf. (A.22) in Appendix A). Thus we can bound the average pairwise error probability, as in (3.54),

$$\mathbb{P}\{\mathbf{X}_A \rightarrow \mathbf{X}_B\} \leq \prod_{\ell=1}^{L} \frac{1}{1 + \text{SNR} \, \lambda_\ell^2/4}. \qquad (3.85)$$

If all the λ_ℓ^2 are strictly positive for all the codeword differences, then the maximal diversity gain of L is achieved. Since the number of positive eigenvalues λ_ℓ^2 equals the rank of the codeword difference matrix, this is possible only if $N \geq L$. If indeed all the λ_ℓ^2 are positive, then,

$$\begin{aligned}
\mathbb{P}\{\mathbf{X}_A \rightarrow \mathbf{X}_B\} &\leq \frac{4^L}{\text{SNR}^L \prod_{\ell=1}^{L} \lambda_\ell^2} \\
&= \frac{4^L}{\text{SNR}^L \, \det[(\mathbf{X}_A - \mathbf{X}_B)(\mathbf{X}_A - \mathbf{X}_B)^*]},
\end{aligned} \qquad (3.86)$$

and a diversity gain of L is achieved. The coding gain is determined by the minimum of the determinant $\det[(\mathbf{X}_A - \mathbf{X}_B)(\mathbf{X}_A - \mathbf{X}_B)^*]$ over all codeword pairs. This is sometimes called the *determinant criterion*.

In the special case when the transmit diversity code comes from a time diversity code, the space-time code matrices are diagonal (cf. (3.79)), and $\lambda_\ell^2 = |d_\ell|^2$, the squared magnitude of the component difference between the corresponding time diversity codewords. The determinant criterion then coincides with the squared product distance criterion (3.68) we already derived for time diversity codes.

We can compare the coding gains obtained by the Alamouti scheme with the repetition scheme. That is, how much less power does the Alamouti scheme consume to achieve the same error probability as the repetition scheme? For the Alamouti scheme with BPSK symbols u_i, the minimum determinant is 4. For the repetition scheme with 4-PAM symbols, the minimum determinant is 16/25. (Verify!) This translates into the Alamouti scheme having a coding

---

[10] A complex square matrix $\mathbf{U}$ is unitary if $\mathbf{U}^* \mathbf{U} = \mathbf{U} \mathbf{U}^* = \mathbf{I}$.

gain of roughly a factor of 6 over the repetition scheme, consistent with the analysis above.

The Alamouti transmit diversity scheme has a particularly simple receiver structure. Essentially, a linear receiver allows us to decouple the two symbols sent over the two transmit antennas in two time slots. Effectively, both symbols pass through non-interfering *parallel* channels, both of which afford a diversity of order 2. In Exercise 3.16, we derive some properties that a code construction must satisfy to mimic this behavior for more than two transmit antennas.

3.3.3 MIMO: a 2 × 2 example

Degrees of freedom

Consider now a MIMO channel with two transmit and two receive antennas (Figure 3.11(c)). Let h_{ij} be the Rayleigh distributed channel gain from transmit antenna j to receive antenna i. Suppose both the transmit antennas and the receive antennas are spaced sufficiently far apart that the fading gains, h_{ij}, can be assumed to be independent. There are four independently faded signal paths between the transmitter and the receiver, suggesting that the maximum diversity gain that can be achieved is 4. The same repetition scheme described in the last section can achieve this performance: transmit the same symbol over the two antennas in two consecutive symbol times (at each time, nothing is sent over the other antenna). If the transmitted symbol is x, the received symbols at the two receive antennas are

$$y_i[1] = h_{i1}x + w_i[1], \qquad i = 1, 2 \tag{3.87}$$

at time 1, and

$$y_i[2] = h_{i2}x + w_i[2], \qquad i = 1, 2 \tag{3.88}$$

at time 2. By performing maximal-ratio combining of the four received symbols, an effective channel with gain $\sum_{i=1}^{2} \sum_{j=1}^{2} |h_{ij}|^2$ is created, yielding a four-fold diversity gain.

However, just as in the case of the 2×1 channel, the repetition scheme utilizes the degrees of freedom in the channel poorly; it only transmits one data symbol per two symbol times. In this regard, the Alamouti scheme performs better by transmitting two data symbols over two symbol times. Exercise 3.20 shows that the Alamouti scheme used over the 2×2 channel provides effectively two independent channels, analogous to (3.76), but with the gain in each channel equal to $\sum_{i=1}^{2} \sum_{j=1}^{2} |h_{ij}|^2$. Thus, *both* the data symbols see a diversity gain of 4, the same as that offered by the repetition scheme.

But does the Alamouti scheme utilize *all* the available degrees of freedom in the 2×2 channel? How many degrees of freedom does the 2×2 channel have anyway?

In Section 2.2.3 we have defined the degrees of freedom of a channel as the dimension of the received signal space. In a channel with two transmit and a single receive antenna, this is equal to *one* for every symbol time. The repetition scheme utilizes only half a degree of freedom per symbol time, while the Alamouti scheme utilizes all of it.

With L receive, but a single transmit antenna, the received signal lies in an L-dimensional vector space, but it does not span the full space. To see this explicitly, consider the channel model from (3.69) (suppressing the symbol time index m):

$$\mathbf{y} = \mathbf{h}x + \mathbf{w}, \tag{3.89}$$

where $\mathbf{y} := [y_1, \ldots, y_L]^t$, $\mathbf{h} = [h_1, \ldots, h_L]^t$ and $\mathbf{w} = [w_1, \ldots, w_L]^t$. The signal of interest, $\mathbf{h}x$, lies in a one-dimensional space.[11] Thus, we conclude that the degrees of freedom of a multiple receive, single transmit antenna channel is still 1 per symbol time.

But in a 2×2 channel, there are potentially *two* degrees of freedom per symbol time. To see this, we can write the channel as

$$\mathbf{y} = \mathbf{h}_1 x_1 + \mathbf{h}_2 x_2 + \mathbf{w}, \tag{3.90}$$

where x_j and $\mathbf{h}_j$ are the transmitted symbol and the vector of channel gains from transmit antenna j respectively, and $\mathbf{y} = [y_1, y_2]^t$ and $\mathbf{w} = [w_1, w_2]^t$ are the vectors of received signals and $\mathcal{CN}(0, N_0)$ noise respectively. As long as $\mathbf{h}_1$ and $\mathbf{h}_2$ are linearly independent, the signal space dimension is 2: the signal from transmit antenna j arrives in its own direction $\mathbf{h}_j$, and with two receive antennas, the receiver can distinguish between the two signals. Compared to a 2×1 channel, there is an additional degree of freedom coming from *space*. Figure 3.12 summarizes the situation.

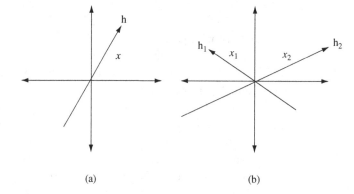

Figure 3.12 (a) In the 1 × 2 channel, the signal space is one-dimensional, spanned by **h**. (b) In the 2 × 2 channel, the signal space is two-dimensional, spanned by **h₁** and **h₂**.

(a)   (b)

---

[11] This is why the *scalar* $(\mathbf{h}^*/\|\mathbf{h}\|)\mathbf{y}$ is a sufficient statistic to detect x (cf. (3.33)).

Spatial multiplexing

Now we see that neither the repetition scheme nor the Alamouti scheme utilizes all the degrees of freedom in a 2×2 channel. A very simple scheme that does is the following: transmit independent uncoded symbols over the different antennas as well as over the different symbol times. This is an example of a *spatial multiplexing* scheme: independent data streams are multiplexed in space. (It is also called V-BLAST in the literature.) To analyze the performance of this scheme, we extend the derivation of the pairwise error probability bound (3.85) from a single receive antenna to multiple receive antennas. Exercise 3.19 shows that with n_{r} receive antennas, the corresponding bound on the probability of confusing codeword $\mathbf{X}_B$ with codeword $\mathbf{X}_A$ is

$$\mathbb{P}\{\mathbf{X}_A \to \mathbf{X}_B\} \le \left[\prod_{\ell=1}^{L} \frac{1}{1 + \mathsf{SNR}\, \lambda_\ell^2/4} \right]^{n_{\mathrm{r}}}. \tag{3.91}$$

where λ_ℓ are the singular values of the codeword difference $\mathbf{X}_A - \mathbf{X}_B$. This bound holds for space-time codes of general block lengths. Our specific scheme does not code across time and is thus "space-only". The block length is 1, the codewords are two-dimensional vectors $\mathbf{x}_1, \mathbf{x}_2$ and the bound simplifies to

$$\begin{aligned} \mathbb{P}\{\mathbf{x}_1 \to \mathbf{x}_2\} &\le \left[\frac{1}{1 + \mathsf{SNR}\, \|\mathbf{x}_1 - \mathbf{x}_2\|^2/4} \right]^2 \\ &\le \frac{16}{\mathsf{SNR}^2\, \|\mathbf{x}_1 - \mathbf{x}_2\|^4}. \end{aligned} \tag{3.92}$$

The exponent of the SNR factor is the diversity gain: the spatial multiplexing scheme achieves a diversity gain of 2. Since there is no coding across the transmit antennas, it is clear that no transmit diversity can be exploited; thus the diversity comes entirely from the dual receive antennas. The factor $\|\mathbf{x}_1 - \mathbf{x}_2\|^4$ plays a role analogous to the determinant $\det[(\mathbf{X}_A - \mathbf{X}_B)(\mathbf{X}_A - \mathbf{X}_B)^*]$ in determining the coding gain (cf. (3.86)).

Compared to the Alamouti scheme, we see that V-BLAST has a smaller diversity gain (2 compared to 4). On the other hand, the full use of the spatial degrees of freedom should allow a more efficient packing of bits, resulting in a better coding gain. To see this concretely, suppose we use BPSK symbols in the spatial multiplexing scheme to deliver 2 bits/s/Hz. Assuming that the average transmit energy per symbol time is normalized to be 1 as before, we can use (3.92) to explicitly calculate a bound on the *worst-case* pairwise error probability:

$$\max_{i \neq j} \mathbb{P}\{\mathbf{x}_i \to \mathbf{x}_j\} \le 4 \cdot \mathsf{SNR}^{-2}. \tag{3.93}$$

On the other hand, the corresponding bound for the Alamouti scheme using 4-PAM symbols to deliver the same 2 bits/s/Hz can be calculated from (3.91) to be

$$\max_{i \neq j} \mathbb{P}\{\mathbf{x}_i \rightarrow \mathbf{x}_j\} \leq 10,000 \cdot \mathsf{SNR}^{-4}. \tag{3.94}$$

We see that indeed the bound for the Alamouti scheme has a much poorer constant before the factor that decays with SNR.

We can draw two lessons from the V-BLAST scheme. First, we see a new role for multiple antennas: in addition to diversity, they can also provide additional degrees of freedom for communication. This is in a sense a more powerful view of multiple antennas, one that will be further explored in Chapter 7. Second, the scheme also reveals limitations in our performance analysis framework for space-time codes. In the earlier sections, our approach has always been to seek schemes which extract the maximum diversity from the channel and then compare them on the basis of the coding gain, which is a function of how efficiently the schemes utilize the available degrees of freedom. This approach falls short in comparing V-BLAST and the Alamouti scheme for the 2×2 channel: V-BLAST has poorer diversity than the Alamouti scheme but is more efficient in exploiting the spatial degrees of freedom, resulting in a better coding gain. A more powerful framework combining the two performance measures into a unified metric is needed; this is one of the main subjects of Chapter 9. There we will also address the issue of what it means by an *optimal* scheme and whether it is possible to find a scheme which achieves the full diversity *and* the full degrees of freedom of the channel.

Low-complexity detection: the decorrelator

One advantage of the Alamouti scheme is its low-complexity ML receiver: the decoding decouples into two orthogonal single-symbol detection problems. ML detection of V-BLAST does not enjoy the same advantage: joint detection of the two symbols is required. The complexity grows exponentially with the number of antennas. A natural question to ask is: what performance can suboptimal single-symbol detectors achieve? We will study MIMO receiver architectures in depth in Chapter 8, but here we will give an example of a simple detector, the *decorrelator*, and analyze its performance in the 2×2 channel.

To motivate the definition of this detector, let us rewrite the channel (3.90) in matrix form:

$$\mathbf{y} = \mathbf{H}\mathbf{x} + \mathbf{w}, \tag{3.95}$$

where $\mathbf{H} = [\mathbf{h}_1, \mathbf{h}_2]$ is the channel matrix. The input $\mathbf{x} := [x_1, x_2]^t$ is composed of two independent symbols x_1, x_2. To decouple the detection of the two symbols, one idea is to invert the effect of the channel:

$$\tilde{\mathbf{y}} = \mathbf{H}^{-1}\mathbf{y} = \mathbf{x} + \mathbf{H}^{-1}\mathbf{w} = \mathbf{x} + \tilde{\mathbf{w}} \tag{3.96}$$

and detect each of the symbols separately. This is in general suboptimal compared to joint ML detection, since the noise samples $\tilde{w}_1$ and $\tilde{w}_2$ are correlated. How much performance do we lose?

Let us focus on the detection of the symbol x_1 from transmit antenna 1. By direct computation, the variance of the noise $\tilde{w}_1$ is

$$\frac{|h_{22}|^2 + |h_{12}|^2}{|h_{11}h_{22} - h_{21}h_{12}|^2} N_0. \tag{3.97}$$

Hence, we can rewrite the first component of the vector equation in (3.96) as

$$\tilde{y}_1 = x_1 + \frac{\sqrt{|h_{22}|^2 + |h_{12}|^2}}{|h_{11}h_{22} - h_{21}h_{12}|} z_1, \tag{3.98}$$

where $z_1 \sim \mathcal{CN}(0, N_0)$, the scaled version of $\tilde{w}_1$, is independent of x_1. Equivalently, the scaled output can be written as

$$\begin{aligned}
y_1' &:= \frac{h_{11}h_{22} - h_{21}h_{12}}{\sqrt{|h_{22}|^2 + |h_{12}|^2}} \tilde{y}_1 \\
&= (\phi_2^* \mathbf{h}_1) x_1 + z_1, \tag{3.99}
\end{aligned}$$

where

$$\mathbf{h}_i := \begin{bmatrix} h_{1i} \\ h_{2i} \end{bmatrix}, \qquad \phi_i := \frac{1}{\sqrt{|h_{2i}|^2 + |h_{1i}|^2}} \begin{bmatrix} h_{2i}^* \\ -h_{1i}^* \end{bmatrix}, \qquad i = 1, 2. \tag{3.100}$$

Geometrically, one can interpret $\mathbf{h}_j$ as the "direction" of the signal from transmit antenna j and ϕ_j as the direction orthogonal to $\mathbf{h}_j$. Equation (3.99) says that when demodulating the symbol from antenna 1, channel inversion eliminates the interference from transmit antenna 2 by projecting the received signal $\mathbf{y}$ in the direction orthogonal to $\mathbf{h}_2$ (Figure 3.13). The signal part is $(\phi_2^* \mathbf{h}_1) x_1$. The scalar gain $\phi_2^* \mathbf{h}_1$ is circular symmetric Gaussian, being the projection of a two-dimensional i.i.d. circular symmetric Gaussian random vector ($\mathbf{h}_1$) onto an independent unit vector (ϕ_2) (cf. (A.22) in Appendix A). The scalar channel (3.99) is therefore Rayleigh faded like a 1×1 channel and has only unit diversity. Note that if there were no interference from antenna 2, the diversity gain would have been 2: the norm $\|\mathbf{h}_1\|^2$ of the entire vector $\mathbf{h}_1$ has to be small for poor reception of x_1. However, here, the component of $\mathbf{h}_1$ perpendicular to $\mathbf{h}_2$ being small already wreaks havoc; this is the price paid for nulling out the interference from antenna 2. In contrast, the ML detector, by jointly detecting the two symbols, retains the diversity gain of 2.

We have discussed V-BLAST in the context of a point-to-point link with two transmit antennas. But since there is no coding across the antennas, we can equally think of the two transmit antennas as two distinct users each with a single antenna. In the multiuser context, the receiver described above is sometimes called the *interference nuller*, *zero-forcing receiver* or

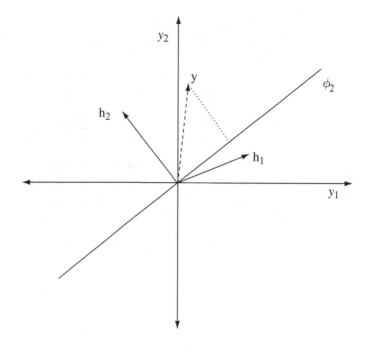

Figure 3.13 Demodulation of x_1: the received vector **y** is projected onto the direction ϕ_2 orthogonal to $\mathbf{h}_2$. The effective channel for x_1 is in deep fade whenever the projection of $\mathbf{h}_1$ onto ϕ_2 is small.

the *decorrelator*. It nulls out the effect of the other user (interferer) while demodulating the symbol of one user. Using this receiver, we see that dual receive antennas can perform one of two functions in a wireless system: they can *either* provide a two-fold diversity gain in a point-to-point link when there is no interference, *or* they can be used to null out the effect of an interfering user but provide no diversity gain more than 1. But they cannot do *both*. This is however not an intrinsic limitation of the *channel* but rather a limitation of the decorrelator; by performing joint ML detection instead, the two users can in fact be simultaneously supported with a two-fold diversity gain each.

Summary 3.2 2 × 2 MIMO schemes

The performance of the various schemes for the 2×2 channel is summarized below.

	Diversity gain	Degrees of freedom utilized per symbol time
Repetition	4	1/2
Alamouti	4	1
V-BLAST (ML)	2	2
V-BLAST (nulling)	1	2
Channel itself	4	2

3.4 Frequency diversity

3.4.1 Basic concept

So far we have focused on narrowband flat fading channels. These channels are modeled by a single-tap filter, as most of the multipaths arrive during one symbol time. In wideband channels, however, the transmitted signal arrives over multiple symbol times and the multipaths can be resolved at the receiver. The frequency response is no longer flat, i.e., the transmission bandwidth W is greater than the coherence bandwidth W_c of the channel. This provides another form of diversity: frequency.

We begin with the discrete-time baseband model of the wireless channel in Section 2.2. Recalling (2.35) and (2.38), the sampled output $y[m]$ can be written as

$$y[m] = \sum_{\ell} h_\ell[m]x[m-\ell] + w[m]. \tag{3.101}$$

Here $h_\ell[m]$ denotes the ℓth channel filter tap at time m. To understand the concept of frequency diversity in the simplest setting, consider first the one-shot communication situation when one symbol $x[0]$ is sent at time 0, and no symbols are transmitted after that. The receiver observes

$$y[\ell] = h_\ell[\ell]x[0] + w[\ell], \qquad \ell = 0, 1, 2, \ldots \tag{3.102}$$

If we assume that the channel response has a finite number of taps L, then the delayed replicas of the signal are providing L branches of diversity in detecting $x[0]$, since the tap gains $h_\ell[\ell]$ are assumed to be independent. This diversity is achieved by the ability of resolving the multipaths at the receiver due to the wideband nature of the channel, and is thus called *frequency diversity*.

A simple communication scheme can be built on the above idea by sending an information symbol every L symbol times. The maximal diversity gain of L can be achieved, but the problem with this scheme is that it is very wasteful of degrees of freedom: only one symbol can be transmitted every delay spread. This scheme can actually be thought of as analogous to the repetition codes used for both time and spatial diversity, where one information symbol is repeated L times. In this setting, once one tries to transmit symbols more frequently, *inter-symbol interference* (ISI) occurs: the delayed replicas of previous symbols interfere with the current symbol. The problem is then how to deal with the ISI while at the same time exploiting the inherent frequency diversity in the channel. Broadly speaking, there are three common approaches:

- **Single-carrier systems with equalization** By using linear and non-linear processing at the receiver, ISI can be mitigated to some extent. Optimal ML detection of the transmitted symbols can be implemented using the Viterbi algorithm. However, the complexity of the Viterbi algorithm grows exponentially with the number of taps, and it is typically used only when the number of significant taps is small. Alternatively, linear equalizers attempt to detect the current symbol while linearly suppressing the interference from the other symbols, and they have lower complexity.

- **Direct-sequence spread-spectrum** In this method, information symbols are modulated by a pseudonoise sequence and transmitted over a bandwidth W much larger than the data rate. Because the symbol rate is very low, ISI is small, simplifying the receiver structure significantly. Although this leads to an inefficient utilization of the total degrees of freedom in the system from the perspective of one user, this scheme allows multiple users to share the total degrees of freedom, with users appearing as pseudonoise to each other.

- **Multi-carrier systems** Here, transmit precoding is performed to convert the ISI channel into a set of non-interfering, orthogonal sub-carriers, each experiencing narrowband flat fading. Diversity can be obtained by coding across the symbols in different sub-carriers. This method is also called Discrete Multi-Tone (DMT) or Orthogonal Frequency Division Multiplexing (OFDM). Frequency-hop spread-spectrum can be viewed as a special case where one carrier is used at a time.

For example, GSM is a single-carrier system, IS-95 CDMA and IEEE 802.11b (a wireless LAN standard) are based on direct-sequence spread-spectrum, and IEEE 802.11a is a multi-carrier system,

Below we study these three approaches in turn. An important conceptual point is that, while frequency diversity is something *intrinsic* in a wideband channel, the presence of ISI is not, as it depends on the modulation technique used. For example, under OFDM, there is no ISI, but sub-carriers that are separated by more than the coherence bandwidth fade more or less independently and hence frequency diversity is still present.

Narrowband systems typically operate in a relatively high SNR regime. In contrast, the energy is spread across many degrees of freedom in many wideband systems, and the impact of the channel uncertainty on the ability of the receiver to extract the inherent diversity in frequency-selective channels becomes more pronounced. This point will be discussed in Section 3.5, but in the present section, we assume that the receiver has a perfect estimate of the channel.

3.4.2 Single-carrier with ISI equalization

Single-carrier with ISI equalization is the classic approach to communication over frequency-selective channels, and has been used in wireless as well as wireline applications such as voiceband modems. Much work has been done in this area but here we focus on the diversity aspects.

Starting at time 1, a sequence of *uncoded* independent symbols $x[1], x[2], \ldots$ is transmitted over the frequency-selective channel (3.101).

Assuming that the channel taps do not vary over these N symbol times, the received symbol at time m is

$$y[m] = \sum_{\ell=0}^{L-1} h_\ell x[m-\ell] + w[m], \qquad (3.103)$$

where $x[m] = 0$ for $m < 1$. For simplicity, we assume here that the taps h_ℓ are i.i.d. Rayleigh with equal variance $1/L$, but the discussion below holds more generally (see Exercise 3.25).

We want to detect each of the transmitted symbols from the received signal. The process of extracting the symbols from the received signal is called *equalization*. In contrast to the simple scheme in the previous section where a symbol is sent every L symbol times, here a symbol is sent *every* symbol time and hence there is significant ISI. Can we still get the maximum diversity gain of L?

Frequency-selective channel viewed as a MISO channel

To analyze this problem, it is insightful to transform the frequency-selective channel into a *flat* fading *MISO* channel with L transmit antennas and a single receive antenna and channel gains $h_0, \ldots, h_{L-1}$. Consider the following transmission scheme on the MISO channel: at time 1, the symbol $x[1]$ is transmitted on antenna 1 and the other antennas are silent. At time 2, $x[1]$ is transmitted at antenna 2, $x[2]$ is transmitted on antenna 1 and the other antennas remain silent. At time m, $x[m-\ell]$ is transmitted on antenna $\ell+1$, for $\ell = 0, \ldots, L-1$. See Figure 3.14. The received symbol at time m in this MISO channel is precisely the same as that in the frequency-selective channel under consideration.

Once we transform the frequency-selective channel into a MISO channel, we can exploit the machinery developed in Section 3.3.2. First, it is clear that if we want to achieve full diversity on a symbol, say $x[N]$, we need to observe the received symbols up to time $N + L - 1$. Over these symbol times, we can write the system in matrix form (as in (3.80)):

$$\mathbf{y}^t = \mathbf{h}^* \mathbf{X} + \mathbf{w}^t, \qquad (3.104)$$

where $\mathbf{y}^t := [y[1], \ldots, y[N+L-1]], \mathbf{h}^* := [h_0, \ldots, h_{L-1}], \mathbf{w}^t := [w[1], \ldots w[N+L-1]]$ and the L by $N+L-1$ space-time code matrix

$$\mathbf{X} = \begin{bmatrix} x[1] & x[2] & \cdot & \cdot & \cdot & x[N] & \cdot & \cdot & x[N+L-1] \\ 0 & x[1] & x[2] & \cdot & \cdot & \cdot & x[N] & \cdot & x[N+L-2] \\ 0 & 0 & x[1] & x[2] & \cdot & \cdot & \cdot & \cdot & \cdot \\ \cdot & \cdot & \cdot & \cdot & \cdot & \cdot & \cdot & \cdot & \cdot \\ 0 & 0 & \cdot & \cdot & x[1] & x[2] & \cdot & \cdot & x[N] \end{bmatrix} \qquad (3.105)$$

corresponds to the transmitted sequence $\mathbf{x} = [x[1], \ldots, x[N+L-1]]^t$.

Figure 3.14 The MISO scenario equivalent to the frequency- selective channel.

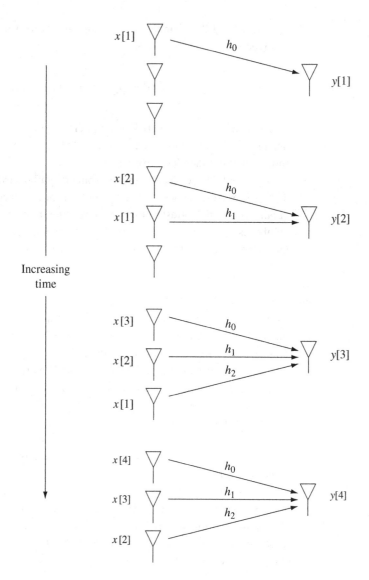

Increasing time

Error probability analysis

Consider the maximum likelihood detection of the *sequence* $\mathbf{x}$ based on the received vector $\mathbf{y}$ (MLSD). With MLSD, the pairwise error probability of confusing $\mathbf{x}_A$ with $\mathbf{x}_B$, when $\mathbf{x}_A$ is transmitted is, as in (3.85),

$$\mathbb{P}\{\mathbf{x}_A \to \mathbf{x}_B\} \le \prod_{\ell=1}^{L} \frac{1}{1 + \mathsf{SNR}\lambda_\ell^2/4}, \tag{3.106}$$

where λ_ℓ^2 are the eigenvalues of the matrix $(\mathbf{X}_A - \mathbf{X}_B)(\mathbf{X}_A - \mathbf{X}_B)^*$ and SNR is the total received SNR per received symbol (summing over all paths). This error probability decays like SNR^{-L} whenever the difference matrix $\mathbf{X}_A - \mathbf{X}_B$ is of rank L.

By a union bound argument, the probability of detecting the particular symbol $x[N]$ incorrectly is bounded by

$$\sum_{\mathbf{x}_B : x_B[N] \neq x_A[N]} \mathbb{P}\{\mathbf{x}_A \to \mathbf{x}_B\}, \tag{3.107}$$

summing over all the transmitted vectors $\mathbf{x}_B$ which differ with $\mathbf{x}_A$ in the Nth symbol.[12] To get full diversity, the difference matrix $\mathbf{X}_A - \mathbf{X}_B$ must be full rank for every such vector $\mathbf{x}_B$ (cf. (3.86)). Suppose m^* is the symbol time in which the vectors $\mathbf{x}_A$ and $\mathbf{x}_B$ *first* differ. Since they differ at least once within the first N symbol times, $m^* \leq N$ and the difference matrix is of the form

$$\mathbf{X}_A - \mathbf{X}_B = \begin{bmatrix} 0 \cdot 0 \; x_A[m^*] - x_B[m^*] & \cdot & \cdot & \cdot & \cdot \\ 0 \cdot \cdot & 0 & x_A[m^*] - x_B[m^*] \cdot & \cdot & \cdot \\ 0 \cdot \cdot & \cdot & 0 & \cdot & \cdot \\ \cdot \cdot \cdot & \cdot & & \cdot & \cdot \\ 0 \cdot \cdot & \cdot & & \cdot & 0 \; x_A[m^*] - x_B[m^*] \cdot \end{bmatrix}.$$
$$\tag{3.108}$$

By inspection, all the rows in the difference matrix are linearly independent. Thus $\mathbf{X}_A - \mathbf{X}_B$ is of full rank (i.e., the rank is equal to L). We can summarize:

Uncoded transmission combined with maximum likelihood sequence detection achieves full diversity on symbol $x[N]$ using the observations up to time $N + L - 1$, i.e., a delay of $L - 1$ symbol times.

Compared to the scheme in which a symbol is transmitted every L symbol times, the same diversity gain of L is achieved and yet an independent symbol can be transmitted every symbol time. This translates into a significant "coding gain" (Exercise 3.26).

In the analysis here it was convenient to transform the frequency-selective channel into a MISO channel. However, we can turn the transformation around: if we transmit the space-time code of the form in (3.105) on a MISO channel, then we have converted the MISO channel into a frequency-selective

---

[12] Strictly speaking, the MLSD only minimizes the *sequence* error probability, not the *symbol* error probability. However, this is the standard detector implemented for ISI equalization via the Viterbi algorithm, to be discussed next. In any case, the symbol error probability performance of the MLSD serves as an upper bound to the optimal symbol error performance.

channel. This is the *delay diversity* scheme and it was one of the first proposed transmit diversity schemes for the MISO channel.

Implementing MLSD: the Viterbi algorithm

Given the received vector **y** of length n, MLSD requires solving the optimization problem

$$\max_{\mathbf{x}} \mathbb{P}\{\mathbf{y}|\mathbf{x}\}. \tag{3.109}$$

A brute-force exhaustive search would require a complexity that grows exponentially with the block length n. An efficient algorithm needs to exploit the structure of the problem and moreover should be recursive in n so that the problem does not have to be solved from scratch for every symbol time. The solution is the ubiquitous *Viterbi algorithm*.

The key observation is that the memory in the frequency-selective channel can be captured by a finite state machine. At time m, define the state (an L-dimensional vector)

$$\mathbf{s}[m] := \begin{bmatrix} x[m-L+1] \\ x[m-L+2] \\ \cdot \\ x[m] \end{bmatrix} \tag{3.110}$$

An example of the finite state machine when the $x[m]$ are BPSK symbols is given in Figure 3.15. The number of states is M^L, where M is the constellation size for each symbol $x[m]$.

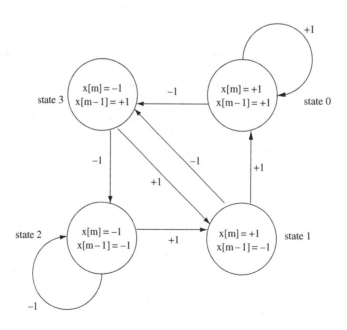

Figure 3.15 A finite state machine when $x[m]$ are ± 1 BPSK symbols and $L = 2$. There is a total of four states.

The received symbol $y[m]$ is given by

$$y[m] = \mathbf{h}^* \mathbf{s}[m] + w[m], \tag{3.111}$$

with $\mathbf{h}$ representing the frequency-selective channel, as in (3.104). The MLSD problem (3.109) can be rewritten as

$$\min_{\mathbf{s}[1], \ldots, \mathbf{s}[n]} -\log \mathbb{P}\{y[1], \ldots, y[n] \mid \mathbf{s}[1], \ldots, \mathbf{s}[n]\}, \tag{3.112}$$

subject to the transition constraints on the state sequence (i.e., the second component of $s[m]$ is the same as the first component of $s[m+1]$). Conditioned on the state sequence $\mathbf{s}[1], \ldots, \mathbf{s}[n]$, the received symbols are independent and the log-likelihood ratio breaks into a sum:

$$\log \mathbb{P}\{y[1], \ldots, y[n] \mid \mathbf{s}[1], \ldots, \mathbf{s}[n]\} = \sum_{m=1}^{n} \log \mathbb{P}\{y[m] \mid \mathbf{s}[m]\}. \tag{3.113}$$

The optimization problem in (3.112) can be represented as the problem of finding the shortest path through an n-stage *trellis*, as shown in Figure 3.16. Each state sequence $(\mathbf{s}[1], \ldots, \mathbf{s}[n])$ is visualized as a path through the trellis, and given the received sequence $y[1], \ldots, y[n]$, the cost associated with the mth transition is

$$c_m(\mathbf{s}[m]) := -\log \mathbb{P}\{y[m] \mid \mathbf{s}[m]\}. \tag{3.114}$$

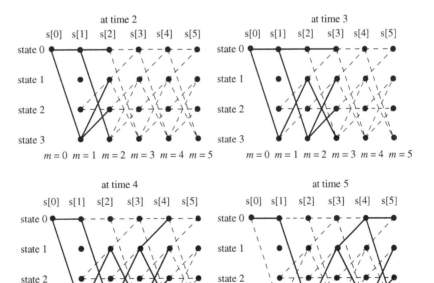

Figure 3.16 The trellis representation of the channel.

The solution is given recursively by the *optimality principle* of dynamic programming. Let $V_m(\mathbf{s})$ be the cost of the shortest path to a given state $\mathbf{s}$ at stage m. Then $V_m(\mathbf{s})$ for all states $\mathbf{s}$ can be computed recursively:

$$V_1(\mathbf{s}) = c_1(\mathbf{s}) \tag{3.115}$$
$$V_m(\mathbf{s}) = \min_{\mathbf{u}}[V_{m-1}(\mathbf{u}) + c_m(\mathbf{s})], \qquad m > 1. \tag{3.116}$$

Here the minimization is over all possible states $\mathbf{u}$, i.e., we only consider the states that the finite state machine can be in at stage $m-1$ and, further, can still end up at state $\mathbf{s}$ at stage m. The correctness of this recursion is based on the following intuitive fact: if the shortest path to state $\mathbf{s}$ at stage m goes through the state $\mathbf{u}^*$ at stage $m-1$, then the part of the path up to stage $m-1$ must itself be the shortest path to state $\mathbf{u}^*$. See Figure 3.17. Thus, to compute the shortest path up to stage m, it suffices to augment only the shortest paths up to stage $m-1$, and these have already been computed.

Once $V_m(\mathbf{s})$ is computed for all states $\mathbf{s}$, the shortest path to stage m is simply the minimum of these values over all states $\mathbf{s}$. Thus, the optimization problem (3.112) is solved. Moreover, the solution is recursive in n.

The complexity of the Viterbi algorithm is linear in the number of stages n. Thus, the cost is constant per symbol, a vast improvement over brute-force exhaustive search. However, its complexity is also proportional to the size of the state space, which is M^L, where M is the constellation size of each symbol. Thus, while MLSD can be done for channels with a small number of taps, it becomes impractical when L becomes large.

The computational complexity of MLSD leads to an interest in seeking suboptimal equalizers which yield comparable performance. Some candidates are *linear* equalizers (such as the zero-forcing and minimum mean square error (MMSE) equalizers, which involve simple linear operations on the received symbols followed by simple hard decoders), and their decision-feedback versions (DFE), where previously detected symbols are removed from the received signal before linear equalization is performed. We will discuss these equalizers further in Discussion 8.1, where we exploit

Figure 3.17 The dynamic programming principle. If the first $m-1$ segments of the shortest path to state s at stage m were not the shortest path to state $\mathbf{u}^*$ at stage $m-1$, then one could have found an even shorter path to state s.

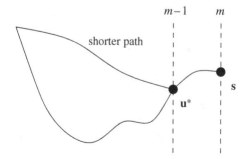

a correspondence between the MIMO channel and the frequency-selective channel.

3.4.3 Direct-sequence spread-spectrum

A common communication system that employs a wide bandwidth is the direct-sequence (DS) spread-spectrum system. Its basic components are shown in Figure 3.18. Information is encoded and modulated by a pseudonoise (PN) sequence and transmitted over a bandwidth W. In contrast to the system we analyzed in the last section where an independent symbol is sent at each symbol time, the data rate R bits/s in a spread-spectrum system is typically much smaller than the transmission bandwidth W Hz. The ratio W/R is sometimes called the *processing gain* of the system. For example, IS-95 (CDMA) is a direct-sequence spread-spectrum system. The bandwidth is 1.2288 MHz and a typical data rate (voice) is 9.6 kbits/s, so the processing gain is 128. Thus, very few bits are transmitted per degree of freedom per user. In spread-spectrum jargon, each sample period is called a *chip*, and another way of describing a spread-spectrum system is that the chip rate is much larger than the data rate.

Because the symbol rate per user is very low in a spread-spectrum system, ISI is typically negligible and equalization is not required. Instead, as we will discuss next, a much simpler receiver called the *Rake* receiver can be used to extract frequency diversity. In the cellular setting, multiple spread-spectrum users would share the large bandwidth so that the aggregate bit rate can be high even though the rate of each user is low. The large processing gain of a user serves to mitigate the interference from other users, which appears as random noise. In addition to providing frequency diversity against multipath fading and allowing multiple access, spread-spectrum systems serve other purposes, such as anti-jamming from intentional interferers, and achieving message privacy in the presence of other listeners. We will discuss the multiple access aspects of spread-spectrum systems in Chapter 4. For now, we focus on how DS spread-spectrum systems can achieve frequency diversity.

The Rake receiver

Suppose we transmit one of two n-chips long pseudonoise sequences $\mathbf{x}_A$ or $\mathbf{x}_B$. Consider the problem of binary detection over a wideband multipath channel. In this context, a binary symbol is transmitted over n chips. The received signal is given by

$$y[m] = \sum_{\ell} h_{\ell}[m]x[m-\ell] + w[m]. \qquad (3.117)$$

We assume that $h_{\ell}[m]$ is non-zero only for $\ell = 0, \ldots, L-1$, i.e., the channel has L taps. One can think of L/W as the delay spread T_{d}. Also, we assume

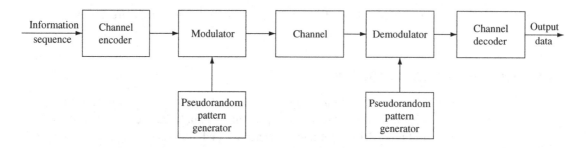

Figure 3.18 Basic elements of a direct sequence spread-spectrum system.

that $h_\ell[m]$ does not vary with m during the transmission of the sequence, i.e., the channel is considered time-invariant. This holds if $n \ll T_c W$, where T_c is the coherence time of the channel. We also assume that there is negligible interference between consecutive symbols, so that we can consider the binary detection problem in isolation for each symbol. This assumption is valid if $n \gg L$, which is quite common in a spread-spectrum system with high processing gain. Otherwise, ISI between consecutive symbols becomes significant, and an equalizer would be needed to mitigate the ISI. Note however we assume that simultaneously $n \gg T_d W$ and $n \ll T_c W$, which is possible only if $T_d \ll T_c$. In a typical cellular system, T_d is of the order of microseconds and T_c of the order of tens of milliseconds, so this assumption is quite reasonable. (Recall from Chapter 2, Table 2.2 that a channel satisfying this condition is called an *underspread* channel.)

With the above assumptions, the output is just a convolution of the input with the LTI channel plus noise

$$y[m] = (h * x)[m] + w[m], \qquad m = 1, \ldots, n + L - 1 \tag{3.118}$$

where h_ℓ is the ℓth tap of the time-invariant channel filter response, with $h_\ell = 0$ for $\ell < 0$ and $\ell > L - 1$. Assuming the channel h is known to the receiver, two sufficient statistics, r_A and r_B, can be obtained by projecting the received vector $\mathbf{y} := [y[1], \ldots, y[n + L - 1]]^t$ onto the $n + L - 1$ dimensional vectors $\mathbf{v}_A$ and $\mathbf{v}_B$, where $\mathbf{v}_A := [(h * x_A)[1], \ldots, (h * x_A)[n + L - 1]]^t$ and $\mathbf{v}_B := [(h * x_B)[1], \ldots, (h * x_B)[n + L - 1]]^t$, i.e.,

$$r_A := \mathbf{v}_A^* \mathbf{y}, \qquad r_B := \mathbf{v}_B^* \mathbf{y}. \tag{3.119}$$

The computation of r_A and r_B can be implemented by first matched filtering the received signal to $\mathbf{x}_A$ and to $\mathbf{x}_B$. The outputs of the matched filters are passed through a filter matched to the channel response h and then sampled at time $n + L - 1$ (Figure 3.19). This is called the *Rake* receiver. What the Rake actually does is taking inner products of the received signal with shifted versions at the candidate transmitted sequences. Each output is then weighted by the channel tap gains at the appropriate delays and summed. The signal path associated with a particular delay is sometimes called a *finger* of the Rake receiver.

Figure 3.19 The Rake receiver. Here, $\tilde{\mathbf{h}}$ is the filter matched to $\mathbf{h}$, i.e., $\tilde{h}_\ell = h^*_{-\ell}$. Each tap of $\tilde{\mathbf{h}}$ represents a finger of the Rake.

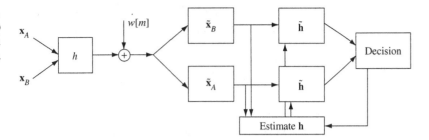

As discussed earlier, we are continuing with the assumption that the channel gains h_ℓ are known at the receiver. In practice, these gains have to be estimated and tracked from either a pilot signal or in a decision-directed mode using the previously detected symbols. (The channel estimation problem will be discussed in Section 3.5.2.) Also, due to hardware limitations, the actual number of fingers used in a Rake receiver may be less than the total number of taps L in the range of the delay spread. In this case, there is also a tracking mechanism in which the Rake receiver continuously searches for the strong paths (taps) to assign the limited number of fingers to.

Performance analysis

Let us now analyze the performance of the Rake receiver. To simplify our notation, we specialize to antipodal modulation (i.e., $\mathbf{x}_A = -\mathbf{x}_B = \mathbf{u}$); the analysis for other modulation schemes is similar. One key aspect of spread-spectrum systems is that the transmitted signal ($\pm\mathbf{u}$) has a *pseudonoise* characteristic. The defining characteristic of a pseudonoise sequence is that its shifted versions are nearly orthogonal to each other. More precisely, if we write $\mathbf{u} = [u[1], \ldots, u[n]]$, and

$$\mathbf{u}^{(\ell)} := [0, \ldots, 0, u[1], \ldots, u[n], 0, \ldots, 0]^t \qquad (3.120)$$

as the $n + L$ dimensional version of $\mathbf{u}$ shifted by ℓ chips (hence there are ℓ zeros preceding $\mathbf{u}$ and $L - \ell$ zeros following $\mathbf{u}$ above), the pseudonoise property means that for every $\ell = 0, \ldots, L-1$,

$$|(\mathbf{u}^{(\ell)})^*(\mathbf{u}^{(\ell')})| \ll \sum_{i=1}^{n} |u[i]|^2, \qquad \ell \neq \ell'. \qquad (3.121)$$

To simplify the analysis, we assume full orthogonality: $(\mathbf{u}^{(\ell)})^*(\mathbf{u}^{(\ell')}) = 0$ if $\ell \neq \ell'$.

We will now show that the performance of the Rake is the same as that in the diversity model with L branches for repetition coding described in Section 3.2. We can see this by looking at a set of sufficient statistics for the detection problem different from the ones we used earlier. First, we rewrite the channel model in vector form

$$\mathbf{y} = \sum_{\ell=0}^{L-1} h_\ell \mathbf{x}^{(\ell)} + \mathbf{w}, \tag{3.122}$$

where $\mathbf{w} := [w[1], \ldots, w[n+L]]^t$ and $\mathbf{x}^{(\ell)} = \pm\mathbf{u}^{(\ell)}$, the version of the transmitted sequence (either $\mathbf{u}$ or $-\mathbf{u}$) shifted by ℓ chips. The received signal (without the noise) therefore lies in the span of the L vectors $\{\mathbf{u}^{(\ell)}/\|\mathbf{u}\|\}_\ell$. By the pseudonoise assumption, all these vectors are orthogonal to each other. A set of L sufficient statistics $\{r^{(\ell)}\}_\ell$ can be obtained by projecting $\mathbf{y}$ onto each of these vectors

$$r^{(\ell)} = h_\ell x + w^{(\ell)}, \qquad \ell = 0, \ldots, L-1, \tag{3.123}$$

where $x = \pm\|\mathbf{u}\|$. Further, the orthogonality of $\mathbf{u}^{(\ell)}$ implies that $w^{(\ell)}$ are i.i.d. $\mathcal{CN}(0, N_0)$. Comparing with (3.32), this is exactly the same as the L-branch diversity model for the case of repetition code interleaved over time. *Thus, we see that the Rake receiver in this case is nothing more than a maximal ratio combiner of the signals from the L diversity branches.* The error probability is given by

$$p_e = \mathbb{E}\left[Q\left(\sqrt{2\|\mathbf{u}\|^2 \sum_{\ell=1}^{L} |h_\ell|^2/N_0} \right) \right]. \tag{3.124}$$

If we assume a Rayleigh fading model such that the tap gains h_ℓ are i.i.d. $\mathcal{CN}(0, 1/L)$, i.e., the energy is spread equally among all the L taps (normalizing such that the $\mathbb{E}[\sum_\ell |h_\ell|^2] = 1$), then the error probability can be explicitly computed (as in (3.37)):

$$p_e = \left(\frac{1-\mu}{2} \right)^L \sum_{\ell=0}^{L-1} \binom{L-1+\ell}{\ell} \left(\frac{1+\mu}{2} \right)^\ell, \tag{3.125}$$

where

$$\mu := \sqrt{\frac{\mathsf{SNR}}{1 + \mathsf{SNR}}} \tag{3.126}$$

and $\mathsf{SNR} := \|\mathbf{u}\|^2/(N_0 L)$ can be interpreted as the average signal-to-noise ratio *per diversity branch*. Noting that $\|\mathbf{u}\|^2$ is the average total energy received per bit of information, we can define $\mathcal{E}_b := \|\mathbf{u}\|^2$. Hence, the SNR per branch is $1/L \cdot \mathcal{E}_b/N_0$. Observe that the factor of $1/L$ accounts for the splitting of energy due to spreading: the larger the spread bandwidth W, the larger L is, and the more diversity one gets, but there is less energy in each branch.[13] As $L \to \infty$, $\sum_{\ell=1}^{L} |h_\ell|^2$ converges to 1 with probability 1 by the law of large numbers, and from (3.124) we see that

$$p_e \to Q\left(\sqrt{2\mathcal{E}_b/N_0}\right),\qquad(3.127)$$

i.e., the performance of the AWGN channel with the same $\mathcal{E}_b/N_0$ is asymptotically achieved.

The above analysis assumes an equal amount of energy in each tap. In a typical multipath delay profile, there is more energy in the taps with shorter delays. The analysis can be extended to the cases when the h_ℓ have unequal variances as well. (See Section 14.5.3 in [96]).

3.4.4 Orthogonal frequency division multiplexing

Both the single-carrier system with ISI equalization and the DS spread-spectrum system with Rake reception are based on a time-domain view of the channel. But we know that if the channel is linear time-invariant, sinusoids are eigenfunctions and they get transformed in a particularly simple way. ISI occurs in a single-carrier system because the transmitted signals are not sinusoids. This suggests that if the channel is underspread (i.e., the coherence time is much larger than the delay spread) and is therefore approximately time-invariant for a sufficiently long time-scale, then transformation into the frequency domain can be a fruitful approach to communication over frequency-selective channels. This is the basic idea behind OFDM.

We begin with the discrete-time baseband model

$$y[m] = \sum_\ell h_\ell[m]x[m-\ell] + w[m].\qquad(3.128)$$

For simplicity, we first assume that for each ℓ, the ℓth tap is not changing with m and hence the channel is linear time-invariant. Again assuming a finite number of non-zero taps $L := T_d W$, we can rewrite the channel model in (3.128) as

$$y[m] = \sum_{\ell=0}^{L-1} h_\ell x[m-\ell] + w[m].\qquad(3.129)$$

Sinusoids are eigenfunctions of LTI systems, but they are of infinite duration. If we transmit over only a finite duration, say N_c symbols, then the sinusoids are no longer eigenfunctions. One way to restore the eigenfunction

---

[13] This is assuming a very rich scattering environment, leading to many paths, all of equal energy. In reality, however, there are just a few paths that are strong enough to matter.

property is by adding a *cyclic prefix* to the symbols. For every block of symbols of length N_c, denoted by

$$\mathbf{d} = [d[0], d[1], \ldots, d[N_c - 1]]^t,$$

we create an $N_c + L - 1$ input block as

$$\mathbf{x} = [d[N_c - L + 1], d[N_c - L + 2], \ldots, d[N_c - 1], d[0], d[1], \ldots, d[N_c - 1]]^t,$$
(3.130)

i.e., we add a *prefix* of length $L - 1$ consisting of data symbols rotated cyclically (Figure 3.20). With this input to the channel (3.129), consider the output

$$y[m] = \sum_{\ell=0}^{L-1} h_\ell x[m - \ell] + w[m], \qquad m = 1, \ldots, N_c + L - 1.$$

The ISI extends over the first $L - 1$ symbols and the receiver ignores it by considering only the output over the time interval $m \in [L, N_c + L - 1]$. Due to the additional cyclic prefix, the output over this time interval (of length N_c) is

$$y[m] = \sum_{\ell=0}^{L-1} h_\ell d[(m - L - \ell) \text{ modulo } N_c] + w[m].$$
(3.131)

See Figure 3.21.

Denoting the output of length N_c by

$$\mathbf{y} = [y[L], \ldots, y[N_c + L - 1]]^t,$$

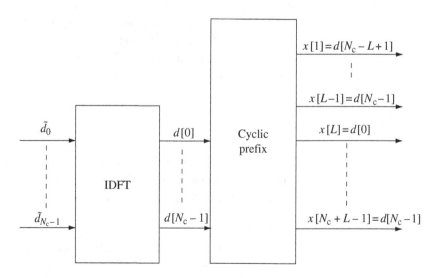

Figure 3.20 The cyclic prefix operation.

Figure 3.21 Convolution between the channel (**h**) and the input (**x**) formed from the data symbols (**d**) by adding a cyclic prefix. The output is obtained by multiplying the corresponding values of **x** and **h** on the circle, and outputs at different times are obtained by rotating the *x*-values with respect to the *h*-values. The current configuration yields the output *y*[*L*].

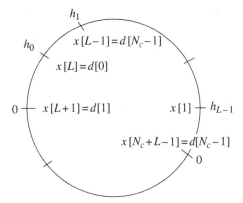

and the channel by a vector of length N_c

$$\mathbf{h} = [h_0, h_1, \ldots, h_{L-1}, 0, \ldots, 0]^t, \tag{3.132}$$

(3.131) can be written as

$$\mathbf{y} = \mathbf{h} \otimes \mathbf{d} + \mathbf{w}. \tag{3.133}$$

Here we denoted

$$\mathbf{w} = [w[L], \ldots, w[N_c + L - 1]]^t, \tag{3.134}$$

as a vector of i.i.d. $\mathcal{CN}(0, N_0)$ random variables. We also used the notation of $\otimes$ to denote the *cyclic convolution* in (3.131). Recall that the discrete Fourier transform (DFT) of **d** is defined to be

$$\tilde{d}_n := \frac{1}{\sqrt{N_c}} \sum_{m=0}^{N_c-1} d[m] \exp\left(\frac{-\mathrm{j}2\pi nm}{N_c}\right), \qquad n = 0, \ldots, N_c - 1. \tag{3.135}$$

Taking the discrete Fourier transform (DFT) of both sides of (3.133) and using the identity

$$\mathrm{DFT}(\mathbf{h} \otimes \mathbf{d})_n = \sqrt{N_c}\mathrm{DFT}(\mathbf{h})_n \cdot \mathrm{DFT}(\mathbf{d})_n, \qquad n = 0, \ldots, N_c - 1, \tag{3.136}$$

we can rewrite (3.133) as

$$\tilde{y}_n = \tilde{h}_n \tilde{d}_n + \tilde{w}_n, \qquad n = 0, \ldots, N_c - 1. \tag{3.137}$$

Here we have denoted $\tilde{w}_0, \ldots, \tilde{w}_{N_c-1}$ as the N_c-point DFT of the noise vector $w[1], \ldots, w[N_c]$. The vector $[\tilde{h}_0, \ldots, \tilde{h}_{N_c-1}]^t$ is defined as the DFT of the L-tap channel **h**, multiplied by $\sqrt{N_c}$,

$$\tilde{h}_n = \sum_{\ell=0}^{L-1} h_\ell \exp\left(\frac{-\mathrm{j}2\pi n\ell}{N_c}\right). \tag{3.138}$$

Note that the nth component $\tilde{h}_n$ is equal to the frequency response of the channel (see (2.20)) at $f = nW/N_c$.

We can redo everything in terms of matrices, a viewpoint which will prove particularly useful in Chapter 7 when we will draw a connection between the frequency-selective channel and the MIMO channel. The circular convolution operation $\mathbf{u} = \mathbf{h} \otimes \mathbf{d}$ can be viewed as a linear transformation

$$\mathbf{u} = \mathbf{Cd}, \tag{3.139}$$

where

$$\mathbf{C} := \begin{bmatrix} h_0 & 0 & \cdot & 0 & h_{L-1} & h_{L-2} & \cdot & h_1 \\ h_1 & h_0 & 0 & \cdot & 0 & h_{L-1} & \cdot & h_2 \\ \cdot & \cdot & \cdot & \cdot & \cdot & \cdot & \cdot & \cdot \\ 0 & \cdot & 0 & h_{L-1} & h_{L-2} & \cdot & h_1 & h_0 \end{bmatrix} \tag{3.140}$$

is a *circulant* matrix, i.e., the rows are cyclic shifts of each other. On the other hand, the DFT of $\mathbf{d}$ can be represented as an N_c-length vector $\mathbf{Ud}$, where $\mathbf{U}$ is the unitary matrix with its (k, n)th entry equal to

$$\frac{1}{\sqrt{N_c}} \exp\left(\frac{-j2\pi kn}{N_c}\right), \qquad k, n = 0, \ldots, N_c - 1. \tag{3.141}$$

This can be viewed as a coordinate change, expressing $\mathbf{d}$ in the basis defined by the rows of $\mathbf{U}$. Equation (3.136) is equivalent to

$$\mathbf{Uu} = \mathbf{\Lambda Ud}, \tag{3.142}$$

where $\mathbf{\Lambda}$ is the diagonal matrix with diagonal entries $\sqrt{N_c}$ times the DFT of $\mathbf{h}$, i.e.,

$$\Lambda_{nn} = \tilde{h}_n := \left(\sqrt{N_c}\mathbf{Uh}\right)_n, \qquad n = 0, \ldots, N_c - 1.$$

Comparing (3.139) and (3.142), we come to the conclusion that

$$\mathbf{C} = \mathbf{U}^{-1}\mathbf{\Lambda U}. \tag{3.143}$$

Equation (3.143) is the matrix version of the key DFT property (3.136). In geometric terms, this means that the circular convolution operation is diagonalized in the coordinate system defined by the rows of $\mathbf{U}$, and the eigenvalues of $\mathbf{C}$ are the DFT coefficients of the channel $\mathbf{h}$. Equation (3.133) can thus be written as

$$\mathbf{y} = \mathbf{Cd} + \mathbf{w} = \mathbf{U}^{-1}\mathbf{\Lambda Ud} + \mathbf{w}. \tag{3.144}$$

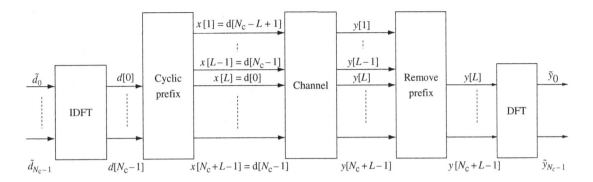

Figure 3.22 The OFDM transmission and reception schemes.

This representation suggests a natural rotation at the input and at the output to convert the channel to a set of non-interfering channels with no ISI. In particular, the actual data symbols (denoted by the length N_c vector $\tilde{\mathbf{d}}$) in the frequency domain are rotated through the IDFT (inverse DFT) matrix $\mathbf{U}^{-1}$ to arrive at the vector $\mathbf{d}$. At the receiver, the output vector of length N_c (obtained by ignoring the first L symbols) is rotated through the DFT matrix $\mathbf{U}$ to obtain the vector $\tilde{\mathbf{y}}$. The final output vector $\tilde{\mathbf{y}}$ and the actual data vector $\tilde{\mathbf{d}}$ are related through

$$\tilde{y}_n = \tilde{h}_n \tilde{d}_n + \tilde{w}_n, \qquad n = 0, \ldots, N_c - 1. \tag{3.145}$$

We have denoted $\tilde{\mathbf{w}} := \mathbf{U}\mathbf{w}$ as the DFT of the random vector $\mathbf{w}$ and we see that since $\mathbf{w}$ is isotropic, $\tilde{\mathbf{w}}$ has the same distribution as $\mathbf{w}$, i.e., a vector of i.i.d. $\mathcal{CN}(0, N_0)$ random variables (cf. (A.26) in Appendix A).

These operations are illustrated in Figure 3.22, which affords the following interpretation. The data symbols modulate N_c *tones* or *sub-carriers*, which occupy the bandwidth W and are uniformly separated by W/N_c. The data symbols on the sub-carriers are then converted (through the IDFT) to time domain. The procedure of introducing the cyclic prefix before transmission allows for the removal of ISI. The receiver ignores the part of the output signal containing the cyclic prefix (along with the ISI terms) and converts the length N_c symbols back to the frequency domain through a DFT. The data symbols on the sub-carriers are maintained to be orthogonal as they propagate through the channel and hence go through narrowband *parallel* sub-channels. This interpretation justifies the name of OFDM for this communication scheme. Finally, we remark that DFT and IDFT can be very efficiently implemented (using Fast Fourier Transform) whenever N_c is a power of 2.

OFDM block length

The OFDM scheme converts communication over a multipath channel into communication over simpler parallel narrowband sub-channels. However, this simplicity is achieved at a cost of underutilizing two resources, resulting in a loss of performance. First, the cyclic prefix occupies an amount of time which cannot be used to communicate data. This loss amounts to a fraction

$L/(N_c + L)$ of the total time. The second loss is in the power transmitted. A fraction $L/(N_c + L)$ of the average power is allocated to the cyclic prefix and cannot be used towards communicating data. Thus, to minimize the overhead (in both time and power) due to the cyclic prefix we prefer to have N_c as large as possible. The time-varying nature of the wireless channel, however, constrains the largest value N_c can reasonably take.

We started the discussion in this section by considering a simple channel model (3.129) that did not vary with time. If the channel is slowly time-varying (as discussed in Section 2.2.1, this is a reasonable assumption) then the coherence time T_c is much larger than the delay spread T_d (the *under-spread* scenario). For underspread channels, the block length of the OFDM communication scheme N_c can be chosen significantly larger than the multi-path length $L = T_d W$, but still much smaller than the coherence block length $T_c W$. Under these conditions, the channel model of linear time invariance approximates a slowly time-varying channel over the block length N_c, while keeping the overhead small.

The constraint on the OFDM block length can also be understood in the frequency domain. A block length of N_c corresponds to an inter-sub-carrier spacing equal to W/N_c. In a wireless channel, the Doppler spread introduces uncertainty in the frequency of the received signal; from Table 2.1 we see that the Doppler spread is inversely proportional to the coherence time of the channel: $D_s = 1/4T_c$. For the inter-sub-carrier spacing to be much larger than the Doppler spread, the OFDM block length N_c should be constrained to be much smaller than $T_c W$. This is the same constraint as above.

Apart from an underutilization of time due to the presence of the cyclic prefix, we also mentioned the additional power due to the cyclic prefix. OFDM schemes that put a zero signal instead of the cyclic prefix have been proposed to reduce this loss. However, due to the abrupt transition in the signal, such schemes introduce harmonics that are difficult to filter in the overall signal. Further, the cyclic prefix can be used for timing and frequency acquisition in wireless applications, and this capability would be lost if a zero signal replaced the cyclic prefix.

Frequency diversity

Let us revert to the non-overlapping narrowband channel representation of the ISI channel in (3.145). The correlation between the channel frequency coefficients $\tilde{h}_0, \ldots, \tilde{h}_{N_c - 1}$ depends on the coherence bandwidth of the channel. From our discussion in Section 2.3, we have learned that the coherence bandwidth is inversely proportional to the multipath spread. In particular, we have from (2.47) that

$$W_c = \frac{1}{2T_d} = \frac{W}{2L},$$

where we use our notation for L as denoting the length of the ISI. Since each sub-carrier is W/N_c wide, we expect approximately

$$\frac{N_c W_c}{W} = \frac{N_c}{2L}$$

as the number of neighboring sub-carriers whose channel coefficients are heavily correlated (Exercise 3.28). One way to exploit the frequency diversity is to consider ideal interleaving across the sub-carriers (analogous to the time-interleaving done in Section 3.2) and consider the model of (3.31)

$$y_\ell = h_\ell x_\ell + w_\ell, \qquad \ell = 1, \dots, L.$$

The difference is that now ℓ represents the sub-carriers while it is used to denote time in (3.31). However, with the ideal frequency interleaving assumption we retain the same independent assumption on the channel coefficients. Thus, the discussion of Section 3.2 on schemes harnessing diversity is directly applicable here. In particular, an L-fold diversity gain (proportional to the number of ISI symbols L) can be obtained. Since the communication scheme is over sub-carriers, the form of diversity is due to the frequency-selective channel and is termed *frequency diversity* (as compared to the time diversity discussed in Section 3.2 which arises due to the time variations of the channel).

Summary 3.3 Communication over frequency-selective channels

We have studied three approaches to extract frequency diversity in a frequency-selective channel (with L taps). We summarize their key attributes and compare their implementational complexity.

1 Single-carrier with ISI equalization
Using maximum likelihood sequence detection (MLSD), full diversity of L can be achieved for uncoded transmission sent at symbol rate.

MLSD can be performed by the Viterbi algorithm. The complexity is constant per symbol time but grows exponentially with the number of taps L.

The complexity is entirely at the receiver.

2 Direct-sequence spread-spectrum
Information is spread, via a pseudonoise sequence, across a bandwidth much larger than the data rate. ISI is typically negligible.

The signal received along the L nearly orthogonal diversity paths is maximal-ratio combined using the Rake receiver. Full diversity is achieved.

Compared to MLSD, complexity of the Rake receiver is much lower. ISI is avoided because of the very low spectral efficiency per user, but the spectrum is typically shared between many interfering users. Complexity is thus shifted to the problem of interference management.

3 Orthogonal frequency division multiplexing
Information is modulated on non-interfering sub-carriers in the frequency domain.

The transformation between the time and frequency domains is done by means of adding/subtracting a cyclic prefix and IDFT/DFT operations. This incurs an overhead in terms of time and power.

Frequency diversity is attained by coding over independently faded sub-carriers. This coding problem is identical to that for time diversity.

Complexity is shared between the transmitter and the receiver in performing the IDFT and DFT operations; the complexity of these operations is insensitive to the number of taps, scales moderately with the number of sub-carriers N_c and is very manageable with current implementation technology.

Complexity of diversity coding across sub-carriers can be traded off with the amount of diversity desired.

3.5 Impact of channel uncertainty

In the past few sections we assumed perfect channel knowledge so that coherent combining can be performed at the receiver. In fast varying channels, it may not be easy to estimate accurately the phases and magnitudes of the tap gains before they change. In this case, one has to understand the impact of estimation errors on performance. In some situations, non-coherent detection, which does not require an estimate of the channel, may be the preferred route. In Section 3.1.1, we have already come across a simple non-coherent detector for fading channels without diversity. In this section, we will extend this to channels with diversity.

When we compared coherent and non-coherent detection for channels without diversity, the difference was seen to be relatively small (cf. Figure 3.2). An important question is what happens to that difference as the number of diversity paths L increases. The answer depends on the specific diversity scenario. We first focus on the situation where channel uncertainty has the most impact: DS spread-spectrum over channels with frequency diversity. Once we understand this case, it is easy to extend the insights to other scenarios.

3.5.1 Non-coherent detection for DS spread-spectrum

We considered this scenario in Section 3.4.3, except now the receiver has no knowledge of the channel gains h_ℓ. As we saw in Section 3.1.1, no information can be communicated in the phase of the transmitted signal in conjunction with non-coherent detection (in particular, antipodal signaling cannot be used). Instead, we consider binary orthogonal modulation,[14] i.e., $\mathbf{x}_A$ and $\mathbf{x}_B$ are orthogonal and $\|\mathbf{x}_A\| = \|\mathbf{x}_B\|$.

Recall that the central pseudonoise property of the transmitted sequences in DS spread-spectrum is that the shifted versions are nearly orthogonal. For simplicity of analysis, we continue with the assumption that shifted versions of the transmitted sequence are exactly orthogonal; this holds for both $\mathbf{x}_A$ and $\mathbf{x}_B$ here. We make the further assumption that versions of the two sequences with different shifts are also orthogonal to each other, i.e., $(\mathbf{x}_A^{(\ell)})^*(\mathbf{x}_B^{(\ell')}) = 0$ for $\ell \neq \ell'$ (the so-called zero cross-correlation property). This approximately holds in many spread-spectrum systems. For example, in the uplink of IS-95, the transmitted sequence is obtained by multiplying the selected codeword of an orthogonal code by a (common) pseudonoise ± 1 sequence, so that the low cross-correlation property carries over from the auto-correlation property of the pseudonoise sequence.

Proceeding as in the analysis of coherent detection, we start with the channel model in vector form (3.122) and observe that the projection of $\mathbf{y}$ onto the $2L$ orthogonal vectors $\{\mathbf{x}_A^{(\ell)}/\|\mathbf{x}_A\|, \mathbf{x}_B^{(\ell)}/\|\mathbf{x}_B\|\}_\ell$ yields $2L$ sufficient statistics:

$$
\begin{aligned}
r_A^{(\ell)} &= h_\ell x_1 + w_A^{(\ell)}, & \ell = 0, \ldots, L-1, \\
r_B^{(\ell)} &= h_\ell x_2 + w_B^{(\ell)}, & \ell = 0, \ldots, L-1,
\end{aligned}
$$

where $w_A^{(\ell)}$ and $w_B^{(\ell)}$ are i.i.d. $\mathcal{CN}(0, N_0)$, and

$$
\begin{pmatrix} x_1 \\ x_2 \end{pmatrix} = \begin{cases} \begin{pmatrix} \|\mathbf{x}_A\| \\ 0 \end{pmatrix} & \text{if } \mathbf{x}_A \text{ is transmitted,} \\ \begin{pmatrix} 0 \\ \|\mathbf{x}_B\| \end{pmatrix} & \text{if } \mathbf{x}_B \text{ is transmitted.} \end{cases} \tag{3.146}
$$

This is essentially a generalization of the non-coherent detection problem in Section 3.1.1 from 1 branch to L branches. Just as in the 1 branch case, a

---

[14] Typically M-ary orthogonal modulation is used. For example, the uplink of IS-95 employs non-coherent detection of 64-ary orthogonal modulation.

square-law type detector is the optimal non-coherent detector: decide in favor of $\mathbf{x}_A$ if

$$\sum_{\ell=0}^{L-1} |r_A^{(\ell)}|^2 \geq \sum_{\ell=0}^{L-1} |r_B^{(\ell)}|^2, \tag{3.147}$$

otherwise decide in favor of $\mathbf{x}_B$. The performance can be analyzed as in the 1 branch case: the error probability has the same form as in (3.125), but with μ given by

$$\mu = \frac{1/L \cdot \mathcal{E}_b/N_0}{2 + 1/L \cdot \mathcal{E}_b/N_0}, \tag{3.148}$$

where $\mathcal{E}_b := \|\mathbf{x}_A\|^2$. (See Exercise 3.31.) As a basis of comparison, the performance of coherent detection of binary orthogonal modulation can be analyzed as for the antipodal case; it is again given by (3.125) but with μ given by (Exercise 3.33):

$$\mu = \sqrt{\frac{1/L \cdot \mathcal{E}_b/N_0}{2 + 1/L \cdot \mathcal{E}_b/N_0}}. \tag{3.149}$$

It is interesting to compare the performance of coherent and non-coherent detection as a function of the number of diversity branches. This is shown in Figures 3.23 and 3.24. For $L = 1$, the gap between the performance of both schemes is small, but they are bad anyway, as there is a lack of diversity. This point has already been made in Section 3.1. As L increases, the performance of coherent combining improves monotonically and approaches the performance of an AWGN channel. In contrast, the performance of non-coherent detection first improves with L but then degrades as L is increased further.

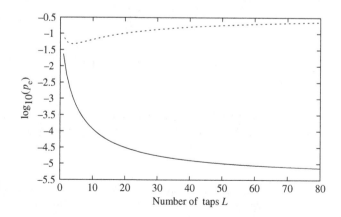

Figure 3.23 Comparison of error probability under coherent detection (—) and non-coherent detection (---), as a function of the number of taps L. Here $\mathcal{E}_b/N_0 = 10\,\text{dB}$.

Figure 3.24 Comparison of error probability under coherent detection (—) and non-coherent detection (- - -), as a function of the number of taps L. Here $\mathcal{E}_b/N_0 = 15\,\text{dB}$.

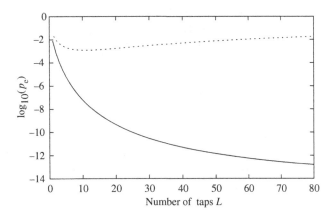

The initial improvement comes from a diversity gain. There is however a law of diminishing return on the diversity gain. At the same time, when L becomes too large, the SNR per branch becomes very poor and non-coherent combining cannot effectively exploit the available diversity. This leads to an ultimate degradation in performance. In fact, it can be shown that as $L \to \infty$ the error probability approaches 1/2.

3.5.2 Channel estimation

The significant performance difference between coherent and non-coherent combining when the number of branches is large suggests the importance of channel knowledge in wideband systems. We assumed perfect channel knowledge when we analyzed the performance of the coherent Rake receiver, but in practice, the channel taps have to be estimated and tracked. It is therefore important to understand the impact of channel measurement errors on the performance of the coherent combiner. We now turn to the issue of channel estimation.

In data detection, the transmitted sequence is one of several possible sequences (representing the data symbol). In channel estimation, the transmitted sequence is assumed to be known at the receiver. In a pilot-based scheme, a known sequence (called a pilot, sounding tone, or training sequence) is transmitted and this is used to estimate the channel.[15] In a decision-feedback scheme, the previously detected symbols are used instead to update the channel estimates. If we assume that the detection is error free, then the development below applies to both pilot-based and decision-directed schemes.

---

[15] The downlink of IS-95 uses a pilot, which is assigned its own pseudonoise sequence and transmitted superimposed on the data.

Focus on one symbol duration, and suppose the transmitted sequence is a known pseudonoise sequence **u**. We return to the channel model in vector form (cf. (3.122))

$$\mathbf{y} = \sum_{\ell=0}^{L-1} h_\ell \mathbf{u}^{(\ell)} + \mathbf{w}. \tag{3.150}$$

We see that since the shifted versions of **u** are orthogonal to each other and the taps are assumed to be independent of each other, projecting **y** onto $\mathbf{u}^{(\ell)}/\|\mathbf{u}^{(\ell)}\|$ will yield a sufficient statistic to estimate h_ℓ (see Summary A.3)

$$r^{(\ell)} := (\mathbf{u}^{(\ell)})^* \mathbf{y} = h_\ell \|\mathbf{u}^{(\ell)}\| + w^{(\ell)} = \sqrt{\mathcal{E}} h_\ell + w^{(\ell)}, \tag{3.151}$$

where $\mathcal{E} := \|\mathbf{u}^{(\ell)}\|^2$. This is implemented by filtering the received signal by a filter matched to **u** and sampling at the appropriate chip time. This operation is the same as the first stage of the Rake receiver, and the channel estimator can in fact be combined with the Rake receiver if done in a decision-directed mode. (See Figure 3.19.)

Typically, channel estimation is obtained by averaging K such measurements over a coherence time period in which the channel is constant:

$$r_k^{(\ell)} := \sqrt{\mathcal{E}} h_\ell + w_k^{(\ell)}, \qquad k = 1, \ldots, K. \tag{3.152}$$

Assuming that $h_\ell \sim \mathcal{CN}(0, 1/L)$, the minimum mean square estimate of h_ℓ given these measurements is (cf. (A.84) in Summary A.3)

$$\hat{h}_\ell = \frac{\sqrt{\mathcal{E}}}{K\mathcal{E} + LN_0} \sum_{k=1}^{K} r_k^{(\ell)}. \tag{3.153}$$

The mean square error associated with this estimate is (cf. (A.85) in Summary A.3)

$$\frac{1}{L} \cdot \frac{1}{1 + K\mathcal{E}/(LN_0)}. \tag{3.154}$$

the same for all branches.

The key parameter affecting the estimation error is

$$\mathsf{SNR}_{\text{est}} := \frac{K\mathcal{E}}{LN_0}. \tag{3.155}$$

When $\mathsf{SNR}_{\text{est}} \gg 1$, the mean square estimation error is much smaller than the variance of h_ℓ (equal to $1/L$) and the impact of the channel estimation error on the performance of the coherent Rake receiver is not significant; perfect channel knowledge is a reasonable assumption in this regime. On the other hand, when $\text{SNR}_{\text{est}} \ll 1$, the mean square error is close to $1/L$, the variance of h_ℓ. In this regime, we hardly have any information about the channel gains and the performance of the coherent combiner cannot be expected to be better than the non-coherent combiner, which we know has poor performance whenever L is large.

How should we interpret the parameter SNR_{est}? Since the channel is constant over the coherence time T_c, we can interpret $K\mathcal{E}$ as the total received energy over the channel coherence time T_c. We can rewrite SNR_{est} as

$$\text{SNR}_{\text{est}} = \frac{PT_c}{LN_0} \qquad (3.156)$$

where P is the received power of the signal from which channel measurements are obtained. Hence, SNR_{est} can be interpreted as the signal-to-noise ratio available to estimate the channel per coherence time per tap. Thus, channel uncertainty has a significant impact on the performance of the Rake receiver whenever this quantity is significantly below $0\,\text{dB}$.

If the measurements are done in a decision-feedback mode, P is the received power of the data stream itself. If the measurements are done from a pilot, then P is the received power of the pilot. On the downlink of a CDMA system, one can have a pilot common to all users, and the power allocated to the pilot can be larger than the power of the signals for the individual users. This results in a larger SNR_{est}, and thus makes coherent combining easier. On the uplink, however, it is not possible to have a common pilot, and the channel estimation will have to be done with a weaker pilot allotted to the individual user. With a lower received power from the individual users, SNR_{est} can be considerably smaller.

3.5.3 Other diversity scenarios

There are two reasons why wideband DS spread-spectrum systems are significantly impacted by channel uncertainty:

- the amount of energy per resolvable path decreases inversely with increasing number of paths, making their gains harder to estimate when there are many paths;
- the number of diversity paths depends both on the bandwidth and the delay spread and, given these parameters, the designer has no control over this number.

What about in other diversity scenarios?

In antenna diversity with L receive antennas, the received energy per antenna is the same regardless of the number of antennas, so the channel measurement problem is the same as with a single receive antenna and does not become harder. The situation is similar in the time diversity scenario. In antenna diversity with L *transmit* antennas, the received energy per diversity path *does* decrease with the number of antennas used, but certainly we can restrict the number L to be the number of different channels that can be reliably learnt by the receiver.

How about in OFDM systems with frequency diversity? Here, the designer has control over how many sub-carriers to spread the signal energy over. Thus, while the number of *available* diversity branches L may increase with the bandwidth, the signal energy can be restricted to a fixed number of sub-carriers $L' < L$ over any one OFDM time block. Such communication can be restricted to concentrated time-frequency blocks and Figure 3.25 visualizes one such scheme (for $L' = 2$), where the choice of the L' sub-carriers is different for different OFDM blocks and is hopped over the entire bandwidth. Since the energy in each OFDM block is concentrated within a fixed number of sub-carriers at any one time, coherent reception is possible. On the other hand, the maximum diversity gain of L can still be achieved by coding across the sub-carriers within one OFDM block as well as across different blocks.

One possible drawback is that since the total power is only concentrated within a subset of sub-carriers, the total degrees of freedom available in the system are not utilized. This is certainly the case in the context of point-to-point communication; in a system with other users sharing the same bandwidth, however, the other degrees of freedom can be utilized by the other users and need not go wasted. In fact, one key advantage of OFDM over DS spread-spectrum is the ability to maintain orthogonality across multiple users in a multiple access scenario. We will return to this point in Chapter 4.

Figure 3.25 An illustration of a scheme that uses only a fixed part of the bandwidth at every time. Here, one small square denotes a single sub-carrier within one OFDM block. The time-axis indexes the different OFDM blocks; the frequency-axis indexes the different sub-carriers.

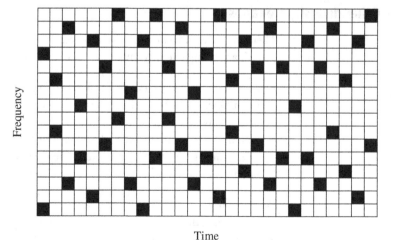

Chapter 3 The main plot

Baseline

We first looked at detection on a narrowband slow fading Rayleigh channel. Under both coherent and non-coherent detection, the error probability behaves like

$$p_e \approx \text{SNR}^{-1} \tag{3.157}$$

at high SNR. In contrast, the error probability decreases *exponentially* with the SNR in the AWGN channel. The typical error event for the fading channel is due to the channel being in deep fade rather than the Gaussian noise being large.

Diversity

Diversity was presented as an effective approach to improve performance drastically by providing redundancy across independently faded branches. Three modes of diversity were considered:

- time – the interleaving of coded symbols over different coherence time periods;
- space – the use of multiple transmit and/or receive antennas;
- frequency – the use of a bandwidth greater than the coherence bandwidth of the channel.

In all cases, a simple scheme that repeats the information symbol across the multiple branches achieves full diversity. With L i.i.d. Rayleigh branches of diversity, the error probability behaves like

$$p_e \approx c \cdot \text{SNR}^{-L} \tag{3.158}$$

at high SNR.

Examples of repetition schemes:

- repeating the same symbol over different coherence periods;
- repeating the same symbol over different transmit antennas one at a time;
- repeating the same symbol across OFDM sub-carriers in different coherence bands;
- transmitting a symbol once every delay spread in a frequency-selective channel so that multiple delayed replicas of the symbol are received without interference.

Code design and degrees of freedom

More sophisticated schemes cannot achieve higher diversity gain but can provide a *coding gain* by improving the constant c in (3.158). This is achieved by utilizing the available *degrees of freedom* better than in the repetition schemes.

Examples:
- rotation and permutation codes for time diversity and for frequency diversity in OFDM;
- Alamouti scheme for transmit diversity;
- uncoded transmission at symbol rate in a frequency-selective channel with ISI equalization.

Criteria to design schemes with good coding gain were derived for the different scenarios by using the union bound (based on pairwise error probabilities) on the actual error probability:
- product distance between codewords for time diversity;
- determinant criterion for space-time codes.

Channel uncertainty

The impact of channel uncertainty is significant in scenarios where there are many diversity branches but only a small fraction of signal energy is received along each branch. Direct-sequence spread-spectrum is a prime example.

The gap between coherent and non-coherent schemes is very significant in this regime. Non-coherent schemes do not work well as they cannot combine the signals along each branch effectively.

Accurate channel estimation is crucial. Given the amount of transmit power devoted to channel estimation, the efficacy of detection performance depends on the key parameter SNR_{est}, the received SNR per coherence time per diversity branch. If $\mathsf{SNR}_{est} \gg 0\,dB$, then detection performance is near coherent. If $\mathsf{SNR}_{est} \ll 0\,dB$, then effective combining is impossible.

Impact of channel uncertainty can be ameliorated in some schemes where the transmit energy can be focused on smaller number of diversity branches. Effectively SNR_{est} is increased. OFDM is an example.

3.6 Bibliographical notes

Reliable communication over fading channels has been studied since the 1960s. Improving the performance via diversity is also an old topic. Standard digital communication texts contain many formulas for the performance of coherent and non-coherent diversity combiners, which we have used liberally in this chapter (see Chapter 14 of Proakis [96], for example).

Early works recognizing the importance of the product distance criterion for improving the coding gain under Rayleigh fading are Wilson and Leung [144] and Divsalar and Simon [30], in the context of trellis-coded modulation. The rotation example is taken from Boutros and Viterbo [13]. Transmit antenna diversity was studied extensively in the late 1990s code design criteria were derived by Tarokh *et al.* [115] and by Guey *et al.* [55]; in particular, the determinant criterion is obtained in Tarokh *et al.* [115]. The delay diversity scheme was introduced by Seshadri and Winters [107]. The Alamouti scheme was introduced by Alamouti [3] and generalized to orthogonal designs by Tarokh *et al.* [117]. The diversity analysis of the decorrelator was performed by Winters *et al.* [145], in the context of a space-division multiple access system with multiple receive antennas.

The topic of equalization has been studied extensively and is covered comprehensively in standard textbooks on communication theory; for example, see the book by Barry *et al.* [4]. The Viterbi algorithm was introduced in [139]. The diversity analysis of MLSD is adopted from Grokop and Tse [54].

The OFDM approach to communicate over a wideband channel was first used in military systems in the 1950s and discussed in early papers in the 1960s by Chang [18] and Saltzberg [104]. Circular convolution and the DFT are classical undergraduate material in digital signal processing (Chapter 8, and Section 8.7.5, in particular, of [87]).

The spread-spectrum approach to harness frequency diversity has been well summarized by Viterbi [140]. The Rake receiver was designed by Price and Green [95]. The impact of channel uncertainty on the performance has been studied by various authors, including Médard and Gallager [85], Telatar and Tse [120] and Subramanian and Hajek [113].

3.7 Exercises

Exercise 3.1 Verify (3.19) and the high SNR approximation (3.21). *Hint*: Write the expression as a double integral and interchange the order of integration.

Exercise 3.2 In Section 3.1.2 we studied the performance of antipodal signaling under coherent detection over a Rayleigh fading channel. In particular, we saw that the error probability p_e decreases like $1/\mathsf{SNR}$. In this question, we study a deeper characterization of the behavior of p_e with increasing SNR.

1. A precise way of saying that p_e decays like $1/\mathsf{SNR}$ with increasing SNR is the following:

$$\lim_{\mathsf{SNR} \to \infty} p_e \cdot \mathsf{SNR} = c,$$

where c is a constant. Identify the value of c for the Rayleigh fading channel.

2. Now we want to test how robust the above result is with respect to the fading distribution. Let h be the channel gain, and suppose $|h|^2$ has an arbitrary continuous pdf f satisfying $f(0) > 0$. Does this give enough information to compute the high SNR error probability like in the previous part? If so, compute it. If not, specify what other information you need. *Hint*: You may need to interchange limit and integration in your calculations. You can assume that this can be done without worrying about making your argument rigorous.

3. Suppose now we have L independent branches of diversity with gains $h_1, \ldots, h_L$, and $|h_\ell|^2$ having an arbitrary distribution as in the previous part. Is there enough

information for you to find the high SNR performance of repetition coding and coherent combining? If so, compute it. If not, what other information do you need?

4. Using the result in the previous part or otherwise, compute the high SNR performance under Rician fading. How does the parameter κ affect the performance?

Exercise 3.3 This exercise shows how the high SNR slope of the probability of error (3.19) versus SNR curve can be obtained using a typical error event analysis, without the need for directly carrying out the integration.

Fix $\epsilon > 0$ and define the ϵ-typical error events $\mathbb{E}_\epsilon$ and $\mathbb{E}_{-\epsilon}$, where

$$\mathbb{E}_\epsilon := \{h : |h|^2 < 1/\mathsf{SNR}^{1-\epsilon}\}. \tag{3.159}$$

1. By conditioning on the event $\mathbb{E}_\epsilon$, show that at high SNR

$$\lim_{\mathsf{SNR}\to\infty} \frac{\log p_\mathrm{e}}{\log \mathsf{SNR}} \leq -(1-\epsilon). \tag{3.160}$$

2. By conditioning on the event $\mathbb{E}_{-\epsilon}$, show that

$$\lim_{\mathsf{SNR}\to\infty} \frac{\log p_\mathrm{e}}{\log \mathsf{SNR}} \geq -(1+\epsilon). \tag{3.161}$$

3. Hence conclude that

$$\lim_{\mathsf{SNR}\to\infty} \frac{\log p_\mathrm{e}}{\log \mathsf{SNR}} = -1. \tag{3.162}$$

This says that the asymptotic slope of the error probability versus SNR plot (in dB/dB scale) is -1.

Exercise 3.4 In Section 3.1.2, we saw that there is a 4-dB energy loss when using 4-PAM on only the I channel rather than using QPSK on both the I and the Q channels, although both modulations convey two bits of information. Compute the corresponding loss when one wants to transmit k bits of information using 2^k-PAM rather than 2^k-QAM. You can assume k to be even. How does the loss depend on k?

Exercise 3.5 Consider the use of the differential BPSK scheme proposed in Section 3.1.3 for the Rayleigh flat fading channel.

1. Find a natural non-coherent scheme to detect $u[m]$ based on $y[m-1]$ and $y[m]$, assuming the channel is constant across the two symbol times. Your scheme does not have to be the ML detector.

2. Analyze the performance of your detector at high SNR. You may need to make some approximations. How does the high SNR performance of your detector compare to that of the coherent detector?

3. Repeat your analysis for differential QPSK.

Exercise 3.6 In this exercise we further study coherent detection in Rayleigh fading.

1. Verify Eq. (3.37).

2. Analyze the error probability performance of coherent detection of binary orthogonal signaling with L branches of diversity, under an i.i.d. Rayleigh fading assumption (i.e., verify Eq. (3.149)).

Exercise 3.7 In this exercise, we study the performance of the rotated code in Section 3.2.2.

1. Give an explicit expression for the *exact* pairwise error probability $\mathbb{P}\{\mathbf{x}_A \to \mathbf{x}_B\}$ in (3.49). *Hint*: The techniques from Exercise 3.1 will be useful here.

2. This pairwise error probability was upper bounded in (3.54). Show that the product of SNR^2 and the difference between the upper bound and the actual pairwise error probability goes to a constant with increasing SNR. In other words, the upper bound in (3.54) is tight up to the leading term in $1/\mathsf{SNR}$.

Exercise 3.8 In the text, we mainly use real symbols to simplify the notation. In practice, complex constellations are used (i.e., symbols are sent along both the I and Q components). The simplest complex constellation is QPSK: the constellation is $\{a(1+\mathrm{j}), a(1-\mathrm{j}), a(-1-\mathrm{j}), a(-1+\mathrm{j})\}$.

1. Compute the error probability of QPSK detection for a Rayleigh fading channel with repetition coding over L branches of diversity. How does the performance compare to a scheme which uses only real symbols?

2. In Section 3.2.2, we developed a diversity scheme based on rotation of real symbols (thus using only the I channel). One can develop an analogous scheme for QPSK complex symbols, using a 2×2 complex unitary matrix instead. Find an analogous pairwise code-design criterion as in the real case.

3. Real orthonormal matrices are special cases of complex unitary matrices. Within the class of real orthonormal matrices, find the optimal rotation to maximize your criterion.

4. Find the optimal unitary matrix to maximize your criterion. (This may be difficult!)

Exercise 3.9 In Section 3.2.2, we rotate two BPSK symbols to demonstrate the possible improvement over repetition coding in a time diversity channel with two diversity paths. Continuing with the same model, now consider transmitting at a higher rate using a $2n$-PAM constellation for each symbol. Consider rotating the resulting 2D constellation by a rotation matrix of the form in (3.46). Using the performance criterion of the minimum squared product distance, construct the optimal rotation matrix.

Exercise 3.10 In Section 3.2.2, we looked at the example of the rotation code to achieve time diversity (with the number of branches, L, equal to 2). In the text, we use real symbols and in Exercise 3.8 we extend to complex symbols. In the latter scenario, another coding scheme is the *permutation code*. Shown in Figure 3.26 are two 16-QAM constellations. Each codeword in the permutation code for $L = 2$ is obtained by picking a pair of points, one from each constellation, which are represented by the same icon. The codeword is transmitted over two (complex) symbol times.

1. Why do you think this is called a permutation code?

2. What is the data rate of this code?

3. Compute the diversity gain and the minimum product distance for this code.

4. How does the performance of this code compare to the rotation code in Exercise 3.8, part (3), in terms of the transmit power required?

Exercise 3.11 In the text, we considered the use of rotation codes to obtain time diversity. Rotation codes are designed specifically for fading channels. Alternatively, one can use standard AWGN codes like binary linear block codes. This question looks at the diversity performance of such codes.

Figure 3.26 A permutation code.

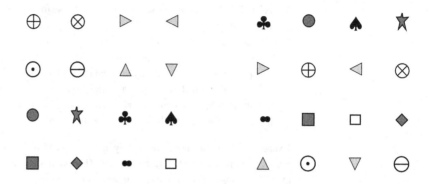

Figure 3.26 A permutation code.

Consider a perfectly interleaved Rayleigh fading channel:

$$y_\ell = h_\ell x_\ell + w_\ell, \qquad \ell = 1, \ldots, L$$

where h_ℓ and w_ℓ are i.i.d. $\mathcal{CN}(0, 1)$ and $\mathcal{CN}(0, N_0)$ random variables respectively. A (L, k) binary linear block code is specified by a k by L generator matrix $\mathbf{G}$ whose entries are 0 or 1. k information bits form a k-dimensional binary-valued vector $\mathbf{b}$ which is mapped into the binary codeword $\mathbf{c} = \mathbf{G}^t \mathbf{b}$ of length L, which is then mapped into L BPSK symbols and transmitted over the fading channel.[16] The receiver is assumed to have a perfect estimate of the channel gains h_ℓ.

1. Compute a bound on the error probability of ML decoding in terms of the SNR and parameters of the code. Hence, compute the diversity gain in terms of code parameter(s).

2. Use your result in (1) to compute the diversity gain of the $(3, 2)$ code with generator matrix:

$$\mathbf{G} = \begin{bmatrix} 1 & 0 & 1 \\ 0 & 1 & 1 \end{bmatrix}. \tag{3.163}$$

How does the performance of this code compare to the rate 1/2 repetition code?

3. The ML decoding is also called *soft decision decoding* as it takes the entire received vector $\mathbf{y}$ and finds the transmitted codeword closest in Euclidean distance to it. Alternatively, a suboptimal but lower-complexity decoder uses *hard decision decoding*, which for each ℓ first makes a hard decision $\hat{c}_\ell$ on the ℓth transmitted coded symbol based only on the corresponding received symbol y_ℓ, and then finds the codeword that is closest in Hamming distance to $\hat{\mathbf{c}}$. Compute the diversity gain of this scheme in terms of basic parameters of the code. How does it compare to the diversity gain achieved by soft decision decoding? Compute the diversity gain of the code in part (2) under hard decision decoding.

4. Suppose now you still do hard decision decoding except that you are allowed to also declare an "erasure" on some of the transmitted symbols (i.e., you can refuse to make a hard decision on some of the symbols). Can you design a scheme that

---

[16] Addition and multiplication are done in the binary field.

yields a better diversity gain than the scheme in part (3)? Can you do as well as soft decision decoding? Justify your answers. Try your scheme out on the example in part (2). *Hint*: the trick is to figure out when to declare an erasure. You may want to start thinking of the problems in terms of the example in part (2). The typical error event view in Exercise 3.3 may also be useful here.

Exercise 3.12 In our study of diversity models (cf. (3.31)), we have modeled the L branches to have independent fading coefficients. Here we explore the impact of correlation between the L diversity branches. In the time diversity scenario, consider the correlated model: $h_1, \ldots, h_L$ are jointly circular symmetric complex Gaussian with zero mean and covariance $\mathbf{K}_h$ ($\mathcal{CN}(0, \mathbf{K}_h)$ in our notation).

1. Redo the diversity calculations for repetition coding (Section 3.2.1) for this correlated channel model by calculating the rate of decay of error probability with SNR. What is the dependence of the asymptotic (in SNR) behavior of the typical error event on the correlation $\mathbf{K}_h$? You can answer this by characterizing the rate of decay of (3.42) at high SNR (as a function of $\mathbf{K}_h$).

2. We arrived at the product distance code design criterion to harvest coding gain along with time diversity in Section 3.2.2. What is the analogous criterion for correlated channels? *Hint*: Jointly complex Gaussian random vectors are related to i.i.d. complex Gaussian vectors via a linear transformation that depends on the covariance matrix.

3. For transmit diversity with independent fading across the transmit antennas, we have arrived at the generalized product distance code design criterion in Section 3.3.2. Calculate the code design criterion for the correlated fading channel here (the channel $\mathbf{h}$ in (3.80) is now $\mathcal{CN}(0, \mathbf{K}_h)$).

Exercise 3.13 The optimal coherent receiver for repetition coding with L branches of diversity is a maximal ratio combiner. For implementation reasons, a simpler receiver one often builds is a *selection combiner*. It does detection based on the received signal along the branch with the strongest gain only, and ignores the rest. For the i.i.d. Rayleigh fading model, analyze the high SNR performance of this scheme. How much of the inherent diversity gain can this scheme get? Quantify the performance loss from optimal combining. *Hint*: You may find the techniques developed in Exercise 3.2 useful for this problem.

Exercise 3.14 It is suggested that full diversity gain can be achieved over a Rayleigh faded MISO channel by simply transmitting the same symbol at each of the transmit antennas simultaneously. Is this correct?

Exercise 3.15 An $L \times 1$ MISO channel can be converted into a time diversity channel with L diversity branches by simply transmitting over one antenna at a time.

1. In this way, any code designed for a time diversity channel with L diversity branches can be used for a MISO (multiple input single output) channel with L transmit antennas. If the code achieves k-fold diversity in the time diversity channel, how much diversity can it obtain in the MISO channel? What is the relationship between the minimum product distance metric of the code when viewed as a time diversity code and its minimum determinant metric when viewed as a transmit diversity code?

2. Using this transformation, the rotation code can be used as a transmit diversity scheme. Compare the performance of this code and the Alamouti scheme in a 2×1 Rayleigh fading channel, using BPSK symbols. Which one is better? How about using QPSK symbols?

3. Use the permutation code (cf. Figure 3.26) from Exercise 3.10 on the 2×1 Rayleigh fading channel and compare (via a numerical simulation) its performance with the Alamouti scheme using QPSK symbols (so the rate is the same in both the schemes).

Exercise 3.16 In this exercise, we derive some properties a code construction must satisfy to mimic the Alamouti scheme behavior for more than two transmit antennas. Consider communication over n time slots on the L transmit antenna channel (cf. (3.80)):

$$\mathbf{y}^t = \mathbf{h}^* \mathbf{X} + \mathbf{w}^t. \tag{3.164}$$

Here $\mathbf{X}$ is the $L \times n$ space-time code. Over n time slots, we want to communicate L independent constellation symbols, $d_1, \ldots, d_L$; the space-time code $\mathbf{X}$ is a deterministic function of these symbols.

1. Consider the following property for every channel realization $\mathbf{h}$ and space-time codeword $\mathbf{X}$

$$(\mathbf{h}^* \mathbf{X})^t = \mathbf{A} \mathbf{d}. \tag{3.165}$$

Here we have written $\mathbf{d} = [d_1, \ldots, d_L]^t$ and $\mathbf{A} = [\mathbf{a}_1, \ldots, \mathbf{a}_L]$, a matrix with *orthogonal* columns. The vector $\mathbf{d}$ depends solely on the codeword $\mathbf{X}$ and the matrix $\mathbf{A}$ depends solely on the channel $\mathbf{h}$. Show that, if the space-time codeword $\mathbf{X}$ satisfies the property in (3.165), the joint receiver to detect $\mathbf{d}$ separates into individual linear receivers, each separately detecting $d_1, \ldots, d_L$.

2. We would like the effective channel (after the linear receiver) to provide each symbol d_m ($m = 1, \ldots, L$) with full diversity. Show that, if we impose the condition that

$$\|\mathbf{a}_m\| = \|\mathbf{h}\|, \qquad m = 1, \ldots L, \tag{3.166}$$

then each data symbol d_m has full diversity.

3. Show that a space-time code $\mathbf{X}$ satisfying (3.165) (the linear receiver property) and (3.166) (the full diversity property) must be of the form

$$\mathbf{X} \mathbf{X}^* = \|\mathbf{d}\|^2 \mathbf{I}_L, \tag{3.167}$$

i.e., the columns of $\mathbf{X}$ must be orthogonal. Such an $\mathbf{X}$ is called an *orthogonal design*. Indeed, we observe that the codeword $\mathbf{X}$ in the Alamouti scheme (cf. (3.77)) is an orthogonal design with $L = n = 2$.

Exercise 3.17 This exercise is a sequel to Exercise 3.16. It turns out that if we require $n = L$, then for $L > 2$ there are no orthogonal designs. (This result is proved in Theorem 5.4.2 in [117].) If we settle for $n > L$ then orthogonal designs exist for

$L > 2$. In particular, Theorem 5.5.2 of [117] constructs orthogonal designs for all L and $n \geq 2L$. This does not preclude the existence of orthogonal designs with rate larger than 0.5. A reading exercise is to study [117] where orthogonal designs with rate larger than 0.5 are constructed.

Exercise 3.18 The pairwise error probability analysis for the i.i.d. Rayleigh fading channel has led us to the product distance (for time diversity) and generalized product distance (for transmit diversity) code design criteria. Extend this analysis for the i.i.d. Rician fading channel.

1. Does the diversity order change for repetition coding over a time diversity channel with the L branches i.i.d. Rician distributed?
2. What is the new code design criterion, analogous to product distance, based on the pairwise error probability analysis?

Exercise 3.19 In this exercise we study the performance of space-time codes (the subject of Section 3.3.2) in the presence of multiple receive antennas.

1. Derive, as an extension of (3.83), the pairwise error probability for space-time codes with n_r receive antennas.
2. Assuming that the channel *matrix* has i.i.d. Rayleigh components, derive, as an extension of (3.86), a simple upper bound for the pairwise error probability.
3. Conclude that the code design criterion remains *unchanged* with multiple receive antennas.

Exercise 3.20 We have studied the performance of the Alamouti scheme in a channel with two transmit and one receive antenna. Suppose now we have an additional receive antenna. Derive the ML detector for the symbols based on the received signals at both receive antennas. Show that the scheme effectively provides two independent scalar channels. What is the gain of each of the channels?

Exercise 3.21 In this exercise we study some expressions for error probabilities that arise in Section 3.3.3.

1. Verify Eqs. (3.93) and (3.94). In which SNR range is (3.93) smaller than (3.94)?
2. Repeat the derivation of (3.93) and (3.94) for a general target rate of R bits/s/Hz (suppose that R is an integer). How does the SNR range in which the spatial multiplexing scheme performs better depend on R?

Exercise 3.22 In Section 3.3.3, the performance comparison between the spatial multiplexing scheme and the Alamouti scheme is done for PAM symbols. Extend the comparison to QAM symbols with the target data rate R bits/s/Hz (suppose that $R \geq 4$ is an even integer).

Exercise 3.23 In the text, we have developed code design criteria for pure time diversity and pure spatial diversity scenarios. In some wireless systems, one can get both time and spatial diversity simultaneously, and we want to develop a code design criterion for that. More specifically, consider a channel with L transmit antennas and 1 receive antenna. The channel remains constant over blocks of k symbol times, but changes to an independent realization every k symbols (as a result of interleaving, say). The channel is assumed to be independent across antennas. All channel gains are Rayleigh distributed.

1. What is the maximal diversity gain that can be achieved by coding over n such blocks?

2. Develop a pairwise code design criterion over this channel. Show how this criterion reduces to the special cases we have derived for pure time and pure spatial diversity.

Exercise 3.24 A mobile having a single receive antenna sees a Rayleigh flat fading channel

$$y[m] = h[m]x[m] + w[m],$$

where $w[m] \sim \mathcal{CN}(0, N_0)$ and i.i.d. and $\{h[m]\}$ is a complex circular symmetric stationary Gaussian process with a given correlation function $R[m]$ which is monotonically decreasing with m. (Recall that $R[m]$ is defined to be $\mathbb{E}[h[0]h[m]^*]$.)

1. Suppose now we want to put an extra antenna on the mobile at a separation d. Can you determine, from the information given so far, the joint distribution of the fading gains the two antennas see at a particular symbol time? If so, compute it. If not, specify any additional information you have to assume and then compute it.

2. We transmit uncoded BPSK symbols from the base-station to the mobile with dual antennas. Give an expression for the average error probability for the ML detector.

3. Give a back-of-the-envelope approximation to the high SNR error probability, making explicit the effect of the correlation of the channel gains across antennas. What is the diversity gain from having two antennas in the correlated case? How does the error probability compare to the case when the fading gains are assumed to be independent across antennas? What is the effect of increasing the antenna separation d?

Exercise 3.25 Show that full diversity can still be obtained with the maximum likelihood sequence equalizer in Section 3.4.2 even when the channel taps h_ℓ have different variances (but are still independent). You can use a heuristic argument based on typical error analysis.

Exercise 3.26 Consider the maximum likelihood sequence detection described in Section 3.4.2. We computed the achieved diversity gain but did not compute an explicit bound on the error probability on detecting each of the symbol $x[m]$. Below you can assume that BPSK modulation is used for the symbols.

1. Suppose $N = 1$. Find a bound on the error probability of the MLSD incorrectly detecting $x[1]$. *Hint*: finding the worst-case pairwise error probability does not require much calculation, but you should be a little careful in applying the union bound.

2. Use your result to estimate the coding gain over the scheme that completely avoids ISI by sending a symbol every L symbol times. How does the coding gain depend on L?

3. Extend your analysis to general block length N and the detection of $x[m]$ for $m \leq N$.

Exercise 3.27 Consider the equalization problem described in Section 3.4.2. We studied the performance of MLSD. In this exercise, we will look at the performance of a linear equalizer. For simplicity, suppose $N = L = 2$.

1. Over the two symbol times (time 0 and time 1), one can think of the ISI channel as a 2×2 MIMO channel between the input and output symbols. Identify the channel matrix $\mathbf{H}$.

2. The MIMO point of view suggests using, as an alternative to MLSD, the zero-forcing (decorrelating) receiver to detect $x[0]$ based on completely inverting the

channel. How much diversity gain can this equalizer achieve? How does it compare to the performance of MLSD?

Exercise 3.28 Consider a multipath channel with L i.i.d. Rayleigh faded taps. Let $\tilde{h}_n$ be the complex gain of the nth carrier in the OFDM modulation at a particular time. Compute the joint statistics of the gains and lend evidence to the statement that the gains of the carriers separated by more than the coherence bandwidths are approximately independent.

Exercise 3.29 Argue that for typical wireless channels, the delay spread is much less than the coherence time. What are the implications of this observation on: (1) an OFDM system; (2) a direct-sequence spread-spectrum system with Rake combining? (There may be multiple implications in each case.)

Exercise 3.30 Communication takes place at passband over a bandwidth W around a carrier frequency of f_c. Suppose the baseband equivalent discrete-time model has a finite number of taps. We use OFDM modulation. Let $\tilde{h}_n[i]$ be the complex gain for the nth carrier and the ith OFDM symbol. We typically assume there are a large number of reflectors so that the tap gains of the discrete-time model can be modeled as Gaussian distributed, but suppose we do not make this assumption here. Only relying on natural assumptions on f_c and W, argue the following. State your assumptions on f_c and W and make your argument as clear as possible.
1. At a fixed symbol time i, the $\tilde{h}_n[i]$ are identically distributed across the carriers.
2. More generally, the *processes* $\{\tilde{h}_n[i]\}_n$ have the same statistics for different n.

Exercise 3.31 Show that the square-law combiner (given by (3.147)) is the optimal non-coherent ML detector for a channel with i.i.d. Rayleigh faded branches, and analyze the non-coherent error probability performance (i.e., verify (3.148)).

Exercise 3.32 Consider the problem of Rake combining under channel measurement uncertainty, discussed in Section 3.4.3. Assume a channel with L i.i.d. Rayleigh faded branches. Suppose the channel estimation is as given in Eqs. (3.152) and (3.153). We communicate using binary orthogonal signaling. The receive is coherent with the channel estimates used in place of the true channel gains h_ℓ. It is not easy to compute explicitly the error probability of this detector, but through either an approximate analysis, numerical computation or simulation, get an idea of its performance as a function of L. In particular, give evidence supporting the intuitive statement that, when $L \gg K\mathcal{E}/N_0$, the performance of this detector is very poor.

Exercise 3.33 We have studied coherent performance of antipodal signaling of the Rake receiver in Section 3.4.3. Now consider binary orthogonal modulation: we either transmit $\mathbf{x}_A$ or $\mathbf{x}_B$, which are both orthogonal and their shifts are also orthogonal with each other. Calculate the error probability with the coherent Rake (i.e., verify (3.149)).

4

Cellular systems: multiple access and interference management

4.1 Introduction

In Chapter 3, our focus was on *point-to-point* communication, i.e., the scenario of a single transmitter and a single receiver. In this chapter, we turn to a *network* of many mobile users interested in communicating with a common wireline network infrastructure.[1] This form of wireless communication is different from radio or TV in two important respects: first, users are interested in messages specific to them as opposed to the common message that is broadcast in radio and TV. Second, there is two-way communication between the users and the network. In particular, this allows feedback from the receiver to the transmitter, which is missing in radio and TV. This form of communication is also different from the all-wireless walkie-talkie communication since an access to a wireline network infrastructure is demanded. *Cellular systems* address such a multiuser communication scenario and form the focus of this chapter.

Broadly speaking, two types of spectra are available for commercial cellular systems. The first is *licensed*, typically nationwide and over a period of a few years, from the spectrum regulatory agency (FCC, in the United States). The second is unlicensed spectrum made available for experimental systems and to aid development of new wireless technologies. While licensing spectrum provides immunity from any kind of interference outside of the system itself, bandwidth is very expensive. This skews the engineering design of the wireless system to be as spectrally efficient as possible. There are no hard constraints on the power transmitted within the licensed spectrum but the power is expected to decay rapidly outside. On the other hand, unlicensed spectrum is very cheap to transmit on (and correspondingly larger

---

[1] A common example of such a network (wireline, albeit) is the public switched telephone network.

than licensed spectrum) but there is a maximum power constraint over the entire spectrum as well as interference to deal with. The emphasis thus is less on spectral efficiency. The engineering design can thus be very different depending on whether the spectrum is licensed or not. In this chapter, we focus on cellular systems that are designed to work on licensed spectrum. Such cellular systems have been deployed nationwide and one of the driving factors for the use of licensed spectrum for such networks is the risk of huge capital investment if one has to deal with malicious interference, as would be the case in unlicensed bands.

A cellular network consists of a number of fixed base-stations, one for each *cell*. The total coverage area is divided into cells and a mobile communicates with the base-station(s) close to it. (See Figure 1.2.) At the physical and medium access layers, there are two main issues in cellular communication: *multiple access* and *interference management*. The first issue addresses how the overall resource (time, frequency, and space) of the system is shared by the users in the same cell (intra-cell) and the second issue addresses the interference caused by simultaneous signal transmissions in different cells (inter-cell). At the network layer, an important issue is that of seamless connectivity to the mobile as it moves from one cell to the other (and thus switching communication from one base-station to the other, an operation known as *handoff*). In this chapter we will focus primarily on the physical-layer issues of multiple access and interference management, although we will see that in some instances these issues are also coupled with how handoff is done.

In addition to resource sharing between different users, there is also an issue of how the resource is allocated between the *uplink* (the communication from the mobile users to the base-station, also called the *reverse link*) and the *downlink* (the communication from the base-station to the mobile users, also called the *forward link*). There are two natural strategies for separating resources between the uplink and the downlink: *time division duplex* (TDD) separates the transmissions in time and *frequency division duplex* (FDD) achieves the separation in frequency. Most commercial cellular systems are based on FDD. Since the powers of the transmitted and received signals typically differ by more than 100 dB at the transmitter, the signals in each direction occupy bands that are separated far apart (tens of MHz), and a device called a *duplexer* is required to filter out any interference between the two bands.

A cellular network provides coverage of the entire area by dividing it into cells. We can carry this idea further by dividing each cell spatially. This is called *sectorization* and involves dividing the cell into, say three, *sectors*. Figure 4.1 shows such a division of a hexagonal cell. One way to think about sectors is to consider them as separate cells, except that the base-station corresponding to the sectors is at the same location. Sectorization is achieved by having a *directional antenna* at the base-station that focuses transmissions

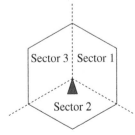

Figure 4.1 A hexagonal cell with three sectors.

into the sector of interest, and is designed to have a null in the other sectors. The ideal end result is an effective creation of new cells without the added burden of new base-stations and network infrastructure. Sectorization is most effective when the base-station is quite tall with few obstacles surrounding it. Even in this ideal situation, there is inter-sector interference. On the other hand, if there is substantial local scattering around the base-station, as is the case when the base-stations are low-lying (such as on the top of lamp posts), sectorization is far less effective because the scattering and reflection would transfer energy to sectors other than the one intended. We will discuss the impact of sectorization on the choice of the system design.

In this chapter, we study three cellular system designs as case studies to illustrate several different approaches to multiple access and interference management. Both the uplink and the downlink designs will be studied. In the first system, which can be termed a *narrowband system*, user transmissions within a cell are restricted to separate narrowband channels. Further, neighboring cells use different narrowband channels for user transmissions. This requires that the total bandwidth be split and reduces the *frequency reuse* in the network. However, the network can now be simplified and approximated by a collection of point-to-point *non-interfering* links, and the physical-layer issues are essentially point-to-point ones. The IS-136 and GSM standards are prime examples of this system. Since the level of interference is kept minimal, the point-to-point links typically have high signal-to-interference-plus-noise ratios (SINRs).[2]

The second and third system designs propose a contrasting strategy: all transmissions are spread to the entire bandwidth and are hence *wideband*. The key feature of these systems is *universal frequency reuse*: the same spectrum is used in every cell. However, simultaneous transmissions can now interfere with each other and links typically operate at low SINRs. The two system designs differ in how the users' signals are spread. The code division multiple access (CDMA) system is based on direct-sequence spread-spectrum. Here, users' information bits are coded at a very low rate and modulated by pseudonoise sequences. In this system, the simultaneous transmissions, intra-cell and inter-cell, cause interference. The IS-95 standard is the main example to highlight the design features of this system. In the orthogonal frequency division multiplexing (OFDM) system, on the other hand, users' information is spread by hopping in the time–frequency grid. Here, the transmissions within a cell can be kept orthogonal but adjacent cells share the same bandwidth and inter-cell interference still exists. This system has the advantage of the full frequency reuse of CDMA while retaining the benefits of the narrowband system where there is no intra-cell interference.

---

[2] Since interference plays an important role in multiuser systems, SINR takes the place of the parameter SNR we used in Chapter 3 when we only talked about point-to-point communication.

We also study the power profiles of the signals transmitted in these systems. This study will be conducted for both the downlink and the uplink to obtain an understanding of the peak and average power profile of the transmissions. We conclude by detailing the impact on power amplifier settings and overall power consumption in the three systems.

Towards implementing the multiple access design, there is an overhead in terms of communicating certain parameters from the base-station to the mobiles and vice versa. They include: authentication of the mobile by the network, allocation of traffic channels, training data for channel measurement, transmit power level, and acknowledgement of correct reception of data. Some of these parameters are one-time communication for a mobile; others continue in time. The amount of overhead this constitutes depends to some extent on the design of the system itself. Our discussions include this topic only when a significant overhead is caused by a specific design choice.

The table at the end of the chapter summarizes the key properties of the three systems.

4.2 Narrowband cellular systems

In this section, we discuss a cellular system design that uses naturally the ideas of reliable point-to-point wireless communication towards constructing a wireless network. The basic idea is to schedule all transmissions so that no two simultaneous transmissions interfere with each other (for the most part). We describe an identical uplink and downlink design of multiple access and interference management that can be termed narrowband to signify that the user transmissions are restricted to a narrow frequency band and the main design goal is to minimize all interference.

Our description of the narrowband system is the same for the uplink and the downlink. The uplink and downlink transmissions are separated, either in time or frequency. For concreteness, let us consider the separation to be in frequency, implemented by adopting an FDD scheme which uses widely separated frequency bands for the two types of transmissions. A bandwidth of W Hz is allocated for the uplink as well as for the downlink. Transmissions of different users are scheduled to be non-overlapping in time and frequency thus eliminating intra-cell interference. Depending on how the overall resource (time and bandwidth) is split among transmissions to the users, the system performance and design implications of the receivers are affected.

We first divide the bandwidth into N narrowband chunks (also denoted as *channels*). Each narrowband channel has width W/N Hz. Each cell is allotted some n of these N channels. These n channels are not necessarily contiguous. The idea behind this allocation is that all transmissions within this cell (in both the uplink and the downlink) are restricted to those n channels. To prevent interference between simultaneous transmissions in neighboring

Figure 4.2 A hexagonal arrangements of cells and a possible reuse pattern of channels 1 through 7 with the condition that a channel cannot be used in one concentric ring of cells around the cell using it. The frequency reuse factor is 1/7.

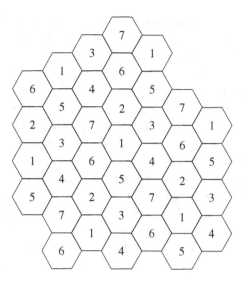

cells, a channel is allocated to a cell only if it is not used by a few concentric rings of neighboring cells. Assuming a regular hexagonal cellular arrangement, Figure 4.2 depicts cells that can use the same channel simultaneously (such cells are denoted by the same number) if we want to avoid any neighboring cell from using the same channel.

The maximum number n of channels that a cell can be allocated depends on the geometry of the cellular arrangement and on the interference avoidance pattern that dictates which cells can share the same channel. The ratio n/N denotes how often a channel can be reused and is termed the *frequency reuse factor*. In the regular hexagonal model of Figure 4.2, for example, the frequency reuse factor is at least 1/7. In other words, $W/7$ is the effective bandwidth used by any base-station. This reduced spectral efficiency is the price paid up front towards satisfying the design goal of reducing all interference from neighboring base-stations. The specific reuse pattern in Figure 4.2 is ad hoc. A more careful analysis of the channel allocation to suit traffic conditions and the effect of reuse patterns among the cells is carried out in Exercises 4.1, 4.2, and 4.3.

Within a cell, different users are allocated transmissions that are non-overlapping, in both time and channels. The nature of this allocation affects various aspects of system design. To get a concrete feel for the issues involved, we treat one specific way of allocation that is used in the GSM system.

4.2.1 Narrowband allocations: GSM system

The GSM system has already been introduced in Example 3.1. Each narrowband channel has bandwidth 200 kHz (i.e. $W/N = 200$ kHz). Time is divided into slots of length $T = 577\,\mu s$. The time slots in the different channels are the finest divisible resources allocated to the users. Over each slot, n simultaneous user transmissions are scheduled within a cell, one in each of the narrowband channels. To minimize the co-channel interference, these n channels have to be chosen as far apart in frequency as possible. Furthermore, each narrowband channel is shared among eight users in a time-division manner. Since voice is a fixed rate application with predictable traffic, each user is periodically allocated a slot out of every eight. Due to the nature of resource allocation (time and frequency), transmissions suffer no interference from within the cell and further see minimal interference from neighboring cells. Hence the network is stitched out of several point-to-point non-interfering wireless links with transmissions over a narrow frequency band, justifying our term "narrowband system" to denote this design paradigm.

Since the allocations are static, the issues of frequency and timing synchronization are the same as those faced by point-to-point wireless communication. The symmetric nature of voice traffic also enables a symmetric design of the uplink and the downlink. Due to the lack of interference, the operating received SINRs can be fairly large (up to 30 dB), and the communication scheme in both the uplink and the downlink is *coherent*. This involves learning the narrowband channel through the use of training symbols (or pilots), which are time-division multiplexed with the data in each slot.

Performance

What is the link reliability? Since the slot length T is fairly small, it is typically within the coherence time of the channel and there is not much time diversity. Further, the transmission is restricted to a contiguous bandwidth 200 kHz that is fairly narrow. In a typical outdoor scenario the delay spread is of the order of 1 μs and this translates to a coherence bandwidth of 500 kHz, significantly larger than the bandwidth of the channel. Thus there is not much frequency diversity either. The tough message of Chapter 3 that the error probability decays very slowly with the SNR is looming large in this scenario. As discussed in Example 3.1 of Chapter 3, GSM solves this problem by coding over eight consecutive time slots to extract a combination of time and frequency diversity (the latter via slow frequency hopping of the frames, each made up of the eight time slots of the users sharing a narrowband channel). Moreover, voice quality not only depends on the average frame error rate but also on how clustered the errors are. A cluster of errors leads to a far more noticeable quality degradation than independent frame errors even though the average frame error rate is the same in both the scenarios. Thus, the frequency hopping serves to break up the cluster of errors as well.

Signal characteristics and receiver design

The mobile user receives signals with energy concentrated in a contiguous, narrow bandwidth (of width (W/N), 200 kHz in the GSM standard). Hence the sample rate can be small and the sampling period is of the order of N/W

(5 μs in the GSM standard). All the signal processing operations are driven off this low rate, simplifying the implementation demands on the receiver design. While the sample rate is small, it might still be enough to resolve multipaths.

Let us consider the signals transmitted by a mobile and by the base-station. The *average* transmit power in the signal determines the performance of the communication scheme. On the other hand, certain devices in the RF chain that carry the transmit signal have to be designed for the *peak power* of the signal. In particular, the current bias setting of the power amplifier is directly proportional to the peak signal power. Typically class AB power amplifiers are used due to the linearity required by the spectrally efficient modulation schemes. Further, class AB amplifiers are very power inefficient and their cost (both capital cost and operating cost) is proportional to the bias setting (the range over which linearity is to be maintained). Thus an engineering constraint is to design transmit signals with reduced peak power for a given average power level. One way to capture this constraint is by studying the *peak to average power ratio* (PAPR) of the transmit signal. This constraint is particularly important in the mobile where power is a very scarce resource, as compared to the base-station.

Let us first turn to the signal transmitted by the mobile user (in the uplink). The signal over a slot is confined to a contiguous narrow frequency band (of width 200 kHz). In GSM, data is modulated on to this single-carrier using constant amplitude modulation schemes. In this context, the PAPR of the transmitted signal is fairly small (see Exercise 4.4), and is not much of a design issue. On the other hand, the signal transmitted from the base-station is a superposition of n such signals, one for each of the 200 kHz channels. The aggregate signal (when viewed in the time domain) has a larger PAPR, but the base-station is usually provided with an AC supply and power consumption is not as much of an issue as in the uplink. Further, the PAPR of the signal at the base-station is of the same order in most system designs.

4.2.2 Impact on network and system design

The specific division of resources here in conjunction with a static allocation among the users simplified the design complexities of multiple access and interference management in the network. There is however no free lunch. Two main types of price have to be paid in this design choice. The first is the physical-layer price of the inefficient use of the total bandwidth (measured through the frequency reuse factor). The second is the complexity of network planning. The orthogonal design entails a frequency division that has to be done up front in a global manner. This includes a careful study of the topology of the base-stations and shadowing conditions to arrive at acceptable interference from a base-station reusing one of the N channels. While Figure 4.2 demonstrated a rather simple setting with a suggestively simple design of reuse pattern, this study is quite involved in a real world system.

Further, the introduction of base-stations is done in an incremental way in real systems. Initially, enough base-stations to provide *coverage* are installed and new ones are added when the existing ones are overloaded. Any new base-station introduced in an area will require reconfiguring the assignment of channels to the base-stations in the neighborhood.

The nature of orthogonal allocations allows a high SINR link to most users, regardless of their location in the cell. Thus, the design is geared to allow the system to operate at about the same SINR levels for mobiles that are close to the base-stations as well as those that are at the edge of the cell. How does sectorization affect this design? Though sectored antennas are designed to isolate the transmissions of neighboring sectors, in practice, inter-sector interference is seen by the mobile users, particularly those at the edge of the sector. One implication of reusing the channels among the sectors of the same cell is that the dynamic range of SINR is reduced due to the inter-sector interference. This means that neighboring sectors cannot reuse the same channels while at the same time following the design principles of this system. To conclude, the gains of sectorization come not so much from frequency reuse as from an antenna gain and the improved capacity of the cell.

4.2.3 Impact on frequency reuse

How robust is this design towards allowing neighboring base-stations to reuse the same set of channels? To answer this question, let us focus on a specific scenario. We consider the uplink of a base-station one of whose neighboring base-stations uses the same set of channels. To study the performance of the uplink with this added interference, let us assume that there are enough users so that all channels are in use. Over one slot, a user transmission interferes directly with another transmission in the neighboring cell that uses the same channel. A simple model for the SINR at the base-station over a slot for one particular user uplink transmission is the following:

$$\text{SINR} = \frac{P|h|^2}{N_0 + I}.$$

The numerator is the received power at the base-station due to the user transmission of interest with P denoting the average transmit power and $|h|^2$ the fading channel gain (with unit mean). The denominator consists of the background noise (N_0) and an extra term due to the interference from the user in the neighboring cell. I denotes the interference and is modeled as a random variable with a mean typically smaller than P (say equal to $0.2P$). The interference from the neighboring cell is random due to two reasons. One of them is small-scale fading and the other is the physical location of the user in the other cell that is reusing the same channel. The mean of I represents the average interference caused, averaged over all locations from which it could originate and the channel variations. But due to the fact that the interfering user can be at a wide range of locations, the variance of I is quite high.

We see that the SINR is a random parameter leading to an undesirably poor performance. There is an appreciably high probability of unreliable transmission of even a small and fixed data rate in the frame. In Chapter 3, we focused on techniques that impart channel diversity to the system; for example, antenna diversity techniques make the channel less variable, improving performance. However, there is an important distinction in the variability of the SINR here that cannot be improved by the diversity techniques of Chapter 3. The randomness in the interference I due to the interferer's location is inherent in this system and remains. Due to this, we can conclude that narrowband systems are unsuitable for universal frequency reuse. To reduce the randomness in the SINR, we would really like the interference to be *averaged* over several simultaneous lower-powered transmissions from the neighboring cell instead of coming from one user only. This is one of the important underlying themes in the design of the next two systems that have universal frequency reuse.

> ## Summary 4.1 Narrowband systems
>
> Orthogonal narrowband channels are assigned to users within a cell.
>
> Users in adjacent cells cannot be assigned the same channel due to the lack of interference averaging across users. This reduces the frequency reuse factor and leads to inefficient use of the total bandwidth.
>
> The network is decomposed into a set of high SINR point-to-point links, simplifying the physical-layer design.
>
> Frequency planning is complex, particularly when new cells have to be added.

4.3 Wideband systems: CDMA

In narrowband systems, users are assigned disjoint time-frequency slots within the cell, and users in adjacent cells are assigned different frequency bands. The network is decomposed into a set of point-to-point non-interfering links. In a code division multiple access (CDMA) system design, the multiple access and interference management strategies are different. Using the direct-sequence spread-spectrum technique briefly mentioned in Section 3.4.3, each user spreads its signal over the entire bandwidth, such that when demodulating any particular user's data, other users' signals appear as pseudo white noise.

Thus, not only all users in the same cell share all the time-frequency degrees of freedom, so do the users in different cells. Universal frequency reuse is a key property of CDMA systems.

Roughly, the design philosophy of CDMA systems can be broken down into two design goals:

- First, the interference seen by any user is made as similar to white Gaussian noise as possible, and the power of that interference is kept to a minimum level and as consistent as possible. This is achieved by:
  - Making the received signal of every user as random looking as possible, via modulating the coded bits onto a long pseudonoise sequence.
  - Tight *power control* among users within the same cell to ensure that the received power of each user is no more than the minimum level needed for demodulation. This is so that the interference from users closer to the base-station will not overwhelm users further away (the so-called *near–far problem*).
  - *Averaging* the interference of many geographically distributed users in nearby cells. This averaging not only makes the aggregate interference look Gaussian, but more importantly reduces the randomness of the interference level due to varying locations of the interferers, thus increasing link reliability. This is the key reason why universal frequency reuse is possible in a wideband system but impossible in a narrowband system.
- Assuming the first design goal is met, each user sees a point-to-point wideband fading channel with additive Gaussian noise. Diversity techniques introduced in Chapter 3, such as coding, time-interleaving, Rake combining and antenna diversity, can be employed to improve the reliability of these point-to-point links.

Thus, CDMA is different from narrowband system design in the sense that all users share all degrees of freedom and therefore interfere with each other: the system is *interference-limited* rather than *degree-of-freedom-limited*. On the other hand, it is similar in the sense that the design philosophy is still to decompose the network problem into a set of independent point-to-point links, only now each link sees both interference as well as the background thermal noise. We do not question this design philosophy here, but we will see that there are alternative approaches in later chapters. In this section, we confine ourselves to discussing the various components of a CDMA system in the quest to meet the two design goals. We use the IS-95 standard to discuss concretely the translation of the design goals into a real system.

Compared to the narrowband systems described in the previous section, CDMA has several potential benefits:

- *Universal frequency reuse* means that users in all cells get the full bandwidth or degrees of freedom of the system. In narrowband systems, the number of degrees of freedom per user is reduced by both the number of users sharing the resources within a cell as well as by the frequency-reuse factor. This increase in degrees of freedom per user of a CDMA system however comes at the expense of a lower signal-to-interference-plus-noise ratio (SINR) per degree of freedom of the individual links.

- Because the performance of a user depends only on the *aggregate* interference level, the CDMA approach automatically takes advantage of the source variability of users; if a user stops transmitting data, the total interference level automatically goes down and benefits all the other users. Assuming that users' activities are independent of each other, this provides a *statistical multiplexing* effect to enable the system to accommodate more users than would be possible if every user were transmitting continuously. Unlike narrowband systems, no explicit re-assignment of time or frequency slots is required.

- In a narrowband system, new users cannot be admitted into a network once the time–frequency slots run out. This imposes a *hard* capacity limit on the system. In contrast, increasing the number of users in a CDMA system increases the total level of interference. This allows a more graceful degradation on the performance of a system and provides a *soft* capacity limit on the system.

- Since all cells share a common spectrum, a user on the edge of a cell can receive or transmit signals to two or more base-stations to improve reception. This is called *soft handoff*, and is yet another diversity technique, but at the network level (sometimes called *macrodiversity*). It is an important mechanism to increase the capacity of CDMA systems.

In addition to these network benefits, there is a further link-level advantage over narrowband systems: every user in a CDMA experiences a wideband fading channel and can therefore exploit the inherent frequency diversity in the system. This is particularly important in a slow fading environment where there is a lack of time diversity. It significantly reduces the *fade margin* of the system (the increased SINR required to achieve the same error probability as in an AWGN channel).

On the cons side, it should be noted that the performance of CDMA systems depends crucially on accurate power control, as the channel attenuation of nearby and cell edge users can differ by many tens of dBs. This requires frequent feedback of power control information and incurs a significant overhead per active user. In contrast, tight power control is not necessary in narrowband systems, and power control is exercised mainly for reducing battery consumption rather than managing interference. Also, it is important in a CDMA system that there be sufficient averaging of out-of-cell interference. While this assumption is rather reasonable in the uplink because the interference comes from many weak users, it is more questionable in the downlink, where the interference comes from a few strong adjacent base-stations.[3]

---

[3] In fact, the downlink of IS-95 is the capacity limiting link.

A comprehensive capacity comparison between CDMA and narrowband systems depends on the specific coding schemes and power control strategies, the channel propagation models, the traffic characteristics and arrival patterns of the users, etc. and is beyond the scope of this book. Moreover, many of the advantages of CDMA outlined above are qualitative and can probably be achieved in the narrowband system, albeit with a more complex engineering design. We focus here on a qualitative discussion on the key features of a CDMA system, backed up by some simple analysis to gain some insights into these features. In Chapter 5, we look at a simplified cellular setting and apply some basic information theory to analyze the tradeoff between the increase in degrees of freedom and the increase in the level of interference due to universal frequency reuse.

In a CDMA system, users interact through the interference they cause each other. We discuss ways to manage that interference and analyze its effect on performance. For concreteness, we first focus on the uplink and then move on to the downlink. Even though there are many similarities in their design, there are several differences worth pointing out.

4.3.1 CDMA uplink

The general schematic of the uplink of a CDMA system with K users in the system is shown in Figure 4.3. A fraction of the K users are in the cell and the rest are outside the cell. The data of the kth user are encoded into two BPSK sequences[4] $\{a_k^I[m]\}$ and $\{a_k^Q[m]\}$, which we assume to have equal amplitude for all m. Each sequence is modulated by a pseudonoise sequence, so that the transmitted complex sequence is

$$x_k[m] = a_k^I[m]s_k^I[m] + ja_k^Q[m]s_k^Q[m], \qquad m = 1, 2, \ldots, \qquad (4.1)$$

where $\{s_k^I[m]\}$ and $\{s_k^Q[m]\}$ are pseudonoise sequences taking values ± 1. Recall that m is called a *chip time*. Typically, the chip rate is much larger than the data rate.[5] Consequently, information bits are heavily coded and the coded sequences $\{a_k^I[m]\}$ and $\{a_k^Q[M]\}$ have a lot of redundancy. The transmitted sequence of user k goes through a discrete-time baseband equivalent multipath channel $h^{(k)}$ and is superimposed at the receiver:

$$y[m] = \sum_{k=1}^{K} \left(\sum_{\ell} h_\ell^{(k)}[m]x_k[m - \ell] \right) + w[m]. \qquad (4.2)$$

The fading channels $\{h^{(k)}\}$ are assumed to be independent across users, in addition to the assumption of independence across taps made in Section 3.4.3.

---

[4] Since CDMA systems operate at very low SINR per degree of freedom, a binary modulation alphabet is always used.
[5] In IS-95, the chip rate is 1.2288 MHz and the data rate is 9.6 kbits/s or less.

Figure 4.3 Schematic of the CDMA uplink.

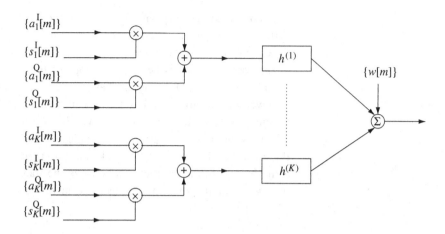

The receiver for user k multiplies the I and Q components of the output sequence $\{y[m]\}$ by the pseudonoise sequences $\{s_k^I[m]\}$ and $\{s_k^Q[m]\}$ respectively to extract the coded streams of user k, which are then fed into a demodulator to recover the information bits. Note that in practice, the users' signals arrive asynchronously at the transmitter but we are making the idealistic assumption that users are *chip-synchronous*, so that the discrete-time model in Chapter 2 can be extended to the multiuser scenario here. Also, we are making the assumption that the receiver is already synchronized with each of the transmitters. In practice, there is a timing acquisition process by which such synchronization is achieved and maintained. Basically, it is a hypothesis testing problem, in which each hypothesis corresponds to a possible relative delay between the transmitter and the receiver. The challenge here is that because timing has to be accurate to the level of a chip, there are many hypotheses to consider and efficient search procedures are needed. Some of these procedures are detailed in Chapter 3 of [140].

Generation of pseudonoise sequences

The pseudonoise sequences are typically generated by *maximum length shift registers*. For a shift register of memory length r, the value of the sequence at time m is a linear function (in the binary field of $\{0, 1\}$) of the values at time $m - 1, m - 2, \ldots, m - r$ (its state). Thus, these binary 0–1 sequences are periodic, and the maximum period length is $p = 2^r - 1$, the number of non-zero states of the register.[6] This occurs when, starting from any non-zero state, the shift register goes through all possible $2^r - 1$ distinct non-zero states before returning to that state. Maximum length shift register (MLSR) sequences have this maximum periodic length, and they exist even for r very

---

[6] Starting from the zero state, the register will remain at the zero state, so the zero state cannot be part of such a period.

large. For CDMA applications, typically, r is somewhere between 20 and 50, thus the period is very long. Note that the generation of the sequence is a deterministic process, and the only randomness is in the initial state. An equivalent way to say this is that realizations of MLSR sequences are random shifts of each other.

The desired pseudonoise sequence $\{s[m]\}$ can be obtained from an MLSR sequence simply by mapping each value from 0 to $+1$ and from 1 to -1. This pseudonoise sequence has the following characteristics which make it look like a typical realization of a Bernoulli coin-flipped sequence ([52, 140]):

-

$$\frac{1}{p} \sum_{m=1}^{p} s[m] = -\frac{1}{p}, \tag{4.3}$$

i.e., the fraction of 0's and 1's is almost half-and-half over the period p.
- For all $\ell \neq 0$:

$$\frac{1}{p} \sum_{m=1}^{p} s[m]s[m+\ell] = -\frac{1}{p}, \tag{4.4}$$

i.e., the shifted versions of the pseudonoise sequence are nearly orthogonal to each other.

For memory $r = 2$, the period is 3 and the MLSR sequence is $110110110\ldots$ The states 11, 10, 01 appear in succession within each period. 00 does not appear, and this is the reason why the sum in (4.3) is not zero. However, this imbalance is very small when the period p is large.

If we randomize the shift of the pseudonoise sequence (i.e., uniformly chosen initial state of the shift register), then it becomes a random process. The above properties suggest that the resulting process is approximately like an i.i.d. Bernoulli sequence over a long time-scale (since p is very large). We will make this assumption below in our analysis of the statistics of the interference.

Statistics of the interference

In a CDMA system, the signal of one user is typically demodulated treating other users' signals as interference. The link level performance then depends on the statistics of the interference. Focusing on the demodulation of user 1, the aggregate interference it sees is

$$I[m] := \sum_{k>1} \left(\sum_{\ell} h_\ell^{(k)}[m]x_k[m-\ell] \right). \tag{4.5}$$

$\{I[m]\}$ has zero mean. Since the fading processes are circular symmetric, the process $\{I[m]\}$ is circular symmetric as well. The second-order statistics are then characterized by $\mathbb{E}[I[m]I[m+\ell]^*]$ for $\ell = 0, 1 \ldots$ They can be computed as

$$\mathbb{E}[|I[m]|^2] = \sum_{k>1} \mathcal{E}_k^c, \qquad \mathbb{E}[I[m]I[m+\ell]^*] = 0 \quad \text{for } \ell \neq 0, \qquad (4.6)$$

where

$$\mathcal{E}_k^c := \mathbb{E}[|x_k[m]|^2] \sum_{\ell} \mathbb{E}[|h_\ell^{(k)}[m]|^2] \qquad (4.7)$$

is the total average energy received per chip from the kth user due to the multipath. In the above variance calculation, we make use of the fact that $\mathbb{E}[x_k[m]x_k[m+\ell]^*] = 0$ (for $\ell \neq 0$), due to the random nature of the spreading sequences. Note that in computing these statistics, we are averaging over both the data and the fading gains of the other users.

When there are many users in the network, and none of them contributes to a significant part of the interference, the Central Limit Theorem can be invoked to justify a Gaussian approximation of the interference process. From the second-order statistics, we see that this process is white. Hence, a reasonable approximation from the point of view of designing the point-to-point link for user 1 is to consider it as a multipath fading channel with white Gaussian noise of power $\sum_{k>1} \mathcal{E}_k^c + N_0$.[7]

We have made the assumption that none of the users contributes a large part of the interference. This is a reasonable assumption due to two important mechanisms in a CDMA system:

- **Power control** The transmit powers of the users within the cell are controlled to solve the near–far problem, and this makes sure that there is no significant intra-cell interferer.
- **Soft handoff** Each base-station that receives a mobile's signal will attempt to decode its data and send them to the MSC (mobile switching center) together with some measure of the quality of the reception. The MSC will select the one with the highest quality of reception. Typically the user's power will be controlled by the base-station which has the best reception. This reduces the chance that some significant out-of-cell interferer is not power controlled.

We will discuss these two mechanisms in more detail later on.

Point-to-point link design

We have already discussed to some extent the design issues of the point-to-point link in a DS spread-spectrum system in Section 3.4.3. In the context

---

[7] This approach is by no means optimal, however. We will see in Chapter 6 that better performance can be achieved by recognizing that the interference consists of the data of the other users that can in fact be decoded.

of the CDMA system, the only difference here is that we are now facing the aggregation of both interference and noise.

The link level performance of user 1 depends on the SINR:

$$\text{SINR}_c := \frac{\mathcal{E}_1^c}{\sum_{k>1} \mathcal{E}_k^c + N_0}. \tag{4.8}$$

Note that this is the SINR *per chip*. The first observation is that typically the SINR per chip is very small. For example, if we consider a system with K perfectly power controlled users in the cell, even ignoring the out-of-cell interference and background noise, SINR_c is $1/(K-1)$. In a cell with 31 users, this is $-15\,\text{dB}$. In IS-95, a typical level of out-of-cell interference is 0.6 of the interference from within the cell. (The background noise, on the other hand, is often negligible in CDMA systems, which are primarily interference-limited.) This reduces the SINR_c further to $-17\,\text{dB}$.

How can we demodulate the transmitted signal at such low SINR? To see this in the simplest setting, let us consider an unfaded channel for user 1 and consider the simple example of BPSK modulation with coherent detection discussed in Section 3.4.3, where each information bit is modulated onto a pseudonoise sequence of length G chips. In the system discussed here which uses a *long* pseudonoise sequence $\{s[m]\}$ (cf. Figure 4.3), this corresponds to repeating every BPSK symbol G times, $a_1^I[Gi+m] = a_1^I[Gi]$, $m = 1, \ldots, G-1$.[8] The detection of the 0th information symbol is accomplished by projecting the in-phase component of the received signal onto the sequence $\mathbf{u} = [s_1^I[0], s_1^I[1], \ldots, s_1^I[G-1]]^t$, and the error probability is

$$p_e = Q\left(\sqrt{\frac{2\|\mathbf{u}\|^2 \mathcal{E}_1^c}{\sum_{k>1} \mathcal{E}_k^c + N_0}}\right) = Q\left(\sqrt{\frac{2G\mathcal{E}_1^c}{\sum_{k>1} \mathcal{E}_k^c + N_0}}\right) = Q\left(\sqrt{\frac{2\mathcal{E}_b}{\sum_{k>1} \mathcal{E}_k^c + N_0}}\right) \tag{4.9}$$

where $\mathcal{E}_b := G\mathcal{E}_1^c$ is the received energy per bit for user 1. Thus, we see that while the SINR *per chip* is low, the SINR *per bit* is increased by a factor of G, due to the averaging of the noise in the G chips over which we repeat the information bits. In terms of system parameters, $G = W/R$, where W Hz is the bandwidth and R bits/s is the data rate. Recall that this parameter is called the *processing gain* of the system, and we see its role here as increasing the effective SINR against a large amount of interference that the user faces. As we scale up the size of a CDMA system by increasing the bandwidth W and the number of users in the system proportionally, but keeping the data rate of each user R fixed, we see that the total interference $\sum_{k>1} \mathcal{E}_k^c$ and the

---

[8] As mentioned, a pseudonoise sequence typically has a period ranging from 2^{20} to 2^{50} chips, much larger than the processing gain G. In contrast, *short* pseudonoise sequences are used in the IS-95 downlink to uniquely identify the individual sector or cell.

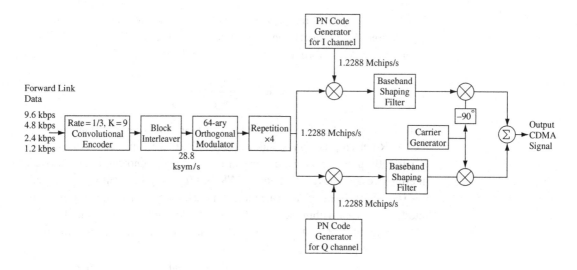

Figure 4.4 The IS-95 uplink.

processing gain G increase proportionally as well. This means that CDMA is an inherently scalable multiple access scheme.[9]

IS-95 link design

The above scheme is based on repetition coding. By using more sophisticated low-rate codes, even better performance can be achieved. Moreover, in practice the actual channel is a multipath fading channel, and so techniques such as time-interleaving and the Rake receiver are important to obtain time and frequency diversity respectively. IS-95, for example, uses a combination of convolutional coding, interleaving and non-coherent demodulation of M-ary orthogonal symbols via a Rake receiver. (See Figure 4.4.) Compressed voice at rate 9.6 kbits/s is encoded using a rate 1/3, constraint length 9, convolutional code. The coded bits are time-interleaved at the level of 6-bit blocks, and each of these blocks is mapped into one of $2^6 = 64$ orthogonal Hadamard sequences,[10] each of length 64. Finally, each symbol of the Hadamard sequence is repeated four times to form the coded sequence $\{a^l[m]\}$. The processing gain is seen to be $3 \cdot 64/6 \cdot 4 = 128$, with a resulting chip rate of $128 \cdot 9.6 = 1.2288$ Mchips/s.

Each of the 6-bit blocks is demodulated non-coherently using a Rake receiver. In the binary orthogonal modulation example in Section 3.5.1, for each orthogonal sequence the non-coherent detector computes the correlation

---

[9] But note that as the bandwidth gets wider and wider, channel uncertainty may eventually become the bottleneck, as we have seen in Section 3.5.

[10] The Hadamard sequences of length $M = 2^J$ are the orthogonal columns of the M by M matrix $\mathbf{H}_M$, defined recursively as $\mathbf{H}_1 = [1]$ and for $M \geq 2$:

$$\mathbf{H}_M = \begin{bmatrix} \mathbf{H}_{M/2} & \mathbf{H}_{M/2} \\ \mathbf{H}_{M/2} & -\mathbf{H}_{M/2} \end{bmatrix}.$$

along each diversity branch (finger) and then forms the sum of the squares. It then decides in favor of the sequence with the largest sum (the square-law detector). (Recall the discussion around (3.147).) Here, each 6-bit block should be thought of as a coded symbol of an outer convolutional code, and we are not interested in hard decision of the block. Instead, we would like to calculate the branch metric for each of the possible values of the 6-bit block, for use by a Viterbi decoder for the outer convolutional code. It happens that the sum of the squares above can be used as a metric, so that the Rake receiver structure can be used for this purpose as well. It should be noted that it is important that the time-interleaving be done at the level of the 6-bit blocks so that the channel remains constant within the chips associated with each such block. Otherwise non-coherent demodulation cannot be performed.

The IS-95 uplink design employs non-coherent demodulation. Another design option is to estimate the channel using a pilot signal and perform coherent demodulation. This option is adopted for CDMA 2000.

Power control

The link-level performance of a user is a function of its SINR. To achieve reliable communication, the SINR, or equivalently the ratio of the energy per bit to the interference and noise per chip (commonly called $\mathcal{E}_b/I_0$ in the CDMA literature), should be above a certain threshold. This threshold depends on the specific code used, as well as the multipath channel statistics. For example, a typical $\mathcal{E}_b/I_0$ threshold in the IS-95 system is 6 to 7 dB. In a mobile communication system, the attenuation of both the user of interest and the interferers varies as the users move, due to varying path loss and shadowing effects. To maintain a target SINR, *transmit power control* is needed.

The power control problem can be formulated in the network setting as follows. There are K users in total in the system and a number of cells (base-stations). Suppose user k is assigned to base-station c_k. Let P_k be the transmit power of user k, and g_{km} be the attenuation of user k's signal to base-station m.

The received energy per chip for user k at base-station m is simply given by $P_k g_{km}/W$. Using the expression (4.8), we see that if each user's target $\mathcal{E}_b/I_0$ is β, then the transmit powers of the users should be controlled such that

$$\frac{G P_k g_{k,c_k}}{\sum_{n \neq k} P_n g_{n,c_k} + N_0 W} \geq \beta, \qquad k = 1, \ldots, K. \qquad (4.10)$$

where $G = W/R$ is the processing gain of the system. Moreover, due to constraints on the dynamic range of the transmitting mobiles, there is a limit of the transmit powers as well:

$$P_k \leq \hat{P}, \qquad k = 1, \ldots, K. \qquad (4.11)$$

These inequalities define the set of all feasible power vectors $\mathbf{P} := (P_1, \ldots, P_K)^t$, and this set is a function of the attenuation of the users. If this set is empty, then the SINR requirements of the users cannot be simultaneously met. The system is said to be in *outage*. On the other hand, whenever this set of feasible powers is non-empty, one is interested in finding a solution which requires as little power as possible to conserve energy. In fact, it can be shown (Exercise 4.8) that whenever the feasible set is non-empty (this characterization is carried out carefully in Exercise 4.5), there exists a *component-wise* minimal solution $\mathbf{P}^*$ in the feasible set, i.e., $P_k^* \leq P_k$ for *every* user k in any other feasible power vector $\mathbf{P}$. This fact follows from a basic monotonicity property of the power control problem: when a user lowers its transmit power, it creates less interference and benefits all other users in the system. At the optimal solution $\mathbf{P}^*$, every user is at the minimal possible power so that their SINR requirements are met with equality and no more. Note that at the optimal point all the users in the same cell have the same received power at the base-station. It can also be shown that a simple distributed power control algorithm will converge to the optimal solution: at each step, each user updates its transmit power so that its own SINR requirement is just met with the current level of the interference. Even if the updates are done asynchronously among the users, convergence is still guaranteed. These results give theoretical justification to the robustness and stability of the power control algorithms implemented in practice. (Exercise 4.12 studies the robustness of the power update algorithm to inaccuracies in controlling the received powers of all the mobiles to be exactly equal.)

Power control in IS-95

The actual power control in IS-95 has an open-loop and a closed-loop component. The open-loop sets the transmit power of the mobile user at roughly the right level by inference from the measurements of the *downlink* channel strength via a pilot signal. (In IS-95, there is a common pilot transmitted in the downlink to all the mobiles.) However, since IS-95 is implemented in the FDD mode, the uplink and downlink channel typically differ in carrier frequency of tens of MHz and are not identical. Thus, open-loop control is typically accurate only up to a few dB. Closed-loop control is needed to adjust the power more precisely.

The closed-loop power control operates at 800 Hz and involves 1 bit feedback from the base-station to the mobile, based on measured SINR values; the command is to increase (decrease) power by 1 dB if the measured SINR is below (above) a threshold. Since there is no pilot in the uplink in IS-95, the SINR is estimated in a decision-directed mode, based on the output of the Rake receiver. In addition to measurement errors, the accuracy of power control is also limited by the 1-bit quantization. Since the SINR threshold β for reliable communication depends on the multipath channel statistics and is therefore not known perfectly in advance, there is also an outer loop which

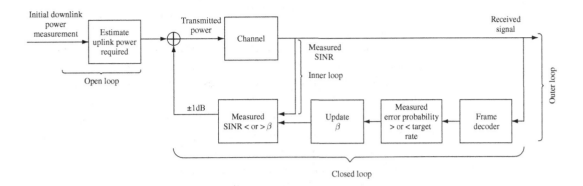

Figure 4.5 Inner and outer loops of power control.

adjusts the SINR threshold as a function of frame error rates (Figure 4.5). An important point, however, is that even though feedback occurs at a high rate (800 Hz), because of the limited resolution of 1 bit per feedback, power control does not track the fast multipath fading of the users when they are at vehicular speeds. It only tracks the slower shadow fading and varying path loss. The multipath fading is dealt with primarily by the diversity techniques discussed earlier.

Soft handoff

Handoff from one cell to the other is an important mechanism in cellular systems. Traditionally, handoffs are *hard*: users are either assigned to one cell or the other but not both. In CDMA systems, since all the cells share the same spectrum, *soft* handoffs are possible: multiple base-stations can simultaneously decode the mobile's data, with the switching center choosing

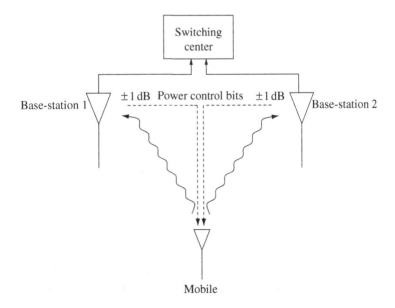

Figure 4.6 Soft handoff.

the best reception among them (Figure 4.6). Soft handoffs provide another level of diversity to the users.

The soft handoff process is mobile-initiated and works like this. While a user is tracking the downlink pilot of the cell it is currently in, it can be searching for pilots of adjacent cells (these pilots are known pseudonoise sequences shifted by known offsets). In general, this involves timing acquisition of the adjacent cell as well. However, we have observed that timing acquisition is a computationally very expensive step. Thus, a practical alternative is for the base-station clocks to be synchronized so that the mobile only has to acquire timing once. Once a pilot is detected and found to have sufficient signal strength relative to the first pilot, the mobile will signal the event to its original base-station. The original base-station will in turn notify the switching center, which enables the second cell's base-station to both send and receive the same traffic to and from the mobile. In the uplink, each base-station demodulates and decodes the frame or packet independently, and it is up to the switching center to arbitrate. Normally, the better cell's decision will be used.

If we view the base-stations as multiple receive antennas, soft handoff is providing a form of receive diversity. We know from Section 3.3.1 that the optimal processing of signals from the multiple antennas is *maximal-ratio combining*; this is however difficult to do in the handoff scenario as the antennas are geographically apart. Instead, what soft handoff achieves is *selection combining* (cf. Exercise 3.13). In IS-95, there is another form of handoff, called *softer handoff*, which takes place between sectors of the same cell. In this case, since the signal from the mobile is received at the sectored antennas which are co-located at the same base-station, maximal-ratio combining can be performed.

How does power control work in conjunction with soft handoff? Soft handoff essentially allows users to choose among several cell sites. In the power control formulation discussed in the previous section, each user is assumed to be assigned to a particular cell, but cell site selection can be easily incorporated in the framework. Suppose user k has an active set S_k of cells among which it is performing soft handoff. Then the transmit powers P_k *and* the cell site assignments $c_k \in S_k$ should be chosen such that the SINR requirements (4.10) are simultaneously met. Again, if there is a feasible solution, it can be shown that there is a component-wise minimal solution for the transmit powers (Exercise 4.5). Moreover, there is an analogous distributed asynchronous algorithm that will converge to the optimal solution: at each step, each user is assigned the cell site that will minimize the transmit power required to meet its SINR requirement, given the current interference levels at the base-stations. Its transmit power is set accordingly (Exercise 4.8). Put it another way, the transmit power is set in such a way that the SINR requirement is just met at the cell with the best reception. This is implemented in the IS-95 system as follows: all the base-stations in the soft handoff set will feedback power control bits to the mobile; the mobile will always decrease its transmit power by 1 dB if at least one of the soft handoff cell sites instructs it to do so. In other words, the minimum transmit power is always used. The advantages of soft handoff are studied in more detail in Exercise 4.10.

Interference averaging and system capacity

Power control and soft handoff minimize the transmit powers required to meet SINR requirements, if there is a feasible solution for the powers at all. If not, then the system is in outage. The *system capacity* is the maximum number of users that can be accommodated in the system for a desired outage probability and a link level $\mathcal{E}_b/I_0$ requirement.

The system can be in outage due to various random events. For example, users can be in certain configurations that create a lot of interference on neighboring cells. Also, voice or data users have periods of activity, and too many users can be active in the system at a given point in time. Another source of randomness is due to imperfect power control. While it is impossible to have a zero probability of outage, one wants to maintain that probability small, below a target threshold. Fortunately, the link level performance of a user in the uplink depends on the *aggregate* interference at the base-station due to many users, and the effect of these sources of randomness tends to average out according to the law of large numbers. This means that one does not have to be too conservative in admitting users into the network and still guarantee a small probability of outage. This translates into a larger system capacity. More specifically,

- **Out-of-cell interference averaging** Users tend to be in random independent locations in the network, and the fluctuations of the aggregate interference created in the adjacent cell are reduced when there are many users in the system.
- **Users' burstiness averaging** Independent users are unlikely to be active all the time, thus allowing the system to admit more users than if it is assumed that every user sends at peak rate all the time.
- **Imperfect power control averaging** Imperfect power control is due to tracking inaccuracy and errors in the feedback loop.[11] However, these errors tend to occur independently across the different users in the system and average out.

These phenomena can be generally termed *interference averaging*, an important property of CDMA systems. Note that the concept of interference averaging is reminiscent of the idea of *diversity* we discussed in Chapter 3: while diversity techniques make a point-to-point link more reliable by averaging over the channel fading, interference averaging makes the link more

---

[11] Since power control bits have to be fed back with a very tight delay constraint, they are usually uncoded which implies quite a high error rate.

reliable by averaging over the effects of different interferers. Thus, interference averaging can also be termed *interference diversity*.

To give a concrete sense of the benefit of interference averaging on system capacity, let us consider the specific example of averaging of users' burstiness. For simplicity, consider a single-cell situation with K users power controlled to a common base-station and no out-of-cell interference. Specializing (4.10) to this case, it can be seen that the $\mathcal{E}_b/I_0$ requirement β of all users is satisfied if

$$\frac{GQ_k}{\sum_{n\neq k} Q_n + N_0 W} \geq \beta, \qquad k = 1, \ldots, K, \tag{4.12}$$

where $Q_k := P_k g_k$ is the received power of user k at the base-station. Equivalently:

$$GQ_k \geq \beta \left(\sum_{n\neq k} Q_n + N_0 W \right) \qquad k = 1, \ldots, K. \tag{4.13}$$

Summing up all the inequalities, we get the following necessary condition for the Q_k:

$$[G - \beta(K-1)]\sum_{k=1}^{K} Q_k \geq K N_0 W \beta. \tag{4.14}$$

Thus a necessary condition for the existence of feasible powers is $G - \beta(K-1) > 0$, or equivalently,

$$K < \frac{G}{\beta} + 1. \tag{4.15}$$

On the other hand, if this condition is satisfied, the powers

$$Q_k = \frac{N_0 W \beta}{G - \beta(K-1)}, \qquad k = 1, \ldots, K \tag{4.16}$$

will meet the $\mathcal{E}_b/I_0$ requirements of all the users. Hence, condition (4.15) is a necessary and sufficient condition for the existence of feasible powers to support a given $\mathcal{E}_b/I_0$ requirement.

Equation (4.15) yields the *interference-limited system capacity* of the single cell. It says that, because of the interference between users, there is a limit on the number of users admissible in the cell. If we substitute $G = W/R$ into (4.15), we get

$$\boxed{\frac{KR}{W} < \frac{1}{\beta} + \frac{1}{G}.} \tag{4.17}$$

The quantity KR/W is the overall spectral efficiency of the system (in bits/s/Hz). Since the processing gain G of a CDMA system is typically large, (4.17) says that the maximal spectral efficiency is approximately $1/\beta$. In IS-95, a typical $\mathcal{E}_b/N_0$ requirement β is 6 dB, which translates into a maximum spectral efficiency of 0.25 bits/s/Hz.

Let us now illustrate the effect of user burstiness on the system capacity and the spectral efficiency in the single cell setting. We have assumed that all K users are active all the time, but suppose now that each user is active and has data to send only with probability p, and users' activities are independent of each other. Voice users, for example, are typically talking 3/8 of the time, and if the voice coder can detect silence, there is no need to send data during the quiet periods. If we let ν_k be the indicator random variable for user k's activity, i.e., $\nu_k = 1$ when user k is transmitting, and $\nu_k = 0$ otherwise, then using (4.15), the $\mathcal{E}_b/I_0$ requirements of the users can be met if and only if

$$\sum_{k=1}^{K} \nu_k < \frac{G}{\beta} + 1. \tag{4.18}$$

Whenever this constraint is not satisfied, the system is in outage. If the system wants to guarantee that no outage can occur, then the maximum number of users admissible in the network is $G/\beta + 1$, the same as the case when users are active all the time. However, more users can be accommodated if a small outage probability p_{out} can be tolerated: this number $K^*(p_{\text{out}})$ is the largest K such that

$$\Pr\left[\sum_{k=1}^{K} \nu_k > \frac{G}{\beta} + 1\right] \leq p_{\text{out}}. \tag{4.19}$$

The random variable $\sum_{k=1}^{K} \nu_k$ is binomially distributed. It has mean Kp and standard deviation $\sqrt{Kp(1-p)}$, where $p(1-p)$ is the variance of ν_k. When $p_{\text{out}} = 0$, $K^*(p_{\text{out}})$ is $G/\beta + 1$. If $p_{\text{out}} > 0$, then $K^*(p_{\text{out}})$ can be chosen larger. It is straightforward to calculate $K^*(p_{\text{out}})$ numerically for a given p_{out}. It is also interesting to see what happens to the spectral efficiency when the bandwidth of the system W scales with the rate R of each user fixed. In this regime, there are many users in the system and it is reasonable to apply a Gaussian approximation to $\sum_{k=1}^{K} \nu_k$. Hence,

$$\Pr\left[\sum_{k=1}^{K} \nu_k > \frac{G}{\beta} + 1\right] \approx Q\left[\frac{G/\beta + 1 - Kp}{\sqrt{Kp(1-p)}}\right]. \tag{4.20}$$

The overall spectral efficiency of the system is given by

$$\rho := \frac{KpR}{W}, \tag{4.21}$$

since the mean rate of each user is pR bits/s. Using the approximation (4.20) in (4.19), we can solve for the constraint on the spectral efficiency ρ:

$$\rho \leq \frac{1}{\beta}\left[1 + Q^{-1}(p_{\text{out}})\sqrt{\frac{1-p}{pK} - \frac{1}{Kp}}\right]^{-1}. \tag{4.22}$$

This bound on the spectral efficiency is plotted in Figure 4.7 as a function of the number of users. As seen in Eq. (4.17), the number $1/\beta$ is the maximum spectral efficiency if each user is non-bursty and transmitting at a constant rate equal to the mean rate pR of the bursty user. However, the *actual* spectral efficiency in the system with burstiness is different from that, by a factor of

$$\left(1 + Q^{-1}(p_{\text{out}})\sqrt{\frac{1-p}{pK} - \frac{1}{Kp}}\right)^{-1}.$$

This loss in spectral efficiency is due to a need to admit fewer users to cater for the burstiness of the traffic. This "safety margin" is larger when the outage probability requirement p_{out} is more stringent. More importantly, for a given outage probability, the spectral efficiency approaches $1/\beta$ as the bandwidth W (and hence the number of users K) scales. When there are many users in the system, interference averaging occurs: the fluctuation of the aggregate interference is smaller relative to the mean interference level. Since the link level performance of the system depends on the aggregate interference, less excess resource needs to be set aside to accommodate the fluctuations. This is a manifestation of the familiar principle of *statistical multiplexing*.

In the above example, we have only considered a single cell, where each active user is assumed to be perfectly power controlled and the only source of interference fluctuation is due to the random number of active users. In a multicell setting, the level of interference from outside of the cell depends on the locations of the interfering users and this contributes to another source

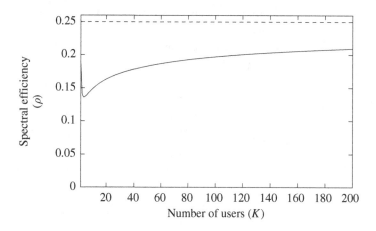

Figure 4.7 Plot of the spectral efficiency as a function of the number of users in a system with burstiness (the right hand side of (4.22)). Here, $p = 3/8$, $p_{\text{out}} = 0.01$ and $\beta = 6\,\text{dB}$.

of fluctuation of the aggregate interference level. Further randomness arises due to imperfect power control. The same principle of interference averaging applies to these settings as well, allowing CDMA systems to benefit from an increase in the system size. These settings are analyzed in Exercises (4.11) and (4.12).

To conclude our discussion, we note that we have made an implicit assumption of *separation of time-scales* in our analysis of the effect of interference in CDMA systems. At a faster time-scale, we average over the pseudorandom characteristics of the signal and the fast multipath fading to compute the statistics of the interference, which determine the bit error rates of the point-to-point demodulators. At a slower time-scale, we consider the burstiness of user traffic and the large-scale motion of the users to determine the outage probability, i.e., the probability that the target bit error rate performance of users cannot be met. Since these error events occur at completely different time-scales and have very different ramifications from a system-level perspective, this way of measuring the performance of the system makes more sense than computing an overall average performance.

4.3.2 CDMA downlink

The design of the one-to-many downlink uses the same basic principles of pseudorandom spreading, diversity techniques, power control and soft handoff we already discussed for the uplink. However, there are several important differences:

- The near–far problem does not exist for the downlink, since all the signals transmitted from a base-station go through the same channel to reach any given user. Thus, power control is less crucial in the downlink than in the uplink. Rather, the problem becomes that of allocating different powers to different users as a function of primarily the amount of out-of-cell interference they see. However, the theoretical formulation of this power allocation problem has the same structure as the uplink power control problem. (See Exercise 4.13.)
- Since signals for the different users in the cell are all transmitted at the base-station, it is possible to make the users orthogonal to each other, something that is more difficult to do in the uplink, as it requires chip-level synchronization between distributed users. This reduces but does not remove intra-cell interference, since the transmitted signal goes through multipath channels and signals with different delays from different users still interfere with each other. Still, if there is a strong line-of-sight component, this technique can significantly reduce the intra-cell interference, since then most of the energy is in the first tap of the channel.
- On the other hand, *inter-cell* interference is more poorly behaved in the downlink than in the uplink. In the uplink, there are many distributed

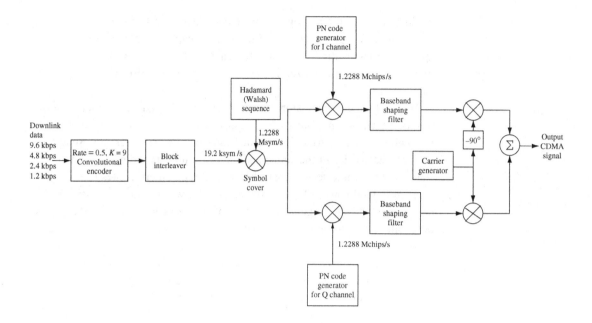

Figure 4.8 The IS-95 downlink.

users transmitting with small power, and significant interference averaging occurs. In the downlink, in contrast, there are only a few neighboring base-stations but each transmits at high power. There is much less interference averaging and the downlink capacity takes a significant hit compared to the uplink.

- In the uplink, soft handoff is accomplished by multiple base-stations listening to the transmitted signal from the mobile. No extra system resource needs to be allocated for this task. In the downlink, however, multiple base-stations have to simultaneously transmit to a mobile in soft handoff. Since each cell has a fixed number of orthogonal codes for the users, this means that a user in soft handoff is consuming double or more system resources. (See Exercise 4.13 for a precise formulation of the downlink soft handoff problem.)

- It is common to use a strong pilot and perform coherent demodulation in the downlink, since the common pilot can be shared by all the users. With the knowledge of the channels from each base-station, a user in soft handoff can also coherently combine the signals from the different base-stations. Synchronization tasks are also made easier in the presence of a strong pilot.

As an example, the IS-95 downlink is shown in Figure 4.8. Note the different roles of the Hadamard sequences in the uplink and in the downlink. In the uplink, the Hadamard sequences serve as an orthogonal modulation for each individual user so that non-coherent demodulation can be performed. In the downlink, in contrast, each user in the cell is assigned a *different* Hadamard sequence to keep them orthogonal (at the transmitter).

4.3.3 System issues

Signal characteristics

Consider the baseband uplink signal of a user given in (4.1). Due to the abrupt transitions (from $+1$ to -1 and vice versa) of the pseudonoise sequences s_n, the bandwidth occupied by this signal is very large. On the other hand, the signal has to occupy an allotted bandwidth. As an example, we see that the IS-95 system uses a bandwidth of 1.2288 MHz and a steep fall off after 1.67 MHz. To fit this allotted bandwidth, the signal in (4.1) is passed through a pulse shaping filter and then modulated on to the carrier. Thus though the signal in (4.1) has a perfect PAPR (equal to 1), the resulting transmit signal has a larger PAPR. The overall signal transmitted from the base-station is the superposition of all the user signals and this aggregate signal has PAPR performance similar to that of the narrowband system described in the previous section.

Sectorization

In the narrowband system we saw that all users can maintain high SINR due to the nature of the allocations. In fact, this was the benefit gained by paying the price of poor (re)use of the spectrum. In the CDMA system, however, due to the intra and inter-cell interferences, the values of SINR possible are very small. Now consider sectorization with universal frequency reuse among the sectors. Ideally (with full isolation among the sectors), this allows us to increase the system capacity by a factor equal to the number of sectors. However, in practice each sector now has to contend with inter-sector interference as well. Since intra-sector and inter-cell interference dominate the noise faced by the user signals, the additional interference caused due to sectorization does not cause a further degradation in SINR. Thus sectors of the same cell reuse the frequency without much of an impact on the performance.

Network issues

We have observed that timing acquisition (at a chip level accuracy) by a mobile is a computationally intensive step. Thus we would like to have this step repeated as infrequently as possible. On the other hand, to achieve soft handoff this acquisition has to be done (synchronously) for all base-stations with which the mobile communicates. To facilitate this step and the eventual handoff, implementations of the IS-95 system use high precision clocks (about 1 ppm (parts per million)) and further, synchronize the clocks at the base-stations through a proprietary wireline network that connects the base-stations. This networking cost is the price paid in the design to ease the handoff process.

Summary 4.2 CDMA

Universal frequency reuse: all users, both within a cell and across different cells, transmit and receive on the entire bandwidth.

The signal of each user is modulated onto a pseudonoise sequence so that it appears as white noise to others.

Interference management is crucial for allowing universal frequency reuse:
- Intra-cell interference is managed via power control. Accurate closed-loop power control is particularly important for combating the near–far problem in the uplink.
- Inter-cell interference is managed via averaging of the effects of multiple interferers. It is more effective in the uplink than in the downlink.

Interference averaging also allows statistical multiplexing of bursty users, thus increasing system capacity.

Diversity of the point-to-point links is achieved by a combination of low-rate coding, time-interleaving and Rake combining.

Soft handoff provides a further level of macrodiversity, allowing users to communicate with multiple base-stations simultaneously.

4.4 Wideband systems: OFDM

The narrowband system design of making transmissions interference-free simplified several aspects of network design. One such aspect was that the performance of a user is insensitive to the received powers of other users. In contrast to the CDMA approach, the requirement for accurate power control is much less stringent in systems where user transmissions in the same cell are kept orthogonal. This is particularly important in systems designed to accommodate many users each with very low average data rate: the fixed overhead needed to perform tight power control for each user may be too expensive for such systems. On the other hand there is a penalty of poor spectral reuse in narrowband systems compared to the CDMA system. Basically, narrowband systems are ill suited for universal frequency reuse since they do not average interference. In this section, we describe a system that combines the desirable features of both these systems: maintaining orthogonality of transmissions within the cell and having universal frequency reuse across cells. Again, the latter feature is made possible through interference averaging.

4.4.1 Allocation design principles

The first step in the design is to decide on the user signals that ensure orthogonality after passing through the wireless channel. Recall from the discussion of the downlink signaling in the CDMA system that though the *transmit signals* of the users are orthogonal, they interfere with each other at the receiver after passing through the multipath channel. Thus any orthogonal set of signals will not suffice. If we model the wireless channel as a linear time invariant multipath channel, then the only eigenfunctions are the *sinusoids*. Thus sinusoid inputs remain orthogonal at the receiver no matter what the multipath channel is. However, due to the channel variations in time, we want to restrict the notion of orthogonality to no more than a coherence time interval. In this context, sinusoids are no longer orthogonal, but the sub-carriers of the OFDM scheme of Section 3.4.4 with the cyclic *prefix* for the multipath channel provide a set of orthogonal signals over an OFDM block length.

We describe an allocation of sets of OFDM sub-carriers as the user signals; this description is identical for both the downlink and the uplink. As in Section 3.4.4, the bandwidth W is divided into N_c sub-carriers. The number of sub-carriers N_c is chosen to be as large as possible. As we discussed earlier, N_c is limited by the coherence time, i.e., the OFDM symbol period $N_c/W < T_c$. In each cell, we would like to distribute these N_c sub-carriers to the users in it (with say n sub-carriers per user). The n sub-carriers should be spread out in frequency to take advantage of frequency diversity. There is no interference among user transmissions within a cell by this allocation.

With universal frequency reuse, there is however inter-cell interference. To be specific, let us focus on the uplink. Two users in neighboring cells sharing the same sub-carrier in any OFDM symbol time interfere with each other directly. If the two users are close to each other, the interference can be very severe and we would like to minimize such overlaps. However, due to full spectral reuse, there is such an overlap at every OFDM symbol time in a fully loaded system. Thus, the best one can do is to ensure that the interference does not come solely from one user (or a small set of users) and the interference seen over a coded sequence of OFDM symbols (forming a *frame*) can be attributed to most of the user transmissions in the neighboring cell. Then the *overall* interference seen over a frame is a function of the average received power of *all* the users in the neighboring cells. This is yet another example of the *interference diversity* concept we already saw in Section 4.3.

How are the designs of the previous two systems geared towards harvesting interference diversity? The CDMA design fully exploits interferer diversity by interference averaging. This is achieved by every user spreading its signals over the entire spectrum. On the other hand, the orthogonal allocation of channels in the GSM system is poorly suited from the point of view of interferer diversity. As we saw in Section 4.2, users in neighboring cells that are close to each other and transmitting on the same channel over the same slot cause severe interference to each other. This leads to a very degraded performance and the reason for it is clear: interference seen by a user comes solely from one interferer and there is no scope to see an average interference from all the users over a slot. If there were no hopping and coding across the sub-carriers, the OFDM system would behave exactly like a narrowband system and suffer the same fate.

Turning to the downlink we see that now all the transmissions in a cell occur from the same place: at the base-station. However, the power in different sub-carriers transmitted from the base-station can be vastly different. For example, the pilots (training symbols) are typically at a much higher power than the signal to a user very close to the base-station. Thus even in the downlink, we would like to hop the sub-carriers allocated to a user every OFDM symbol time so that over a frame the interference seen by a mobile is a function of the average transmit power of the neighboring base-stations.

4.4.2 Hopping pattern

We have arrived at two design rules for the sub-carrier allocations to the users. Allocate the n sub-carriers for the user as spread out as possible and further, hop the n sub-carriers every OFDM symbol time. We would like the hop patterns to be as "apart" as possible for neighboring base-stations. We now delve into the design of *periodic* hopping patterns that meet these broad design rules that repeat, say, every N_c OFDM symbol intervals. As we will see, the choice of the period to be equal to N_c along with the assumption that N_c be prime (which we now make) simplifies the construction of the hopping pattern.

The periodic hopping pattern of the N_c sub-carriers can be represented by a square matrix (of dimension N_c) with entries from the set of *virtual channels*, namely $0, 1, \ldots, N_c - 1$. Each virtual channel hops over different sub-carriers at different OFDM symbol times. Each row of the hopping matrix corresponds to a sub-carrier and each column represents an OFDM symbol time, and the entries represent the virtual channels that use that sub-carrier in different OFDM symbol times. In particular, the (i, j) entry of the matrix corresponds to the virtual channel number the ith sub-carrier is taken on by, at OFDM symbol time j. We require that every virtual channel hop over all the sub-carriers in each period for maximal frequency diversity. Further, in any OFDM symbol time the virtual channels occupy different sub-carriers. These two requirements correspond to the constraint that each row and column of the hopping matrix contains every virtual channel number $(0, \ldots, N_c - 1)$, exactly once. Such a matrix is called a *Latin square*. Figure 4.9 shows hopping patterns of the 5 virtual channels over the 5 OFDM symbol times (i.e., $N_c = 5$). The horizontal axis corresponds to OFDM symbol times and the vertical axis denotes the 5 physical sub-carriers (as in Figure 3.25), and the sub-carriers the virtual channels adopt are denoted by darkened squares. The corresponding hopping pattern matrix is

$$\begin{bmatrix} 0 & 1 & 2 & 3 & 4 \\ 2 & 3 & 4 & 0 & 1 \\ 4 & 0 & 1 & 2 & 3 \\ 1 & 2 & 3 & 4 & 0 \\ 3 & 4 & 0 & 1 & 2 \end{bmatrix}.$$

Figure 4.9 Virtual channel hopping patterns for $N_c = 5$.

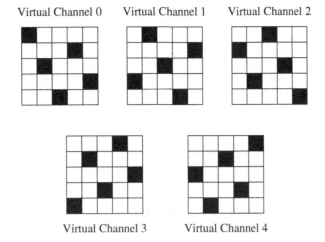

Virtual Channel 0   Virtual Channel 1   Virtual Channel 2

Virtual Channel 3   Virtual Channel 4

For example, we see that the virtual channel 0 is assigned the OFDM symbol time and sub-carrier pairs $(0, 0)$, $(1, 2)$, $(2, 4)$, $(3, 1)$, $(4, 3)$. Now users could be allocated n virtual channels, accommodating $\lfloor N_c/n \rfloor$ users.

Each base-station has its own hopping matrix (Latin square) that determines the physical structure of the virtual channels. Our design rule to maximize interferer diversity requires us to have minimal overlap between virtual channels of neighboring base-stations. In particular, we would like to have exactly one time/sub-carrier collision for every pair of virtual channels of two base-stations that employ these hopping patterns. Two Latin squares that have this property are said to be *orthogonal*.

When N_c is prime, there is a simple construction for a family of $N_c - 1$ mutually orthogonal Latin squares. For $a = 1, \ldots, N_c - 1$ we define an $N_c \times N_c$ matrix R^a with (i, j)th entry

$$R_{ij}^a = ai + j \quad \text{modulo } N_c. \tag{4.23}$$

Here we index rows and columns from 0 through $N_c - 1$. In Exercise 4.14, you are asked to verify that R^a is a Latin square and further that for every $a \neq b$ the Latin squares R^a and R^b are orthogonal. Observe that Figure 4.9 depicts a Latin square hopping pattern of this type with $a = 2$ and $N_c = 5$.

With these Latin squares as the hopping patterns, we can assess the performance of data transmission over a single virtual channel. First, due to the hopping over the entire band, the frequency diversity in the channel is harnessed. Second, the interference seen due to inter-cell transmissions comes from different virtual channels (and repeats after N_c symbol times). Coding over several OFDM symbols allows the full interferer diversity to be harnessed: coding ensures that no one single strong interference from a virtual channel can cause degradation in performance. If sufficient interleaving is permitted, then the time diversity in the system can also be obtained.

To implement these design goals in a cellular system successfully, the users within the cell must be synchronized to their corresponding base-station. This way, the simultaneous uplink transmissions are still orthogonal at the base-station. Further, the transmissions of neighboring base-stations also have to be synchronized. This way the design of the hopping patterns to average the interference is fully utilized. Observe that the synchronization needs to be done only at the level of OFDM symbols, which is much coarser than at the level of chips.

4.4.3 Signal characteristics and receiver design

Let us consider the signal transmission corresponding to a particular user (either in the uplink or the downlink). The signal consists of n virtual channels, which over a slot constitute a set of n OFDM sub-carriers that are hopped over OFDM symbol times. Thus, though the signal *information content* can be "narrow" (for small ratios n/N_c), the signal bandwidth itself is wide. Further, since the bandwidth range occupied varies from symbol to symbol, each (mobile) receiver has to be wideband. That is, the sampling rate is proportional to $1/W$. Thus this signal constitutes a (frequency hopped) spread-spectrum signal just as the CDMA signal is: the ratio of data rate to bandwidth occupied by the signal is small. However, unlike the CDMA signal, which spreads the energy over the entire bandwidth, here the energy of the signal is only in certain sub-carriers (n of a total N_c). As discussed in Chapter 3, fewer channel parameters have to be measured and channel estimation with this signal is superior to that with the CDMA signal.

The major advantages of the third system design are the frequency and interferer diversity features. There are a few engineering drawbacks to this choice. The first is that the mobile sampling rate is quite high (same as that of the CDMA system design but much higher than that of the first system). All signal processing operations (such as the FFT and IFFT) are driven off this basic rate and this dictates the processing power required at the mobile receiver. The second drawback is with respect to the transmit signal on the uplink. In Exercise 4.15, we calculate the PAPR of a canonical transmit signal in this design and observe that it is significantly high, as compared to the signal in the GSM and CDMA systems. As we discussed in the first system earlier, this higher PAPR translates into a larger bias in the power amplifier settings and a correspondingly lower average efficiency. Several engineering solutions have been proposed to this essentially engineering problem (as opposed to the more central communication problem which deals with the uncertainties in the channel) and we review some of these in Exercise 4.16.

4.4.4 Sectorization

What range of SINRs is possible for the users in this system? We observed that while the first (narrowband) system provided high SINRs to all the mobiles, almost no user was in a high SINR scenario in the CDMA system due to the intra-cell interference. The range of SINRs possible in this system is midway between these two extremes. First, we observe that the only source of interference is inter-cell. So, users close to the base-station will be able to have high SINRs since they are impacted less from inter-cell interference. On the other hand, users at the edge of the cell are interference limited and cannot support high SINRs. If there is a feedback of the received SINRs then users closer by the base-station can take advantage of the higher SINR by transmitting and receiving at higher data rates.

What is the impact of sectorization? If we universally reuse the frequency among the sectors, then there is inter-sector interference. We can now observe an important difference between inter-sector and inter-cell interference. While inter-cell interference affects mostly the users at the edge of the cell, inter-sector interference affects users regardless of whether they are at the edge of the cell or close to the base-station (the impact is pronounced on those at the edge of the sectors). This interference now reduces the dynamic range of SINRs this system is capable of providing.

Example 4.1 Flash-OFDM

A technology that partially implements the design features of the wideband OFDM system is Flash-OFDM, developed by Flarion Technologies [38]. Over 1.25 MHz, there are 113 sub-carriers, i.e., $N_c = 113$. The 113 virtual channels are created from these sub-carriers using the Latin square hopping patterns (in the downlink the hops are done every OFDM symbol but once in every 7 OFDM symbols in the uplink). The sampling rate (or equivalently, chip rate) is 1.25 MHz and a cyclic prefix of 16 samples (or chips) covers for a delay spread of approximately 11 μs. This means that the OFDM symbol is 128 samples, or approximately 100 μs long.

There are four *traffic channels* of different granularity: there are five in the uplink (comprising 7, 14, 14, 14 and 28 virtual channels) and four in the downlink (comprising 48, 24, 12, 12 virtual channels). Users are scheduled on different traffic channels depending on their traffic requirements and channel conditions (we study the desired properties of the scheduling algorithm in greater detail in Chapter 6). The scheduling algorithm operates once every *slot*: a slot is about 1.4 ms long, i.e., it consists of 14 OFDM symbols. So, if a user is scheduled (say, in the downlink) the traffic channel consisting of 48 virtual channels, it can transmit 672 OFDM symbols over the slot when it is scheduled. An appropriate rate LDPC (low-density parity check) code combined with a simple modulation scheme (such as

QPSK or 16-QAM) is used to convert the raw information bits into the 672 OFDM symbols.

The different levels of granularity of the traffic channels are ideally suited to carry bursty traffic. Indeed, Flash-OFDM is designed to act in a data network where it harnesses the statistical multiplexing gains of the user's bursty data traffic by its packet-switching operation.

The mobiles are in three different states in the network. When they are inactive, they go to a "sleep" mode monitoring the base-station signal every once in a while: this mode saves power by turning off most of the mobile device functionalities. On the other hand, when the mobile is actively receiving and/or sending data it is in the "ON" mode: this mode requires the network to assign resources to the mobile to perform periodic power control updates and timing and frequency synchronization. Apart from these two states, there is an in-between "HOLD" mode: here mobiles that have been recently active are placed without power control updates but still maintaining timing and frequency synchronization with the base-station. Since the intra-cell users are orthogonal and the accuracy of power control can be coarse, users in a HOLD state can be quickly moved to an ON state when there is a need to send or receive data. Flash-OFDM has the ability to hold approximately 30, 130 and 1000 mobiles in the ON, HOLD and sleep modes.

For many data applications, it is important to be able to keep a large number of users in the HOLD state, since each user may send traffic only once in a while and in short bursts (requests for http transfers, acknowledgements, etc.) but when they do want to send, they require short latency and quick access to the wireless resource. It is difficult to support this HOLD state in a CDMA system. Since accurate power control is crucial because of the near–far problem, a user who is not currently power-controlled is required to slowly ramp up its power before it can send traffic. This incurs a very significant delay.[12] On the other hand, it is very expensive to power control a large number of users who only transmit infrequently. In an orthogonal system like OFDM, this overhead can be largely avoided. The issue does not arise in a voice system since each user sends constantly and the power control overhead is only a small percentage of the payload (about 10% in IS-95).

Chapter 4 The main plot

The focus of this chapter is on multiple access, interference management and the system issues in the design of cellular networks. To highlight the

---

[12] Readers from the San Francisco Bay area may be familiar with the notorious "Fast Track" lanes for the Bay Bridge. Once a car gets on one of these lanes, it can cross the toll plaza very quickly. But the problem is that most of the delay is in getting to them through the traffic jam!

issues, we looked at three different system designs. Their key characteristics are compared and contrasted in the table below.

	Narrowband system	Wideband CDMA	Wideband OFDM
Signal	Narrowband	Wideband	Wideband
Intra-cell BW allocation	Orthogonal	Pseudorandom	Orthogonal
Intra-cell interference	None	Significant	None
Inter-cell BW allocation	Partial reuse	Universal reuse	Universal reuse
Inter-cell uplink interference	Bursty	Averaged	Averaged
Accuracy of power control	Low	High	Low
Operating SINR	High	Low	Range: low to high
PAPR of uplink signal	Low	Medium	High
Example system	GSM	IS-95	Flash-OFDM

4.5 Bibliographical notes

The two important aspects that have to be addressed by a wireless system designer are how resource is allocated within a cell among the users and how interference (both intra- and inter-cell) is handled. Three topical wireless technologies have been used as case studies to bring forth the tradeoffs the designer has to make. The standards IS-136 [60] and GSM [99] have been the substrate on which the discussion of the narrowband system design is built. The wideband CDMA design is based on the widely implemented second-generational technology IS-95 [61]. A succinct description of the the technical underpinnings of the IS-95 design has been done by Viterbi [140] with emphasis on a system view, and our discussion here has been influenced by it. The frequency hopping OFDM system based on Latin squares was first suggested by Wyner [150] and Pottie and Calderbank [94]. This basic physical-layer construct has been built into a technology (Flash-OFDM [38]).

4.6 Exercises

Exercise 4.1 In Figure 4.2 we set a specific reuse pattern. A channel used in a cell precludes its use in all the neighboring cells. With this allocation policy the reuse factor is at least 1/7. This is a rather ad hoc allocation of channels to the cells and the reuse ratio can be improved; for example, the four-color theorem [102] asserts that a planar graph can be colored with four colors with no two vertices joined by an edge sharing the same channel. Further, we may have to allocate more channels to cells which are crowded. In this question, we consider modeling this problem.

Let us represent the cells by a finite set (of vertices) $V := \{v_1, \ldots, v_C\}$; one vertex for each cell, so there are C cells. We want to be able to say that only a certain collection of vertices can share the same channel. We do this by defining an *allowable set* $S \subseteq V$ such that all the vertices in S can share the same channel. We are only interested in *maximal allowable sets*: these are allowable sets with no strict superset also an allowable set. Suppose the maximal allowable sets are M in number, denoted as $S_1, \ldots, S_M$. Each of these maximal allowable sets can be thought of as a *hyper-edge* (the traditional definition of *edge* means a pair of vertices) and the collection of V and the hyper-edges forms a *hyper-graph*. You can learn more about hyper-graphs from [7].

1. Consider the hexagonal cellular system in Figure 4.10. Suppose we do not allow any two neighboring cells to share the same channel and further not allow the same channel to be allocated to cells 1, 3 and 5. Similarly, cells 2, 4 and 6 cannot share the same channel. For this example, what are C and M? Enumerate the maximal allowable sets $S_1, \ldots, S_M$.

2. The hyper-edges can also be represented as an *adjacency matrix* of size $C \times M$: the (i, j)th entry is

$$a_{ij} := \begin{cases} 1 & \text{if } v_i \in S_j, \\ 0 & \text{if } v_i \notin S_j. \end{cases} \tag{4.24}$$

For the example in Figure 4.10, explicitly construct the adjacency matrix.

Exercise 4.2 [84] In Exercise 4.1, we considered a graphical model of the cellular system and constraints on channel allocation. In this exercise, we consider modeling the *dynamic traffic and channel allocation algorithms*.

Suppose there are N channels to be allocated. Further, the allocation has to satisfy the reuse conditions: in the graphical model this means that each channel is mapped to one of the maximal allowable sets. The traffic comprises calls originating and terminating in the cells. Consider the following statistical model. The average number of overall calls in all the cells is B. This number accounts for *new* call arrivals and calls leaving the cell due to termination. The *traffic intensity* is the number of call arrivals per available channel, $r := B/N$ (in Erlangs per channel). A fraction p_i of these calls occur in cell i (so that $\sum_{i=1}^{C} p_i = 1$). So, the long-term average number of calls per channel to be handled in cell i is $p_i r$. We need a channel to service a call, so to meet this traffic we need on an average at least $p_i r$ channels allocated to cell i. We fix the traffic profile $p_1, \ldots, p_C$ over the time-scale for which the number of calls averaging is done. If a cell has used up all its allocated channels, then a new call cannot be serviced and is dropped.

A dynamic channel allocation algorithm allocates the N channels to the C cells to meet the instantaneous traffic requirements and further satisfies the reuse pattern. Let us focus on the average performance of a dynamic channel allocation algorithm: this is the sum of the average traffic per channel supported by each cell, denoted by $T(r)$.

1. Show that

$$T(r) \leq \max_{j=1 \ldots M} \sum_{i=1}^{C} a_{ij}. \tag{4.25}$$

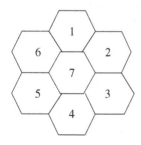

Figure 4.10 A narrowband system with seven cells. Adjacent cells cannot share the same channel and cells $\{1, 3, 5\}$ and $\{2, 4, 6\}$ cannot share the same channel either.

Hint: The quantity on the right hand side is the cardinality of the largest maximal allowable set.

2. Show that

$$T(r) \leq \sum_{i=1}^{C} p_i r = r, \qquad (4.26)$$

i.e., the total arrival rate is also an upper bound.

3. Let us combine the two simple upper bounds in (4.25) and (4.26). For every fixed list of C numbers $y_i \in [0,1], i = 1, \ldots, C$, show that

$$T(r) \leq \sum_{i=1}^{C} y_i p_i r + \max_{j=1\ldots M} \sum_{i=1}^{C} (1-y_i) a_{ij}. \qquad (4.27)$$

Exercise 4.3 This exercise is a sequel to Exercises 4.1 and 4.2. Consider the cellular system example in Figure 4.10, with the arrival rates $p_i = 1/8$ for $i = 1, \ldots, 6$ (all the cells at the edge) and $p_7 = 1/4$ (the center cell).

1. Derive a good upper bound on $T(r)$, the traffic carried per channel for any dynamic channel allocation algorithm for this system. In particular, use the upper bound derived in (4.27), but optimized over all choices of $y_1, \ldots, y_C$. *Hint*: The upper bound on $T(r)$ in (4.27) is *linear* in the variables $y_1, \ldots, y_C$. So, you can use software such as MATLAB (with the function `linprog`) to arrive at your answer.

2. In general, a channel allocation policy is dynamic: i.e., the number of channels allocated to a cell varies with time as a function of the traffic. Since we are interested in the average behavior of a policy over a large amount of time, it is possible that *static* channel allocation policies also do well. (Static policies allocate channels to the cells in the beginning and do not alter this allocation to suit the varying traffic levels.) Consider the following static allocation policy defined by the probability vector $\mathbf{x} := (x_1, \ldots, x_M)$, i.e., $\sum_{j=1}^{M} x_j = 1$. Each maximal allowable set S_j is allocated $\lfloor Nx_j \rfloor$ channels, in the sense that each cell in S_j is allocated these $\lfloor Nx_j \rfloor$ channels. Observe that cell i is allocated

$$\sum_{j=1}^{M} \lfloor Nx_j \rfloor a_{ij}$$

channels. Denote $T_{\mathbf{x}}(r)$ as the carried traffic by using this static channel allocation algorithm.

If the incoming traffic is smooth enough that the carried traffic in each cell is the minimum of arrival traffic in that cell and the number of channels allocated to that cell,

$$\lim_{N \to \infty} T_{\mathbf{x}}(r) = \sum_{i=1}^{C} \min\left(rp_i, \sum_{j=1}^{M} x_j a_{ij}\right), \qquad \forall r > 0. \qquad (4.28)$$

What are good static allocation policies? For the cellular system model in Figure 4.10, try out simple static channel allocation algorithms that you can think of. You can evaluate the performance of your algorithm numerically by simulating a smooth traffic arrival process (common models are uniform arrivals and independent and exponential inter-arrival times). How does your answer compare to the upper bound derived in part (1)?

In [84], the authors show that there exists a static allocation policy that can actually achieve (for large N, because the integer truncation effects have to be smoothed out) the upper bound in part (1) for every graphical model and traffic arrival rate.

Exercise 4.4 In this exercise we study the PAPR of the uplink transmit signal in narrowband systems. The uplink transmit signal is confined to a small bandwidth (200 kHz in the GSM standard). Consider the folowing simple model of the transmit signal using the idealized pulse shaping filter:

$$s(t) = \Re \left[\sum_{n=0}^{\infty} x[n] \, \mathrm{sinc} \left(\frac{t - nT}{T} \right) \exp(j2\pi f_c t) \right], \qquad t \geq 0. \tag{4.29}$$

Here T is approximately the inverse of the bandwidth (5 μs in the GSM standard) and $\{x[n]\}$ is the sequence of (complex) data symbols. The carrier frequency is denoted by f_c; for simplicity let us assume that $f_c T$ is an integer.

1. The raw information bits are coded and modulated resulting in the data symbols $x[n]$. Modeling the data symbols as i.i.d. uniformly distributed on the complex unit circle, calculate the average power in the transmit signal $s(t)$, averaged over the data symbols. Let us denote the average power by P_{av}.

2. The statistical behavior of the transmit signal $s(t)$ is periodic with period T. Thus we can focus on the peak power within the time interval $[0, T]$, denoted as

$$PP(x) = \max_{0 \leq t \leq T} |s(t)|^2. \tag{4.30}$$

The peak power is a random variable since the data symbols are random. Obtain an estimate for the average peak power. How does your estimate depend on T? What does this imply about the PAPR (ratio of PP to P_{av}) of the narrowband signal $s(t)$?

Exercise 4.5 [56] In this problem we study the uplink power control problem in the CDMA system in some detail. Consider the uplink of a CDMA system with a total of K mobiles trying to communicate with L base-stations. Each mobile k communicates with just one among a subset S_k of the L base-stations; this base-station assignment is denoted by c_k (i.e., we do not model diversity combining via soft handoff in this problem). Observe that by restricting S_k to have just one element, we are ruling out soft handoff as well. As in Section 4.3.1, we denote the transmit power of mobile k by P_k and the channel attenuation from mobile k to base-station m by g_{km}. For successful communication we require the $\mathcal{E}_b/I_0$ to be at least a target level β, i.e., successful uplink communication of the mobiles entails the constraints (cf. (4.10)):

$$\frac{\mathcal{E}_b}{I_0} = \frac{G P_k g_{k,c_k}}{\sum_{n \neq k} P_n g_{n,c_k} + N_0 W} \geq \beta_k, \qquad k = 1, 2, \ldots, K. \tag{4.31}$$

Here we have let the target level be potentially different for each mobile and denoted $G = W/R$ as the processing gain of the CDMA system. Writing the transmit powers as the vector $\mathbf{p} = (P_1, \ldots, P_K)^t$, show that (4.31) can be written as

$$(\mathbf{I}_K - \mathbf{F})\mathbf{p} \geq \mathbf{b}, \tag{4.32}$$

where $\mathbf{F}$ is the $K \times K$ matrix with strictly positive off-diagonal entries

$$f_{ij} = \begin{cases} 0 & \text{if } i = j, \\ g_{jc_i} \dfrac{\beta_i}{G} & \\ \dfrac{}{g_{ic_i}} & \text{if } i \neq j, \end{cases} \tag{4.33}$$

and

$$\mathbf{b} := N_0 W \left(\frac{\frac{\beta_1}{G}}{g_{1,c_1}}, \ldots, \frac{\frac{\beta_K}{G}}{g_{K,c_K}} \right)^t. \tag{4.34}$$

It can be shown (see Exercise 4.6) that there exist positive powers to make $\mathcal{E}_b/I_0$ meet the target levels, exactly when all the eigenvalues of $\mathbf{F}$ have absolute value strictly less than 1. In this case, there is in fact a *component-wise minimal* vector of powers that allows successful communication and is simply given by

$$\mathbf{p}^* = (\mathbf{I}_K - \mathbf{F})^{-1}\mathbf{b}. \tag{4.35}$$

Exercise 4.6 Consider the set of linear inequalities in (4.32) that correspond to the $\mathcal{E}_b/I_0$ requirements in the uplink of a CDMA system. In this exercise we investigate the mathematical constraints on the physical parameters of the CDMA system (i.e., the channel gains and desired target levels) which allow reliable communication.

We begin by observing that $\mathbf{F}$ is a non-negative matrix (i.e., it has non-negative entries). A non-negative matrix F is said to be *irreducible* if there exists a positive integer m such that $\mathbf{F}^m$ has all entries strictly positive.

1. Show that $\mathbf{F}$ in (4.33) is irreducible. (The number of mobiles K is greater than two.)
2. Non-negative matrices also show up as the probability transition matrices of finite state *Markov chains*. An important property of irreducible non-negative matrices is the *Perron–Frobenius* theorem: There exists a strictly positive eigenvalue (called the Perron–Frobenius eigenvalue) which is strictly bigger than the absolute value of any of the other eigenvalues. Further, there is a unique right eigenvector corresponding to the Perron–Frobenius eigenvalue, and this has strictly positive entries. Recall this result from a book on non-negative matrices such as [106].
3. Consider the vector form of the $\mathcal{E}_b/I_0$ constraints of the mobiles in (4.32) with $\mathbf{F}$ a non-negative irreducible matrix and $\mathbf{b}$ having strictly positive entries. Show that the following statements are equivalent.
   (a) There exists $\mathbf{p}$ satisfying (4.32) and having strictly positive entries.
   (b) The Perron–Frobenius eigenvalue of $\mathbf{F}$ is strictly smaller than 1.
   (c) $(\mathbf{I}_K - \mathbf{F})^{-1}$ exists and has strictly positive entries.

The upshot is that the existence or non-existence of a power vector that permits successful uplink communication from all the mobiles to their corresponding base-stations (with the assignment $k \mapsto c_k$) can be characterized in terms of the Perron–Frobenius eigenvalue of an irreducible non-negative matrix $\mathbf{F}$.

Exercise 4.7 In this problem, a sequel to Exercise 4.5, we allow the assignment of mobiles to base-stations to be in our control. Let $\mathbf{t} := (\beta_1, \ldots, \beta_K)$ denote the vector of the desired target thresholds on the $\mathcal{E}_{\mathrm{b}}/I_0$ of the mobiles. Given an assignment of mobiles to base-stations $k \mapsto c_k$ (with $c_k \in S_k$), we say that the pair $(c, \mathbf{t})$ is *feasible* if there is a power vector that permits successful communication from all the mobiles to their corresponding base-stations (i.e., user k's $\mathcal{E}_{\mathrm{b}}/I_0$ meets the target level β_k).

1. Show that if $(c, \mathbf{t}^{(1)})$ is feasible and $\mathbf{t}^{(2)}$ is another vector of desired target levels such that $\beta_k^{(1)} \geq \beta_k^{(2)}$ for each mobile $1 \leq k \leq K$, then $(c, \mathbf{t}^{(2)})$ is also feasible.
2. Suppose $(c^{(1)}, \mathbf{t})$ and $(c^{(2)}, \mathbf{t})$ are feasible. Let $\mathbf{p}^{(1)*}$ and $\mathbf{p}^{(2)*}$ denote the corresponding minimal vectors of powers allowing successful communication, and define

$$p_k^{(3)} := \min\left(p_k^{(1)*}, p_k^{(2)*}\right).$$

Define the new assignment

$$c_k^{(3)} := \begin{cases} c_k^{(1)} & \text{if } p_k^{(1)*} \leq p_k^{(2)*}, \\ c_k^{(2)} & \text{if } p_k^{(1)*} > p_k^{(2)*}. \end{cases}$$

Define the new target levels

$$\beta_k^{(3)} := \frac{g_{k c_k^{(3)}} p_k^{(3)*}}{N_0 W + \sum_{n \neq k} g_{n c_n^{(3)}} p_n^{(3)*}}, \qquad k = 1, \ldots, K,$$

and the vector $\mathbf{t}^{(3)} = (\beta_1^{(3)}, \ldots, \beta_K^{(3)})$. Show that $(c^{(3)}, \mathbf{t}^{(3)})$ is feasible and further that $\beta_k^{(3)} \geq \beta_k$ for all mobiles $1 \leq k \leq K$ (i.e., $\mathbf{t}^{(3)} \geq \mathbf{t}$ component-wise).

3. Using the results of the previous two parts, show that if uplink communication is feasible, then there is a unique component-wise minimum vector of powers that allows for successful uplink communication of all the mobiles, by appropriate assignment of mobiles to base-stations allowing successful communication. Further show that for any other assignment of mobiles to base-stations allowing successful communication the corresponding minimal power vector is component-wise at least as large as this power vector.

Exercise 4.8 [56, 151] In this problem, a sequel to Exercise 4.7, we will see an adaptive algorithm that updates the transmit powers of the mobiles in the uplink and the assignment of base-stations to the mobiles. The key property of this adaptive algorithm is that it converges to the component-wise minimal power among all assignments of base-stations to the mobiles (if there exists some assignment that is feasible, as discussed in Exercise 4.7(3)).

Users begin with an arbitrary power vector $\mathbf{p}^{(1)}$ and base-station assignment $c^{(1)}$ at the starting time 1. At time m, let the transmit powers of the mobiles be denoted by (the vector) $\mathbf{p}^{(m)}$ and the base-station assignment function be denoted by $c^{(m)}$. Let us first calculate the interference seen by mobile n at each of the base-stations $l \in S_n$; here S_n is the set of base-stations that can be assigned to mobile n.

$$I_{nl}^{(m)} := \sum_{k \neq n} g_{kl} p_k^{(m)} + N_0 W. \tag{4.36}$$

Now, we choose *greedily* to assign mobile n to that base-station which requires the least transmit power on the part of mobile n to meet its target level β_n. That is,

$$p_n^{(m+1)} := \min_{l \in S_n} \frac{\beta_n I_{nl}^{(m)}}{G g_{nl}}, \tag{4.37}$$

$$c_n^{(m+1)} := \arg\min_{l \in S_n} \frac{\beta_n I_{nl}^{(m)}}{g_{nl}}. \tag{4.38}$$

Consider this greedy update to each mobile being done *synchronously*: i.e., the updates of transmit power and base-station assignment for every mobile at time $m+1$ is made based on the transmit powers of all other the mobiles at time m. Let us denote this greedy update algorithm by the map $I: \mathbf{p}^{(m)} \mapsto \mathbf{p}^{(m+1)}$.

1. Show the following properties of I. Vector inequalities are defined to be component-wise inequalities.
   (a) $I(\mathbf{p}) > 0$ for every $\mathbf{p} \geq \mathbf{0}$.
   (b) $I(\mathbf{p}) \geq I(\tilde{\mathbf{p}})$, whenever $\mathbf{p} \geq \tilde{\mathbf{p}}$.
   (c) $I(\alpha \mathbf{p}) < \alpha I(\mathbf{p})$ whenever $\alpha > 1$.
2. Using the previous part, or otherwise, show that if I has a fixed point (denoted by $\mathbf{p}^*$) then it is unique.
3. Using the previous two parts, show that if I has a fixed point then $\mathbf{p}^{(m)} \to \mathbf{p}^*$ component-wise as $m \to \infty$ where $\mathbf{p}^{(m)} := I(\mathbf{p}^{(m-1)})$ and $\mathbf{p}^{(1)}$ and $c^{(1)}$ are an arbitrary initial allocation of transmit powers and assignments of base-stations.
4. If I has a fixed point, then show that the uplink communication problem must be feasible and further, the fixed point $\mathbf{p}^*$ must be the same as the component-wise minimal power vector derived in Exercise 4.7(3).

Exercise 4.9 Consider the following *asynchronous* version of the update algorithm in Exercise 4.8. Each mobile's update (of power and base-station assignment) occurs asynchronously based on some previous knowledge of all the other users' transmit powers. Say the update of mobile n at time m is based on mobile k's transmit power at time $\tau_{nk}(m)$. Clearly, $\tau_{nk}(m) \leq m$ and we require that each user eventually has an update of the other users' powers, i.e., for every time m_0 there exists time $m_1 \geq m_0$ such that $\tau_{nk}(m) \geq m_0$ for every time $m \geq m_1$. We further require that each user's power and base-station assignment is allocated infinitely often. Then, starting from any initial condition of powers of the users, show that the asynchronous power update algorithm converges to the optimal power vector $\mathbf{p}^*$ (assuming the problem is feasible, so that $\mathbf{p}^*$ exists in the first place).

Exercise 4.10 Consider the uplink of a CDMA system. Suppose there is only a single cell with just two users communicating to the base-station in the cell.

1. Express mathematically the set of all feasible power vectors to support given $\mathcal{E}_b/I_0$ requirements (assumed to be both equal to β).
2. Sketch examples of sets of feasible power vectors. Give one example where the feasible set is non-empty and give one example where the feasible set is empty. For the case where the feasible set is non-empty, identify the component-wise minimum power vector.
3. For the example in part (2) where the feasible set is non-empty, start from an arbitrary initial point and run the power control algorithm described in Section 4.3.1 (and studied in detail in Exercise 4.8). Exhibit the trajectory of power updates and

how it converges to the component-wise minimum solution. (You can either do this by hand or use MATLAB.)

4. Now suppose there are two cells with two base-stations and each of the two users can be connected to either one of them, i.e. the users are in soft handoff. Extend parts (1) and (2) to this scenario.

5. Extend the iterative power control algorithm in part (3) to the soft handoff scenario and redo part (3).

6. For a general number of users, do you think that it is always true that, in the optimal solution, each user is always connected to the base-station to which it has the strongest channel gain? Explain.

Exercise 4.11 (Out-of-cell interference averaging) Consider a cellular system with two adjacent single-dimensional cells along a highway, each of length d. The base-stations are at the midpoint of their respective cell. Suppose there are K users in each cell, and the location of each user is uniformly and independently located in its cell. Users in cell i are power controlled to the base-station in cell i, and create interference at the base-station in the adjacent cell. The power attenuation is proportional to $r^{-\alpha}$ where r is the distance. The system bandwidth is W Hz and the $\mathcal{E}_b/I_0$ requirement of each user is β. You can assume that the background noise is small compared to the interference and that users are maintained orthogonal within a cell with the out-of-cell interference from each of the interferers spread across the entire bandwidth. (This is an approximate model for the OFDM system in the text.)

1. Outage occurs when the users are located such that the out-of-cell interference is too large. For a given outage probability p_{out}, give an approximate expression for the spectral efficiency of the system as a function of K, α and β.

2. What is the limiting spectral efficiency as K and W grow? How does this depend on α?

3. Plot the spectral efficiency as a function of K for $\alpha = 2$ and $\beta = 7\,\mathrm{dB}$. Is the spectral efficiency an increasing or decreasing function of K? What is the limiting value?

4. We have assumed orthogonal users within a cell. But in a CDMA system, there is intra-cell interference as well. Assuming that all users within a cell are perfectly power controlled at their base-station, repeat the analysis in the first three parts of the question. From your plots, what qualitative differences between the CDMA and orthogonal systems can you observe? Intuitively explain your observations. *Hint*: Consider first what happens when the number of users increases from $K = 1$ to $K = 2$.

Exercise 4.12 Consider the uplink of a single-cell CDMA system with N users active all the time. In the text we have assumed the received powers are controlled such that they are exactly equal to the target level needed to deliver the desired SINR requirement for each user. In practice, the received powers are controlled imperfectly due to various factors such as tracking errors and errors in the feedback links. Suppose that when the target received power level is P, the actual received power of user i is $\epsilon_i P$, where ϵ_i are i.i.d. random variables whose statistics do not depend on P. Experimental data and theoretical analysis suggest that a good model for ϵ_i is a log normal distribution, i.e., $\log(\epsilon_i)$ follows a Gaussian distribution with mean μ and variance σ^2.

1. Assuming there is no power constraint on the users, give an approximate expression for the achievable spectral efficiency (bits/s/Hz) to support N users for a given outage probability p_{out} and $\mathcal{E}_b/I_0$ requirement β for each user.

2. Plot this expression as a function of N for reasonable values of the parameters and compare this to the perfect power control case. Do you see any interference averaging effect?

3. How does this scenario differ from the users' activity averaging example considered in the text?

Exercise 4.13 In the downlink of a CDMA system, each users' signal is spread onto a pseudonoise sequence.[13] Uncoded BPSK modulation is used, with a processing gain of G. Soft handoff is performed by sending the same symbol to the mobile from multiple base-stations, the symbol being spread onto independently chosen pseudonoise sequences. The mobile receiver has knowledge of all the sequences used to spread the data intended for it as well as the channel gains and can detect the transmitted symbol in the optimal way. We ignore fading and assume an AWGN channel between the mobile and each of the base-stations.

1. Give an expression for the detection error probability for a mobile in soft handoff between two base-stations. You may need to make several simplifying assumptions here. Feel free to make them but state them explicitly.

2. Now consider a whole network where each mobile is already assigned to a set of base-stations among which it is in soft handoff. Formulate the power control problem to meet the error probability requirement for each mobile in the downlink.

Exercise 4.14 In this problem we consider the design of hopping patterns of neighboring cells in the OFDM system. Based on the design principles in Section 4.4.2, we want the hopping patterns to be Latin squares and further require these Latin squares to be orthogonal. Another way to express the orthogonality of a pair of Latin squares is the following. For the two Latin squares, the N_c^2 ordered pairs (n_1, n_2), where n_1 and n_2 are the entries (sub-carrier index) from the same position in the respective Latin squares, exhaust the N_c^2 possibilities, i.e., every ordered pair occurs exactly once.

1. Show that the $N_c - 1$ Latin squares constructed in Section 4.4.2 (denoted by R^a in (4.23)) are mutually orthogonal.

2. Show that there cannot be more than $N_c - 1$ mutually orthogonal Latin squares. You can learn more about Latin squares from a book on combinatorial theory such as [16].

Exercise 4.15 In this exercise we derive some insight into the PAPR of the uplink transmit signal in the OFDM system. The uplink signal is restricted to n of the N_c sub-carriers and the specific choice of n depends on the allocation and further hops from one OFDM symbol to the other. So, for concreteness, we assume that n divides N_c and assume that sub-carriers are uniformly separated. Let us take the carrier frequency to be f_c and the inter-sub-carrier spacing to be $1/T$ Hz. This means that the passband transmit signal over one OFDM symbol (of length T) is

$$s(t) = \Re\left[\frac{1}{\sqrt{N_c}} \sum_{i=0}^{n-1} \tilde{d}_i \exp\left(j2\pi \left(f_c + \frac{iN_c}{nT} \right)t \right) \right], \qquad t \in [0, T].$$

---

[13] Note that this is different from the downlink of IS-95, where each user is assigned an orthogonal sequence.

Here we have denoted $\tilde{d}_0, \ldots, \tilde{d}_{n-1}$ to be the data (constellation) symbols chosen according to the (coded) data bits. We also denote the product $f_c T$ by ζ, which is typically a very large number. For example, with carrier frequency $f_c = 2\,\text{GHz}$ and bandwidth $W = 1\,\text{MHz}$ with $N_c = 512$ tones, the length of the OFDM symbol is approximately $T = N_c/W$. Then ζ is of the order of 10^6.

1. What is the (average) power of $s(t)$ as a function of the data symbols $\tilde{d}_i$, $i = 0, \ldots, n-1$? In the uplink, the constellation is usually small in size (due to low SINR values and transmit power constraints). A typical example is equal energy constellation such as (Q)PSK. For this problem, we assume that the data symbols are uniform over the circle in the complex plane with unit radius. With this assumption, compute the average of the power of $s(t)$, averaged over the data symbols. We denote this average by P_{av}.

2. We define the peak power of the signal $s(t)$ as a function of the data symbols as the square of the largest absolute value $s(t)$ can take in the time interval $[0, T]$. We denote this by $PP(\tilde{\mathbf{d}})$, the peak power as a function of the data symbols $\mathbf{d}$. Observe that the peak power can be written in our notation as

$$PP(\tilde{\mathbf{d}}) = \max_{0 \leq t \leq 1} \left(\Re \left[\frac{1}{\sqrt{N_c}} \sum_{i=0}^{n-1} \tilde{d}_i \exp\left(j2\pi \left(\zeta + \frac{iN_c}{n} \right) t \right) \right] \right)^2.$$

The peak to average power ratio (PAPR) is the ratio of $PP(\tilde{\mathbf{d}})$ to P_{av}.

We would like to understand how $PP(\tilde{\mathbf{d}})$ behaves with the data symbols $\tilde{\mathbf{d}}$. Since ζ is a large number, $s(t)$ is wildly fluctuating with time and is rather hard to analyze in a clean way. To get some insight, let us take a look at the values of $s(t)$ at the sample times: $t = l/W, l = 0, \ldots, N_c - 1$:

$$s(l/W) = \Re[d[l] \exp(j2\pi\zeta l)],$$

where $(d[0], \ldots, d[N_c - 1])$ is the N_c point IDFT (see Figure 3.20) of the vector with ith component equal to

$$\begin{cases} \tilde{d}_l & \text{when } i = lN_c/n \text{ for integer } l, \\ 0 & \text{otherwise.} \end{cases}$$

The worst amplitude of $s(l/W)$ is equal to the amplitude of $d[l]$, so let us focus on $d[0], \ldots, d[N_c - 1]$. With the assumption that the data symbols $\tilde{\mathbf{d}}_0, \ldots, \tilde{\mathbf{d}}_{n-1}$ are uniformly distributed on the circle in the complex plane of radius $1/\sqrt{N_c}$, what can you say about the marginal distributions of $d[0], \ldots, d[N_c - 1]$? In particular, what happens to these marginal distributions as $n, N_c \to \infty$ with n/N_c equal to a non-zero constant? The random variable $|d[0]|^2/P_{av}$ can be viewed as a lower bound to the PAPR.

3. Thus, even though the constellation symbols were all of equal energy, the PAPR of the resultant time domain signal is quite large. In practice, we can tolerate some codewords having large PAPRs as long as the majority of the codewords (say a fraction equal to $1 - \eta$) have well-behaved PAPRs. Using the distribution

$|d[0]|^2/P_{av}$ for large n, N_c as a lower bound substitute for the PAPR, calculate $\theta(\eta)$ defined as

$$\mathbb{P}\left\{\frac{|d[0]|^2}{P_{av}} < \theta(\eta)\right\} = 1 - \eta.$$

Calculate $\theta(\eta)$ for $\eta = 0.05$. When the power amplifier bias is set to the average power times θ, then on the average 95% of the codewords do not get clipped. This large value of $\theta(\eta)$ is one of the main implementational obstacles to using OFDM in the uplink.

Exercise 4.16 Several techniques have been proposed to reduce the PAPR in OFDM transmissions. In this exercise, we take a look at a few of these.

1. A standard approach to reduce the large PAPR of OFDM signals is to restrict signals transmitted to those that have guaranteed small PAPRs. One approach is based on Golay's complementary sequences [48, 49, 50]. These sequences possess an extremely low PAPR of 2 but their rate rapidly approaches zero with the number of sub-carriers (in the binary case, there are roughly $n \log n$ Golay sequences of length n). A reading exercise is to go through [14] and [93] which first suggested the applicability of Golay sequences in multitone communication.

2. However, in many communication systems codes are designed to have maximal rate. For example, LDPC and Turbo codes operate very close to the Shannon limits on many channels (including the AWGN channel). Thus it is useful to have strategies that improve the PAPR behavior of *existing* code sets. In this context, [64] proposes the following interesting idea: Introduce fixed phase rotations, say $\theta_0, \ldots, \theta_{n-1}$, to each of the data symbols $\tilde{d}_0, \ldots, \tilde{d}_{n-1}$. The choice of these fixed rotations is made such that the overall PAPR behavior of the signal set (corresponding to the code set) is improved. Focusing on the worst case PAPR (the largest signal power at any time for any signal among the code set), [116] introduces a geometric viewpoint and a computationally efficient algorithm to find the good choice of phase rotations. This reading exercise takes you through [64] and [116] and introduces these developments.

3. The worst case PAPR may be too conservative in predicting the bias setting. As an alternative, one can allow large peaks to occur but they should do so with small probability. When a large peak does occur, the signal will not be faithfully reproduced by the power amplifier thereby introducing noise into the signal. Since communication systems are designed to tolerate a certain amount of noise, one can attempt to control the probability that peak values are exceeded and then ameliorate the effects of the additional noise through the error control codes. A probabilistic approach to reduce PAPR of existing codesets is proposed in [70]. The idea is to remove the worst (say half) of the codewords based on the PAPR performance. This reduces the code rate by a negligible amount but the probability (η) that a certain threshold is exceeded by the transmit signal can be reduced a lot (as small as η^2). Since the peak threshold requirement of the amplifiers is typically chosen so as to set this probability to a sufficiently small level, such a scheme will permit the threshold to be set lower. A reading exercise takes you through the unpublished manuscript [70] where a scheme that is specialized to OFDM systems is detailed.

5 Capacity of wireless channels

In the previous two chapters, we studied *specific* techniques for communication over wireless channels. In particular, Chapter 3 is centered on the point-to-point communication scenario and there the focus is on diversity as a way to mitigate the adverse effect of fading. Chapter 4 looks at cellular wireless networks as a whole and introduces several multiple access and interference management techniques.

The present chapter takes a more fundamental look at the problem of communication over wireless fading channels. We ask: what is the *optimal* performance achievable on a given channel and what are the techniques to achieve such optimal performance? We focus on the point-to-point scenario in this chapter and defer the multiuser case until Chapter 6. The material covered in this chapter lays down the theoretical basis of the modern development in wireless communication to be covered in the rest of the book.

The framework for studying performance limits in communication is *information theory*. The basic measure of performance is the *capacity* of a channel: the maximum rate of communication for which arbitrarily small error probability can be achieved. Section 5.1 starts with the important example of the AWGN (additive white Gaussian noise) channel and introduces the notion of capacity through a heuristic argument. The AWGN channel is then used as a building block to study the capacity of wireless fading channels. Unlike the AWGN channel, there is no single definition of capacity for fading channels that is applicable in all scenarios. Several notions of capacity are developed, and together they form a systematic study of performance limits of fading channels. The various capacity measures allow us to see clearly the different types of resources available in fading channels: power, diversity and degrees of freedom. We will see how the diversity techniques studied in Chapter 3 fit into this big picture. More importantly, the capacity results suggest an alternative technique, *opportunistic communication*, which will be explored further in the later chapters.

5.1 AWGN channel capacity

Information theory was invented by Claude Shannon in 1948 to characterize the limits of reliable communication. Before Shannon, it was widely believed that the only way to achieve reliable communication over a noisy channel, i.e., to make the error probability as small as desired, was to reduce the data rate (by, say, repetition coding). Shannon showed the surprising result that this belief is incorrect: by more intelligent coding of the information, one can in fact communicate at a strictly *positive* rate but at the same time with as small an error probability as desired. However, there is a maximal rate, called the *capacity* of the channel, for which this can be done: if one attempts to communicate at rates above the channel capacity, then it is *impossible* to drive the error probability to zero.

In this section, the focus is on the familiar (real) AWGN channel:

$$y[m] = x[m] + w[m], \tag{5.1}$$

where $x[m]$ and $y[m]$ are real input and output at time m respectively and $w[m]$ is $\mathcal{N}(0, \sigma^2)$ noise, independent over time. The importance of this channel is two-fold:

- It is a building block of all of the wireless channels studied in this book.
- It serves as a motivating example of what capacity means operationally and gives some sense as to why arbitrarily reliable communication is possible at a strictly positive data rate.

5.1.1 Repetition coding

Using uncoded BPSK symbols $x[m] = \pm\sqrt{P}$, the error probability is $Q\left(\sqrt{P/\sigma^2}\right)$. To reduce the error probability, one can repeat the same symbol N times to transmit the one bit of information. This is a repetition code of block length N, with codewords $\mathbf{x}_A = \sqrt{P}[1, \ldots, 1]^t$ and $\mathbf{x}_B = \sqrt{P}[-1, \ldots, -1]^t$. The codewords meet a power constraint of P joules/symbol. If $\mathbf{x}_A$ is transmitted, the received vector is

$$\mathbf{y} = \mathbf{x}_A + \mathbf{w}, \tag{5.2}$$

where $\mathbf{w} = (w[1], \ldots, w[N])^t$. Error occurs when $\mathbf{y}$ is closer to $\mathbf{x}_B$ than to $\mathbf{x}_A$, and the error probability is given by

$$Q\left(\frac{\|\mathbf{x}_A - \mathbf{x}_B\|}{2\sigma}\right) = Q\left(\sqrt{\frac{NP}{\sigma^2}}\right), \tag{5.3}$$

which decays exponentially with the block length N. The good news is that communication can now be done with arbitrary reliability by choosing a large enough N. The bad news is that the data rate is only $1/N$ bits per symbol time and with increasing N the data rate goes to zero.

The reliably communicated data rate with repetition coding can be marginally improved by using multilevel PAM (generalizing the two-level BPSK scheme from earlier). By repeating an M-level PAM symbol, the levels equally spaced between $\pm\sqrt{P}$, the rate is $\log M/N$ bits per symbol time[1] and the error probability for the inner levels is equal to

$$Q\left(\frac{\sqrt{NP}}{(M-1)\sigma}\right). \tag{5.4}$$

As long as the number of levels M grows at a rate less than $\sqrt{N}$, reliable communication is guaranteed at large block lengths. But the data rate is bounded by $(\log\sqrt{N})/N$ and this still goes to zero as the block length increases. Is that the price one must pay to achieve reliable communication?

5.1.2 Packing spheres

Geometrically, repetition coding puts all the codewords (the M levels) in just one dimension (Figure 5.1 provides an illustration; here, all the codewords are on the same line). On the other hand, the signal space has a large number of dimensions N. We have already seen in Chapter 3 that this is a very inefficient way of packing codewords. To communicate more efficiently, the codewords should be spread in all the N dimensions.

We can get an estimate on the maximum number of codewords that can be packed in for the given power constraint P, by appealing to the classic sphere-packing picture (Figure 5.2). By the law of large numbers, the N-dimensional received vector $\mathbf{y} = \mathbf{x} + \mathbf{w}$ will, with high probability, lie within

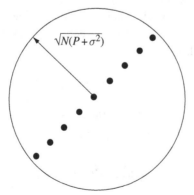

Figure 5.1 Repetition coding packs points inefficiently in the high-dimensional signal space.

---

[1] In this chapter, all logarithms are taken to be to the base 2 unless specified otherwise.

Figure 5.2 The number of noise spheres that can be packed into the y-sphere yields the maximum number of codewords that can be reliably distinguished.

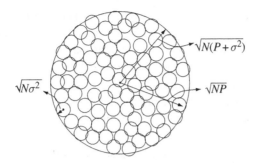

a y-sphere of radius $\sqrt{N(P+\sigma^2)}$; so without loss of generality we need only focus on what happens inside this y-sphere. On the other hand

$$\frac{1}{N} \sum_{m=1}^{N} w^2[m] \to \sigma^2 \qquad (5.5)$$

as $N \to \infty$, by the law of large numbers again. So, for N large, the received vector **y** lies, with high probability, near the surface of a *noise sphere* of radius $\sqrt{N}\sigma$ around the transmitted codeword (this is sometimes called the *sphere hardening* effect). Reliable communication occurs as long as the noise spheres around the codewords do not overlap. The maximum number of codewords that can be packed with non-overlapping noise spheres is the ratio of the volume of the y-sphere to the volume of a noise sphere:[2]

$$\frac{\left(\sqrt{N(P+\sigma^2)}\right)^N}{\left(\sqrt{N\sigma^2}\right)^N}. \qquad (5.6)$$

This implies that the maximum number of bits per symbol that can be reliably communicated is

$$\frac{1}{N} \log \left(\frac{\left(\sqrt{N(P+\sigma^2)}\right)^N}{\left(\sqrt{N\sigma^2}\right)^N} \right) = \frac{1}{2} \log \left(1 + \frac{P}{\sigma^2}\right). \qquad (5.7)$$

This is indeed the capacity of the AWGN channel. (The argument might sound very heuristic. Appendix B.5 takes a more careful look.)

The sphere-packing argument only yields the maximum number of codewords that can be packed while ensuring reliable communication. How to construct codes to achieve the promised rate is another story. In fact, in Shannon's argument, he never explicitly constructed codes. What he showed is that if

---

[2] The volume of an N-dimensional sphere of radius r is proportional to r^N and an exact expression is evaluated in Exercise B.10.

one picks the codewords randomly and independently, with the components of each codeword i.i.d. $\mathcal{N}(0, P)$, then with very high probability the randomly chosen code will do the job at any rate $R < C$. This is the so-called *i.i.d. Gaussian code*. A sketch of this random coding argument can be found in Appendix B.5.

From an engineering standpoint, the essential problem is to identify easily encodable and decodable codes that have performance close to the capacity. The study of this problem is a separate field in itself and Discussion 5.1 briefly chronicles the success story: codes that operate very close to capacity have been found and can be implemented in a relatively straightforward way using current technology. In the rest of the book, these codes are referred to as "capacity-achieving AWGN codes".

Discussion 5.1 Capacity-achieving AWGN channel codes

Consider a code for communication over the real AWGN channel in (5.1). The ML decoder chooses the nearest codeword to the received vector as the most likely transmitted codeword. The closer two codewords are to each other, the higher the probability of confusing one for the other: this yields a geometric design criterion for the set of codewords, i.e., place the codewords as far apart from each other as possible. While such a set of maximally spaced codewords are likely to perform very well, this in itself does not constitute an *engineering* solution to the problem of code construction: what is required is an arrangement that is "easy" to describe and "simple" to decode. In other words, the computational complexity of encoding and decoding should be practical.

Many of the early solutions centered around the theme of ensuring efficient ML decoding. The search of codes that have this property leads to a rich class of codes with nice algebraic properties, but their performance is quite far from capacity. A significant breakthrough occurred when the stringent ML decoding was relaxed to an *approximate* one. An iterative decoding algorithm with near ML performance has led to *turbo* and *low density parity check* codes.

A large ensemble of linear parity check codes can be considered in conjunction with the iterative decoding algorithm. Codes with good performance can be found offline and they have been verified to perform very close to capacity. To get a feel for their performance, we consider some sample performance numbers. The capacity of the AWGN channel at 0 dB SNR is 0.5 bits per symbol. The error probability of a carefully designed LDPC code in these operating conditions (rate 0.5 bits per symbol, and the signal-to-noise ratio is equal to 0.1 dB) with a block length of 8000 bits is approximately 10^{-4}. With a larger block length, much smaller error probabilities have been achieved. These modern developments are well surveyed in [100].

The capacity of the AWGN channel is probably the most well-known result of information theory, but it is in fact only a special case of Shannon's general theory applied to a specific channel. This general theory is outlined in Appendix B. All the capacity results used in the book can be derived from this general framework. To focus more on the implications of the results in the main text, the derivation of these results is relegated to Appendix B. In the main text, the capacities of the channels looked at are justified by either

---

Summary 5.1 Reliable rate of communication and capacity

- Reliable communication at rate R bits/symbol means that one can design codes at that rate with arbitrarily small error probability.
- To get reliable communication, one *must* code over a long block; this is to exploit the law of large numbers to average out the randomness of the noise.
- Repetition coding over a long block can achieve reliable communication, but the corresponding data rate goes to zero with increasing block length.
- Repetition coding does not pack the codewords in the available degrees of freedom in an efficient manner. One can pack a number of codewords that is exponential in the block length and still communicate reliably. This means the data rate can be strictly positive even as reliability is increased arbitrarily by increasing the block length.
- The maximum data rate at which reliable communication is possible is called the capacity C of the channel.
- The capacity of the (real) AWGN channel with power constraint P and noise variance σ^2 is:

$$C_{\text{awgn}} = \frac{1}{2} \log\left(1 + \frac{P}{\sigma^2}\right), \tag{5.8}$$

and the engineering problem of constructing codes close to this performance has been successfully addressed.

Figure 5.3 summarizes the three communication schemes discussed.

Figure 5.3 The three communication schemes when viewed in N-dimensional space: (a) uncoded signaling: error probability is poor since large noise in any dimension is enough to confuse the receiver; (b) repetition code: codewords are now separated in all dimensions, but there are only a few codewords packed in a single dimension; (c) capacity-achieving code: codewords are separated in all dimensions and there are many of them spread out in the space.

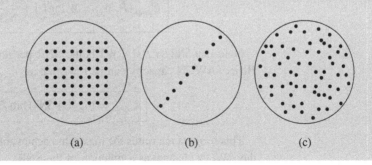

(a)   (b)   (c)

transforming the channels back to the AWGN channel, or by using the type of heuristic sphere-packing arguments we have just seen.

5.2 Resources of the AWGN channel

The AWGN capacity formula (5.8) can be used to identify the roles of the key resources of *power* and *bandwidth*.

5.2.1 Continuous-time AWGN channel

Consider a continuous-time AWGN channel with bandwidth W Hz, power constraint $\bar{P}$ watts, and additive white Gaussian noise with power spectral density $N_0/2$. Following the passband–baseband conversion and sampling at rate $1/W$ (as described in Chapter 2), this can be represented by a discrete-time complex baseband channel:

$$y[m] = x[m] + w[m], \tag{5.9}$$

where $w[m]$ is $\mathcal{CN}(0, N_0)$ and is i.i.d. over time. Note that since the noise is independent in the I and Q components, each use of the complex channel can be thought of as two independent uses of a real AWGN channel. The noise variance and the power constraint per real symbol are $N_0/2$ and $\bar{P}/(2W)$ respectively. Hence, the capacity of the channel is

$$\frac{1}{2} \log \left(1 + \frac{\bar{P}}{N_0 W} \right) \text{ bits per real dimension,} \tag{5.10}$$

or

$$\log \left(1 + \frac{\bar{P}}{N_0 W} \right) \text{ bits per complex dimension.} \tag{5.11}$$

This is the capacity in bits per complex dimension or degree of freedom. Since there are W complex samples per second, the capacity of the continuous-time AWGN channel is

$$\boxed{C_{\mathrm{awgn}}(\bar{P}, W) = W \log \left(1 + \frac{\bar{P}}{N_0 W} \right) \text{ bits/s.}} \tag{5.12}$$

Note that $\mathsf{SNR} := \bar{P}/(N_0 W)$ is the SNR per (complex) degree of freedom. Hence, AWGN capacity can be rewritten as

$$\boxed{C_{\mathrm{awgn}} = \log(1 + \mathsf{SNR}) \text{ bits/s/Hz.}} \tag{5.13}$$

This formula measures the maximum achievable *spectral efficiency* through the AWGN channel as a function of the SNR.

5.2.2 Power and bandwidth

Let us ponder the significance of the capacity formula (5.12) to a communication engineer. One way of using this formula is as a benchmark for evaluating the performance of channel codes. For a system engineer, however, the main significance of this formula is that it provides a high-level way of thinking about how the performance of a communication system depends on the basic *resources* available in the channel, without going into the details of specific modulation and coding schemes used. It will also help identify the bottleneck that limits performance.

The basic resources of the AWGN channel are the received power $\bar{P}$ and the bandwidth W. Let us first see how the capacity depends on the received power. To this end, a key observation is that the function

$$f(\mathsf{SNR}) := \log(1 + \mathsf{SNR}) \qquad (5.14)$$

is *concave*, i.e., $f''(x) \leq 0$ for all $x \geq 0$ (Figure 5.4). This means that increasing the power $\bar{P}$ suffers from a law of diminishing marginal returns: the higher the SNR, the smaller the effect on capacity. In particular, let us look at the low and the high SNR regimes. Observe that

$$\log_2(1 + x) \approx x \log_2 e \qquad \text{when } x \approx 0, \qquad (5.15)$$

$$\log_2(1 + x) \approx \log_2 x \qquad \text{when } x \gg 1. \qquad (5.16)$$

Thus, when the SNR is low, the capacity increases linearly with the received power $\bar{P}$: every 3 dB increase in (or, doubling) the power doubles the capacity. When the SNR is high, the capacity increases logarithmically with $\bar{P}$: every 3 dB increase in the power yields only one additional bit per dimension. This phenomenon should not come as a surprise. We have already seen in

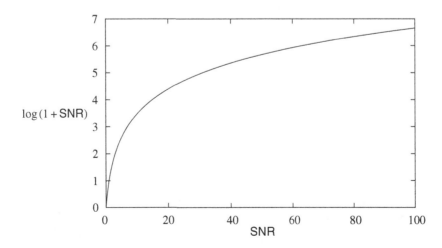

Figure 5.4 Spectral efficiency $\log(1 + \mathsf{SNR})$ of the AWGN channel.

Chapter 3 that packing many bits per dimension is very power-inefficient. The capacity result says that this phenomenon not only holds for specific schemes but is in fact fundamental to *all* communication schemes. In fact, for a fixed error probability, the data rate of uncoded QAM also increases logarithmically with the SNR (Exercise 5.7).

The dependency of the capacity on the bandwidth W is somewhat more complicated. From the formula, the capacity depends on the bandwidth in two ways. First, it increases the degrees of freedom available for communication. This can be seen in the linear dependency on W for a fixed $\mathsf{SNR} = \bar{P}/(N_0 W)$. On the other hand, for a given received power $\bar{P}$, the SNR per dimension decreases with the bandwidth as the energy is spread more thinly across the degrees of freedom. In fact, it can be directly calculated that the capacity is an increasing, concave function of the bandwidth W (Figure 5.5). When the bandwidth is small, the SNR per degree of freedom is high, and then the capacity is insensitive to small changes in SNR. Increasing W yields a rapid increase in capacity because the increase in degrees of freedom more than compensates for the decrease in SNR. The system is in the *bandwidth-limited* regime. When the bandwidth is large such that the SNR per degree of freedom is small,

$$W \log\left(1 + \frac{\bar{P}}{N_0 W}\right) \approx W\left(\frac{\bar{P}}{N_0 W}\right) \log_2 e = \frac{\bar{P}}{N_0} \log_2 e. \qquad (5.17)$$

In this regime, the capacity is proportional to the *total* received power across the entire band. It is insensitive to the bandwidth, and increasing the bandwidth has a small impact on capacity. On the other hand, the capacity is now linear in the received power and increasing power has a significant effect. This is the *power-limited* regime.

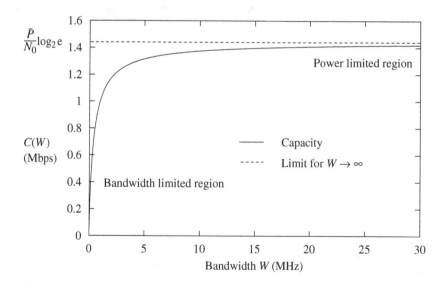

Figure 5.5 Capacity as a function of the bandwidth W. Here $\bar{P}/N_0 = 10^6$.

As W increases, the capacity increases monotonically (why must it?) and reaches the asymptotic limit

$$C_\infty = \frac{\bar{P}}{N_0} \log_2 e \text{ bits/s.} \tag{5.18}$$

This is the infinite bandwidth limit, i.e., the capacity of the AWGN channel with only a power constraint but no limitation on bandwidth. It is seen that even if there is no bandwidth constraint, the capacity is finite.

In some communication applications, the main objective is to minimize the required energy per bit $\mathcal{E}_b$ rather than to maximize the spectral efficiency. At a given power level $\bar{P}$, the minimum required energy per bit $\mathcal{E}_b$ is $\bar{P}/C_{awgn}(\bar{P}, W)$. To minimize this, we should be operating in the most power-efficient regime, i.e., $\bar{P} \to 0$. Hence, the minimum $\mathcal{E}_b/N_0$ is given by

$$\left(\frac{\mathcal{E}_b}{N_0}\right)_{min} = \lim_{\bar{P} \to 0} \frac{\bar{P}}{C_{awgn}(\bar{P}, W)N_0} = \frac{1}{\log_2 e} = -1.59 \, \text{dB.} \tag{5.19}$$

To achieve this, the SNR per degree of freedom goes to zero. The price to pay for the energy efficiency is *delay*: if the bandwidth W is fixed, the communication rate (in bits/s) goes to zero. This essentially mimics the infinite bandwidth regime by spreading the total energy over a long time interval, instead of spreading the total power over a large bandwidth.

It was already mentioned that the success story of designing capacity-achieving AWGN codes is a relatively recent one. In the infinite bandwidth regime, however, it has long been known that *orthogonal codes*[3] achieve the capacity (or, equivalently, achieve the minimum $\mathcal{E}_b/N_0$ of -1.59 dB). This is explored in Exercises 5.8 and 5.9.

Example 5.2 Bandwidth reuse in cellular systems

The capacity formula for the AWGN channel can be used to conduct a simple comparison of the two orthogonal cellular systems discussed in Chapter 4: the narrowband system with frequency reuse versus the wideband system with universal reuse. In both systems, users within a cell are orthogonal and do not interfere with each other. The main parameter of interest is the reuse ratio $\rho(\rho \leq 1)$. If W denotes the bandwidth per user within a cell, then each user transmission occurs over a bandwidth of ρW. The parameter $\rho = 1$ yields the full reuse of the wideband OFDM system and $\rho < 1$ yields the narrowband system.

---

[3] One example of orthogonal coding is the Hadamard sequences used in the IS-95 system (Section 4.3.1). Pulse position modulation (PPM), where the position of the on–off pulse (with large duty cycle) conveys the information, is another example.

Here we consider the uplink of this cellular system; the study of the downlink in orthogonal systems is similar. A user at a distance r is heard at the base-station with an attenuation of a factor $r^{-\alpha}$ in power; in free space the decay rate α is equal to 2 and the decay rate is 4 in the model of a single reflected path off the ground plane, cf. Section 2.1.5.

The uplink user transmissions in a neighboring cell that reuses the same frequency band are averaged and this constitutes the interference (this averaging is an important feature of the wideband OFDM system; in the narrowband system in Chapter 4, there is no interference averaging but that effect is ignored here). Let us denote by f_ρ the amount of total out-of-cell interference at a base-station as a *fraction* of the received signal power of a user at the edge of the cell. Since the amount of interference depends on the number of neighboring cells that reuse the same frequency band, the fraction f_ρ depends on the reuse ratio and also on the topology of the cellular system.

For example, in a one-dimensional linear array of base-stations (Figure 5.6), a reuse ratio of ρ corresponds to one in every $1/\rho$ cells using the same frequency band. Thus the fraction f_ρ decays roughly as ρ^α. On the other hand, in a two-dimensional hexagonal array of base-stations, a reuse ratio of ρ corresponds to the nearest reusing base-station roughly a distance of $\sqrt{1/\rho}$ away: this means that the fraction f_ρ decays roughly as $\rho^{\alpha/2}$. The exact fraction f_ρ takes into account geographical features of the cellular system (such as shadowing) and the geographic averaging of the interfering uplink transmissions; it is usually arrived at using numerical simulations (Table 6.2 in [140] has one such enumeration for a full reuse system). In a simple model where the interference is considered to come from the center of the cell reusing the same frequency band, f_ρ can be taken to be $2(\rho/2)^\alpha$ for the linear cellular system and $6(\rho/4)^{\alpha/2}$ for the hexagonal planar cellular system (see Exercises 5.2 and 5.3).

The received SINR at the base-station for a cell edge user is

$$\text{SINR} = \frac{\text{SNR}}{\rho + f_\rho \text{SNR}}, \tag{5.20}$$

where the SNR for the cell edge user is

$$\text{SNR} := \frac{P}{N_0 W d^\alpha}, \tag{5.21}$$

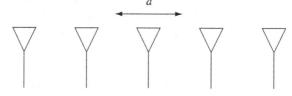

Figure 5.6 A linear cellular system with base-stations along a line (representing a highway).

with d the distance of the user to the base-station and P the uplink transmit power. The operating value of the parameter SNR is decided by the *coverage* of a cell: a user at the edge of a cell has to have a minimum SNR to be able to communicate reliably (at aleast a fixed minimum rate) with the nearest base-station. Each base-station comes with a capital installation cost and recurring operation costs and to minimize the number of base-stations, the cell size d is usually made as large as possible; depending on the uplink transmit power capability, coverage decides the cell size d.

Using the AWGN capacity formula (cf. (5.14)), the rate of reliable communication for a user at the edge of the cell, as a function of the reuse ratio ρ, is

$$R_\rho = \rho W \log_2(1 + \text{SINR}) = \rho W \log_2 \left(1 + \frac{\text{SNR}}{\rho + f_\rho \text{SNR}} \right) \text{bits/s}. \qquad (5.22)$$

The rate depends on the reuse ratio through the available degrees of freedom and the amount of out-of-cell interference. A large ρ increases the available bandwidth per cell but also increases the amount of out-of-cell interference. The formula (5.22) allows us to study the optimal reuse factor. At low SNR, the system is not degree of freedom limited *and* the interference is small relative to the noise; thus the rate is insensitive to the reuse factor and this can be verified directly from (5.22). On the other hand, at large SNR the interference grows as well and the SINR peaks at $1/f_\rho$. (A general rule of thumb in practice is to set SNR such that the interference is of the same order as the background noise; this will guarantee that the operating SINR is close to the largest value.) The largest rate is

$$\rho W \log_2 \left(1 + \frac{1}{f_\rho} \right). \qquad (5.23)$$

This rate goes to zero for small values of ρ; thus sparse reuse is not favored. It can be verified that universal reuse yields the largest rate in (5.23) for the hexagonal cellular system (Exercise 5.3). For the linear cellular model, the corresponding optimal reuse is $\rho = 1/2$, i.e., reusing the frequency every other cell (Exercise 5.5). The reduction in interference due to less reuse is more dramatic in the linear cellular system when compared to the hexagonal cellular system. This difference is highlighted in the optimal reuse ratios for the two systems at high SNR: universal reuse is preferred for the hexagonal cellular system while a reuse ratio of $1/2$ is preferred for the linear cellular system.

This comparison also holds for a range of SNR between the small and the large values: Figures 5.7 and 5.8 plot the rates in (5.22) for different reuse ratios for the linear and hexagonal cellular systems respectively. Here the power decay rate α is fixed to 3 and the rates are plotted as a function of the SNR for a user at the edge of the cell, cf. (5.21). In the

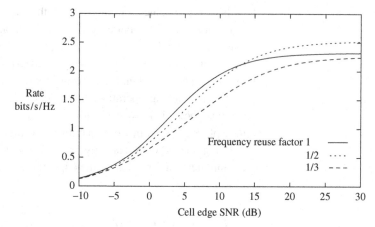

Figure 5.7 Rates in bits/s/Hz as a function of the SNR for a user at the edge of the cell for universal reuse and reuse ratios of 1/2 and 1/3 for the linear cellular system. The power decay rate α is set to 3.

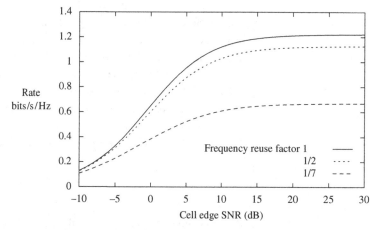

Figure 5.8 Rates in bits/s/Hz as a function of the SNR for a user at the edge of the cell for universal reuse, reuse ratios 1/2 and 1/7 for the hexagonal cellular system. The power decay rate α is set to 3.

hexagonal cellular system, universal reuse is clearly preferred at all ranges of SNR. On the other hand, in a linear cellular system, universal reuse and a reuse of 1/2 have comparable performance and if the operating SNR value is larger than a threshold (10 dB in Figure 5.7), then it pays to reuse, i.e., $R_{1/2} > R_1$. Otherwise, universal reuse is optimal. If this SNR threshold is within the rule of thumb setting mentioned earlier (i.e., the gain in rate is worth operating at this SNR), then reuse is preferred. This Preference has to be traded off with the size of the cell dictated by (5.21) due to a transmit power constraint on the mobile device.

5.3 Linear time-invariant Gaussian channels

We give three examples of channels which are closely related to the simple AWGN channel and whose capacities can be easily computed. Moreover, optimal codes for these channels can be constructed directly from an optimal code for the basic AWGN channel. These channels are *time-invariant*, known to both the transmitter and the receiver, and they form a bridge to the fading channels which will be studied in the next section.

5.3.1 Single input multiple output (SIMO) channel

Consider a SIMO channel with one transmit antenna and L receive antennas:

$$y_\ell[m] = h_\ell x[m] + w_\ell[m] \qquad \ell = 1, \ldots, L, \qquad (5.24)$$

where h_ℓ is the *fixed* complex channel gain from the transmit antenna to the ℓth receive antenna, and $w_\ell[m]$ is $\mathcal{CN}(0, N_0)$ is additive Gaussian noise independent across antennas. A sufficient statistic for detecting $x[m]$ from $\mathbf{y}[m] := [y_1[m], \ldots, y_L[m]]^t$ is

$$\tilde{y}[m] := \mathbf{h}^* \mathbf{y}[m] = \|\mathbf{h}\|^2 x[m] + \mathbf{h}^* \mathbf{w}[m], \qquad (5.25)$$

where $\mathbf{h} := [h_1, \ldots, h_L]^t$ and $\mathbf{w}[m] := [w_1[m], \ldots, w_L[m]]^t$. This is an AWGN channel with received SNR $P\|\mathbf{h}\|^2/N_0$ if P is the average energy per transmit symbol. The capacity of this channel is therefore

$$\boxed{C = \log\left(1 + \frac{P\|\mathbf{h}\|^2}{N_0}\right) \text{ bits/s/Hz.}} \qquad (5.26)$$

Multiple receive antennas increase the effective SNR and provide a *power gain*. For example, for $L = 2$ and $|h_1| = |h_2| = 1$, dual receive antennas provide a 3 dB power gain over a single antenna system. The linear combining (5.25) maximizes the output SNR and is sometimes called *receive beamforming*.

5.3.2 Multiple input single output (MISO) channel

Consider a MISO channel with L transmit antennas and a single receive antenna:

$$y[m] = \mathbf{h}^* \mathbf{x}[m] + w[m], \qquad (5.27)$$

where $\mathbf{h} = [h_1, \ldots, h_L]^t$ and h_ℓ is the (fixed) channel gain from transmit antenna ℓ to the receive antenna. There is a total power constraint of P across the transmit antennas.

In the SIMO channel above, the sufficient statistic is the projection of the L-dimensional received signal onto $\mathbf{h}$: the projections in orthogonal directions contain noise that is not helpful to the detection of the transmit signal. A natural reciprocal transmission strategy for the MISO channel would send information only in the direction of the channel vector $\mathbf{h}$; information sent in any orthogonal direction will be nulled out by the channel anyway. Therefore, by setting

$$\mathbf{x}[m] = \frac{\mathbf{h}}{\|\mathbf{h}\|} \tilde{x}[m], \tag{5.28}$$

the MISO channel is reduced to the scalar AWGN channel:

$$y[m] = \|\mathbf{h}\| \tilde{x}[m] + w[m], \tag{5.29}$$

with a power constraint P on the scalar input. The capacity of this scalar channel is

$$\log\left(1 + \frac{P\|\mathbf{h}\|^2}{N_0}\right) \text{ bits/s/Hz.} \tag{5.30}$$

Can one do better than this scheme? Any reliable code for the MISO channel can be used as a reliable code for the scalar AWGN channel $y[m] = x[m] + w[m]$: if $\{\mathbf{X}_i\}$ are the transmitted $L \times N$ (space-time) code matrices for the MISO channel, then the received $1 \times N$ vectors $\{\mathbf{h}^*\mathbf{X}_i\}$ form a code for the scalar AWGN channel. Hence, the rate achievable by a reliable code for the MISO channel must be at most the capacity of a scalar AWGN channel with the same received SNR. Exercise 5.11 shows that the received SNR, $P\|\mathbf{h}\|^2/N_0$, of the transmission strategy above is in fact the *largest* possible SNR given the transmit power constraint of P. Any other scheme has a lower received SNR and hence its reliable rate must be less than (5.30), the rate achieved by the proposed transmission strategy. We conclude that the capacity of the MISO channel is indeed

$$\boxed{C = \log\left(1 + \frac{P\|\mathbf{h}\|^2}{N_0}\right) \text{ bits/s/Hz.}} \tag{5.31}$$

Intuitively, the transmission strategy maximizes the received SNR by having the received signals from the various transmit antennas add up in-phase (coherently) and by allocating more power to the transmit antenna with the better gain. This strategy, "aligning the transmit signal in the direction of the transmit antenna array pattern", is called *transmit beamforming*. Through beamforming, the MISO channel is converted into a scalar AWGN channel and thus any code which is optimal for the AWGN channel can be used directly.

In both the SIMO and the MISO examples the benefit from having multiple antennas is a *power* gain. To get a gain in degrees of freedom, one has to use both multiple transmit and multiple receive antennas (MIMO). We will study this in depth in Chapter 7.

5.3.3 Frequency-selective channel

Transformation to a parallel channel

Consider a *time-invariant* L-tap frequency-selective AWGN channel:

$$y[m] = \sum_{\ell=0}^{L-1} h_\ell x[m-\ell] + w[m], \qquad (5.32)$$

with an average power constraint P on each input symbol. In Section 3.4.4, we saw that the frequency-selective channel can be converted into N_c independent sub-carriers by adding a cyclic prefix of length $L-1$ to a data vector of length N_c, cf. (3.137). Suppose this operation is repeated over *blocks* of data symbols (of length N_c each, along with the corresponding cyclic prefix of length $L-1$); see Figure 5.9. Then communication over the ith OFDM block can be written as

$$\tilde{y}_n[i] = \tilde{h}_n \tilde{d}_n[i] + \tilde{w}_n[i] \qquad n = 0, 1, \ldots, N_c - 1. \qquad (5.33)$$

Here,

$$\tilde{\mathbf{d}}[i] := [\tilde{d}_0[i], \ldots, \tilde{d}_{N_c-1}[i]]^t, \qquad (5.34)$$

$$\tilde{\mathbf{w}}[i] := [\tilde{w}_0[i], \ldots, \tilde{w}_{N_c-1}[i]]^t, \qquad (5.35)$$

$$\tilde{\mathbf{y}}[i] := [\tilde{y}_0[i], \ldots, \tilde{y}_{N_c-1}[i]]^t \qquad (5.36)$$

are the DFTs of the input, the noise and the output of the ith OFDM block respectively. $\tilde{\mathbf{h}}$ is the DFT of the channel scaled by $\sqrt{N_c}$ (cf. (3.138)). Since the overhead in the cyclic prefix relative to the block length N_c can be made arbitrarily small by choosing N_c large, the capacity of the original frequency-selective channel is the same as the capacity of this transformed channel as $N_c \to \infty$.

The transformed channel (5.33) can be viewed as a collection of sub-channels, one for each sub-carrier n. Each of the sub-channels is an AWGN channel. The

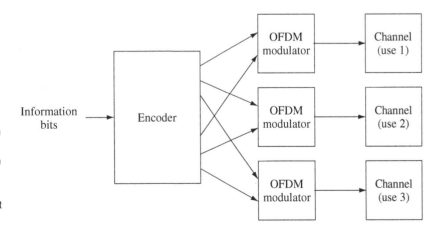

Figure 5.9 A coded OFDM system. Information bits are coded and then sent over the frequency-selective channel via OFDM modulation. Each channel use corresponds to an OFDM block. Coding can be done across different OFDM blocks as well as over different sub-carriers.

transformed noise $\tilde{\mathbf{w}}[i]$ is distributed as $\mathcal{CN}(0, N_0\mathbf{I})$, so the noise is $\mathcal{CN}(0, N_0)$ in each of the sub-channels and, moreover, the noise is independent across sub-channels. The power constraint on the input symbols in time translates to one on the data symbols on the sub-channels (Parseval theorem for DFTs):

$$\mathbb{E}\left[\|\tilde{\mathbf{d}}[i]\|^2\right] \le N_c P. \tag{5.37}$$

In information theory jargon, a channel which consists of a set of non-interfering sub-channels, each of which is corrupted by independent noise, is called a *parallel channel*. Thus, the transformed channel here is a parallel AWGN channel, with a total power constraint across the sub-channels. A natural strategy for reliable communication over a parallel AWGN channel is illustrated in Figure 5.10. We allocate power to each sub-channel, P_n to the nth sub-channel, such that the total power constraint is met. Then, a separate capacity-achieving AWGN code is used to communicate over each of the sub-channels. The maximum rate of reliable communication using this scheme is

$$\sum_{n=0}^{N_c-1} \log\left(1 + \frac{P_n|\tilde{h}_n|^2}{N_0}\right) \text{ bits/OFDM symbol.} \tag{5.38}$$

Further, the power allocation can be chosen appropriately, so as to maximize the rate in (5.38). The "optimal power allocation", thus, is the solution to the optimization problem:

$$\boxed{C_{N_c} := \max_{P_0,\ldots,P_{N_c-1}} \sum_{n=0}^{N_c-1} \log\left(1 + \frac{P_n|\tilde{h}_n|^2}{N_0}\right),} \tag{5.39}$$

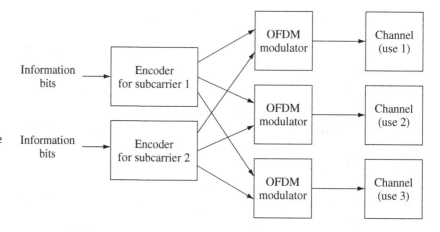

Figure 5.10 Coding independently over each of the sub-carriers. This architecture, with appropriate power and rate allocations, achieves the capacity of the frequency-selective channel.

subject to

$$\sum_{n=0}^{N_c-1} P_n = N_c P, \qquad P_n \geq 0, \quad n = 0, \ldots, N_c - 1. \qquad (5.40)$$

Waterfilling power allocation

The optimal power allocation can be explicitly found. The objective function in (5.39) is jointly concave in the powers and this optimization problem can be solved by Lagrangian methods. Consider the Lagrangian

$$\mathcal{L}(\lambda, P_0, \ldots, P_{N_c-1}) := \sum_{n=0}^{N_c-1} \log\left(1 + \frac{P_n|\tilde{h}_n|^2}{N_0}\right) - \lambda \sum_{n=0}^{N_c-1} P_n, \qquad (5.41)$$

where λ is the Lagrange multiplier. The Kuhn–Tucker condition for the optimality of a power allocation is

$$\frac{\partial \mathcal{L}}{\partial P_n} \begin{cases} = 0 & \text{if } P_n > 0 \\ \leq 0 & \text{if } P_n = 0. \end{cases} \qquad (5.42)$$

Define $x^+ := \max(x, 0)$. The power allocation

$$P_n^* = \left(\frac{1}{\lambda} - \frac{N_0}{|\tilde{h}_n|^2}\right)^+, \qquad (5.43)$$

satisfies the conditions in (5.42) and is therefore optimal, with the Lagrange multiplier λ chosen such that the power constraint is met:

$$\frac{1}{N_c} \sum_{n=0}^{N_c-1} \left(\frac{1}{\lambda} - \frac{N_0}{|\tilde{h}_n|^2}\right)^+ = P. \qquad (5.44)$$

Figure 5.11 gives a pictorial view of the optimal power allocation strategy for the OFDM system. Think of the values $N_0/|\tilde{h}_n|^2$ plotted as a function of the sub-carrier index $n = 0, \ldots, N_c - 1$, as tracing out the bottom of a vessel. If P units of water per sub-carrier are filled into the vessel, the depth of the water at sub-carrier n is the power allocated to that sub-carrier, and $1/\lambda$ is the height of the water surface. Thus, this optimal strategy is called *waterfilling* or *waterpouring*. Note that there are some sub-carriers where the bottom of the vessel is above the water and no power is allocated to them. In these sub-carriers, the channel is too poor for it to be worthwhile to transmit information. In general, the transmitter allocates more power to the stronger sub-carriers, taking advantage of the better channel conditions, and less or even no power to the weaker ones.

Figure 5.11 Waterfilling power allocation over the N_c sub-carriers.

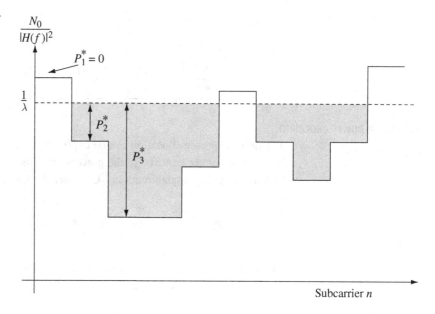

Observe that

$$\tilde{h}_n = \sum_{\ell=0}^{L-1} h_\ell \exp\left(-\frac{j2\pi\ell n}{N_c}\right), \tag{5.45}$$

is the discrete-time Fourier transform $H(f)$ evaluated at $f = nW/N_c$, where (cf. (2.20))

$$H(f) := \sum_{\ell=0}^{L-1} h_\ell \exp\left(-\frac{j2\pi\ell f}{W}\right), \qquad f \in [0, W]. \tag{5.46}$$

As the number of sub-carriers N_c grows, the frequency width W/N_c of the sub-carriers goes to zero and they represent a finer and finer sampling of the continuous spectrum. So, the optimal power allocation converges to

$$P^*(f) = \left(\frac{1}{\lambda} - \frac{N_0}{|H(f)|^2}\right)^+, \tag{5.47}$$

where the constant λ satisfies (cf. (5.44))

$$\int_0^W P^*(f)df = P. \tag{5.48}$$

The power allocation can be interpreted as waterfilling over frequency (see Figure 5.12). With N_c sub-carriers, the largest reliable communication rate

Figure 5.12 Waterfilling power allocation over the frequency spectrum of the two-tap channel (high-pass filter): $h[0] = 1$ and $h[1] = 0.5$.

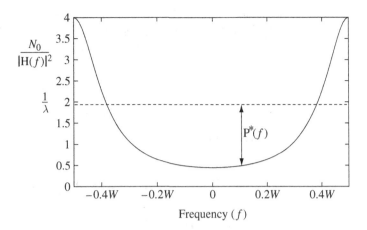

with independent coding is C_{N_c} bits per OFDM symbol or C_{N_c}/N_c bits/s/Hz (C_{N_c} given in (5.39)). So as $N_c \to \infty$, the WC_{N_c}/N_c converges to

$$C = \int_0^W \log\left(1 + \frac{P^*(f)|H(f)|^2}{N_0}\right) df \text{ bits/s.} \qquad (5.49)$$

Does coding across sub-carriers help?

So far we have considered a very simple scheme: coding independently over each of the sub-carriers. By coding *jointly* across the sub-carriers, presumably better performance can be achieved. Indeed, over a finite block length, coding jointly over the sub-carriers yields a smaller error probability than can be achieved by coding separately over the sub-carriers at the same rate. However, somewhat surprisingly, the *capacity* of the parallel channel is equal to the largest reliable rate of communication with independent coding within each sub-carrier. In other words, if the block length is very large then coding jointly over the sub-carriers cannot increase the rate of reliable communication any more than what can be achieved simply by allocating power and rate over the sub-carriers but not coding across the sub-carriers. So indeed (5.49) is the capacity of the time-invariant frequency-selective channel.

To get some insight into why coding *across* the sub-carriers with large block length does not improve capacity, we turn to a geometric view. Consider a code, with block length $N_c N$ symbols, coding over all N_c of the sub-carriers with N symbols from each sub-carrier. In high dimensions, i.e., $N \gg 1$, the $N_c N$-dimensional *received* vector after passing through the parallel channel (5.33) lives in an ellipsoid, with different axes stretched and shrunk by the different channel gains $\tilde{h}_n$. The volume of the ellipsoid is proportional to

$$\prod_{n=0}^{N_c-1} \left(|\tilde{h}_n|^2 P_n + N_0\right)^N, \qquad (5.50)$$

see Exercise 5.12. The volume of the noise sphere is, as in Section 5.1.2, proportional to $N_0^{N_c N}$. The maximum number of distinguishable codewords that can be packed in the ellipsoid is therefore

$$\prod_{n=0}^{N_c-1} \left(1 + \frac{P_n |\tilde{h}_n|^2}{N_0}\right)^N.$$ (5.51)

The maximum reliable rate of communication is

$$\frac{1}{N} \log \prod_{n=0}^{N_c-1} \left(1 + \frac{P_n |\tilde{h}_n|^2}{N_0}\right)^N = \sum_{n=0}^{N_c-1} \log\left(1 + \frac{P_n |\tilde{h}_n|^2}{N_0}\right) \text{ bits/OFDM symbol.}$$
(5.52)

This is precisely the rate (5.38) achieved by separate coding and this suggests that coding across sub-carriers can do no better. While this sphere-packing argument is heuristic, Appendix B.6 gives a rigorous derivation from information theoretic first principles.

Even though coding across sub-carriers cannot improve the reliable *rate* of communication, it can still improve the *error probability* for a given data rate. Thus, coding across sub-carriers can still be useful in practice, particularly when the block length for each sub-carrier is small, in which case the coding effectively increases the overall block length.

In this section we have used parallel channels to model a frequency-selective channel, but parallel channels will be seen to be very useful in modeling many other wireless communication scenarios as well.

5.4 Capacity of fading channels

The basic capacity results developed in the last few sections are now applied to analyze the limits to communication over wireless fading channels.

Consider the complex baseband representation of a flat fading channel:

$$y[m] = h[m]x[m] + w[m],$$ (5.53)

where $\{h[m]\}$ is the fading process and $\{w[m]\}$ is i.i.d. $\mathcal{CN}(0, N_0)$ noise. As before, the symbol rate is W Hz, there is a power constraint of P joules/symbol, and $\mathbb{E}[|h[m]|^2] = 1$ is assumed for normalization. Hence $\text{SNR} := P/N_0$ is the average received SNR.

In Section 3.1.2, we analyzed the performance of uncoded transmission for this channel. What is the ultimate performance limit when information can be coded over a sequence of symbols? To answer this question, we make the simplifying assumption that the receiver can perfectly track the fading process, i.e., coherent reception. As we discussed in Chapter 2, the coherence time of typical wireless channels is of the order of hundreds of symbols and so the channel varies slowly relative to the symbol rate and can be estimated by say a pilot signal. For now, the *transmitter* is not assumed to have any knowledge of the channel realization other than the statistical characterization. The situation when the transmitter has access to the channel realizations will be studied in Section 5.4.6.

5.4.1 Slow fading channel

Let us first look at the situation when the channel gain is random but remains constant for all time, i.e., $h[m] = h$ for all m. This models the *slow fading* situation where the delay requirement is short compared to the channel coherence time (cf. Table 2.2). This is also called the *quasi-static* scenario.

Conditional on a realization of the channel h, this is an AWGN channel with received signal-to-noise ratio $|h|^2\mathsf{SNR}$. The maximum rate of reliable communication supported by this channel is $\log(1 + |h|^2\mathsf{SNR})$ bits/s/Hz. This quantity is a function of the random channel gain h and is therefore random (Figure 5.13). Now suppose the transmitter encodes data at a rate R bits/s/Hz. If the channel realization h is such that $\log(1 + |h|^2\mathsf{SNR}) < R$, then whatever the code used by the transmitter, the decoding error probability cannot be made arbitrarily small. The system is said to be *in outage*, and the outage probability is

$$p_{\text{out}}(R) := \mathbb{P}\{\log(1 + |h|^2\mathsf{SNR}) < R\}. \qquad (5.54)$$

Thus, the best the transmitter can do is to encode the data assuming that the channel gain is strong enough to support the desired rate R. Reliable communication can be achieved whenever that happens, and outage occurs otherwise.

A more suggestive interpretation is to think of the channel as allowing $\log(1 + |h|^2\mathsf{SNR})$ bits/s/Hz of information through when the fading gain is h.

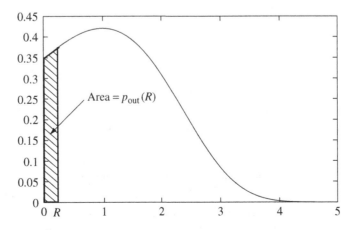

Figure 5.13 Density of $\log(1 + |h|^2\mathsf{SNR})$, for Rayleigh fading and $\mathsf{SNR} = 0$ dB. For any target rate R, there is a non-zero outage probability.

Reliable decoding is possible as long as this amount of information exceeds the target rate.

For Rayleigh fading (i.e., h is $\mathcal{CN}(0, 1)$), the outage probability is

$$p_{\text{out}}(R) = 1 - \exp\left(\frac{-(2^R - 1)}{\text{SNR}}\right). \tag{5.55}$$

At high SNR,

$$p_{\text{out}}(R) \approx \frac{(2^R - 1)}{\text{SNR}}, \tag{5.56}$$

and the outage probability decays as $1/\text{SNR}$. Recall that when we discussed uncoded transmission in Section 3.1.2, the detection error probability also decays like $1/\text{SNR}$. Thus, we see that coding *cannot* significantly improve the error probability in a slow fading scenario. The reason is that while coding can average out the Gaussian white noise, it cannot average out the channel fade, which affects *all* the coded symbols. Thus, deep fade, which is the typical error event in the uncoded case, is also the typical error event in the coded case.

There is a conceptual difference between the AWGN channel and the slow fading channel. In the former, one can send data at a positive rate (in fact, any rate less than C) while making the error probability as small as desired. This cannot be done for the slow fading channel as long as the probability that the channel is in deep fade is non-zero. Thus, the capacity of the slow fading channel in the strict sense is zero. An alternative performance measure is the ϵ-*outage capacity* C_ϵ. This is the largest rate of transmission R such that the outage probability $p_{\text{out}}(R)$ is less than ϵ. Solving $p_{\text{out}}(R) = \epsilon$ in (5.54) yields

$$\boxed{C_\epsilon = \log(1 + F^{-1}(1 - \epsilon)\,\text{SNR})\,\text{bits/s/Hz},} \tag{5.57}$$

where F is the complementary cumulative distribution function of $|h|^2$, i.e., $F(x) := \mathbb{P}\{|h|^2 > x\}$.

In Section 3.1.2, we looked at uncoded transmission and there it was natural to focus only on the high SNR regime; at low SNR, the error probability of uncoded transmission is very poor. On the other hand, for coded systems, it makes sense to consider both the high and the low SNR regimes. For example, the CDMA system in Chapter 4 operates at very low SINR and uses very low-rate orthogonal coding. A natural question is: in which regime does fading have a more significant impact on outage performance? One can answer this question in two ways. Eqn (5.57) says that, to achieve the same rate as the AWGN channel, an extra $10\log(1/F^{-1}(1 - \epsilon))$ dB of power is needed. This is true regardless of the operating SNR of the environment. Thus the *fade margin* is the same at all SNRs. If we look at the outage capacity at a *given* SNR, however, the impact of fading depends very much on the operating regime. To get a sense, Figure 5.14 plots the ϵ-outage capacity as

Figure 5.14 ϵ-outage capacity as a fraction of AWGN capacity under Rayleigh fading, for $\epsilon = 0.1$ and $\epsilon = 0.01$.

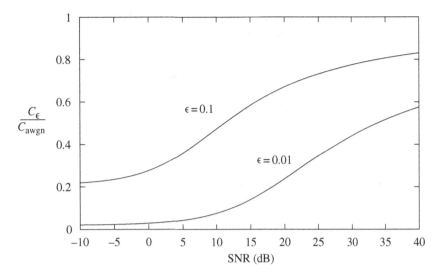

a function of SNR for the Rayleigh fading channel. To assess the impact of fading, the ϵ-outage capacity is plotted as a fraction of the AWGN capacity at the same SNR. It is clear that the impact is much more significant in the low SNR regime. Indeed, at high SNR,

$$C_\epsilon \approx \log \mathsf{SNR} + \log(F^{-1}(1-\epsilon)) \tag{5.58}$$

$$\approx C_{\mathrm{awgn}} - \log\left(\frac{1}{F^{-1}(1-\epsilon)}\right), \tag{5.59}$$

a constant *difference* irrespective of the SNR. Thus, the relative loss gets smaller at high SNR. At low SNR, on the other hand,

$$C_\epsilon \approx F^{-1}(1-\epsilon)\mathsf{SNR}\log_2 \mathrm{e} \tag{5.60}$$

$$\approx F^{-1}(1-\epsilon)C_{\mathrm{awgn}}. \tag{5.61}$$

For reasonably small outage probabilities, the outage capacity is only a small fraction of the AWGN capacity at low SNR. For Rayleigh fading, $F^{-1}(1-\epsilon) \approx \epsilon$ for small ϵ and the impact of fading is very significant. At an outage probability of 0.01, the outage capacity is only 1% of the AWGN capacity! Diversity has a significant effect at high SNR (as already seen in Chapter 3), but can be more important at low SNR. Intuitively, the impact of the randomness of the channel is in the received SNR, and the reliable rate supported by the AWGN channel is much more sensitive to the received SNR at low SNR than at high SNR. Exercise 5.10 elaborates on this point.

5.4.2 Receive diversity

Let us increase the diversity of the channel by having L receive antennas instead of one. For given channel gains $\mathbf{h} := [h_1, \ldots, h_L]'$, the capacity was calculated in Section 5.3.1 to be $\log(1 + \|\mathbf{h}\|^2 \text{SNR})$. Outage occurs whenever this is below the target rate R:

$$\boxed{p_{\text{out}}^{\text{rx}}(R) := \mathbb{P}\{\log(1 + \|\mathbf{h}\|^2 \text{SNR}) < R\}.} \tag{5.62}$$

This can be rewritten as

$$p_{\text{out}}(R) = \mathbb{P}\left\{\|\mathbf{h}\|^2 < \frac{2^R - 1}{\text{SNR}}\right\}. \tag{5.63}$$

Under independent Rayleigh fading, $\|\mathbf{h}\|^2$ is a sum of the squares of $2L$ independent Gaussian random variables and is distributed as Chi-square with $2L$ degrees of freedom. Its density is

$$f(x) = \frac{1}{(L-1)!} x^{L-1} e^{-x}, \qquad x \geq 0. \tag{5.64}$$

Approximating e^{-x} by 1 for x small, we have (cf. (3.44)),

$$\mathbb{P}\{\|\mathbf{h}\|^2 < \delta\} \approx \frac{1}{L!}\delta^L, \tag{5.65}$$

for δ small. Hence at high SNR the outage probability is given by

$$p_{\text{out}}(R) \approx \frac{(2^R - 1)^L}{L!\,\text{SNR}^L}. \tag{5.66}$$

Comparing with (5.55), we see a diversity gain of L: the outage probability now decays like $1/\text{SNR}^L$. This parallels the performance of uncoded transmission discussed in Section 3.3.1: thus, coding cannot increase the diversity gain.

The impact of receive diversity on the ϵ-outage capacity is plotted in Figure 5.15. The ϵ-outage capacity is given by (5.57) with F now the cumulative distribution function of $\|\mathbf{h}\|^2$. Receive antennas yield a diversity gain and an L-fold power gain. To emphasize the impact of the diversity gain, let us normalize the outage capacity C_ϵ by $C_{\text{awgn}} = \log(1 + L\text{SNR})$. The dramatic salutary effect of diversity on outage capacity can now be seen. At low SNR and small ϵ, (5.61) and (5.65) yield

$$C_\epsilon \approx F^{-1}(1-\epsilon)\text{SNR}\log_2 e \tag{5.67}$$

$$\approx (L!)^{\frac{1}{L}}(\epsilon)^{\frac{1}{L}}\text{SNR}\log_2 e \text{ bits/s/Hz} \tag{5.68}$$

and the loss with respect to the AWGN capacity is by a factor of $\epsilon^{1/L}$ rather than by ϵ when there is no diversity. At $\epsilon = 0.01$ and $L = 2$, the outage capacity is increased to 14% of the AWGN capacity (as opposed to 1% for $L = 1$).

Figure 5.15 ϵ-outage capacity with L-fold receive diversity, as a fraction of the AWGN capacity $\log(1 + L\mathsf{SNR})$, for $\epsilon = 0.01$ and different L.

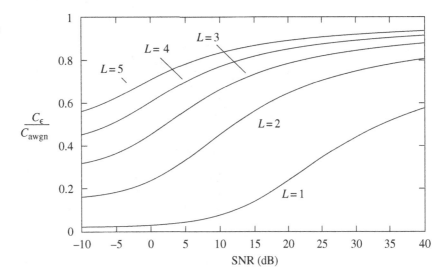

5.4.3 Transmit diversity

Now suppose there are L transmit antennas but only one receive antenna, with a total power constraint of P. From Section 5.3.2, the capacity of the channel conditioned on the channel gains $\mathbf{h} = [h_1, \ldots, h_L]^t$ is $\log(1 + \|\mathbf{h}\|^2 \mathsf{SNR})$. Following the approach taken in the SISO and the SIMO cases, one is tempted to say that the outage probability for a fixed rate R is

$$p_{\text{out}}^{\text{full-csi}}(R) = \mathbb{P}\{\log(1 + \|\mathbf{h}\|^2 \mathsf{SNR}) < R\}, \tag{5.69}$$

which would have been exactly the same as the corresponding SIMO system with 1 transmit and L receive antennas. However, this outage performance is achievable only if the transmitter knows the phases and magnitudes of the gains $\mathbf{h}$ so that it can perform transmit beamforming, i.e., allocate more power to the stronger antennas and arrange the signals from the different antennas to align in phase at the receiver. When the transmitter does not know the channel gains $\mathbf{h}$, it has to use a fixed transmission strategy that does not depend on $\mathbf{h}$. (This subtlety does not arise in either the SISO or the SIMO case because the transmitter need not know the channel realization to achieve the capacity for those channels.) How much performance loss does not knowing the channel entail?

Alamouti scheme revisited

For concreteness, let us focus on $L = 2$ (dual transmit antennas). In this situation, we can use the Alamouti scheme, which extracts transmit diversity without transmitter channel knowledge (introduced in Section 3.3.2). Recall from (3.76) that, under this scheme, both the transmitted symbols u_1, u_2 over a block of 2 symbol times see an equivalent scalar fading channel with gain $\|\mathbf{h}\|$

equivalent scalar channel

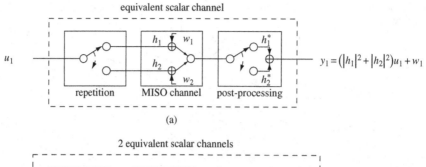

(a)

2 equivalent scalar channels

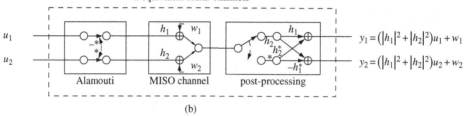

(b)

Figure 5.16 A space-time coding scheme combined with the MISO channel can be viewed as an equivalent scalar channel: (a) repetition coding; (b) the Alamouti scheme. The outage probability of the scheme is the outage probability of the equivalent channel.

and additive noise $\mathcal{CN}(0, N_0)$ (Figure 5.16(b)). The energy in the symbols u_1 and u_2 is $P/2$. Conditioned on h_1, h_2, the capacity of the equivalent scalar channel is

$$\log\left(1 + \|\mathbf{h}\|^2 \frac{\mathsf{SNR}}{2}\right) \text{ bits/s/Hz}. \tag{5.70}$$

Thus, if we now consider successive blocks and use an AWGN capacity-achieving code of rate R over each of the streams $\{u_1[m]\}$ and $\{u_2[m]\}$ separately, then the outage probability of each stream is

$$p_{\text{out}}^{\text{Ala}}(R) = \mathbb{P}\left\{\log\left(1 + \|\mathbf{h}\|^2 \frac{\mathsf{SNR}}{2}\right) < R\right\}. \tag{5.71}$$

Compared to (5.69) when the transmitter knows the channel, the Alamouti scheme performs strictly worse: the loss is 3 dB in the received SNR. This can be explained in terms of the efficiency with which energy is transferred to the receiver. In the Alamouti scheme, the symbols sent at the two transmit antennas in each time are *independent* since they come from two separately coded streams. Each of them has power $P/2$. Hence, the total SNR at the receive antenna at any given time is

$$\left(|h_1|^2 + |h_2|^2\right) \frac{\mathsf{SNR}}{2}. \tag{5.72}$$

In contrast, when the transmitter knows the channel, the symbols transmitted at the two antennas are *completely correlated* in such a way that the signals add up in phase at the receive antenna and the SNR is now

$$\left(|h_1|^2 + |h_2|^2\right) \mathsf{SNR},$$

a 3-dB power gain over the independent case.[4] Intuitively, there is a power loss because, without channel knowledge, the transmitter is sending signals that have energy in *all* directions instead of focusing the energy in a specific direction. In fact, the Alamouti scheme radiates energy in a perfectly *isotropic* manner: the signal transmitted from the two antennas has the same energy when projected in any direction (Exercise 5.14).

A scheme radiates energy isotropically whenever the signals transmitted from the antennas are uncorrelated and have equal power (Exercise 5.14). Although the Alamouti scheme does not perform as well as transmit beamforming, it is optimal in one important sense: it has the *best* outage probability among all schemes that radiate energy isotropically. Indeed, any such scheme must have a received SNR equal to (5.72) and hence its outage performance must be no better than that of a scalar slow fading AWGN channel with that received SNR. But this is precisely the performance achieved by the Alamouti scheme.

Can one do even better by radiating energy in a non-isotropic manner (but in a way that does not depend on the random channel gains)? In other words, can one improve the outage probability by correlating the signals from the transmit antennas and/or allocating unequal powers on the antennas? The answer depends of course on the distribution of the gains h_1, h_2. If h_1, h_2 are i.i.d. Rayleigh, Exercise 5.15 shows, using symmetry considerations, that correlation never improves the outage performance, but it is not necessarily optimal to use all the transmit antennas. Exercise 5.16 shows that uniform power allocation across antennas is always optimal, but the number of antennas used depends on the operating SNR. For reasonable values of target outage probabilities, it is optimal to use all the antennas. This implies that in most cases of interest, *the Alamouti scheme has the optimal outage performance for the i.i.d. Rayleigh fading channel.*

What about for $L > 2$ transmit antennas? An information theoretic argument in Appendix B.8 shows (in a more general framework) that

$$p_{\text{out}}(R) = \mathbb{P}\left\{\log\left(1 + \|\mathbf{h}\|^2 \frac{\text{SNR}}{L}\right) < R\right\} \tag{5.73}$$

is achievable. This is the natural generalization of (5.71) and corresponds again to isotropic transmission of energy from the antennas. Again, Exercises 5.15 and 5.16 show that this strategy is optimal for the i.i.d. Rayleigh fading channel and for most target outage probabilities of interest. However, there is no natural generalization of the Alamouti scheme for a larger number of transmit antennas (cf. Exercise 3.17). We will return to the problem of outage-optimal code design for $L > 2$ in Chapter 9.

---

[4] The addition of two in-phase signals of equal power yields a sum signal that has double the amplitude and four times the power of each of the signals. In contrast, the addition of two independent signals of equal power only *doubles* the power.

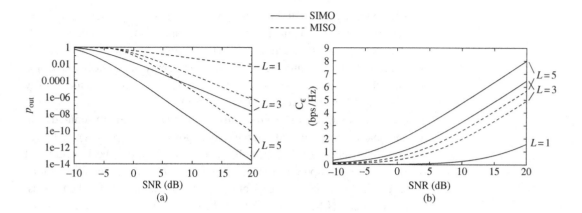

Figure 5.17 Comparison of outage performance between SIMO and MISO channels for different L: (a) outage probability as a function of SNR, for fixed $R = 1$; (b) outage capacity as a function of SNR, for a fixed outage probability of 10^{-2}.

The outage performances of the SIMO and the MISO channels with i.i.d. Rayleigh gains are plotted in Figure 5.17 for different numbers of transmit antennas. The difference in outage performance clearly outlines the asymmetry between receive and transmit antennas caused by the transmitter lacking knowledge of the channel.

Suboptimal schemes: repetition coding

In the above, the Alamouti scheme is viewed as an *inner code* that converts the MISO channel into a scalar channel. The outage performance (5.71) is achieved when the Alamouti scheme is used in conjunction with an *outer code* that is capacity-achieving for the scalar AWGN channel. Other space-time schemes can be similarly used as inner codes and their outage probability analyzed and compared to the channel outage performance.

Here we consider the simplest example, the repetition scheme: the same symbol is transmitted over the L different antennas over L symbol periods, using only one antenna at a time to transmit. The receiver does maximal ratio combining to demodulate each symbol. As a result, each symbol sees an equivalent scalar fading channel with gain $\|\mathbf{h}\|$ and noise variance N_0 (Figure 5.16(a)). Since only one symbol is transmitted every L symbol periods, a rate of LR bits/symbol is required on this scalar channel to achieve a target rate of R bits/symbol on the original channel. The outage probability of this scheme, when combined with an outer capacity-achieving code, is therefore:

$$p_{\text{out}}^{\text{rep}}(R) = \mathbb{P}\left\{ \frac{1}{L}\log(1 + \|\mathbf{h}\|^2 \text{SNR}) < R \right\}. \tag{5.74}$$

Compared to the outage probability (5.73) of the *channel*, this scheme is suboptimal: the SNR has to be increased by a factor of

$$\frac{2^{LR} - 1}{L(2^R - 1)}, \tag{5.75}$$

to achieve the same outage probability for the same target rate R. Equivalently, the reciprocal of this ratio can be interpreted as the maximum achievable *coding gain* over the simple repetition scheme. For a fixed R, the performance loss increases with L: the repetition scheme becomes increasingly inefficient in using the degrees of freedom of the channel. For a fixed L, the performance loss increases with the target rate R. On the other hand, for R small, $2^R - 1 \approx R \ln 2$ and $2^{RL} - 1 \approx RL \ln 2$, so

$$\frac{L(2^R - 1)}{2^{LR} - 1} \approx \frac{LR \ln 2}{LR \ln 2} = 1, \tag{5.76}$$

and there is hardly any loss in performance. Thus, while the repetition scheme is very suboptimal in the high SNR regime where the target rate can be high, it is nearly optimal in the low SNR regime. This is not surprising: the system is degree-of-freedom limited in the high SNR regime and the inefficiency of the repetition scheme is felt more there.

> ### Summary 5.2 Transmit and receive diversity
>
> With receive diversity, the outage probability is
>
> $$p_{\text{out}}^{\text{rx}}(R) := \mathbb{P}\{\log(1 + \|\mathbf{h}\|^2 \mathsf{SNR}) < R\}. \tag{5.77}$$
>
> With transmit diversity and isotropic transmission, the outage probability is
>
> $$p_{\text{out}}^{\text{tx}}(R) := \mathbb{P}\left\{\log\left(1 + \|\mathbf{h}\|^2 \frac{\mathsf{SNR}}{L}\right) < R\right\}, \tag{5.78}$$
>
> a loss of a factor of L in the received SNR because the transmitter has no knowledge of the channel direction and is unable to beamform in the specific channel direction.
>
> With two transmit antennas, capacity-achieving AWGN codes in conjunction with the Alamouti scheme achieve the outage probability.

5.4.4 Time and frequency diversity

Outage performance of parallel channels

Another way to increase channel diversity is to exploit the time-variation of the channel: in addition to coding over symbols within one coherence period, one can code over symbols from L such periods. Note that this is a generalization of the schemes considered in Section 3.2, which take *one* symbol from each coherence period. When coding can be performed over *many* symbols from each period, as well as between symbols from different periods, what is the performance limit?

One can model this situation using the idea of *parallel channels* introduced in Section 5.3.3: each of the sub-channels, $\ell = 1, \ldots, L$, represents a coherence period of duration T_c symbols:

$$y_\ell[m] = h_\ell x_\ell[m] + w_\ell[m], \qquad m = 1, \ldots, T_c. \tag{5.79}$$

Here h_ℓ is the (non-varying) channel gain during the ℓth coherence period. It is assumed that the coherence time T_c is large such that one can code over many symbols in each of the sub-channels. An average transmit power constraint of P on the original channel translates into a total power constraint of LP on the parallel channel.

For a given realization of the channel, we have already seen in Section 5.3.3 that the optimal power allocation across the sub-channels is *waterfilling*. However, since the transmitter does not know what the channel gains are, a reasonable strategy is to allocate equal power P to each of the sub-channels. In Section 5.3.3, it was mentioned that the maximum rate of reliable communication given the fading gains h_ℓ is

$$\sum_{\ell=1}^{L} \log(1 + |h_\ell|^2 \mathsf{SNR}) \text{ bits/s/Hz}, \tag{5.80}$$

where $\mathsf{SNR} = P/N_0$. Hence, if the target rate is R bits/s/Hz per sub-channel, then outage occurs when

$$\sum_{\ell=1}^{L} \log(1 + |h_\ell|^2 \mathsf{SNR}) < LR. \tag{5.81}$$

Can one design a code to communicate reliably whenever

$$\sum_{\ell=1}^{L} \log(1 + |h_\ell|^2 \mathsf{SNR}) > LR? \tag{5.82}$$

If so, an L-fold diversity is achieved for i.i.d. Rayleigh fading: outage occurs only if each of the terms in the sum $\sum_{\ell=1}^{L} \log(1 + |h_\ell|^2 \mathsf{SNR})$ is small.

The term $\log(1 + |h_\ell|^2 \mathsf{SNR})$ is the capacity of an AWGN channel with received SNR equal to $|h_\ell|^2 \mathsf{SNR}$. Hence, a seemingly straightforward strategy, already used in Section 5.3.3, would be to use a capacity-achieving AWGN code with rate

$$\log(1 + |h_\ell|^2 \mathsf{SNR})$$

for the ℓth coherence period, yielding an average rate of

$$\frac{1}{L} \sum_{\ell=1}^{L} \log(1 + |h_\ell|^2 \mathsf{SNR}) \text{ bits/s/Hz}$$

and meeting the target rate whenever condition (5.82) holds. The caveat is that this strategy requires the transmitter to know in advance the channel state during each of the coherence periods so that it can adapt the rate it allocates to each period. This knowledge is not available. *However, it turns out that such transmitter adaptation is unnecessary: information theory guarantees that one can design a single code that communicates reliably at rate R whenever the condition (5.82) is met.* Hence, the outage probability of the time diversity channel is precisely

$$p_{\text{out}}(R) = \mathbb{P}\left\{ \frac{1}{L}\sum_{\ell=1}^{L}\log(1+|h_\ell|^2\text{SNR}) < R \right\}. \qquad (5.83)$$

Even though this outage performance can be achieved with or without transmitter knowledge of the channel, the coding strategy is vastly different. With transmitter knowledge of the channel, dynamic rate allocation and separate coding for each sub-channel suffices. Without transmitter knowledge, separate coding would mean using a fixed-rate code for each sub-channel and poor diversity results: errors occur whenever one of the sub-channels is bad. Indeed, coding *across* the different coherence periods is now necessary: if the channel is in deep fade during one of the coherence periods, the information bits can still be protected if the channel is strong in other periods.

A geometric view

Figure 5.18 gives a geometric view of our discussion so far. Consider a code with rate R, coding over all the sub-channels and over one coherence time-interval; the block length is LT_c symbols. The codewords lie in an LT_c-dimensional sphere. The received LT_c-dimensional signal lives in an ellipsoid, with (L groups of) different axes stretched and shrunk by the different sub-channel gains (cf. Section 5.3.3). The ellipsoid is a function of the sub-channel gains, and hence random. The no-outage condition (5.82) has a geometric interpretation: it says that the volume of the ellipsoid is large enough to contain $2^{LT_c R}$ noise spheres, one for each codeword. (This was already seen in the sphere-packing argument in Section 5.3.3.) An outage-optimal code is one that communicates reliably whenever the random ellipsoid is at least this large. The subtlety here is that the *same* code must work for all such ellipsoids. Since the shrinking can occur in any of the L groups of dimensions, a robust code needs to have the property that the codewords are simultaneously well-separated in *each* of the sub-channels (Figure 5.18(a)). A set of independent codes, one for each sub-channel, is not robust: errors will be made when even only one of the sub-channels fades (Figure 5.18(b)).

We have already seen, in the simple context of Section 3.2, codes for the parallel channel which are designed to be well-separated in all the sub-channels. For example, the repetition code and the rotation code in Figure 3.8 have the property that the codewords are separated in both the sub-channels

Figure 5.18 Effect of the fading gains on codes for the parallel channel. Here there are $L = 2$ sub-channels and each axis represents T_c dimensions within a sub-channel. (a) Coding across the sub-channels. The code works as long as the volume of the ellipsoid is big enough. This requires good codeword separation in both the sub-channels. (b) Separate, non-adaptive code for each sub-channel. Shrinking of one of the axes is enough to cause confusion between the codewords.

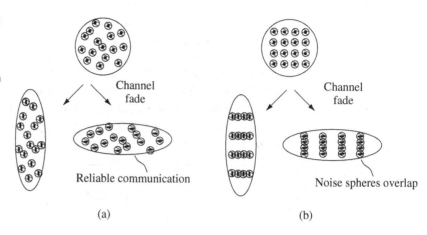

Channel fade

Reliable communication

Channel fade

Noise spheres overlap

(a)   (b)

(here $T_c = 1$ symbol and $L = 2$ sub-channels). More generally, the code design criterion of maximizing the product distance for all pairs of codewords naturally favors codes that satisfy this property. Coding over long blocks affords a larger coding gain; information theory guarantees the existence of codes with large enough coding gain to achieve the outage probability in (5.83).

To achieve the outage probability, one wants to design a code that communicates reliably over *every* parallel channel that is not in outage (i.e., parallel channels that satisfy (5.82)). In information theory jargon, a code that communicates reliably for a class of channels is said to be *universal* for that class. In this language, we are looking for universal codes for parallel channels that are not in outage. In the slow fading scalar channel without diversity ($L = 1$), this problem is the same as the code design problem for a *specific* channel. This is because all scalar channels are ordered by their received SNR; hence a code that works for the channel that is just strong enough to support the target rate will automatically work for all better channels. For parallel channels, each channel is described by a vector of channel gains and there is no natural ordering of channels; the universal code design problem is now non-trivial. In Chapter 9, a universal code design criterion will be developed to construct universal codes that come close to achieving the outage probability.

Extensions

In the above development, a uniform power allocation across the sub-channels is assumed. Instead, if we choose to allocate power P_ℓ to sub-channel ℓ, then the outage probability (5.83) generalizes to

$$p_{\text{out}}(R) = \mathbb{P}\left\{\sum_{\ell=1}^{L}\log(1 + |h_\ell|^2\text{SNR}_\ell) < LR\right\}, \qquad (5.84)$$

where $\text{SNR}_\ell = P_\ell/N_0$. Exercise 5.17 shows that for the i.i.d. Rayleigh fading model, a non-uniform power allocation that does not depend on the channel gains cannot improve the outage performance.

The parallel channel is used to model time diversity, but it can model frequency diversity as well. By using the usual OFDM transformation, a slow frequency-selective fading channel can be converted into a set of parallel sub-channels, one for each sub-carrier. This allows us to characterize the outage capacity of such channels as well (Exercise 5.22).

We summarize the key idea in this section using more suggestive language.

Summary 5.3 Outage for parallel channels

Outage probability for a parallel channel with L sub-channels and the ℓth channel having random gain h_ℓ:

$$p_{\text{out}}(R) = \mathbb{P}\left\{\frac{1}{L}\sum_{\ell=1}^{L}\log(1+|h_\ell|^2\text{SNR}) < R\right\}, \tag{5.85}$$

where R is in bits/s/Hz per sub-channel.

The ℓth sub-channel allows $\log(1+|h_\ell|^2\text{SNR})$ bits of information per symbol through. Reliable decoding can be achieved as long as the *total* amount of information allowed through exceeds the target rate.

5.4.5 Fast fading channel

In the slow fading scenario, the channel remains constant over the transmission duration of the codeword. If the codeword length spans several coherence periods, then time diversity is achieved and the outage probability improves. When the codeword length spans *many* coherence periods, we are in the so-called *fast fading* regime. How does one characterize the performance limit of such a fast fading channel?

Capacity derivation

Let us first consider a very simple model of a fast fading channel:

$$y[m] = h[m]x[m] + w[m], \tag{5.86}$$

where $h[m] = h_\ell$ remains constant over the ℓth coherence period of T_c symbols and is i.i.d. across different coherence periods. This is the so-called *block fading* model; see Figure 5.19(a). Suppose coding is done over L such coherence periods. If $T_c \gg 1$, we can effectively model this as L parallel sub-channels that fade independently. The outage probability from (5.83) is

$$p_{\text{out}}(R) = \mathbb{P}\left\{\frac{1}{L}\sum_{\ell=1}^{L}\log(1+|h_\ell|^2\text{SNR}) < R\right\}. \tag{5.87}$$

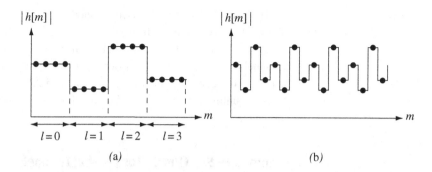

(a)   (b)

For finite L, the quantity

$$\frac{1}{L}\sum_{\ell=1}^{L}\log(1+|h_\ell|^2\mathsf{SNR})$$

is random and there is a non-zero probability that it will drop below any target rate R. Thus, there is no meaningful notion of capacity in the sense of maximum rate of arbitrarily reliable communication and we have to resort to the notion of outage. However, as $L \to \infty$, the law of large numbers says that

$$\frac{1}{L}\sum_{\ell=1}^{L}\log(1+|h_\ell|^2\mathsf{SNR}) \to \mathbb{E}[\log(1+|h|^2\mathsf{SNR})]. \qquad (5.88)$$

Now we can average over many independent fades of the channel by coding over a large number of coherence time intervals and a reliable rate of communication of $\mathbb{E}[\log(1+|h|^2\mathsf{SNR})]$ can indeed be achieved. In this situation, it is now meaningful to assign a positive capacity to the fast fading channel:

$$\boxed{C = \mathbb{E}[\log(1+|h|^2\mathsf{SNR})]\,\text{bits/s/Hz}} \qquad (5.89)$$

Impact of interleaving

In the above, we considered codes with block lengths LT_c symbols, where L is the number of coherence periods and T_c is the number of symbols in each coherence block. To approach the capacity of the fast fading channel, L has to be large. Since T_c is typically also a large number, the overall block length may become prohibitively large for implementation. In practice, shorter codes are used but they are interleaved so that the symbols of each codeword are spaced far apart in time and lie in different coherence periods. (Such interleaving is used for example in the IS-95 CDMA system, as illustrated in Figure 4.4.) Does interleaving impart a performance loss in terms of capacity?

Going back to the channel model (5.86), ideal interleaving can be modeled by assuming the $h[m]$ are now i.i.d., i.e., successive interleaved symbols go through independent fades. (See Figure 5.19(b).) In Appendix B.7.1, it is shown that for a large block length N and a given realization of the fading gains $h[1], \ldots, h[N]$, the maximum achievable rate through this interleaved channel is

$$\frac{1}{N} \sum_{m=1}^{N} \log(1 + |h[m]|^2 \mathsf{SNR}) \text{ bits/s/Hz.} \qquad (5.90)$$

By the law of large numbers,

$$\frac{1}{N} \sum_{m=1}^{N} \log(1 + |h[m]|^2 \mathsf{SNR}) \to \mathbb{E}[\log(1 + |h|^2 \mathsf{SNR})] \qquad (5.91)$$

as $N \to \infty$, for almost all realizations of the random channel gains. Thus, even with interleaving, the capacity (5.89) of the fast fading channel can be achieved. The important benefit of interleaving is that this capacity can now be achieved with a much shorter block length.

A closer examination of the above argument reveals why the capacity under interleaving (with $\{h[m]\}$ i.i.d.) and the capacity of the original block fading model (with $\{h[m]\}$ block-wise constant) are the same: the convergence in (5.91) holds for both fading processes, allowing the same long-term average rate through the channel. If one thinks of $\log(1 + |h[m]|^2 \mathsf{SNR})$ as the rate of information flow allowed through the channel at time m, the only difference is that in the block fading model, the rate of information flow is constant over each coherence period, while in the interleaved model, the rate varies from symbol to symbol. See Figure 5.19 again.

This observation suggests that the capacity result (5.89) holds for a much broader class of fading processes. Only the convergence in (5.91) is needed. This says that the time average should converge to the same limit for almost all realizations of the fading process, a concept called *ergodicity*, and it holds in many models. For example, it holds for the Gaussian fading model mentioned in Section 2.4. What matters from the point of view of capacity is only the long-term time average rate of flow allowed, and not on how fast that rate fluctuates over time.

Discussion

In the earlier parts of the chapter, we focused exclusively on deriving the capacities of *time-invariant* channels, particularly the AWGN channel. We have just shown that *time-varying* fading channels also have a well-defined capacity. However, the operational significance of capacity in the two cases is quite different. In the AWGN channel, information flows at a *constant* rate of $\log(1 + \mathsf{SNR})$ through the channel, and reliable communication can take place as long as the coding block length is large enough to average out the white Gaussian noise. The resulting coding/decoding delay is typically much smaller than the delay requirement of applications and this is not a big concern. In the fading channel, on the other hand, information flows at a *variable* rate of $\log(1 + |h[m]|^2\mathsf{SNR})$ due to variations of the channel strength; the coding block length now needs to be large enough to average out *both* the Gaussian noise *and* the fluctuations of the channel. To average out the latter, the coded symbols must span many coherence time periods, and this coding/decoding delay can be quite significant. Interleaving reduces the block length but not the coding/decoding delay: one still needs to wait many coherence periods before the bits get decoded. For applications that have a tight delay constraint relative to the channel coherence time, this notion of capacity is not meaningful, and one will suffer from outage.

The capacity expression (5.89) has the following interpretation. Consider a family of codes, one for each possible fading state h, and the code for state h achieves the capacity $\log(1 + |h|^2\mathsf{SNR})$ bits/s/Hz of the AWGN channel at the corresponding received SNR level. From these codes, we can build a variable-rate coding scheme that adaptively selects a code of appropriate rate depending on what the current channel condition is. This scheme would then have an average throughput of $\mathbb{E}[\log(1 + |h|^2\mathsf{SNR})]$ bits/s/Hz. For this variable-rate scheme to work, however, the transmitter needs to know the current channel state. The significance of the fast fading capacity result (5.89) is that one can communicate reliably at this rate even when the transmitter is blind and cannot track the channel.[5]

The nature of the information theoretic result that guarantees a code which achieves the capacity of the fast fading channel is similar to what we have already seen in the outage performance of the slow fading channel (cf. (5.83)). In fact, information theory guarantees that a fixed code with the rate in (5.89) is *universal* for the class of ergodic fading processes (i.e., (5.91) is satisfied with the same limiting value). This class of processes includes the AWGN channel (where the channel is fixed for all time) and, at the other extreme, the interleaved fast fading channel (where the channel varies i.i.d. over time). This suggests that capacity-achieving AWGN channel codes (cf. Discussion 5.1) could be suitable for the fast fading channel as well. While this is still an active research area, LDPC codes have been adapted successfully to the fast Rayleigh fading channel.

Performance comparison

Let us explore a few implications of the capacity result (5.89) by comparing it with that for the AWGN channel. The capacity of the fading channel is always less than that of the AWGN channel with the same SNR. This follows directly from Jensen's inequality, which says that if f is a strictly concave function and u is any random variable, then $\mathbb{E}[f(u)] \leq f(\mathbb{E}[u])$, with equality if and only if u is deterministic (Exercise B.2). Intuitively, the gain from

---

[5] Note however that if the transmitter can really track the channel, one can do even better than this rate. We will see this next in Section 5.4.6.

the times when the channel strength is above the average cannot compensate for the loss from the times when the channel strength is below the average. This again follows from the law of diminishing marginal return on capacity from increasing the received power.

At low SNR, the capacity of the fading channel is

$$C = \mathbb{E}[\log(1 + |h|^2\mathsf{SNR})] \approx \mathbb{E}[|h|^2\mathsf{SNR}]\log_2 e = \mathsf{SNR}\log_2 e \approx C_{\text{awgn}}, \quad (5.92)$$

where C_{awgn} is the capacity of the AWGN channel and is measured in bits per symbol. Hence at low SNR the "Jensen's loss" becomes negligible; this is because the capacity is approximately linear in the received SNR in this regime. At high SNR,

$$C \approx \mathbb{E}[\log(|h|^2\mathsf{SNR})] = \log\mathsf{SNR} + \mathbb{E}[\log|h|^2] \approx C_{\text{awgn}} + \mathbb{E}[\log|h|^2], \quad (5.93)$$

i.e., a constant difference with the AWGN capacity at high SNR. This difference is -0.83 bits/s/Hz for the Rayleigh fading channel. Equivalently, 2.5 dB more power is needed in the fading case to achieve the same capacity as in the AWGN case. Figure 5.20 compares the capacity of the Rayleigh fading channel with the AWGN capacity as a function of the SNR. The difference is not that large for the entire plotted range of SNR.

5.4.6 Transmitter side information

So far we have assumed that only the receiver can track the channel. But let us now consider the case when the *transmitter* can track the channel as well. There are several ways in which such channel information can be obtained at the transmitter. In a TDD (time-division duplex) system, the transmitter

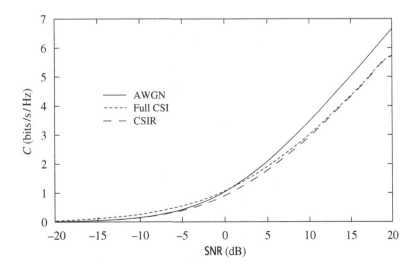

Figure 5.20 Plot of AWGN capacity, fading channel capacity with receiver tracking the channel only (CSIR) and capacity with both transmitter and the receiver tracking the channel (full CSI). (A discussion of the latter is in Section 5.4.6.)

can exploit channel reciprocity and make channel measurements based on the signal received along the opposite link. In an FDD (frequency-division duplex) system, there is no reciprocity and the transmitter will have to rely on feedback information from the receiver. For example, power control in the CDMA system implicitly conveys some channel state information through the feedback in the uplink.

Slow fading: channel inversion

When we discussed the slow fading channel in Section 5.4.1, it was seen that with no channel knowledge at the transmitter, outage occurs whenever the channel cannot support the target data rate R. With transmitter knowledge, one option is now to control the transmit power such that the rate R can be delivered no matter what the fading state is. This is the *channel inversion* strategy: the received SNR is kept constant irrespective of the channel gain. (This strategy is reminiscent of the power control used in CDMA systems, discussed in Section 4.3.) With exact channel inversion, there is zero outage probability. The price to pay is that huge power has to be consumed to invert the channel when it is very bad. Moreover, many systems are also peak-power constrained and cannot invert the channel beyond a certain point. Systems like IS-95 use a combination of channel inversion and diversity to achieve a target rate with reasonable power consumption (Exercise 5.24).

Fast fading: waterfilling

In the slow fading scenario, we are interested in achieving a target data rate within a coherence time period of the channel. In the fast fading case, one is now concerned with the rate averaged over *many* coherence time periods. With transmitter channel knowledge, what is the capacity of the fast fading channel? Let us again consider the simple block fading model (cf. (5.86)):

$$y[m] = h[m]x[m] + w[m], \qquad (5.94)$$

where $h[m] = h_\ell$ remains constant over the ℓ th coherence period of $T_c (T_c \gg 1)$ symbols and is i.i.d. across different coherence periods. The channel over L such coherence periods can be modeled as a *parallel channel* with L sub-channels that fade independently. For a given realization of the channel gains $h_1, \ldots, h_L$, the capacity (in bits/symbol) of this parallel channel is (cf. (5.39), (5.40) in Section 5.3.3)

$$\max_{P_1, \ldots, P_L} \frac{1}{L} \sum_{\ell=1}^{L} \log\left(1 + \frac{P_\ell |h_\ell|^2}{N_0}\right) \qquad (5.95)$$

subject to

$$\frac{1}{L} \sum_{\ell=1}^{L} P_\ell = P, \qquad (5.96)$$

where P is the average power constraint. It was seen (cf. (5.43)) that the optimal power allocation is *waterfilling*:

$$P_\ell^* = \left(\frac{1}{\lambda} - \frac{N_0}{|h_\ell|^2} \right)^+, \tag{5.97}$$

where λ satisfies

$$\frac{1}{L} \sum_{\ell=1}^{L} \left(\frac{1}{\lambda} - \frac{N_0}{|h_\ell|^2} \right)^+ = P. \tag{5.98}$$

In the context of the frequency-selective channel, waterfilling is done over the OFDM sub-carriers; here, waterfilling is done over time. In both cases, the basic problem is that of power allocation over a parallel channel.

The optimal power P_ℓ allocated to the ℓth coherence period depends on the channel gain in that coherence period and λ, which in turn depends on all the other channel gains through the constraint (5.98). So it seems that implementing this scheme would require knowledge of the future channel states. Fortunately, as $L \to \infty$, this non-causality requirement goes away. By the law of large numbers, (5.98) converges to

$$\boxed{\mathbb{E} \left[\left(\frac{1}{\lambda} - \frac{N_0}{|h|^2} \right)^+ \right] = P} \tag{5.99}$$

for almost all realizations of the fading process $\{h[m]\}$. Here, the expectation is taken with respect to the stationary distribution of the channel state. The parameter λ now converges to a constant, depending only on the channel *statistics* but not on the specific *realization* of the fading process. Hence, the optimal power at any time depends only on the channel gain h at that time:

$$\boxed{P^*(h) = \left(\frac{1}{\lambda} - \frac{N_0}{|h|^2} \right)^+.} \tag{5.100}$$

The capacity of the fast fading channel with transmitter channel knowledge is

$$\boxed{C = \mathbb{E} \left[\log \left(1 + \frac{P^*(h)|h|^2}{N_0} \right) \right] \text{ bits/s/Hz.}} \tag{5.101}$$

Equations (5.101), (5.100) and (5.99) together allow us to compute the capacity.

We have derived the capacity assuming the block fading model. The generalization to any ergodic fading process can be done exactly as in the case with no transmitter channel knowledge.

Discussion

Figure 5.21 gives a pictorial view of the waterfilling power allocation strategy. In general, the transmitter allocates more power when the channel is good, taking advantage of the better channel condition, and less or even no power when the channel is poor. This is precisely the opposite of the channel inversion strategy. Note that only the magnitude of the channel gain is needed to implement the waterfilling scheme. In particular, phase information is not required (in contrast to transmit beamforming, for example).

The derivation of the waterfilling capacity suggests a natural variable-rate coding scheme (see Figure 5.22). This scheme consists of a set of codes of different rates, one for each channel state h. When the channel is in state h, the code for that state is used. This can be done since both the transmitter and the receiver can track the channel. A transmit power of $P^*(h)$ is used when

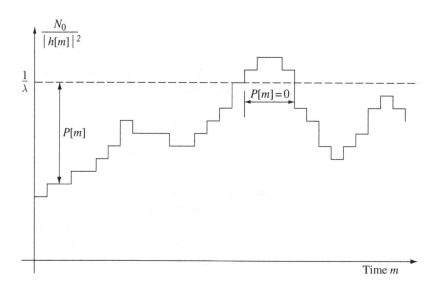

Figure 5.21 Pictorial representation of the waterfilling strategy.

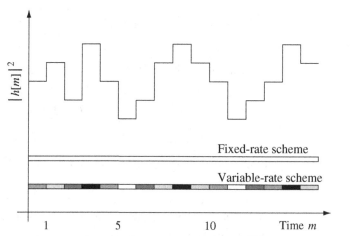

Figure 5.22 Comparison of the fixed-rate and variable-rate schemes. In the fixed-rate scheme, there is only one code spanning many coherence periods. In the variable-rate scheme, different codes (distinguished by different shades) are used depending on the channel quality at that time. For example, the code in white is a low-rate code used only when the channel is weak.

the channel gain is h. The rate of that code is therefore $\log(1 + P^*(h)|h|^2/N_0)$ bits/s/Hz. No coding across channel states is necessary. This is in contrast to the case without transmitter channel knowledge, where a single fixed-rate code with the coded symbols spanning across different coherence time periods is needed (Figure 5.22). Thus, knowledge of the channel state at the transmitter not only allows dynamic power allocation but simplifies the code design problem as one can now use codes designed for the AWGN channel.

Waterfilling performance

Figure 5.20 compares the waterfilling capacity and the capacity with channel knowledge only at the receiver, under Rayleigh fading. Figure 5.23 focuses on the low SNR regime. In the literature the former is also called the capacity with full channel side information (CSI) and the latter is called the capacity with channel side information at the receiver (CSIR). Several observations can be made:

- At low SNR, the capacity with full CSI is significantly larger than the CSIR capacity.
- At high SNR, the difference between the two goes to zero.
- Over a wide range of SNR, the gain of waterfilling over the CSIR capacity is very small.

The first two observations are in fact generic to a wide class of fading models, and can be explained by the fact that the benefit of dynamic power allocation is a *received power gain*: by spending more power when the channel is good, the received power gets boosted up. At high SNR, however, the capacity is insensitive to the received power per degree of freedom and varying the amount of transmit power as a function of the channel state yields a minimal gain (Figure 5.24(a)). At low SNR, the capacity is quite sensitive to the received power (linear, in fact) and so the boost in received power from optimal transmit power allocation provides significant gain. Thus, dynamic

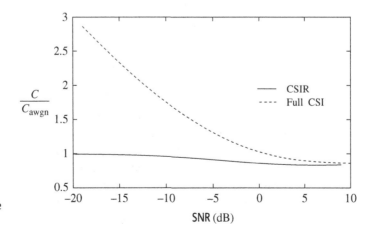

Figure 5.23 Plot of capacities with and without CSI at the transmitter, as a fraction of the AWGN capacity.

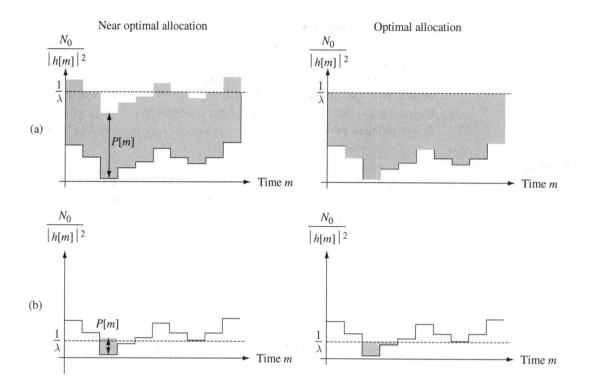

Figure 5.24 (a) High SNR: allocating equal powers at all times is almost optimal. (b) Low SNR: allocating all the power when the channel is strongest is almost optimal.

power allocation is more important in the power-limited (low SNR) regime than in the bandwidth-limited (high SNR) regime.

Let us look more carefully at the low SNR regime. Consider first the case when the channel gain $|h|^2$ has a peak value G_{max}. At low SNR, the waterfilling strategy transmits information only when the channel is very good, near G_{max}: when there is very little water, the water ends up at the bottom of the vessel (Figure 5.24(b)). Hence at low SNR

$$C \approx \mathbb{P}\left\{|h|^2 \approx G_{max}\right\} \log\left(1 + G_{max} \cdot \frac{\mathsf{SNR}}{\mathbb{P}\{|h|^2 \approx G_{max}\}}\right)$$

$$\approx G_{max} \cdot \mathsf{SNR}\log_2 \mathrm{e}\ \mathrm{bits/s/Hz}. \tag{5.102}$$

Recall that at low SNR the CSIR capacity is $\mathsf{SNR}\log_2 \mathrm{e}$ bits/s/Hz. Hence, transmitter CSI increases the capacity by G_{max} times, or a $10\log_{10} G_{max}$ dB gain. Moreover, since the AWGN capacity is the same as the CSIR capacity at low SNR, this leads to the interesting conclusion that with full CSI, *the capacity of the fading channel can be much larger than when there is no fading*. This is in contrast to the CSIR case where the fading channel capacity is always less than the capacity of the AWGN channel with the same average SNR. The gain is coming from the fact that in a fading channel, channel fluctuations create peaks and deep nulls, but when the energy per degree of freedom is small, the sender *opportunistically* transmits only when the channel is near its peak. In a non-fading AWGN channel, the channel stays constant at the average level and there are no peaks to take advantage of.

For models like Rayleigh fading, the channel gain is actually unbounded. Hence, theoretically, the gain of the fading channel waterfilling capacity over the AWGN channel capacity is also unbounded. (See Figure 5.23.) However, to get very large relative gains, one has to operate at very low SNR. In this regime, it may be difficult for the receiver to track and feed back the channel state to the transmitter to implement the waterfilling strategy.

Overall, the performance gain from full CSI is not that large compared to CSIR, unless the SNR is very low. On the other hand, full CSI potentially simplifies the code design problem, as no coding across channel states is necessary. In contrast, one has to interleave and code across many channel states with CSIR.

Waterfilling versus channel inversion

The capacity of the fading channel with full CSI (by using the waterfilling power allocation) should be interpreted as a long-term average rate of flow of information, averaged over the fluctuations of the channel. While the waterfilling strategy increases the long-term throughput of the system by transmitting when the channel is good, an important issue is the *delay* entailed. In this regard, it is interesting to contrast the waterfilling power allocation strategy with the channel inversion strategy. Compared to waterfilling, channel inversion is much less power-efficient, as a huge amount of power is consumed to invert the channel when it is bad. On the other hand, the rate of flow of information is now the same in all fading states, and so the associated delay is *independent* of the time-scale of channel variations. Thus, one can view the channel inversion strategy as a *delay-limited* power allocation strategy. Given an average power constraint, the maximum achievable rate by this strategy can be thought of as a *delay-limited* capacity. For applications with very tight delay constraints, this delay-limited capacity may be a more appropriate measure of performance than the waterfilling capacity.

Without diversity, the delay-limited capacity is typically very small. With increased diversity, the probability of encountering a bad channel is reduced and the average power consumption required to support a target delay-limited rate is reduced. Put another way, a larger delay-limited capacity is achieved for a given average power constraint (Exercise 5.24).

Example 5.3 Rate adaptation in IS-856

IS-856 downlink
IS-856, also called CDMA 2000 1× EV-DO (Enhanced Version Data Optimized) is a cellular data standard operating on the 1.25-MHz bandwidth.

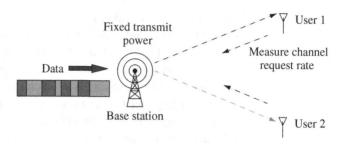

Figure 5.25 Downlink of IS-856 (CDMA 2000 1 x EV-DO). Users measure their channels based on the downlink pilot and feed back requested rates to the base-station. The base-station schedules users in a time-division manner.

The uplink is CDMA-based, not too different from IS-95, but the downlink is quite different (Figure 5.25):

- Multiple access is TDMA, with one user transmission at a time. The finest granularity for scheduling the user transmissions is a *slot* of duration 1.67 ms.

- Each user is *rate*-controlled rather than *power-* controlled. The transmit power at the base-station is fixed at all times and the rate of transmission to a user is adapted based on the current channel condition.

In contrast, the downlink of IS-95 (cf. Section 4.3.2) is CDMA-based, with the total power dynamically allocated among the users to meet their individual SIR requirements. The multiple access and scheduling aspects of IS-856 are discussed in Chapter 6; here the focus is only on rate adaptation.

Rate versus power control

The contrast between power control in IS-95 and rate control in IS-856 is roughly analogous to that between the channel inversion and the waterfilling strategies discussed above. In the former, power is allocated dynamically to a user to maintain a constant target rate at all times; this is suitable for voice, which has a stringent delay requirement and requires a consistent throughput. In the latter, rate is adapted to transmit more information when the channel is strong; this is suitable for data, which have a laxer delay requirement and can take better advantage of a variable transmission rate. The main difference between IS-856 and the waterfilling strategy is that there is no dynamic power adaptation in IS-856, only rate adaption.

Rate control in IS-856

Like IS-95, IS-856 is an FDD system. Hence, rate control has to be performed based on channel state feedback from the mobile to the base-station. The mobile measures its own channel based on a common strong pilot broadcast by the base-station. Using the measured values, the mobile predicts the SINR for the next time slot and uses that to predict the rate the base-station can send information to it. This *requested rate* is fed back to the base-station on the uplink. The transmitter then sends a packet at the requested rate to the mobile starting at the next time slot (if the mobile is scheduled). The table below describes the possible requested rates, the SINR thresholds for those rates, the modulation used and the number of time slots the transmission takes.

Requested rate (kbits/s)	SINR threshold (dB)	Modulation	Number of slots
38.4	−11.5	QPSK	16
76.8	−9.2	QPSK	8
153.6	−6.5	QPSK	4
307.2	−3.5	QPSK	2 or 4
614.4	−0.5	QPSK	1 or 2
921.6	2.2	8-PSK	2
1228.8	3.9	QPSK or 16-QAM	1 or 2
1843.2	8.0	8-PSK	1
2457.6	10.3	16-QAM	1

To simplify the implementation of the encoder, the codes at the different rates are all derived from a basic 1/5-rate turbo code. The low-rate codes are obtained by repeating the turbo-coded symbols over a number of time slots; as demonstrated in Exercise 5.25, such repetition loses little spectral efficiency in the low SNR regime. The higher-rate codes are obtained by using higher-order constellations in the modulation.

Rate control is made possible by the presence of the strong pilot to measure the channel and the rate request feedback from the mobile to the base-station. The pilot is shared between all users in the cell and is also used for many other functions such as coherent reception and synchronization. The rate request feedback is solely for the purpose of rate control. Although each request is only 4 bits long (to specify the various rate levels), this is sent by every active user at every slot and moreover considerable power and coding is needed to make sure the information gets fed back accurately and with little delay. Typically, sending this feedback consumes about 10% of the uplink capacity.

Impact of prediction uncertainty
Proper rate adaptation relies on the accurate tracking and prediction of the channel at the transmitter. This is possible only if the coherence time of the channel is much longer than the lag between the time the channel is measured at the mobile and the time when the packet is actually transmitted at the base-station. This lag is at least two slots ($2 \times 1.67 \, \text{ms}$) due to the delay in getting the requested rate fed back to the base-station, but can be considerably more at the low rates since the packet is transmitted over multiple slots and the predicted channel has to be valid during this time.

At a walking speed of 3 km/h and a carrier frequency $f_c = 1.9$ GHz, the coherence time is of the order of 25 ms, so the channel can be quite accurately predicted. At a driving speed of 30 km/h, the coherence time is only 2.5 ms and accurate tracking of the channel is already very difficult. (Exercise 5.26 explicitly connects the prediction error to the physical parameters of the channel.) At an even faster speed of 120 km/h, the coherence time is less than 1 ms and tracking of the channel is impossible; there is now no transmitter CSI. On the other hand, the multiple slot low rate packets essentially go through a fast fading channel with significant time diversity over the duration of the packet. Recall that the fast fading capacity is given by (5.89):

$$C = \mathbb{E}\left[\log\left(1 + |h|^2 \mathsf{SNR}\right)\right] \approx \mathbb{E}[|h|^2]\mathsf{SNR}\log_2 e \text{ bits/s/Hz} \qquad (5.103)$$

in the low SNR regime, where h follows the stationary distribution of the fading. Thus, to determine an appropriate transmission rate across this fast fading channel, it suffices for the mobile to predict the *average* SINR over the transmission time of the packet, and this average is quite easy to predict. Thus, the difficult regime is actually in between the very slow and very fast fading scenarios, where there is significant uncertainty in the channel prediction and yet not very much time diversity over the packet transmission time. This channel uncertainty has to be taken into account by being more conservative in predicting the SINR and in requesting a rate. This is similar to the outage scenario considered in Section 5.4.1, except that the randomness of the channel is conditional on the predicted value. The requested rate should be set to meet a target outage probability (Exercise 5.27).

The various situations are summarized in Figure 5.26. Note the different roles of coding in the three scenarios. In the first scenario, when the predicted SINR is accurate, the main role of coding is to combat the additive Gaussian noise; in the other two scenarios, coding combats the residual randomness in the channel by exploiting the available time diversity.

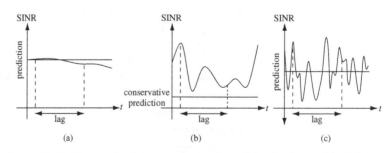

Figure 5.26 (a) Coherence time is long compared to the prediction time lag; predicted SINR is accurate. Near perfect CSI at transmitter. (b) Coherence time is comparable to the prediction time lag, predicted SINR has to be conservative to meet an outage criterion. (c) Coherence time is short compared to the prediction time lag; prediction of average SINR suffices. No CSI at the transmitter.

> To reduce the loss in performance due to the conservativeness of the channel prediction, IS-856 employs an *incremental* ARQ (or hybrid-ARQ) mechanism for the repetition-coded multiple slot packets. Instead of waiting until the end of the transmission of all slots before decoding, the mobile will attempt to decode the information incrementally as it receives the repeated copies over the time slots. When it succeeds in decoding, it will send an acknowledgement back to the base-station so that it can stop the transmission of the remaining slots. This way, a rate higher than the requested rate can be achieved if the actual SINR is higher than the predicted SINR.

5.4.7 Frequency-selective fading channels

So far, we have considered flat fading channels (cf. (5.53)). In Section 5.3.3, the capacity of the *time-invariant* frequency-selective channel (5.32) was also analyzed. It is simple to extend the understanding to *underspread* time-varying frequency-selective fading channels: these are channels with the coherence time much larger than the delay spread. We model the channel as a time-invariant L-tap channel as in (5.32) over each coherence time interval and view it as N_c parallel sub-channels (in frequency). For underspread channels, N_c can be chosen large so that the cyclic prefix loss is negligible. This model is a generalization of the flat fading channel in (5.53): here there are N_c (frequency) sub-channels over each coherence time interval and multiple (time) sub-channels over the different coherence time intervals. Overall it is still a parallel channel. We can extend the capacity results from Sections 5.4.5 and 5.4.6 to the frequency-selective fading channel. In particular, the fast fading capacity with full CSI (cf. Section 5.4.6) can be generalized here to a combination of waterfilling over time and frequency: the coherence time intervals provide sub-channels in time and each coherence time interval provides sub-channels in frequency. This is carried out in Exercise 5.30.

5.4.8 Summary: a shift in point of view

Let us summarize our investigation on the performance limits of fading channels. In the slow fading scenario without transmitter channel knowledge, the amount of information that is allowed through the channel is random, and no positive rate of communication can be reliably supported (in the sense of arbitrarily small error probability). The outage probability is the main performance measure, and it behaves like $1/\mathsf{SNR}$ at high SNR. This is due to a lack of diversity and, equivalently, the outage capacity is very small. With L branches of diversity, either over space, time or frequency, the outage probability is improved and decays like $1/\text{SNR}^L$. The fast fading scenario can be viewed as the limit of infinite time diversity and has a capacity of $\mathbb{E}[\log(1 + |h|^2\text{SNR})]$ bits/s/Hz. This however incurs a coding delay much longer than the coherence time of the channel. Finally, when the transmitter and the receiver can both track the channel, a further performance gain can be obtained by dynamically allocating power and opportunistically transmitting when the channel is good.

The slow fading scenario emphasizes the *detrimental* effect of fading: a slow fading channel is very unreliable. This unreliability is *mitigated* by providing more diversity in the channel. This is the traditional way of viewing the fading phenomenon and was the central theme of Chapter 3. In a narrowband channel with a single antenna, the only source of diversity is through time. The capacity of the fast fading channel (5.89) can be viewed as the performance limit of any such time diversity scheme. Still, the capacity is less than the AWGN channel capacity as long as there is no channel knowledge at the transmitter. With channel knowledge at the transmitter, the picture changes. Particularly at low SNR, the capacity of the fading channel with full CSI can be larger than that of the AWGN channel. Fading can be *exploited* by transmitting near the peak of the channel fluctuations. Channel fading is now turned from a foe to a friend.

This new theme on fading will be developed further in the multiuser context in Chapter 6, where we will see that opportunistic communication will have a significant impact at *all* SNRs, and not only at low SNR.

Chapter 5 The main plot

Channel capacity
The maximum rate at which information can be communicated across a noisy channel with arbitrary reliability.

Linear time-invariant Gaussian channels
Capacity of the AWGN channel with SNR per degree of freedom is

$$C_{\text{awgn}} = \log(1 + \text{SNR})\text{bits/s/Hz}. \tag{5.104}$$

Capacity of the continuous-time AWGN channel with bandwidth W, average received power $\bar{P}$ and white noise power spectral density N_0 is

$$C_{\text{awgn}} = W\log\left(1 + \frac{\bar{P}}{N_0 W}\right)\text{bits/s}. \tag{5.105}$$

Bandwidth-limited regime: $\text{SNR} = \bar{P}/(N_0 W)$ is high and capacity is logarithmic in the SNR.

Power-limited regime: SNR is low and capacity is linear in the SNR.

Capacities of the SIMO and the MISO channels with time-invariant channel gains $h_1, \ldots, h_L$ are the same:

$$C = \log(1 + \mathsf{SNR}\|\mathbf{h}\|^2) \text{ bits/s/Hz.} \qquad (5.106)$$

Capacity of frequency-selective channel with response $H(f)$ and power constraint P per degree of freedom:

$$C = \int_0^W \log\left(1 + \frac{P^*(f)|H(f)|^2}{N_0}\right) df \text{ bits/s} \qquad (5.107)$$

where $P^*(f)$ is waterfilling:

$$P^*(f) = \left(\frac{1}{\lambda} - \frac{N_0}{|H(f)|^2}\right)^+, \qquad (5.108)$$

and λ satisfies:

$$\int_0^W \left(\frac{1}{\lambda} - \frac{N_0}{|H(f)|^2}\right)^+ df = P. \qquad (5.109)$$

Slow fading channels with receiver CSI only
Setting: coherence time is much longer than constraint on coding delay.

Performance measures:

Outage probability $p_{\text{out}}(R)$ at a target rate R.

Outage capacity C_ϵ at a target outage probability ϵ.

Basic flat fading channel:

$$y[m] = hx[m] + w[m]. \qquad (5.110)$$

Outage probability is

$$p_{\text{out}}(R) = \mathbb{P}\left\{\log\left(1 + |h|^2 \mathsf{SNR}\right) < R\right\}, \qquad (5.111)$$

where SNR is the average signal-to-noise ratio at each receive antenna.

Outage probability with receive diversity is

$$p_{\text{out}}(R) := \mathbb{P}\left\{\log\left(1 + \|\mathbf{h}\|^2 \text{SNR}\right) < R\right\}. \tag{5.112}$$

This provides power and diversity gains.

Outage probability with L-fold transmit diversity is

$$p_{\text{out}}(R) := \mathbb{P}\left\{\log\left(1 + \|\mathbf{h}\|^2 \frac{\text{SNR}}{L}\right) < R\right\}. \tag{5.113}$$

This provides diversity gain only.

Outage probability with L-fold time diversity is

$$p_{\text{out}}(R) = \mathbb{P}\left\{\frac{1}{L}\sum_{\ell=1}^{L}\log\left(1 + |h_\ell|^2\text{SNR}\right) < R\right\}. \tag{5.114}$$

This provides diversity gain only.

Fast fading channels
Setting: coherence time is much shorter than coding delay.

Performance measure: capacity.

Basic model:

$$y[m] = h[m]x[m] + w[m]. \tag{5.115}$$

$\{h[m]\}$ is an ergodic fading process.

Receiver CSI only:

$$C = \mathbb{E}\left[\log\left(1 + |h|^2\text{SNR}\right)\right]. \tag{5.116}$$

Full CSI:

$$C = \mathbb{E}\left[\log\left(1 + \frac{P^*(h)|h|^2}{N_0}\right)\right] \text{ bits/s/Hz} \tag{5.117}$$

where $P^*(h)$ waterfills over the fading states:

$$P^*(h) = \left(\frac{1}{\lambda} - \frac{N_0}{|h|^2}\right)^+, \tag{5.118}$$

and λ satisfies:

$$\mathbb{E}\left[\left(\frac{1}{\lambda} - \frac{N_0}{|h|^2}\right)^+\right] = P. \tag{5.119}$$

Power gain over the receiver CSI only case. Significant at low SNR.

5.5 Bibliographical notes

Information theory and the formulation of the notions of reliable communication and channel capacity were introduced in a path-breaking paper by Shannon [109]. The underlying philosophy of using simple models to understand the essence of an engineering problem has pervaded the development of the communication field ever since. In that paper, as a consequence of his general theory, Shannon also derived the capacity of the AWGN channel. He returned to a more in-depth geometric treatment of this channel in a subsequent paper [110]. Sphere-packing arguments were used extensively in the text by Wozencraft and Jacobs [148].

The linear cellular model was introduced by Shamai and Wyner [108]. One of the early studies of wireless channels using information theoretic techniques is due to Ozarow. *et al.* [88], where they introduced the concept of outage capacity. Telatar [119] extended the formulation to multiple antennas. The capacity of fading channels with full CSI was analyzed by Goldsmith and Varaiya [51]. They observed the optimality of the waterfilling power allocation with full CSI and the corollary that full CSI over CSI at the receiver alone is beneficial only at low SNRs. A comprehensive survey of information theoretic results on fading channels was carried out by Biglieri, Proakis and Shamai [9].

The design issues in IS-856 have been elaborately discussed in Bender *et al.* [6] and by Wu and Esteves [149].

5.6 Exercises

Exercise 5.1 What is the maximum reliable rate of communication over the (complex) AWGN channel when only the I channel is used? How does that compare to the capacity of the complex channel at low and high SNR, with the same average power constraint? Relate your conclusion to the analogous comparison between uncoded schemes in Section 3.1.2 and Exercise 3.4, focusing particularly on the high SNR regime.

Exercise 5.2 Consider a linear cellular model with equi-spaced base-stations at distance $2d$ apart. With a reuse ratio of ρ, base-stations at distances of integer multiples of $2d/\rho$ reuse the same frequency band. Assuming that the interference emanates from the center of the cell, calculate the fraction f_ρ defined as the ratio of the interference to the received power from a user at the edge of the cell. You can assume that all uplink transmissions are at the same transmit power P and that the dominant interference comes from the nearest cells reusing the same frequency.

Exercise 5.3 Consider a regular hexagonal cellular model (cf. Figure 4.2) with a frequency reuse ratio of ρ.

1. Identify "appropriate" reuse patterns for different values of ρ, with the design goal of minimizing inter-cell interference. You can use the assumptions made in Exercise 5.2 on how the interference originates.
2. For the reuse patterns identified, show that $f_\rho = 6(\sqrt{\rho}/2)^\alpha$ is a good approximation to the fraction of the received power of a user at the edge of the cell that the interference represents. *Hint*: You can explicitly construct reuse patterns for $\rho = 1, 1/3, 1/4, 1/7, 1/9$ with exactly these fractions.

3. What reuse ratio yields the largest symmetric uplink rate at high SNR (an expression for the symmetric rate is in (5.23))?

Exercise 5.4 In Exercise 5.3 we computed the interference as a fraction of the signal power of interest assuming that the interference emanated from the center of the cell using the same frequency. Re-evaluate f_ρ using the assumption that the interference emanates uniformly in the cells using the same frequency. (You might need to do numerical computations varying the power decay rate α.)

Exercise 5.5 Consider the expression in (5.23) for the rate in the uplink at very high SNR values.
1. Plot the rate as a function of the reuse parameter ρ.
2. Show that $\rho = 1/2$, i.e., reusing the frequency every other cell, yields the largest rate.

Exercise 5.6 In this exercise, we study *time sharing*, as a means to communicate over the AWGN channel by using different codes over different intervals of time.
1. Consider a communication strategy over the AWGN channel where for a fraction of time α a capacity-achieving code at power level P_1 is used, and for the rest of the time a capacity-achieving code at power level P_2 is used, meeting the overall average power constraint P. Show that this strategy is strictly suboptimal, i.e., it is not capacity-achieving for the power constraint P.
2. Consider an additive noise channel:

$$y[m] = x[m] + w[m]. \qquad (5.120)$$

The noise is still i.i.d. over time but not necessarily Gaussian. Let $C(P)$ be the capacity of this channel under an average power constraint of P. Show that $C(P)$ must be a concave function of P. *Hint*: Hardly any calculation is needed. The insight from part (1) will be useful.

Exercise 5.7 In this exercise we use the formula for the capacity of the AWGN channel to see the contrast with the performance of certain communication schemes studied in Chapter 3. At high SNR, the capacity of the AWGN channel scales like $\log_2 \text{SNR}$ bits/s/Hz. Is this consistent with how the rate of an uncoded QAM system scales with the SNR?

Exercise 5.8 For the AWGN channel with general SNR, there is no known explicitly constructed capacity-achieving code. However, it is known that *orthogonal* codes can achieve the minimum $\mathcal{E}_b/N_0$ in the power-limited regime. This exercise shows that orthogonal codes can get arbitrary reliability with a finite $\mathcal{E}_b/N_0$. Exercise 5.9 demonstrates how the Shannon limit can actually be achieved. We focus on the discrete-time complex AWGN channel with noise variance N_0 per dimension.
1. An orthogonal code consists of M orthogonal codewords, each with the same energy $\mathcal{E}_s$. What is the energy per bit $\mathcal{E}_b$ for this code? What is the block length required? What is the data rate?
2. Does the ML error probability of the code depend on the specific choice of the orthogonal set? Explain.
3. Give an expression for the pairwise error probability, and provide a good upper bound for it.
4. Using the union bound, derive a bound on the overall ML error probability.

5. To achieve reliable communication, we let the number of codewords M grow and adjust the energy $\mathcal{E}_s$ per codeword such that the $\mathcal{E}_b/N_0$ remains fixed. What is the minimum $\mathcal{E}_b/N_0$ such that your bound in part (4) vanishes with M increasing? How far are you from the Shannon limit of $-1.59\,\text{dB}$?

6. What happens to the data rate? Reinterpret the code as consuming more and more bandwidth but at a fixed data rate (in bits/s).

7. How do you contrast the orthogonal code with a repetition code of longer and longer block length (as in Section 5.1.1)? In what sense is the orthogonal code better?

Exercise 5.9 (Orthogonal codes achieve $\mathcal{E}_b/N_0 = -1.59\,\text{dB}$.) The minimum $\mathcal{E}_b/N_0$ derived in Exercise 5.8 does not meet the Shannon limit, not because the orthogonal code is not good but because the union bound is not tight enough when $\mathcal{E}_b/N_0$ is close to the Shannon limit. This exercise explores how the union bound can be tightened in this range.

1. Let u_i be the real part of the inner product of the received signal vector with the ith orthogonal codeword. Express the ML detection rule in terms of $u_1, u_2 \ldots, u_M$.

2. Suppose codeword 1 is transmitted. Conditional on u_1 large, the ML detector can get confused with very few other codewords, and the union bound on the conditional error probability is quite tight. On the other hand, when u_1 is small, the ML detector can get confused with many other codewords and the union bound is lousy and can be much larger than 1. In the latter regime, one might as well bound the conditional error by 1. Compute then a bound on the ML error probability in terms of γ, a threshold that determines whether u_1 is "large" or "small". Simplify your bound as much as possible.

3. By an appropriate choice of γ, find a good bound on the ML error probability in terms of $\mathcal{E}_b/N_0$ so that you can demonstrate that orthogonal codes can approach the Shannon limit of $-1.59\,\text{dB}$. *Hint*: a good choice of γ is when the union bound on the conditional error is approximately 1. Why?

4. In what range of $\mathcal{E}_b/N_0$ does your bound in the previous part coincide with the union bound used in Exercise 5.8?

5. From your analysis, what insights about the typical error events in the various ranges of $\mathcal{E}_b/N_0$ can you derive?

Exercise 5.10 The outage performance of the slow fading channel depends on the randomness of $\log(1 + |h|^2\text{SNR})$. One way to quantify the randomness of a random variable is by the ratio of the standard deviation to the mean. Show that this parameter goes to zero at high SNR. What about low SNR? Does this make sense to you in light of your understanding of the various regimes associated with the AWGN channel?

Exercise 5.11 Show that the transmit beamforming strategy in Section 5.3.2 maximizes the received SNR for a given total transmit power constraint. (Part of the question involves making precise what this means!)

Exercise 5.12 Consider coding over N OFDM blocks in the parallel channel in (5.33), i.e., $i = 1, \ldots, N$, with power P_n over the nth sub-channel. Suppose that $\tilde{\mathbf{y}}_n := [\tilde{y}_n[1], \ldots, \tilde{y}_n[N]]^t$, with $\tilde{\mathbf{d}}_n$ and $\tilde{\mathbf{w}}_n$ defined similarly. Consider the entire received vector with $2NN_c$ real dimensions:

$$\tilde{\mathbf{y}} := \text{diag}\{\tilde{h}_1\mathbf{I}_N, \ldots, \tilde{h}_{N_c}\mathbf{I}_N\}\tilde{\mathbf{d}} + \tilde{\mathbf{w}}, \qquad (5.121)$$

where $\tilde{\mathbf{d}} := \left[\tilde{\mathbf{d}}_1^t, \ldots, \tilde{\mathbf{d}}_{N_c}^t\right]^t$ and $\tilde{\mathbf{w}} := [\tilde{\mathbf{w}}_1^t, \ldots, \tilde{\mathbf{w}}_{N_c}^t]^t$.

1. Fix $\epsilon > 0$ and consider the ellipsoid $E^{(\epsilon)}$ defined as

$$\left\{ \mathbf{a} : \mathbf{a}^* \left(\text{diag}\left\{ P_1|\tilde{h}_1|^2 \mathbf{I}_N, \ldots, P_{N_c}|\tilde{h}_{N_c}|^2 \mathbf{I}_N \right\} + N_0 \mathbf{I}_{NN_c} \right)^{-1} \mathbf{a} \leq N(N_c + \epsilon) \right\}.$$

$$(5.122)$$

Show for every ϵ that

$$\mathbb{P}\{\tilde{\mathbf{y}} \in E^{(\epsilon)}\} \rightarrow 1, \quad \text{as } N \rightarrow \infty. \tag{5.123}$$

Thus we can conclude that the received vector lives in the ellipsoid $E^{(0)}$ for large N with high probability.

2. Show that the volume of the ellipsoid $E^{(0)}$ is equal to

$$\left(\prod_{n=1}^{N_c} \left(|\tilde{h}_n|^2 P_n + N_0 \right)^N \right) \tag{5.124}$$

times the volume of a $2NN_c$-dimensional real sphere with radius $\sqrt{NN_c}$. This justifies the expression in (5.50).

3. Show that

$$\mathbb{P}\{\|\tilde{\mathbf{w}}\|^2 \leq N_0 N(N_c + \epsilon)\} \rightarrow 1, \quad \text{as } N \rightarrow \infty. \tag{5.125}$$

Thus $\tilde{\mathbf{w}}$ lives, with high probability, in a $2NN_c$-dimensional real sphere of radius $\sqrt{N_0 NN_c}$. Compare the volume of this sphere to the volume of the ellipsoid in (5.124) to justify the expression in (5.51).

Exercise 5.13 Consider a system with 1 transmit antenna and L receive antennas. Independent $\mathcal{CN}(0, N_0)$ noise corrupts the signal at each of the receive antennas. The transmit signal has a power constraint of P.

1. Suppose the gain between the transmit antenna and each of the receive antennas is constant, equal to 1. What is the capacity of the channel? What is the performance gain compared to a single receive antenna system? What is the nature of the performance gain?

2. Suppose now the signal to each of the receive antennas is subject to independent Rayleigh fading. Compute the capacity of the (fast) fading channel with channel information only at the receiver. What is the nature of the performance gain compared to a single receive antenna system? What happens when $L \rightarrow \infty$?

3. Give an expression for the capacity of the fading channel in part (2) with CSI at both the transmitter and the receiver. At low SNR, do you think the benefit of having CSI at the transmitter is more or less significant when there are multiple receive antennas (as compared to having a single receive antenna)? How about when the operating SNR is high?

4. Now consider the slow fading scenario when the channel is random but constant. Compute the outage probability and quantify the performance gain of having multiple receive antennas.

Exercise 5.14 Consider a MISO slow fading channel.

1. Verify that the Alamouti scheme radiates energy in an isotropic manner.
2. Show that a transmit diversity scheme radiates energy in an isotropic manner if and only if the signals transmitted from the antennas have the same power and are uncorrelated.

Exercise 5.15 Consider the MISO channel with L transmit antennas and channel gain vector $\mathbf{h} = [h_1, \ldots, h_L]^t$. The noise variance is N_0 per symbol and the total power constraint across the transmit antennas is P.

1. First, think of the channel gains as fixed. Suppose someone uses a transmission strategy for which the input symbols at any time have zero mean and a covariance matrix $\mathbf{K}_x$. Argue that the maximum achievable reliable rate of communication under this strategy is no larger than

$$\log\left(1 + \frac{\mathbf{h}^* \mathbf{K}_x \mathbf{h}}{N_0}\right) \text{bits/symbol}. \tag{5.126}$$

2. Now suppose we are in a slow fading scenario and $\mathbf{h}$ is random and i.i.d. Rayleigh. The outage probability of the scheme in part (1) is given by

$$p_{\text{out}}(R) = \mathbb{P}\left\{\log\left(1 + \frac{\mathbf{h}^* \mathbf{K}_x \mathbf{h}}{N_0}\right) < R\right\}. \tag{5.127}$$

Show that correlation never improves the outage probability: i.e., given a total power constraint P, one can do no worse by choosing $\mathbf{K}_x$ to be diagonal. *Hint*: Observe that the covariance matrix $\mathbf{K}_x$ admits a decomposition of the form $\mathbf{U} \operatorname{diag}\{P_1, \ldots, P_L\}\mathbf{U}^*$.

Exercise 5.16 Exercise 5.15 shows that for the i.i.d. Rayleigh slow fading MISO channel, one can always choose the input to be uncorrelated, in which case the outage probability is

$$\mathbb{P}\left\{\log\left(1 + \frac{\sum_{\ell=1}^{L} P_\ell |h_\ell|^2}{N_0}\right) < R\right\}, \tag{5.128}$$

where P_ℓ is the power allocated to antenna ℓ. Suppose the operating SNR is high relative to the target rate and satisfies

$$\log\left(1 + \frac{P}{N_0}\right) \geq R, \tag{5.129}$$

with P equal to the total transmit power constraint.

1. Show that the outage probability (5.128) is a symmetric function of $P_1, \ldots, P_L$.
2. Show that the partial double derivative of the outage probability (5.128) with respect to P_j is non-negative as long as $\sum_{\ell=1}^{L} P_\ell = P$, for each $j = 1, \ldots, L$. These two conditions imply that the isotropic strategy, i.e., $P_1 = \cdots = P_L = P/L$ minimizes the outage probability (5.128) subject to the constraint $P_1 + \cdots + P_L = P$. This result is adapted from Theorem 1 of [11], where the justification for the last step is provided.
3. For different values of L, calculate the range of outage probabilities for which the isotropic strategy is optimal, under condition (5.129).

Exercise 5.17 Consider the expression for the outage probability of the parallel fading channel in (5.84). In this exercise we consider the Rayleigh model, i.e., the channel entries $h_1, \ldots, h_L$ to be i.i.d. $\mathcal{CN}(0, 1)$, and show that uniform power allocation, i.e., $P_1 = \cdots = P_L = P/L$ achieves the minimum in (5.84). Consider the outage probability:

$$\mathbb{P}\left\{\sum_{\ell=1}^{L} \log\left(1 + \frac{P_\ell |h_\ell|^2}{N_0}\right) < LR\right\}. \tag{5.130}$$

1. Show that (5.130) is a symmetric function of $P_1, \ldots, P_L$.
2. Show that (5.130) is a convex function of P_ℓ, for each $\ell = 1, \ldots, L$.[6]

With the sum power constraint $\sum_{\ell=1}^{L} P_\ell = P$, these two conditions imply that the outage probability in (5.130) is minimized when $P_1 = \cdots = P_L = P/L$. This observation follows from a result in the theory of *majorization*, a partial order on vectors. In particular, Theorem 3.A.4 in [80] provides the required justification.

Exercise 5.18 Compute a high-SNR approximation of the outage probability for the parallel channel with L i.i.d. Rayleigh faded branches.

Exercise 5.19 In this exercise we study the slow fading parallel channel.
1. Give an expression for the outage probability of the repetition scheme when used on the parallel channel with L branches.
2. Using the result in Exercise 5.18, compute the extra SNR required for the repetition scheme to achieve the same outage probability as capacity, at high SNR. How does this depend on L, the target rate R and the SNR?
3. Redo the previous part at low SNR.

Exercise 5.20 In this exercise we study the outage capacity of the parallel channel in further detail.
1. Find an approximation for the ϵ-outage capacity of the parallel channel with L branches of time diversity in the low SNR regime.
2. Simplify your approximation for the case of i.i.d. Rayleigh faded branches and small outage probability ϵ.
3. IS-95 operates over a bandwidth of 1.25 MHz. The delay spread is 1 μs, the coherence time is 50 ms, the delay constraint (on voice) is 100 ms. The SINR each user sees is -17 dB per chip. Estimate the 1%-outage capacity for each user. How far is that from the capacity of an unfaded AWGN channel with the same SNR? *Hint*: You can model the channel as a parallel channel with i.i.d. Rayleigh faded sub-channels.

Exercise 5.21 In Chapter 3, we have seen that one way to communicate over the MISO channel is to convert it into a parallel channel by sending symbols over the different transmit antennas one at a time.
1. Consider first the case when the channel is fixed (known to both the transmitter and the receiver). Evaluate the capacity loss of using this strategy at high and low SNR. In which regime is this transmission scheme a good idea?

---

[6] Observe that this condition is weaker than saying that (5.130) is *jointly* convex in the arguments $(P_1, \ldots, P_L)$.

2. Now consider the slow fading MISO channel. Evaluate the loss in performance of using this scheme in terms of (i) the outage probability $p_{\text{out}}(R)$ at high SNR; (ii) the ϵ-outage capacity C_ϵ at low SNR.

Exercise 5.22 Consider the frequency-selective channel with CSI only at the receiver with L i.i.d. Rayleigh faded paths.

1. Compute the capacity of the fast fading channel. Give approximate expressions at the high and low SNR regimes.
2. Provide an expression for the outage probability of the slow fading channel. Give approximate expressions at the high and low SNR regimes.
3. In Section 3.4, we introduced a suboptimal scheme which transmits one symbol every L symbol times and uses maximal ratio combining at the receiver to detect each symbol. Find the outage and fast fading performance achievable by this scheme if the transmitted symbols are ideally coded and the outputs from the maximal-ratio are soft combined. Calculate the loss in performance (with respect to the optimal outage and fast fading performance) in using this scheme for a GSM system with two paths operating at average SNR of 15 dB. In what regime do we *not* lose much performance by using this scheme?

Exercise 5.23 In this exercise, we revisit the CDMA system of Section 4.3 in the light of our understanding of capacity of wireless channels.

1. In our analysis in Chapter 4 of the performance of CDMA systems, it was common for us to assume a $\mathcal{E}_b/N_0$ requirement for each user. This requirement depends on the data rate R of each user, the bandwidth W Hz, and also the code used. Assuming an AWGN channel and the use of capacity-achieving codes, compute the $\mathcal{E}_b/N_0$ requirement as a function of the data rate and bandwidth. What is this number for an IS-95 system with $R = 9.6$ kbits/s and $W = 1.25$ MHz? At the low SNR, power-limited regime, what happens to this $\mathcal{E}_b/N_0$ requirement?
2. In IS-95, the code used is not optimal: each coded symbol is repeated four times in the last stage of the spreading. With only this constraint on the code, find the maximum achievable rate of reliable communication over an AWGN channel. *Hint*: Exercise 5.13(1) may be useful here.
3. Compare the performance of the code used in IS-95 with the capacity of the AWGN channel. Is the performance loss greater in the low SNR or high SNR regime? Explain intuitively.
4. With the repetition constraint of the code as in part (2), quantify the resulting increase in $\mathcal{E}_b/N_0$ requirement compared to that in part (1). Is this penalty serious for an IS-95 system with $R = 9.6$ kbits/s and $W = 1.25$ MHz?

Exercise 5.24 In this exercise we study the price of channel inversion.

1. Consider a narrowband Rayleigh flat fading SISO channel. Show that the average power (averaged over the channel fading) needed to implement the channel inversion scheme is infinite for any positive target rate.
2. Suppose now there are $L > 1$ receive antennas. Show that the average power for channel inversion is now finite.
3. Compute numerically and plot the average power as a function of the target rate for different L to get a sense of the amount of gain from having multiple receive antennas. Qualitatively describe the nature of the performance gain.

Exercise 5.25 This exercise applies basic capacity results to analyze the IS-856 system. You should use the parameters of IS-865 given in the text.

1. The table in the IS-865 example in the text gives the SINR thresholds for using the various rates. What would the thresholds have been if capacity-achieving codes were used? Are the codes used in IS-856 close to optimal? (You can assume that the interference plus noise is Gaussian and that the channel is time-invariant over the time-scale of the coding.)

2. At low rates, the coding is performed by a turbo code followed by a repetition code to reduce the complexity. How much is the sub- optimality of the IS-865 codes due to the repetition structure? In particular, at the lowest rate of 38.4 kbits/s, coded symbols are repeated 16 times. With only this constraint on the code, find the minimum SINR needed for reliable communication. Comparing this to the corresponding threshold calculated in part (1), can you conclude whether one loses a lot from the repetition?

Exercise 5.26 In this problem we study the nature of the error in the channel estimate fed back to the transmitter (to adapt the transmission rate, as in the IS-856 system). Consider the following time-varying channel model (called the Gauss–Markov model):

$$h[m+1] = \sqrt{1-\delta}\, h[m] + \sqrt{\delta}\, w[m+1], \quad m \geq 0, \tag{5.131}$$

with $\{w[m]\}$ a sequence of i.i.d. $\mathcal{CN}(0,1)$ random variables independent of $h[0] \sim \mathcal{CN}(0,1)$. The coherence time of the channel is controlled by the parameter δ.

1. Calculate the auto-correlation function of the channel process in (5.131).

2. Defining the coherence time as the largest time for which the auto-correlation is larger than 0.05 (cf. Section 2.4.3), derive an expression for δ in terms of the coherence time and the sample rate. What are some typical values of δ for the IS-856 system at different vehicular speeds?

3. The channel is estimated at the receiver using training symbols. The estimation error (evaluated in Section 3.5.2) is small at high SNR and we will ignore it by assuming that $h[0]$ is estimated exactly. Due to the delay, the fed back $h[0]$ reaches the transmitter at time n. Evaluate the predictor $\hat{h}[n]$ of $h[n]$ from $h[0]$ that minimizes the mean squared error.

4. Show that the minimum mean squared error predictor can be expressed as

$$h[n] = \hat{h}[n] + h_e[n], \tag{5.132}$$

with the error $h_e[n]$ independent of $\hat{h}[n]$ and distributed as $\mathcal{CN}(0, \sigma_e^2)$. Find an expression for the variance of the prediction error σ_e^2 in terms of the delay n and the channel variation parameter δ. What are some typical values of σ_e^2 for the IS-856 system with a 2-slot delay in the feedback link?

Exercise 5.27 Consider the slow fading channel (cf. Section 5.4.1)

$$y[m] = hx[m] + w[m], \tag{5.133}$$

with $h \sim \mathcal{CN}(0, 1)$. If there is a feedback link to the transmitter, then an estimate of the channel quality can be relayed back to the transmitter (as in the IS-856 system). Let us suppose that the transmitter is aware of $\hat{h}$, which is modeled as

$$h = \hat{h} + h_e, \tag{5.134}$$

where the error in the estimate h_e is independent of the estimate $\hat{h}$ and is $\mathcal{CN}(0, \sigma_e^2)$ (see Exercise 5.26 and (5.132) in particular). The rate of communication R is chosen as a function of the channel estimate $\hat{h}$. If the estimate is perfect, i.e., $\sigma_e^2 = 0$, then the slow fading channel is simply an AWGN channel and R can be chosen to be less than the capacity and an arbitrarily small error probability is achieved. On the other hand, if the estimate is very noisy, i.e., $\sigma_e^2 \gg 1$, then we have the original slow fading channel studied in Section 5.4.1.

1. Argue that the outage probability, conditioned on the estimate of the channel $\hat{h}$, is

$$\mathbb{P}\left\{\log(1 + |h|^2 \mathsf{SNR}) < R(\hat{h}) | \hat{h}\right\}. \tag{5.135}$$

2. Let us fix the outage probability in (5.135) to be less than ϵ for every realization of the channel estimate $\hat{h}$. Then the rate can be adapted as a function of the channel estimate $\hat{h}$. To get a feel for the amount of loss in the rate due to the imperfect channel estimate, carry out the following numerical experiment. Fix $\epsilon = 0.01$ and evaluate numerically (using a software such as MATLAB) the average difference between the rate with perfect channel feedback and the rate R with imperfect channel feedback for different values of the variance of the channel estimate error σ_e^2 (the average is carried out over the joint distribution of the channel and its estimate).

   What is the average difference for the IS-856 system at different vehicular speeds? You can use the results from the calculation in Exercise 5.26(3) that connect the vehicular speeds to σ_e^2 in the IS-856 system.

3. The numerical example gave a feel for the amount of loss in transmission rate due to the channel uncertainty. In this part, we study approximations to the optimal transmission rate as a function of the channel estimate.

   (a) If $\hat{h}$ is small, argue that the optimal rate adaptation is of the form

$$R(\hat{h}) \approx \log\left(1 + a_1 |\hat{h}|^2 + b_1\right), \tag{5.136}$$

   by finding appropriate constants a_1, b_1 as functions of ϵ and σ_e^2.

   (b) When $\hat{h}$ is large, argue that the optimal rate adaptation is of the form

$$R(\hat{h}) \approx \log\left(1 + a_2 |\hat{h}| + b_2\right), \tag{5.137}$$

   and find appropriate constants a_2, b_2.

Exercise 5.28 In the text we have analyzed the performance of fading channels under the assumption of receiver CSI. The CSI is obtained in practice by transmitting training symbols. In this exercise, we will study how the loss in degrees of freedom from sending training symbols compares with the actual capacity of the *non-coherent* fading channel. We will conduct this study in the context of a *block fading* model: the channel remains constant over a block of time equal to the coherence time and jumps to independent realizations over different coherent time intervals. Formally,

$$y[m + nT_c] = h[n]x[m + nT_c] + w[m + nT_c], \quad m = 1, \ldots, T_c, \ n \geq 1, \quad (5.138)$$

where T_c is the coherence time of the channel (measured in terms of the number of samples). The channel variations across the blocks $h[n]$ are i.i.d. Rayleigh.

1. For the IS-856 system, what are typical values of T_c at different vehicular speeds?
2. Consider the following pilot (or training symbol) based scheme that converts the non-coherent communication into a coherent one by providing receiver CSI. The first symbol of the block is a known symbol and information is sent in the remaining symbols ($T_c - 1$ of them). At high SNR, the pilot symbol allows the receiver to estimate the channel ($h[n]$, over the nth block) with a high degree of accuracy. Argue that the reliable rate of communication using this scheme at high SNR is approximately

$$\frac{T_c - 1}{T_c} C(\text{SNR}) \, \text{bits/s/Hz}, \quad (5.139)$$

where $C(\text{SNR})$ is the capacity of the channel in (5.138) *with* receiver CSI. In what mathematical sense can you make this approximation precise?
3. A reading exercise is to study [83] where the authors show that the capacity of the original non-coherent block fading channel in (5.138) is comparable (in the same sense as the approximation in the previous part) to the rate achieved with the pilot based scheme (cf. (5.139)). Thus there is little loss in performance with pilot based reliable communication over fading channels at high SNR.

Exercise 5.29 Consider the block fading model (cf. (5.138)) with a very short coherent time T_c. In such a scenario, the pilot based scheme does not perform very well as compared to the capacity of the channel with receiver CSI (cf. (5.139)). A reading exercise is to study the literature on the capacity of the non-coherent i.i.d. Rayleigh fading channel (i.e., the block fading model in (5.138) with $T_c = 1$) [68, 114, 1]. The main result is that the capacity is approximately

$$\log \log \text{SNR} \quad (5.140)$$

at high SNR, i.e., communication at high SNR is very inefficient. An intuitive way to think about this result is to observe that a logarithmic transform converts the multiplicative noise (channel fading) into an additive Gaussian one. This allows us to use techniques from the AWGN channel, but now the effective SNR is only $\log \text{SNR}$.

Exercise 5.30 In this problem we will derive the capacity of the underspread frequency-selective fading channel modeled as follows. The channel is time invariant over each coherence time interval (with length T_c). Over the ith coherence time interval the channel has L_n taps with coefficients[7]

$$h_0[i], \ldots, h_{L_i-1}[i]. \quad (5.141)$$

---

[7] We have slightly abused our notation here: in the text $h_\ell[m]$ was used to denote the ℓth tap at symbol time m, but here $h_\ell[i]$ is the ℓth tap at the ith coherence interval.

The underspread assumption ($T_c \gg L_i$) means that the edge effect of having the next coherent interval overlap with the last $L_i - 1$ symbols of the current coherent interval is insignificant. One can then jointly code over coherent time intervals with the same (or nearly the same) channel tap values to achieve the corresponding largest reliable communication rate afforded by that frequency-selective channel. To simplify notation we use this operational reasoning to make the following assumption: over the finite time interval T_c, the reliable rate of communication can be well approximated as equal to the capacity of the corresponding time-invariant frequency-selective channel.

1. Suppose a power $P[i]$ is allocated to the ith coherence time interval. Use the discussion in Section 5.4.7 to show that the largest rate of reliable communication over the ith coherence time interval is

$$\max_{P_0[i],\ldots,P_{T_c-1}[i]} \frac{1}{T_c} \sum_{n=0}^{T_c-1} \log\left(1 + \frac{P_n[i]|\tilde{h}_n[i]|^2}{N_0}\right), \tag{5.142}$$

subject to the power constraint

$$\sum_{n=0}^{T_c-1} P_n[i] \leq T_c P[i]. \tag{5.143}$$

It is optimal to choose $P_n[i]$ to waterfill $N_0/|\tilde{h}_n[i]|^2$ where $\tilde{h}_0[i],\ldots,\tilde{h}_{T_c-1}[i]$ is the T_c-point DFT of the channel $h_0[i],\ldots,h_{L_i-1}[i]$ scaled by $\sqrt{T_c}$.

2. Now consider M coherence time intervals over which the powers $P[1],\ldots,P[M]$ are to be allocated subject to the constraint

$$\sum_{i=1}^{M} P[i] \leq MP.$$

Determine the optimal power allocation $P_n[i]$, $n = 0,\ldots,T_c - 1$ and $i = 1,\ldots,M$ as a function of the frequency-selective channels in each of the coherence time intervals.

3. What happens to the optimal power allocation as M, the number of coherence time intervals, grows large? State precisely any assumption you make about the ergodicity of the frequency-selective channel sequence.

6

Multiuser capacity and opportunistic communication

In Chapter 4, we studied several specific multiple access *techniques* (TDMA/FDMA, CDMA, OFDM) designed to share the channel among several users. A natural question is: what are the "optimal" multiple access schemes? To address this question, one must now step back and take a fundamental look at the multiuser *channels* themselves. Information theory can be generalized from the point-to-point scenario, considered in Chapter 5, to the multiuser ones, providing limits to multiuser communications and suggesting optimal multiple access strategies. New techniques and concepts such as *successive cancellation*, *superposition coding* and *multiuser diversity* emerge.

The first part of the chapter focuses on the uplink (many-to-one) and downlink (one-to-many) AWGN channel without fading. For the uplink, an optimal multiple access strategy is for all users to spread their signal across the entire bandwidth, much like in the CDMA system in Chapter 4. However, rather than decoding every user treating the interference from other users as noise, a *successive interference cancellation* (SIC) receiver is needed to achieve capacity. That is, after one user is decoded, its signal is stripped away from the aggregate received signal before the next user is decoded. A similar strategy is optimal for the downlink, with signals for the users superimposed on top of each other and SIC done at the mobiles: each user decodes the information intended for all of the weaker users and strips them off before decoding its own. It is shown that in situations where users have very disparate channels to the base-station, CDMA together with successive cancellation can offer significant gains over the conventional multiple access techniques discussed in Chapter 4.

In the second part of the chapter, we shift our focus to multiuser *fading* channels. One of the main insights learnt in Chapter 5 is that, for fast fading channels, the ability to track the channel at the transmitter can increase point-to-point capacity by *opportunistic communication*: transmitting at high rates when the channel is good, and at low rates or not at all when the channel is poor. We extend this insight to the multiuser setting, both for the uplink and for the downlink. The performance gain of opportunistic communication comes from exploiting the fluctuations of the fading channel. Compared to the point-to-point setting, the multiuser settings offer more opportunities to exploit. In addition to the choice of *when* to transmit, there is now an additional choice of *which user(s)* to transmit from (in the uplink) or to transmit to (in the downlink) and the amount of power to allocate between the users. This additional choice provides a further performance gain not found in the point-to-point scenario. It allows the system to benefit from a *multiuser diversity* effect: at any time in a large network, with high probability there is a user whose channel is near its peak. By allowing such a user to transmit at that time, the overall multiuser capacity can be achieved.

In the last part of the chapter, we will study the system issues arising from the implementation of opportunistic communication in a cellular system. We use as a case study IS-856, the third-generation standard for wireless data already introduced in Chapter 5. We show how multiple antennas can be used to further boost the performance gain that can be extracted from opportunistic communication, a technique known as *opportunistic beamforming*. We distill the insights into a new design principle for wireless systems based on opportunistic communication and multiuser diversity.

6.1 Uplink AWGN channel

6.1.1 Capacity via successive interference cancellation

The baseband discrete-time model for the uplink AWGN channel with two users (Figure 6.1) is

$$y[m] = x_1[m] + x_2[m] + w[m], \tag{6.1}$$

where $w[m] \sim \mathcal{CN}(0, N_0)$ is i.i.d. complex Gaussian noise. User k has an average power constraint of P_k joules/symbol (with $k = 1, 2$).

In the point-to-point case, the *capacity* of a channel provides the performance limit: reliable communication can be attained at any rate $R < C$; reliable communication is impossible at rates $R > C$. In the multiuser case, we should extend this concept to a *capacity region* $\mathcal{C}$: this is the set of all pairs (R_1, R_2) such that *simultaneously* user 1 and 2 can reliably communicate at rate R_1 and R_2, respectively. Since the two users share the same bandwidth, there is naturally a tradeoff between the reliable communication rates of the users: if one wants to communicate at a higher rate, the other user may need to lower its rate. For example, in orthogonal multiple access schemes, such as OFDM, this tradeoff can be achieved by varying the number of sub-carriers allocated to each user. The capacity region $\mathcal{C}$ characterizes the *optimal* tradeoff achievable by *any* multiple access scheme. From this

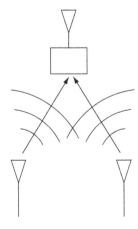
Figure 6.1 Two-user uplink.

capacity region, one can derive other scalar performance measures of interest. For example:

- The symmetric capacity:

$$C_{\text{sym}} := \max_{(R,R)\in\mathcal{C}} R \tag{6.2}$$

is the maximum common rate at which both the users can simultaneously reliably communicate.
- The sum capacity:

$$C_{\text{sum}} := \max_{(R_1,R_2)\in\mathcal{C}} R_1 + R_2 \tag{6.3}$$

is the maximum total throughput that can be achieved.

Just like the capacity of the AWGN channel, there is a very simple characterization of the capacity region $\mathcal{C}$ of the uplink AWGN channel: this is the set of all rates (R_1, R_2) satisfying the three constraints (Appendix B.9 provides a formal justification):

$$R_1 < \log\left(1 + \frac{P_1}{N_0}\right), \tag{6.4}$$

$$R_2 < \log\left(1 + \frac{P_2}{N_0}\right), \tag{6.5}$$

$$R_1 + R_2 < \log\left(1 + \frac{P_1+P_2}{N_0}\right). \tag{6.6}$$

The capacity region is the pentagon shown in Figure 6.2. All the three constraints are natural. The first two say that the rate of the individual user cannot exceed the capacity of the point-to-point link with the other user absent from

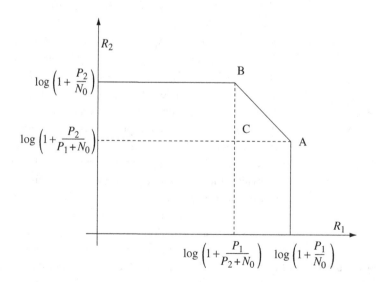

Figure 6.2 Capacity region of the two-user uplink AWGN channel.

the system (these are called single-user bounds). The third says that the total throughput cannot exceed the capacity of a point-to-point AWGN channel with the sum of the received powers of the two users. This is indeed a valid constraint since the signals the two users send are independent and hence the power of the aggregate received signal is the sum of the powers of the individual received signals.[1] Note that without the third constraint, the capacity region would have been a rectangle, and each user could simultaneously transmit at the point-to-point capacity as if the other user did not exist. This is clearly too good to be true and indeed the third constraint says this is not possible: there must be a tradeoff between the performance of the two users.

Nevertheless, something surprising does happen: user 1 can achieve its single-user bound while at the same time user 2 can get a non-zero rate; in fact as high as its rate at point A, i.e.,

$$R_2^* = \log\left(1 + \frac{P_1 + P_2}{N_0}\right) - \log\left(1 + \frac{P_1}{N_0}\right) = \log\left(1 + \frac{P_2}{P_1 + N_0}\right). \quad (6.7)$$

How can this be achieved? Each user encodes its data using a capacity-achieving AWGN channel code. The receiver decodes the information of both the users in two stages. In the first stage, it decodes the data of user 2, treating the signal from user 1 as Gaussian interference. The maximum rate user 2 can achieve is precisely given by (6.7). Once the receiver decodes the data of user 2, it can reconstruct user 2's signal and subtract it from the aggregate received signal. The receiver can then decode the data of user 1. Since there is now only the background Gaussian noise left in the system, the maximum rate user 1 can transmit at is its single-user bound $\log(1 + P_1/N_0)$. This receiver is called a *successive interference cancellation* (SIC) receiver or simply a successive cancellation decoder. If one reverses the order of cancellation, then one can achieve point B, the other corner point. All the other rate points on the segment AB can be obtained by time-sharing between the multiple access strategies in point A and point B. (We see in Exercise 6.7 another technique called *rate-splitting* that also achieves these intermediate points.)

The segment AB contains all the "optimal" operating points of the channel, in the sense that any other point in the capacity region is component-wise dominated by some point on AB. Thus one can always increase *both* users' rates by moving to a point on AB, and there is no reason not to.[2] No such domination exists *among* the points on AB, and the preferred operating point depends on the system objective. If the goal of the system is to maximize the sum rate, then any point on AB is equally fine. On the other hand, some operating points are not *fair*, especially if the received power of one user is

---

[1] This is the same argument we used for deriving the capacity of the MISO channel in Section 5.3.2.

[2] In economics terms, the points on AB are called *Pareto optimal*.

much larger than the other. In this case, consider operating at the corner point in which the strong user is decoded *first*: now the weak user gets the best possible rate.[3] In the case when the weak user is the one further away from the base-station, it is shown in Exercise 6.10 that this decoding order has the property of minimizing the total *transmit* power to meet given target rates for the two users. Not only does this lead to savings in the battery power of the users, it also translates to an increase in the system capacity of an interference-limited cellular system (Exercise 6.11).

6.1.2 Comparison with conventional CDMA

There is a certain similarity between the multiple access technique that achieves points A and B, and the CDMA technique discussed in Chapter 4. The only difference is that in the CDMA system described there, *every* user is decoded treating the other users as interference. This is sometimes called a *conventional* or a *single-user* CDMA receiver. In contrast, the SIC receiver is a *multiuser* receiver: one of the users, say user 1, is decoded treating user 2 as interference, but user 2 is decoded with the benefit of the signal of user 1 already removed. Thus, we can immediately conclude that the performance of the conventional CDMA receiver is suboptimal; in Figure 6.2, it achieves the point C which is strictly in the interior of the capacity region.

The benefit of SIC over the conventional CDMA receiver is particularly significant when the received power of one user is much larger than that of the other: by decoding and subtracting the signal of the strong user first, the weaker user can get a much higher data rate than when it has to contend with the interference of the strong user (Figure 6.3). In the context of a cellular system, this means that rather than having to keep the received powers of all users equal by transmit power control, users closer to the base-station can be allowed to take advantage of the stronger channel and transmit at a higher rate while not degrading the performance of the users in the edge of the cell. With a conventional receiver, this is not possible due to the *near–far problem*. With the SIC, we are turning the near–far problem into a near–far advantage. This advantage is less apparent in providing voice service where the required data rate of a user is constant over time, but it can be important for providing data services where users can take advantage of the higher data rates when they are closer to the base-station.

6.1.3 Comparison with orthogonal multiple access

How about orthogonal multiple access techniques? Can they be information theoretically optimal? Consider an orthogonal scheme that allocates a fraction

---

[3] This operating point is said to be *max–min fair*.

Figure 6.3 In the case when the received powers of the users are very disparate, successive cancellation (point A) can provide a significant advantage to the weaker user compared to conventional CDMA decoding (point C). The conventional CDMA solution is to control the received power of the strong user to equal that of the weak user (point D), but then the rates of both users are much lower. Here, $P_1/N_0 = 0$ dB, $P_2/N_0 = 20$ dB.

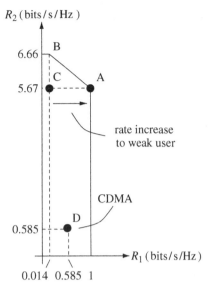

α of the degrees of freedom to user 1 and the rest, $1 - \alpha$, to user 2 (note that it is irrelevant for the capacity analysis whether the partitioning is across frequency or across time, since the power constraint is on the average across the degrees of freedom). Since the received power of user 1 is P_1, the amount of received energy is P_1/α joules per degree of freedom. The maximum rate user 1 can achieve over the total bandwidth W is

$$\alpha W \log\left(1 + \frac{P_1}{\alpha N_0}\right) \text{ bits/s.} \tag{6.8}$$

Similarly, the maximum rate user 2 can achieve is

$$(1 - \alpha)W \log\left(1 + \frac{P_2}{(1 - \alpha)N_0}\right) \text{ bits/s.} \tag{6.9}$$

Varying α from 0 to 1 yields all the rate pairs achieved by orthogonal schemes. See Figure 6.4.

Comparing these rates with the capacity region, one can see that the orthogonal schemes are in general suboptimal, except for one point: when $\alpha = P_1/(P_1 + P_2)$, i.e., the amount of degrees of freedom allocated to each user is proportional to its received power (Exercise 6.2 explores the reason why). However, when there is a large disparity between the received powers of the two users (as in the example of Figure 6.4), this operating point is highly unfair since most of the degrees of freedom are given to the strong user and the weak user has hardly any rate. On the other hand, by decoding the strong user first and then the weak user, the weak user can achieve the highest possible rate and this is therefore the most fair possible operating point (point A in Figure 6.4). In contrast, orthogonal multiple access techniques

Figure 6.4 Performance of orthogonal multiple access compared to capacity. The SNRs of the two users are: $P_1/N_0 = 0$ dB and $P_2/N_0 = 20$ dB. Orthogonal multiple access achieves the sum capacity at exactly one point, but at that point the weak user 1 has hardly any rate and it is therefore a highly unfair operating point. Point A gives the highest possible rate to user 1 and is most fair.

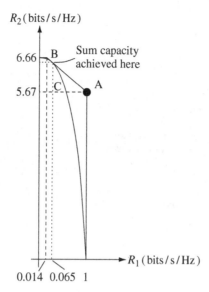

can approach this performance for the weak user only by nearly sacrificing all the rate of the strong user. Here again, as in the comparison with CDMA, SIC's advantage is in exploiting the proximity of a user to the base-station to give it high rate while protecting the far-away user.

6.1.4 General K-user uplink capacity

We have so far focused on the two-user case for simplicity, but the results extend readily to an arbitrary number of users. The K-user capacity region is described by $2^K - 1$ constraints, one for each possible non-empty subset $\mathcal{S}$ of users:

$$\sum_{k \in \mathcal{S}} R_k < \log\left(1 + \frac{\sum_{k \in \mathcal{S}} P_k}{N_0}\right) \qquad \text{for all } \mathcal{S} \subset \{1, \ldots, K\}. \qquad (6.10)$$

The right hand side corresponds to the maximum sum rate that can be achieved by a single transmitter with the total power of the users in $\mathcal{S}$ and with no other users in the system. The sum capacity is

$$C_{\text{sum}} = \log\left(1 + \frac{\sum_{k=1}^{K} P_k}{N_0}\right) \text{bits/s/Hz}. \qquad (6.11)$$

It can be shown that there are exactly $K!$ corner points, each one corresponding to a successive cancellation order among the users (Exercise 6.9).

The equal received power case ($P_1 = \ldots = P_K = P$) is particularly simple. The sum capacity is

$$C_{\text{sum}} = \log\left(1 + \frac{KP}{N_0}\right). \qquad (6.12)$$

The symmetric capacity is

$$C_{\text{sym}} = \frac{1}{K} \cdot \log\left(1 + \frac{KP}{N_0}\right). \qquad (6.13)$$

This is the maximum rate for each user that can be obtained if every user operates at the same rate. Moreover, this rate can be obtained via orthogonal multiplexing: each user is allocated a fraction $1/K$ of the total degrees of freedom.[4] In particular, we can immediately conclude that under equal received powers, the OFDM scheme considered in Chapter 4 has a better performance than the CDMA scheme (which uses conventional receivers.)

Observe that the sum capacity (6.12) is unbounded as the number of users grows. In contrast, if the conventional CDMA receiver (decoding every user treating all other users as noise) is used, each user will face an interference from $K-1$ users of total power $(K-1)P$, and thus the sum rate is only

$$K \cdot \log\left(1 + \frac{P}{(K-1)P + N_0}\right) \text{bits/s/Hz}, \qquad (6.14)$$

which approaches

$$K \cdot \frac{P}{(K-1)P + N_0} \log_2 e \approx \log_2 e = 1.442 \,\text{bits/s/Hz}, \qquad (6.15)$$

as $K \to \infty$. Thus, the total spectral efficiency is bounded in this case: the growing interference is eventually the limiting factor. Such a rate is said to be *interference-limited*.

The above comparison pertains effectively to a single-cell scenario, since the only external effect modeled is white Gaussian noise. In a cellular network, the out-of-cell interference must be considered, and as long as the out-of-cell signals cannot be decoded, the system would still be interference-limited, no matter what the receiver is.

6.2 Downlink AWGN channel

The downlink communication features a single transmitter (the base-station) sending separate information to multiple users (Figure 6.5). The baseband downlink AWGN channel with two users is

$$y_k[m] = h_k x[m] + w_k[m], \qquad k = 1, 2, \qquad (6.16)$$

where $w_k[m] \sim \mathcal{CN}(0, N_0)$ is i.i.d. complex Gaussian noise and $y_k[m]$ is the received signal at user k at time m, for both the users $k = 1, 2$. Here h_k is

---

[4] This fact is specific to the AWGN channel and does not hold in general. See Section 6.3.

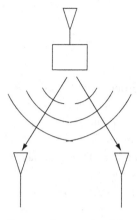

Figure 6.5 Two-user downlink.

the fixed (complex) channel gain corresponding to user k. We assume that h_k is known to both the transmitter and the user k (for $k = 1, 2$). The transmit signal $\{x[m]\}$ has an average power constraint of P joules/symbol. Observe the difference from the uplink of this overall constraint: there the power restrictions are separate for the signals of each user. The users separately decode their data using the signals they receive.

As in the uplink, we can ask for the capacity region $\mathcal{C}$, the region of the rates (R_1, R_2), at which the two users can simultaneously reliably communicate. We have the single-user bounds, as in (6.4) and (6.5),

$$R_k < \log\left(1 + \frac{P|h_k|^2}{N_0}\right), \qquad k = 1, 2. \tag{6.17}$$

This upper bound on R_k can be attained by using all the power and degrees of freedom to communicate to user k (with the other user getting zero rate). Thus, we have the two extreme points (with rate of one user being zero) in Figure 6.6. Further, we can share the degrees of freedom (time and bandwidth) between the users in an orthogonal manner to obtain any rate pair on the line joining these two extreme points. Can we achieve a rate pair outside this triangle by a more sophisticated communication strategy?

6.2.1 Symmetric case: two capacity-achieving schemes

To get more insight, let us first consider the symmetric case where $|h_1| = |h_2|$. In this symmetric situation, the SNR of both the users is the same. This means that if user 1 can successfully decode its data, then user 2 should also be

Figure 6.6 The capacity region of the downlink with two users having symmetric AWGN channels, i.e., $|h_1| = |h_2|$.

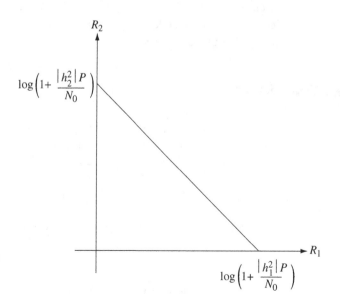

able to decode successfully the data of user 1 (and vice versa). Thus the sum information rate must also be bounded by the single-user capacity:

$$R_1 + R_2 < \log\left(1 + \frac{P|h_1|^2}{N_0}\right). \tag{6.18}$$

Comparing this with the single-user bounds in (6.17) and recalling the symmetry assumption $|h_1| = |h_2|$, we have shown the triangle in Figure 6.6 to be the capacity region of the symmetric downlink AWGN channel.

Let us continue our thought process within the realm of the symmetry assumption. The rate pairs in the capacity region can be achieved by strategies used on point-to-point AWGN channels and sharing the degrees of freedom (time and bandwidth) between the two users. However, the symmetry between the two channels (cf. (6.16)) suggests a natural, and alternative, approach. The main idea is that if user 1 can successfully decode its data from y_1, then user 2, which has the same SNR, should also be able to decode the data of user 1 from y_2. Then user 2 can *subtract* the codeword of user 1 from its received signal y_2 to better decode its own data, i.e., it can perform *successive interference cancellation*. Consider the following strategy that *superposes* the signals of the two users, much like in a spread-spectrum CDMA system. The transmit signal is the sum of two signals,

$$x[m] = x_1[m] + x_2[m], \tag{6.19}$$

where $\{x_k[m]\}$ is the signal intended for user k. The transmitter encodes the information for each user using an i.i.d. Gaussian code spread on the entire bandwidth (and powers P_1, P_2, respectively, with $P_1 + P_2 = P$). User 1 treats the signal for user 2 as noise and can hence be communicated to reliably at a rate of

$$\begin{aligned}R_1 &= \log\left(1 + \frac{P_1|h_1|^2}{P_2|h_1|^2 + N_0}\right) \\ &= \log\left(1 + \frac{(P_1 + P_2)|h_1|^2}{N_0}\right) - \log\left(1 + \frac{P_2|h_1|^2}{N_0}\right). \tag{6.20}\end{aligned}$$

User 2 performs successive interference cancellation: it first decodes the data of user 1 by treating x_2 as noise, subtracts the exactly determined (with high probability) user 1 signal from y_2 and extracts its data. Thus user 2 can support reliably a rate

$$R_2 = \log\left(1 + \frac{P_2|h_2|^2}{N_0}\right). \tag{6.21}$$

This superposition strategy is schematically represented in Figures 6.7 and 6.8. Using the power constraint $P_1 + P_2 = P$ we see directly from (6.20) and (6.21) that the rate pairs in the capacity region (Figure 6.6) can be achieved by this strategy as well. We have hence seen two coding schemes for the

Figure 6.7 Superposition encoding example. The QPSK constellation of user 2 is superimposed on that of user 1.

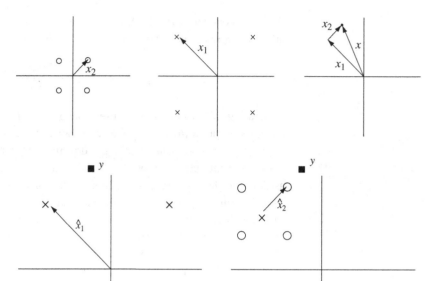

Figure 6.8 Superposition decoding example. The transmitted constellation point of user 1 is decoded first, followed by decoding of the constellation point of user 2.

symmetric downlink AWGN channel that are both optimal: single-user codes followed by orthogonalization of the degrees of freedom among the users, and the superposition coding scheme.

6.2.2 General case: superposition coding achieves capacity

Let us now return to the general downlink AWGN channel without the symmetry assumption and take $|h_1| < |h_2|$. Now user 2 has a better channel than user 1 and hence can decode any data that user 1 can successfully decode. Thus, we can use the superposition coding scheme: First the transmit signal is the (linear) superposition of the signals of the two users. Then, user 1 treats the signal of user 2 as noise and decodes its data from y_1. Finally, user 2, which has the better channel, performs SIC: it decodes the data of user 1 (and hence the transmit signal corresponding to user 1's data) and then proceeds to subtract the transmit signal of user 1 from y_2 and decode its data. As before, with each possible power split of $P = P_1 + P_2$, the following rate pair can be achieved:

$$
\begin{aligned}
R_1 &= \log\left(1 + \frac{P_1|h_1|^2}{P_2|h_1|^2 + N_0}\right) \text{ bits/s/Hz,} \\
R_2 &= \log\left(1 + \frac{P_2|h_2|^2}{N_0}\right) \text{ bits/s/Hz.}
\end{aligned}
\tag{6.22}
$$

On the other hand, orthogonal schemes achieve, for each power split $P = P_1 + P_2$ and degree-of-freedom split $\alpha \in [0, 1]$, as in the uplink (cf. (6.8) and (6.9)),

$$R_1 = \alpha \log \left(1 + \frac{P_1 |h_1|^2}{\alpha N_0} \right) \text{ bits/s/Hz,}$$

$$R_2 = (1 - \alpha) \log \left(1 + \frac{P_2 |h_2|^2}{(1 - \alpha) N_0} \right) \text{ bits/s/Hz.} \tag{6.23}$$

Here, α represents the fraction of the bandwidth devoted to user 1. Figure 6.9 plots the boundaries of the rate regions achievable with superposition coding and optimal orthogonal schemes for the asymmetric downlink AWGN channel (with $\mathsf{SNR}_1 = 0\,\mathrm{dB}$ and $\mathsf{SNR}_2 = 20\,\mathrm{dB}$). We observe that the performance of the superposition coding scheme is better than that of the orthogonal scheme.

One can show that the superposition decoding scheme is strictly better than the orthogonalization schemes (except for the two corner points where only one user is being communicated to). That is, for any rate pair achieved by orthogonalization schemes there is a power split for which the successive decoding scheme achieves rate pairs that are strictly larger (see Exercise 6.25). This gap in performance is more pronounced when the asymmetry between the two users deepens. In particular, superposition coding can provide a very reasonable rate to the strong user, while achieving close to the single-user bound for the weak user. In Figure 6.9, for example, while maintaining the rate of the weaker user R_1 at 0.9 bits/s/Hz, superposition coding can provide a rate of around $R_2 = 3$ bits/s/Hz to the strong user while an orthogonal scheme can provide a rate of only around 1 bits/s/Hz. Intuitively, the strong user, being at high SNR, is degree-of-freedom limited and superposition coding allows it to use the full degrees of freedom of the channel while being allocated only a small amount of transmit power, thus causing small amount

Figure 6.9 The boundary of rate pairs (in bits/s/Hz) achievable by superposition coding (solid line) and orthogonal schemes (dashed line) for the two-user asymmetric downlink AWGN channel with the user SNRs equal to 0 and 20 dB (i.e., $P|h_1|^2/N_0 = 1$ and $P|h_2|^2/N_0 = 100$). In the orthogonal schemes, both the power split $P = P_1 + P_2$ and split in degrees of freedom α are jointly optimized to compute the boundary.

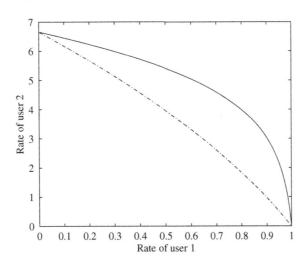

of interference to the weak user. In contrast, an orthogonal scheme has to allocate a significant fraction of the degrees of freedom to the weak user to achieve near single-user performance, and this causes a large degradation in the performance of the strong user.

So far we have considered a specific signaling scheme: linear superposition of the signals of the two users to form the transmit signal (cf. (6.19)). With this specific encoding method, the SIC decoding procedure is optimal. However, one can show that *this scheme in fact achieves the capacity* and the boundary of the capacity region of the downlink AWGN channel is given by (6.22) (Exercise 6.26).

While we have restricted ourselves to two users in the presentation, these results have natural extensions to the general K-user downlink channel. In the symmetric case $|h_k| = |h|$ for all k, the capacity region is given by the single constraint

$$\sum_{k=1}^{K} R_k < \log\left(1 + \frac{P|h|^2}{N_0}\right). \tag{6.24}$$

In general with the ordering $|h_1| \leq |h_2| \leq \cdots \leq |h_K|$, the boundary of the capacity region of the downlink AWGN channel is given by the parameterized rate tuple

$$R_k = \log\left(1 + \frac{P_k|h_k|^2}{N_0 + \left(\sum_{j=k+1}^{K} P_j\right)|h_k|^2}\right), \qquad k = 1\ldots K, \tag{6.25}$$

where $P = \sum_{k=1}^{K} P_k$ is the power split among the users. Each rate tuple on the boundary, as in (6.25), is achieved by superposition coding.

Since we have a full characterization of the tradeoff between the rates at which users can be reliably communicated to, we can easily derive specific scalar performance measures. In particular, we focused on sum capacity in the *uplink* analysis; to achieve the sum capacity we required all the users to transmit simultaneously (using the SIC receiver to decode the data). In contrast, we see from (6.25) that the sum capacity of the downlink is achieved by transmitting to a *single* user, the user with the highest SNR.

Summary 6.1 Uplink and downlink AWGN capacity

Uplink:

$$y[m] = \sum_{k=1}^{K} x_k[m] + w[m] \tag{6.26}$$

with user k having power constraint P_k.

Achievable rates satisfy:

$$\sum_{k \in \mathcal{S}} R_k \le \log \left(1 + \frac{\sum_{k \in \mathcal{S}} P_k}{N_0} \right) \qquad \text{for all } \mathcal{S} \subset \{1, \ldots, K\} \qquad (6.27)$$

The $K!$ corner points are achieved by SIC, one corner point for each cancellation order. They all achieve the same optimal sum rate.

A natural ordering would be to decode starting from the strongest user first and move towards the weakest user.

Downlink:

$$y_k[m] = h_k x[m] + w_k[m], \qquad k = 1, \ldots K \qquad (6.28)$$

with $|h_1| \le |h_2| \le \cdots \le |h_K|$.

The boundary of the capacity region is given by the rate tuples:

$$R_k = \log \left(1 + \frac{P_k |h_k|^2}{N_0 + (\sum_{j=k+1}^{K} P_j) |h_k|^2} \right), \qquad k = 1 \ldots K, \qquad (6.29)$$

for all possible splits $P = \sum_k P_k$ of the total power at the base-station.

The optimal points are achieved by superposition coding at the transmitter and SIC at each of the receivers.

The cancellation order at every receiver is *always* to decode the weaker users before decoding its own data.

Discussion 6.1 SIC: implementation issues

We have seen that successive interference cancellation plays an important role in achieving the capacities of both the uplink and the downlink channels. In contrast to the receivers for the multiple access systems in Chapter 4, SIC is a multiuser receiver. Here we discuss several potential practical issues in using SIC in a wireless system.

- **Complexity scaling with the number of users** In the uplink, the base-station has to decode the signals of every user in the cell, whether it uses the conventional single-user receiver or the SIC. In the downlink, on the other hand, the use of SIC at the mobile means that it now has to decode information intended for some of the other users, something it would not be doing in a conventional system. Then the complexity at each mobile scales with the number of users in the cell; this is not very acceptable. However, we have seen that superposition coding in conjunction with

SIC has the largest performance gain when the users have very disparate channels from the base-station. Due to the spatial geometry, typically there are only a few users close to the base-station while most of the users are near the edge of the cell. This suggests a practical way of limiting complexity: break the users in the cell into groups, with each group containing a small number of users with disparate channels. Within each group, superposition coding/SIC is performed, and across the groups, transmissions are kept orthogonal. This should capture a significant part of the performance gain.

- **Error propagation** Capacity analysis assumes error-free decoding but of course, with actual codes, errors are made. Once an error occurs for a user, all the users later in the SIC decoding order will very likely be decoded incorrectly. Exercise 6.12 shows that if $p_e^{(i)}$ is the probability of decoding the ith user incorrectly, assuming that all the previous users are decoded correctly, then the actual error probability for the kth user under SIC is at most

$$\sum_{i=1}^{k} p_e^{(i)}. \tag{6.30}$$

So, if all the users are coded with the same target error probability assuming no propagation, the effect of error propagation degrades the error probability by a factor of at most the number of users K. If K is reasonably small, this effect can easily be compensated by using a slightly stronger code (by, say, increasing the block length by a small amount).

- **Imperfect channel estimates** To remove the effect of a user from the aggregate received signal, its contribution must be reconstructed from the decoded information. In a wireless multipath channel, this contribution depends also on the impulse response of the channel. Imperfect estimate of the channel will lead to *residual* cancellation errors. One concern is that, if the received powers of the users are very disparate (as in the example in Figure 6.3 where they differ by 20 dB), then the residual error from cancelling the stronger user can still swamp the weaker user's signal. On the other hand, it is also easier to get an accurate channel estimate when the user is strong. It turns out that these two effects compensate each other and the effect of residual errors does not grow with the power disparity (Exercise 6.13).

- **Analog-to-digital quantization error** When the received powers of the users are very disparate, the analog-to-digital (A/D) converter needs to have a very large dynamic range, and at the same time, enough resolution to quantize accurately the contribution from the weak signal. For example, if the power disparity is 20 dB, even 1-bit accuracy for the weak signal would require an 8-bit A/D converter. This may well pose an implementation constraint on how much gain SIC can offer.

6.3 Uplink fading channel

Let us now include fading. Consider the complex baseband representation of the uplink flat fading channel with K users:

$$y[m] = \sum_{k=1}^{K} h_k[m] x_k[m] + w[m], \tag{6.31}$$

where $\{h_k[m]\}_m$ is the fading process of user k. We assume that the fading processes of different users are independent of each other and $\mathbb{E}[|h_k[m]|^2] = 1$. Here, we focus on the symmetric case when each user is subject to the same average power constraint, P, and the fading processes are identically distributed. In this situation, the sum and the symmetric capacities are the key performance measures. We will see later in Section 6.7 how the insights obtained from this idealistic symmetric case can be applied to more realistic asymmetric situations. To understand the effect of the channel fluctuations, we make the simplifying assumption that the base-station (receiver) can perfectly track the fading processes of all the users.

6.3.1 Slow fading channel

Let us start with the slow fading situation where the time-scale of communication is short relative to the coherence time interval for all the users, i.e., $h_k[m] = h_k$ for all m. Suppose the users are transmitting at the same rate R bits/s/Hz. Conditioned on each realization of the channels $h_1, \ldots, h_K$, we have the standard uplink AWGN channel with received SNR of user k equal to $|h_k|^2 P/N_0$. If the symmetric capacity of this uplink AWGN channel is less than R, then the base-station can never recover *all* of the users' information accurately; this results in outage. From the expression for the capacity region of the general K-user uplink AWGN channel (cf. (6.10)), the probability of the outage event can be written as

$$p_{\text{out}}^{\text{ul}} := \mathbb{P} \left\{ \log \left(1 + \text{SNR} \sum_{k \in \mathcal{S}} |h_k|^2 \right) < |\mathcal{S}| R, \quad \text{for some } \mathcal{S} \subset \{1, \ldots, K\} \right\}. \tag{6.32}$$

Here $|\mathcal{S}|$ denotes the cardinality of the set $\mathcal{S}$ and $\text{SNR} := P/N_0$. The corresponding ϵ-outage symmetric capacity, C_ϵ^{sym}, is then the largest rate R such that the outage probability in (6.32) is smaller than or equal to ϵ.

In Section 5.4.1, we have analyzed the behavior of the outage capacity, $C_\epsilon(\text{SNR})$, of the point-to-point slow fading channel. Since this corresponds to the performance of just a single user, it is equal to C_ϵ^{sym} with $K = 1$. With more than one user, C_ϵ^{sym} is only smaller: now each user has to deal not only with a random channel realization but also inter-user interference. Orthogonal multiple access is designed to completely eliminate inter-user interference at the cost of lesser (by a factor of $1/K$) degrees of freedom to each user (but the SNR is boosted by a factor of K). Since the users experience independent fading, an individual outage probability of ϵ for each user translates into

$$1 - (1 - \epsilon)^K \approx K\epsilon$$

outage probability when we require *each* user's information to be successfully decoded. We conclude that the largest symmetric ϵ-outage rate with orthogonal multiple access is equal to

$$\frac{C_{\epsilon/K}(K\mathsf{SNR})}{K}. \tag{6.33}$$

How much improved are the outage performances of more sophisticated multiple access schemes, as compared to orthogonal multiple access?

At low SNRs, the outage performance for any K is just as poor as the point-to-point case (with the outage probability, p_{out}, in (5.54)): indeed, at low SNRs we can approximate (6.32) as

$$\begin{aligned}
p_{\mathrm{out}}^{\mathrm{ul}} &\approx \mathbb{P}\left\{ \frac{|h_k|^2 P}{N_0} < R\log_e 2, \quad \text{for some } k \in \{1, \dots, K\} \right\} \\
&\approx K p_{\mathrm{out}}.
\end{aligned} \tag{6.34}$$

So we can write

$$\begin{aligned}
C_\epsilon^{\mathrm{sym}} &\approx C_{\epsilon/K}(\mathsf{SNR}) \\
&\approx F^{-1}\left(1 - \frac{\epsilon}{K}\right) C_{\mathrm{awgn}}.
\end{aligned} \tag{6.35}$$

Here we used the approximation for C_ϵ at low SNR in (5.61). Since C_{awgn} is linear in SNR at low SNR,

$$C_\epsilon^{\mathrm{sym}} \approx \frac{C_{\epsilon/K}(K\mathsf{SNR})}{K}, \tag{6.36}$$

the same performance as orthogonal multiple access (cf. (6.33)).

The analysis at high SNR is more involved, so to get a feel for the role of inter-user interference on the outage performance of optimal multiple access schemes, we plot $C_\epsilon^{\mathrm{sym}}$ for $K = 2$ users as compared to C_ϵ, for Rayleigh fading, in Figure 6.10. As SNR increases, the ratio of $C_\epsilon^{\mathrm{sym}}$ to C_ϵ increases; thus the effect of the inter-user interference is becoming smaller. However, as SNR becomes very large, the ratio starts to decrease; the inter-user interference begins to dominate. In fact, at very large SNRs the ratio drops back to $1/K$ (Exercise 6.14). We will obtain a deeper understanding of this behavior when we study outage in the uplink with multiple antennas in Section 10.1.4.

Figure 6.10 Plot of the symmetric ϵ-outage capacity of the two-user Rayleigh slow fading uplink as compared to C_ϵ, the corresponding performance of a point-to-point Rayleigh slow fading channel.

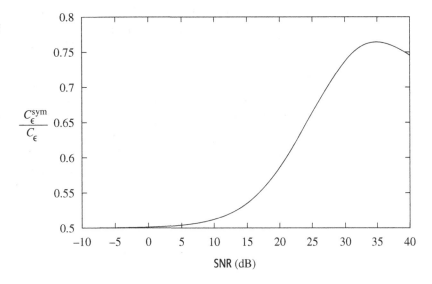

6.3.2 Fast fading channel

Let us now turn to the fast fading scenario, where each $\{h_k[m]\}_m$ is modelled as a time-varying ergodic process. With the ability to code over multiple coherence time intervals, we can have a meaningful definition of the capacity region of the uplink fading channel. With only receiver CSI, the transmitters cannot track the channel and there is no dynamic power allocation. Analogous to the discussion in the point-to-point case (cf. Section 5.4.5 and, in particular, (5.89)), the sum capacity of the uplink fast fading channel can be expressed as:

$$C_{\text{sum}} = \mathbb{E}\left[\log\left(1 + \frac{\sum_{k=1}^{K} |h_k|^2 P}{N_0}\right)\right]. \tag{6.37}$$

Here h_k is the random variable denoting the fading of user k at a particular time and the time averages are taken to converge to the same limit for all realizations of the fading process (i.e., the fading processes are ergodic). A formal derivation of the capacity region of the fast fading uplink (with potentially multiple antenna elements) is carried out in Appendix B.9.3.

How does this compare to the sum capacity of the uplink channel without fading (cf. (6.12))? Jensen's inequality implies that

$$\mathbb{E}\left[\log\left(1 + \frac{\sum_{k=1}^{K} |h_k|^2 P}{N_0}\right)\right] \leq \log\left(1 + \frac{\mathbb{E}[\sum_{k=1}^{K} |h_k|^2]P}{N_0}\right)$$

$$= \log\left(1 + \frac{KP}{N_0}\right).$$

Hence, without channel state information at the transmitter, fading always hurts, just as in the point-to-point case. However, when the number of users becomes large, $1/K \cdot \sum_{k=1}^{K} |h_k|^2 \to 1$ with probability 1, and the penalty due to fading vanishes.

To understand why the effect of fading goes away as the number of users grows, let us focus on a specific decoding strategy to achieve the sum capacity. With each user spreading their information on the entire bandwidth simultaneously, the successive interference cancellation (SIC) receiver, which is optimal for the uplink AWGN channel, is also optimal for the uplink fading channel. Consider the kth stage of the cancellation procedure, where user k is being decoded and users $k+1, \ldots, K$ are not canceled. The effective channel that user k sees is

$$y[m] = h_k[m]x_k[m] + \sum_{i=k+1}^{K} h_i[m]x_i[m] + w[m]. \qquad (6.38)$$

The rate that user k gets is

$$R_k = \mathbb{E}\left[\log\left(1 + \frac{|h_k|^2 P}{\sum_{i=k+1}^{K} |h_i|^2 P + N_0}\right)\right]. \qquad (6.39)$$

Since there are many users sharing the spectrum, the SINR for user k is low. Thus, the capacity penalty due to the fading of user k is small (cf. (5.92)). Moreover, there is also *averaging* among the interferers. Thus, the effect of the fading of the interferers also vanishes. More precisely,

$$\begin{aligned}
R_k &\approx \mathbb{E}\left[\frac{|h_k|^2 P}{\sum_{i=k+1}^{K} |h_i|^2 P + N_0}\right]\log_2 e \\
&\approx \mathbb{E}\left[\frac{|h_k|^2 P}{(K-k)P + N_0}\right]\log_2 e \\
&= \frac{P}{(K-k)P + N_0}\log_2 e,
\end{aligned}$$

which is the rate that user k would have got in the (unfaded) AWGN channel. The first approximation comes from the linearity of $\log(1 + \mathsf{SNR})$ for small SNR, and the second approximation comes from the law of large numbers.

In the AWGN case, the sum capacity can be achieved by an orthogonal multiple access scheme which gives a fraction, $1/K$, of the total degrees of freedom to each user. How about the fading case? The sum rate achieved by this orthogonal scheme is

$$\sum_{k=1}^{K} \frac{1}{K}\mathbb{E}\left[\log\left(1 + \frac{K|h_k|^2 P}{N_0}\right)\right] = \mathbb{E}\left[\log\left(1 + \frac{K|h_k|^2 P}{N_0}\right)\right], \qquad (6.40)$$

which is strictly less than the sum capacity of the uplink fading channel (6.37) for $K \geq 2$. In particular, the penalty due to fading persists even when there is a large number of users.

6.3.3 Full channel side information

We now come to a case of central interest in this chapter, the fast fading channel with tracking of the channels of all the users at the receiver and all the transmitters.[5] As opposed to the case with only receiver CSI, we can now dynamically allocate powers to the users as a function of the channel states. Analogous to the point-to-point case, we can without loss of generality focus on the simple block fading model

$$y[m] = \sum_{k=1}^{K} h_k[m] x_k[m] + w[m], \tag{6.41}$$

where $h_k[m] = h_{k,\ell}$ remains constant over the ℓth coherence period of $T_c(T_c \gg 1)$ symbols and is i.i.d. across different coherence periods. The channel over L such coherence periods can be modeled as a parallel uplink channel with L sub-channels which fade independently. Each sub-channel is an uplink AWGN channel. For a given realization of the channel gains $h_{k,\ell}$, $k = 1, \ldots, K, \ell = 1, \ldots, L$, the sum capacity (in bits/symbol) of this parallel channel is, as for the point-to-point case (cf. (5.95)),

$$\max_{P_{k,\ell}:k=1,\ldots,K,\ell=1,\ldots,L} \frac{1}{L} \sum_{\ell=1}^{L} \log \left(1 + \frac{\sum_{k=1}^{K} P_{k,\ell} |h_{k,\ell}|^2}{N_0} \right), \tag{6.42}$$

subject to the powers being non-negative and the average power constraint on each user:

$$\frac{1}{L} \sum_{\ell=1}^{L} P_{k,\ell} = P, \qquad k = 1, \ldots, K. \tag{6.43}$$

The solution to this optimization problem as $L \to \infty$ yields the appropriate power allocation policy to be followed by the users.

As discussed in the point-to-point communication context with full CSI (cf. Section 5.4.6), we can use a variable rate coding scheme: in the ℓth sub-channel, the transmit powers dictated by the solution to the optimization problem above (6.42) are used by the users and a code designed for this fading state is used. For this code, each codeword sees a *time-invariant* uplink

---

[5] As we will see, the transmitters will not need to explicitly keep track of the channel variations of *all* the users. Only an appropriate function of the channels of all the users needs to be tracked, which the receiver can compute and feed back to the users.

AWGN channel. Thus, we can use the encoding and decoding procedures for the code designed for the uplink AWGN channel. In particular, to achieve the maximum sum rate, we can use orthogonal multiple access: this means that the codes designed for the *point-to-point* AWGN channel can be used. Contrast this with the case when only the receiver has CSI, where we have shown that orthogonal multiple access is *strictly* suboptimal for fading channels. Note that this argument on the optimality of orthogonal multiple access holds regardless of whether the users have symmetric fading statistics.

In the case of the *symmetric* uplink considered here, the optimal power allocation takes on a particularly simple structure. To derive it, let us consider the optimization problem (6.42), but with the individual power constraints in (6.43) relaxed and replaced by a total power constraint:

$$\frac{1}{L}\sum_{\ell=1}^{L}\sum_{k=1}^{K}P_{k,\ell} = KP. \tag{6.44}$$

The sum rate in the ℓth sub-channel is

$$\log\left(1 + \frac{\sum_{k=1}^{K}P_{k,\ell}|h_{k,\ell}|^2}{N_0}\right), \tag{6.45}$$

and for a given total power $\sum_{k=1}^{K}P_{k,\ell}$ allocated to the ℓth sub-channel, this quantity is maximized by giving all that power to the user with the strongest channel gain. Thus, the solution of the optimization problem (6.42) subject to the constraint (6.44) is that at each time, allow only the user with the *best* channel to transmit. Since there is just one user transmitting at any time, we have reduced to a point-to-point problem and can directly infer from our discussion in Section 5.4.6 that the best user allocates its power according to the *waterfilling* policy. More precisely, the optimal power allocation policy is

$$P_{k,\ell} = \begin{cases} \left(\dfrac{1}{\lambda} - \dfrac{N_0}{\max_i |h_{i,\ell}|^2}\right)^{+} & \text{if } |h_{k,\ell}| = \max_i |h_{i,\ell}|, \\ \qquad\qquad 0 & \text{else,} \end{cases} \tag{6.46}$$

where λ is chosen to meet the sum power constraint (6.44). Taking the number of coherence periods $L \to \infty$ and appealing to the ergodicity of the fading process, we get the optimal capacity-achieving power allocation strategy, which allocates powers to the users as a function of the joint channel state $\mathbf{h} := (h_1, \ldots, h_K)$:

$$\boxed{P_k^*(\mathbf{h}) = \begin{cases} \left(\dfrac{1}{\lambda} - \dfrac{N_0}{\max_i |h_i|^2}\right)^{+} & \text{if } |h_k|^2 = \max_i |h_i|^2, \\ \qquad\qquad 0 & \text{else,} \end{cases}} \tag{6.47}$$

with λ chosen to satisfy the power constraint

$$\sum_{k=1}^{K} \mathbb{E}[P_k^*(\mathbf{h})] = KP. \tag{6.48}$$

(Rigorously speaking, this formula is valid only when there is exactly one user with the strongest channel. See Exercise 6.16 for the generalization to the case when multiple users can have the same fading state.) The resulting sum capacity is

$$C_{\text{sum}} = \mathbb{E}\left[\log\left(1 + \frac{P_{k^*}(\mathbf{h})|h_{k^*}|^2}{N_0}\right)\right], \tag{6.49}$$

where $k^*(\mathbf{h})$ is the index of the user with the strongest channel at joint channel state $\mathbf{h}$.

We have derived this result assuming a *total* power constraint on all the users, but by symmetry, the power consumption of all the users is the same under the optimal solution (recall that we are assuming independent and identical fading processes across the users here). Therefore the individual power constraints in (6.43) are automatically satisfied and we have solved the original problem as well.

This result is the multiuser generalization of the idea of opportunistic communication developed in Chapter 5: resource is allocated at the times and to the user whose channel is good.

When one attempts to generalize the optimal power allocation solution from the point-to-point setting to the multiuser setting, it may be tempting to think of "users" as a new dimension, in addition to the time dimension, over which dynamic power allocation can be performed. This may lead us to guess that the optimal solution is waterfilling over the joint time/user space. This, as we have already seen, is not the correct solution. The flaw in this reasoning is that having multiple users *does not* provide additional degrees of freedom in the system: the users are just sharing the time/frequency degrees of freedom already existing in the channel. Thus, the optimal power allocation problem should really be thought of as how to partition the total resource (power) across the time/frequency degrees of freedom and how to share the resource across the users in each of those degrees of freedom. The above solution says that from the point of view of maximizing the sum capacity, the optimal sharing is just to allocate all the power to the user with the strongest channel on that degree of freedom.

We have focused on the sum capacity in the symmetric case where users have identical channel statistics and power constraints. It turns out that in the asymmetric case, the optimal strategy to achieve sum capacity is still to have one user transmitting at a time, but the criterion of choosing which user is different. This problem is analyzed in Exercise 6.15. However, in the asymmetric case, maximizing the sum rate may not be the appropriate objective, since the user with the statistically better channel may get a much higher rate at the expense of the other users. In this case, one may be interested in operating at points in the multiuser capacity region of the uplink fading channel other than the point maximizing the sum rate. This problem is analyzed in Exercise 6.18. It turns out that, as in the time-invariant uplink, orthogonal multiple access is not optimal. Instead, users transmit simultaneously and are jointly decoded (using SIC, for example), even though the rates and powers are still dynamically allocated as a function of the channel states.

Summary 6.2 Uplink fading channel

Slow Rayleigh fading At low SNR, the symmetric outage capacity is equal to the outage capacity of the point-to-point channel, but scaled down by the number of users. At high SNR, the symmetric outage capacity for moderate number of users is approximately equal to the outage capacity of the point-to-point channel. Orthogonal multiple access is close to optimal at low SNR.

Fast fading, receiver CSI With a large number of users, each user gets the same performance as in an uplink AWGN channel with the same average SNR. Orthogonal multiple access is strictly suboptimal.

Fast fading, full CSI Orthogonal multiple access can still achieve the sum capacity. In a symmetric uplink, the policy of allowing only the best user to transmit at each time achieves the sum capacity.

6.4 Downlink fading channel

We now turn to the downlink fading channel with K users:

$$y_k[m] = h_k[m]x[m] + w_k[m], \qquad k = 1, \ldots, K, \tag{6.50}$$

where $\{h_k[m]\}_m$ is the channel fading process of user k. We retain the average power constraint of P on the transmit signal and $w_k[m] \sim \mathcal{CN}(0, N_0)$ to be i.i.d. in time m (for each user $k = 1, \ldots, K$).

As in the uplink, we consider the symmetric case: $\{h_k[m]\}_m$ are identically distributed processes for $k = 1 \ldots K$. Further, let us also make the same assumption we did in the uplink analysis: the processes $\{h_k[m]\}_m$ are ergodic (i.e., the time average of every realization equals the statistical average).

6.4.1 Channel side information at receiver only

Let us first consider the case when the receivers can track the channel but the transmitter does not have access to the channel realizations (but has access to a statistical characterization of the channel processes of the users). To get a feel for good strategies to communicate on this fading channel and to understand the capacity region, we can argue as in the downlink AWGN channel. We have the single-user bounds, in terms of the point-to-point fading channel capacity in (5.89):

$$R_k < \mathbb{E}\left[\log\left(1 + \frac{|h|^2 P}{N_0}\right)\right], \qquad k = 1, \dots, K, \qquad (6.51)$$

where h is a random variable distributed as the stationary distribution of the ergodic channel processes. In the symmetric downlink AWGN channel, we argued that the users have the same channel quality and hence could decode each other's data. Here, the fading statistics are symmetric and by the assumption of ergodicity, we can extend the argument of the AWGN case to say that, if user k can decode its data reliably, then all the other users can also successfully decode user k's data. Analogous to (6.18) in the AWGN downlink analysis, we obtain

$$\sum_{k=1}^{K} R_k < \mathbb{E}\left[\log\left(1 + \frac{|h|^2 P}{N_0}\right)\right]. \qquad (6.52)$$

An alternative way to see that the right hand side in (6.52) is the best sum rate one can achieve is outlined in Exercise 6.27. The bound (6.52) is clearly achievable by transmitting to one user only or by time-sharing between any number of users. Thus in the symmetric fading channel, we obtain the same conclusion as in the symmetric AWGN downlink: the rate pairs in the capacity region can be achieved by both orthogonalization schemes and superposition coding.

How about the downlink fading channel with *asymmetric* fading statistics of the users? While we can use the orthogonalization scheme in this asymmetric model as well, the applicability of superposition decoding is not so clear. Superposition coding was successfully applied in the downlink AWGN channel because there is an *ordering* of the channel strength of the users from weak to strong. In the asymmetric fading case, users in general have different fading distributions and there is no longer a *complete* ordering of the users. In this case, we say that the downlink channel is *non-degraded* and little is known about good strategies for communication. Another interesting situation when the downlink channel is non-degraded arises when the transmitter has an array of multiple antennas; this is studied in Chapter 10.

6.4.2 Full channel side information

We saw in the uplink that the communication scenario becomes more interesting when the transmitters can track the channel as well. In this case, the transmitters can vary their powers as a function of the channel. Let us now turn to the analogous situation in the downlink where the single transmitter tracks all the channels of the users it is communicating to (the users continue to track their individual channels). As in the uplink, we can allocate powers to the users as a function of the channel fade level. To see the effect, let us continue to focus on sum capacity. We have seen that without fading, the sum capacity is achieved by transmitting only to the best user. Now as the channels vary, we can pick the best user at each time and further allocate it an appropriate power, subject to a constraint on the average power. Under this strategy, the downlink channel reduces to a point-to-point channel with the channel gain distributed as

$$\max_{k=1\ldots K} |h_k|^2.$$

The optimal power allocation is the, by now familiar, waterfilling solution:

$$P^*(\mathbf{h}) = \left(\frac{1}{\lambda} - \frac{N_0}{\max_{k=1\ldots K} |h_k|^2} \right)^+, \tag{6.53}$$

where $\mathbf{h} = (h_1, \ldots, h_K)^t$ is the joint fading state and $\lambda > 0$ is chosen such that the average power constraint is met. The optimal strategy is exactly the same as in the sum capacity of the uplink. The sum capacity of the downlink is:

$$\mathbb{E}\left[\log\left(1 + \frac{P^*(\mathbf{h})(\max_{k=1\ldots K} |h_k^2|)}{N_0} \right) \right]. \tag{6.54}$$

6.5 Frequency-selective fading channels

The extension of the flat fading analysis in the uplink and the downlink to underspread frequency-selective fading channels is conceptually straightforward. As we saw in Section 5.4.7 in the point-to-point setting, we can think of the underspread channel as a set of parallel sub-carriers over each coherence time interval and varying independently from one coherence time interval to the other. We can see this constructively by imposing a cyclic prefix to all the transmit signals; the cyclic prefix should be of length that is larger than the largest multipath delay spread that we are likely to encounter among the different users. Since this overhead is fixed, the loss is amortized when communicating over a long block length.

We can apply exactly the same OFDM transformation to the multiuser channels. Thus on the nth sub-carrier, we can write the uplink channel as

$$\tilde{y}_n[i] = \sum_{k=1}^{K} \tilde{h}_n^{(k)}[i]\, \tilde{d}_n^{(k)}[i] + \tilde{w}_n[i], \tag{6.55}$$

where $\widetilde{\mathbf{d}}^{(k)}[i]$, $\widetilde{\mathbf{h}}^{(k)}[i]$ and $\widetilde{\mathbf{y}}[i]$, respectively, represent the DFTs of the transmitted sequence of user k, of the channel and of the received sequence at OFDM symbol time i.

The flat fading uplink channel can be viewed as a set of parallel multiuser sub-channels, one for each coherence time interval. With full CSI, the optimal strategy to maximize the sum rate in the symmetric case is to allow only the user with the best channel to transmit at each coherence time interval. The frequency-selective fading uplink channel can also be viewed as a set of parallel multiuser sub-channels, one for each sub-carrier and each coherence time interval. Thus, the optimal strategy is to allow the best user to transmit on each of these sub-channels. The power allocated to the best user is waterfilling over time and frequency. As opposed to the flat fading case, multiple users can now transmit at the same time, but over different sub-carriers. Exactly the same comments apply to the downlink.

6.6 Multiuser diversity

6.6.1 Multiuser diversity gain

Let us consider the sum capacity of the uplink and downlink flat fading channels (see (6.49) and (6.54), respectively). Each can be interpreted as the waterfilling capacity of a point-to-point link with a power constraint equal to the total transmit power (in the uplink this is equal to KP and in the downlink it is equal to P), and a fading process whose magnitude varies as $\{\max_k |h_k[m]|\}$. Compared to a system with a single transmitting user, the multiuser gain comes from two effects:

1. the increase in total transmit power in the case of the uplink;
2. the effective channel gain at time m that is improved from $|h_1[m]|^2$ to $\max_{1 \leq k \leq K} |h_k[m]|^2$.

The first effect already appeared in the uplink AWGN channel and also in the fading channel with channel side information only at the receiver. The second effect is entirely due to the ability to dynamically schedule resources among the users as a function of the channel state.

The sum capacity of the uplink Rayleigh fading channel with full CSI is plotted in Figure 6.11 for different numbers of users. The performance curves are plotted as a function of the total $\mathsf{SNR} := KP/N_0$ so as to focus on the second effect. The sum capacity of the channel with only CSI at the receiver is also plotted for different numbers of users. The capacity of the point-to-point AWGN channel with received power KP (which is also the sum capacity of a K-user uplink AWGN channel) is shown as a baseline. Figure 6.12 focuses on the low SNR regime.

Figure 6.11 Sum capacity of
the uplink Rayleigh fading
channel plotted as a function
of SNR $= KP/N_0$.

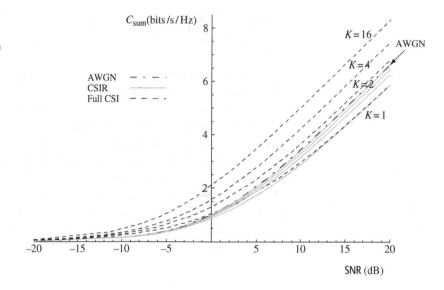

Figure 6.12 Sum capacity of
the uplink Rayleigh fading
channel plotted as a function
of SNR $= KP/N_0$ in the low
SNR regime. Everything is
plotted as a fraction of the
AWGN channel capacity.

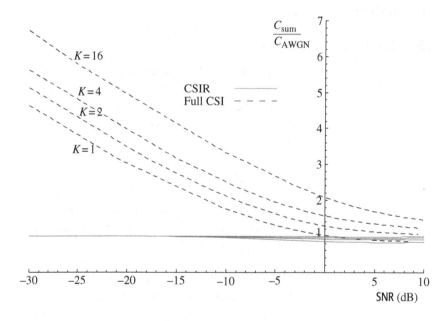

Several observations can be made from the plots:

- The sum capacity without transmitter CSI increases with the number of the
  users, but not significantly. This is due to the multiuser averaging effect
  explained in the last section. This sum capacity is always bounded by the
  capacity of the AWGN channel.
- The sum capacity with full CSI increases significantly with the number of
  users. In fact, with even two users, this sum capacity already exceeds that

of the AWGN channel. At 0 dB, the capacity with $K = 16$ users is about a factor of 2.5 of the capacity with $K = 1$. The corresponding power gain is about 7 dB. Compared to the AWGN channel, the capacity gain for $K = 16$ is about a factor of 2.2 and an SNR gain of 5.5 dB.

- For $K = 1$, the capacity benefit of transmitter CSI only becomes apparent at quite low SNR levels; at high SNR there is no gain. For $K > 1$ the benefit is apparent throughout the entire SNR range, although the relative gain is still more significant at low SNR. This is because the gain is still primarily a *power gain*.

The increase in the full CSI sum capacity comes from a *multiuser diversity* effect: when there are many users that fade *independently*, at any one time there is a high probability that one of the users will have a strong channel. By allowing only that user to transmit, the shared channel resource is used in the most efficient manner and the total system throughput is maximized. The larger the number of users, the stronger tends to be the strongest channel, and the more the multiuser diversity gain.

The amount of multiuser diversity gain depends crucially on the *tail* of the fading distribution $|h_k|^2$: the heavier the tail, the more likely there is a user with a very strong channel, and the larger the multiuser diversity gain. This is shown in Figure 6.13, where the sum capacity is plotted as a function of the number of users for both Rayleigh and Rician fading with κ-factor equal to 5, with the total SNR, equal to KP/N_0, fixed at 0 dB. Recall from

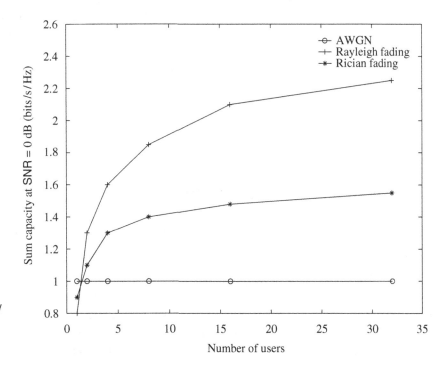

Figure 6.13 Multiuser diversity gain for Rayleigh and Rician fading channels ($\kappa = 5$); $KP/N_0 = 0$ dB.

Section 2.4 that, Rician fading models the situation when there is a strong specular line-of-sight path plus many small reflected paths. The parameter κ is defined as the ratio of the energy in the specular line-of-sight path to the energy in the diffused components. Because of the line-of-sight component, the Rician fading distribution is less "random" and has a lighter tail than the Rayleigh distribution with the same average channel gain. As a consequence, it can be seen that the multiuser diversity gain is significantly smaller in the Rician case, as compared to the Rayleigh case (Exercise 6.21).

6.6.2 Multiuser versus classical diversity

We have called the above explained phenomenon multiuser diversity. Like the diversity techniques discussed in Chapter 3, multiuser diversity also arises from the existence of independently faded signal paths, in this case from the multiple users in the network. However, there are several important differences. First, the main objective of the diversity techniques in Chapter 3 is to improve the *reliability* of communication in slow fading channels; in contrast, the role of multiuser diversity is to increase the total throughput over fast fading channels. Under the sum-capacity-achieving strategy, a user has no guarantee of a high rate in any particular slow fading state; only by averaging over the variations of the channel is a high long-term average throughput attained. Second, while the diversity techniques are designed to *counteract* the adverse effect of fading, multiuser diversity improves system performance by *exploiting* channel fading: channel fluctuations due to fading ensure that with high probability there is a user with a channel strength much larger than the mean level; by allocating all the system resources to that user, the benefit of this strong channel is fully capitalized. Third, while the diversity techniques in Chapter 3 pertain to a point-to-point link, the benefit of multiuser diversity is *system-wide*, across the users in the network. This aspect of multiuser diversity has ramifications on the implementation of multiuser diversity in a cellular system. We will discuss this next.

6.7 Multiuser diversity: system aspects

The cellular system requirements to extract the multiuser diversity benefits are:

- the base-station has access to channel quality measurements: in the downlink, we need each receiver to track its own channel SNR, through say a common downlink pilot, and feed back the instantaneous channel quality to the base-station (assuming an FDD system); and in the uplink, we need transmissions from the users so that their channel qualities can be tracked;

- the ability of the base-station to schedule transmissions among the users as well as to adapt the data rate as a function of the instantaneous channel quality.

These features are already present in the designs of many third-generation systems. Nevertheless, in practice there are several considerations to take into account before realizing such gains. In this section, we study three main hurdles towards a system implementation of the multiuser diversity idea and some prominent ways of addressing these issues.

1. **Fairness and delay** To implement the idea of multiuser diversity in a real system, one is immediately confronted with two issues: fairness and delay. In the ideal situation when users' fading *statistics* are the same, the strategy of communicating with the user having the best channel maximizes not only the total throughput of the system but also that of individual users. In reality, the statistics are not symmetric; there are users who are closer to the base-station with a better average SNR; there are users who are stationary and some that are moving; there are users who are in a rich scattering environment and some with no scatterers around them. More-over, the strategy is only concerned with maximizing long-term average throughputs; in practice there are latency requirements, in which case the average throughput over the delay time-scale is the performance metric of interest. The challenge is to address these issues while at the same time exploiting the multiuser diversity gain inherent in a system with users having independent, fluctuating channel conditions. As a case study, we will look at one particular scheduler that harnesses multiuser diversity while addressing the real-world fairness and delay issues.

2. **Channel measurement and feedback** One of the key system requirements to harness multiuser diversity is to have scheduling decisions by the base-station be made as a function of the channel states of the users. In the uplink, the base-station has access to the user transmissions (over trickle channels which are used to convey control information) and has an estimate of the user channels. In the downlink, the users have access to their channel states but need to feedback these values to the base-station. Both the error in channel state measurement and the delay in feeding it back constitute a significant bottleneck in extracting the multiuser diversity gains.

3. **Slow and limited fluctuations** We have observed that the multiuser diversity gains depend on the distribution of channel fluctuations. In particular, larger and faster variations in a channel are preferred over slow ones. However, there may be a line-of-sight path and little scattering in the environment, and hence the dynamic range of channel fluctuations may be small. Further, the channel may fade very slowly compared to the delay constraints of the application so that transmissions cannot wait until the channel reaches its peak. Effectively, the dynamic range of channel fluctuations is small within the time-scale of interest. Both are important

sources of hindrance to implementing multiuser diversity in a real system. We will see a simple and practical scheme using an antenna array at the base-station that creates fast and large channel fluctuations even when the channel is originally slow fading with a small range of fluctuation.

6.7.1 Fair scheduling and multiuser diversity

As a case study, we describe a simple scheduling algorithm, called the *proportional fair* scheduler, designed to meet the challenges of delay and fairness constraints while harnessing multiuser diversity. This is the baseline scheduler for the downlink of IS-856, the third-generation data standard, introduced in Chapter 5. Recall that the downlink of IS-856 is TDMA-based, with users scheduled on time slots of length 1.67 ms based on the requested rates from the users (Figure 5.25). We have already discussed the rate adaptation mechanism in Chapter 5; here we will study the scheduling aspect.

Proportional fair scheduling: hitting the peaks

The *scheduler* decides which user to transmit information to at each time slot, based on the requested rates the base-station has previously received from the mobiles. The simplest scheduler transmits data to each user in a *round-robin* fashion, regardless of the channel conditions of the users. The scheduling algorithm used in IS-856 schedules in a *channel-dependent* manner to exploit multiuser diversity. It works as follows. It keeps track of the average throughput $T_k[m]$ of each user in an exponentially weighted window of length t_c. In time slot m, the base-station receives the "requested rates" $R_k[m]$, $k = 1, \ldots, K$, from all the users and the scheduling algorithm simply transmits to the user k^* with the largest

$$\frac{R_k[m]}{T_k[m]}$$

among all active users in the system. The average throughputs $T_k[m]$ are updated using an exponentially weighted low-pass filter:

$$T_k[m+1] = \begin{cases} (1 - 1/t_c)T_k[m] + (1/t_c)R_k[m] & k = k^*, \\ (1 - 1/t_c)T_k[m] & k \neq k^*. \end{cases} \quad (6.56)$$

One can get an intuitive feel of how this algorithm works by inspecting Figures 6.14 and 6.15. We plot the sample paths of the requested data rates of two users as a function of time slots (each time slot is 1.67 ms in IS-856). In Figure 6.14, the two users have identical fading *statistics*. If the scheduling time-scale t_c is much larger than the coherence time of the channels, then by symmetry the throughput of each user $T_k[m]$ converges to the same quantity. The scheduling algorithm reduces to always picking the user with the highest

Figure 6.14 For symmetric channel statistics of users, the scheduling algorithm reduces to serving each user with the largest requested rate.

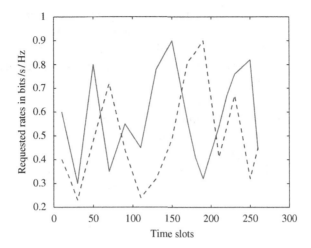

Figure 6.15 In general, with asymmetric user channel statistics, the scheduling algorithm serves each user when it is near its peak within the latency time-scale t_c.

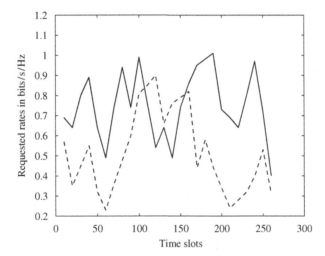

requested rate. Thus, each user is scheduled when its channel is good and at the same time the scheduling algorithm is perfectly fair in the long-term.

In Figure 6.15, due perhaps to different distances from the base-station, one user's channel is much stronger than that of the other user on average, even though both channels fluctuate due to multipath fading. Always picking the user with the highest requested rate means giving all the system resources to the statistically stronger user, and would be highly unfair. In contrast, under the scheduling algorithm described above, users compete for resources not directly based on their requested rates but based on the rates normalized by their respective average throughputs. The user with the statistically stronger channel will have a higher average throughput.

Thus, the algorithm schedules a user when its instantaneous channel quality is high *relative* to its own average channel condition over the time-scale t_c.

In short, data are transmitted to a user when its channel is *near its own peaks*. Multiuser diversity benefit can still be extracted because channels of different users fluctuate independently so that if there is a sufficient number of users in the system, most likely there will be a user near its peak at any one time.

The parameter t_c is tied to the latency time-scale of the application. Peaks are defined with respect to this time-scale. If the latency time-scale is large, then the throughput is averaged over a longer time-scale and the scheduler can afford to wait longer before scheduling a user when its channel hits a really high peak.

The main theoretical property of this algorithm is the following: With a very large t_c (approaching ∞), the algorithm maximizes

$$\sum_{k=1}^{K} \log T_k, \tag{6.57}$$

among all schedulers (see Exercise 6.28). Here, T_k is the long-term average throughput of user k.

Multiuser diversity and superposition coding

Proportional fair scheduling is an approach to deal with fairness among asymmetric users within the orthogonal multiple access constraint (TDMA in the case of IS-856). But we understand from Section 6.2.2 that for the AWGN channel, superposition coding in conjunction with SIC can yield significantly better performance than orthogonal multiple access in such asymmetric environments. One would expect similar gains in fading channels, and it is therefore natural to combine the benefits of superposition coding with multiuser diversity scheduling.

One approach is to divide the users in a cell into, say, two classes depending on whether they are near the base-station or near the cell edge, so that users in each class have statistically comparable channel strengths. Users whose current channel is instantaneously strongest in their own class are scheduled for simultaneous transmission via superposition coding (Figure 6.16). The user near the base-station can decode its own signal after stripping off the signal destined for the far-away user. By transmitting to the strongest user in each class, multiuser diversity benefits are captured. On the other hand, the nearby user has a very strong channel and the full degrees of freedom available (as opposed to only a fraction under orthogonal multiple access), and thus only needs to be allocated a small fraction of the power to enjoy very good rates. Allocating a small fraction of power to the nearby user has a salutary effect: the presence of this user will minimally affect the performance of the cell edge user. Hence, fairness can be maintained by a suitable allocation of power. The efficiency of this approach over proportional fair TDMA scheduling is quantified in Exercise 6.20. Exercise 6.19 shows that this strategy is in fact optimal in achieving *any* point on the boundary of

Figure 6.16 Superposition coding in conjunction with multiuser diversity scheduling. The strongest user from each cluster is scheduled and they are simultaneously transmitted to, via superposition coding.

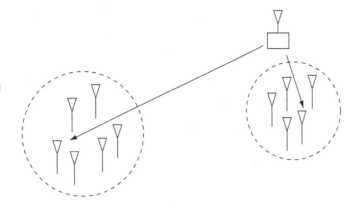

the downlink fading channel capacity region (as opposed to the strategy of transmitting to the user with the best channel overall, which is only optimal for the sum rate and which is an unfair operating point in this asymmetric scenario).

Multiuser diversity gain in practice

We can use the proportional fair algorithm to get some more insights into the issues involved in realizing multiuser diversity benefits in practice. Consider the plot in Figure 6.17, showing the total simulated throughput of the 1.25 MHz IS-856 downlink under the proportional fair scheduling algorithm in three environments:

- **Fixed** Users are fixed, but there are movements of objects around them (2 Hz Rician, $\kappa := E_{\text{direct}}/E_{\text{specular}} = 5$). Here E_{direct} is the energy in the direct

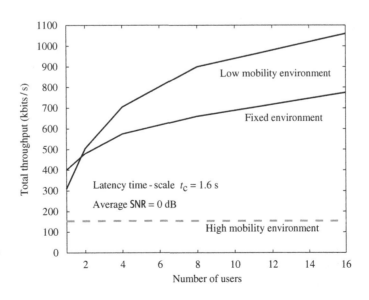

Figure 6.17 Multiuser diversity gain in fixed and mobile environments.

path that is not varying, while E_{specular} refers to the energy in the specular or time-varying component that is assumed to be Rayleigh distributed. The Doppler spectrum of this component follows Clarke's model with a Doppler spread of 2 Hz.

- **Low mobility** Users move at walking speeds (3 km/hr, Rayleigh).
- **High mobility** Users move at 30 km/hr, Rayleigh.

The average channel gain $\mathbb{E}[|h|^2]$ is kept the same in all the three scenarios for fairness of comparison. The total throughput increases with the number of users in both the fixed and low mobility environments, but the increase is more dramatic in the low mobility case. While the channel varies in both cases, the dynamic range and the rate of the variations is larger in the mobile environment than in the fixed one (Figure 6.18). This means that over the latency time-scale ($t_c = 1.67$ s in these examples) the peaks of the channel fluctuations are likely to be higher in the mobile environment, and the peaks are what determines the performance of the scheduling algorithm. Thus, the inherent multiuser diversity is more limited in the fixed environment.

Should one then expect an even higher throughput gain in the high mobility environment? In fact quite the opposite is true. The total throughput hardly increases with the number of users! It turns out that at this speed the receiver has trouble tracking and predicting the channel variations, so that the predicted channel is a low-pass smoothed version of the actual fading process. Thus, even though the actual channel fluctuates, opportunistic communication is impossible without knowing when the channel is actually good.

In the next section, we will discuss how the tracking of the channel can be improved in high mobility environments. In Section 6.7.3, we will discuss a scheme that *boosts* the inherent multiuser diversity in fixed environments.

6.7.2 Channel prediction and feedback

The prediction error is due to two effects: the error in measuring the channel from the pilot and the delay in feeding back the information to the base-station.

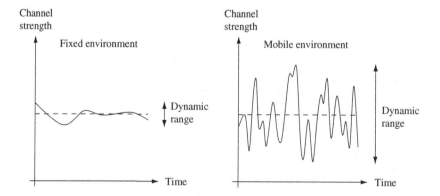

Figure 6.18 The channel varies much faster and has larger dynamic range in the mobile environment.

In the downlink, the pilot is shared between many users and is strong; so, the measurement error is quite small and the prediction error is mainly due to the feedback delay. In IS-856, this delay is about two time slots, i.e., 3.33 ms. At a vehicular speed of 30 km/h and carrier frequency of 1.9 GHz, the coherence time is approximately 2.5 ms; the channel coherence time is comparable to the delay and this makes prediction difficult.

One remedy to reduce the feedback delay is to shrink the size of the scheduling time slot. However, this increases the requested rate feedback frequency in the uplink and thus increases the system overhead. There are ways to reduce this feedback though. In the current system, *every* user feeds back the requested rates, but in fact only users whose channels are near their peaks have any chance of getting scheduled. Thus, an alternative is for each user to feed back the requested rate only when its current requested rate to average throughput ratio, $R_k[m]/T_k[m]$, exceeds a threshold γ. This threshold, γ, can be chosen to trade off the average aggregate amount of feedback the users send with the probability that none of the users sends any feedback in a given time slot (thus wasting the slot) (Exercise 6.22).

In IS-856, multiuser diversity scheduling is implemented in the downlink, but the same concept can be applied to the uplink. However, the issues of prediction error and feedback are different. In the uplink, the base-station would be measuring the channels of the users, and so a separate pilot would be needed for each user. The downlink has a single pilot and this amortization among the users is used to have a strong pilot. However, in the uplink, the fraction of power devoted to the pilot is typically small. Thus, it is expected that the *measurement* error will play a larger role in the uplink. Moreover, the pilot will have to be sent continuously even if the user is not currently scheduled, thus causing some interference to other users. On the other hand, the base-station only needs to broadcast which user is scheduled at that time slot, so the amount of feedback is much smaller than in the downlink (unless the selective feedback scheme is implemented).

The above discussion pertains to an FDD system. You are asked to discuss the analogous issues for a TDD system in Exercise 6.23.

6.7.3 Opportunistic beamforming using dumb antennas

The amount of multiuser diversity depends on the rate and dynamic range of channel fluctuations. In environments where the channel fluctuations are small, a natural idea comes to mind: why not amplify the multiuser diversity gain by *inducing* faster and larger fluctuations? Focusing on the downlink, we describe a technique that does this using multiple transmit antennas at the base-station as illustrated in Figure 6.19.

Consider a system with n_t transmit antennas at the base-station. Let $h_{lk}[m]$ be the complex channel gain from antenna l to user k in time m. In time m, the same symbol $x[m]$ is transmitted from all of the antennas except that it is

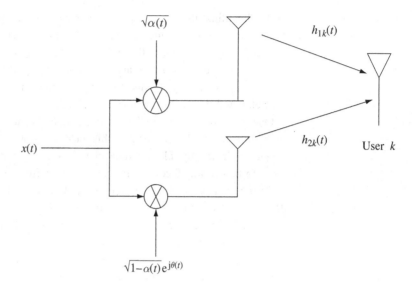

multiplied by a complex number $\sqrt{\alpha_l[m]}\, e^{j\theta_l[m]}$ at antenna l, for $l = 1, \ldots, n_t$, such that $\sum_{l=1}^{n_t} \alpha_l[m] = 1$, preserving the total transmit power. The received signal at user k (see the basic downlink fading channel model in (6.50) for comparison) is given by

$$y_k[m] = \left(\sum_{l=1}^{n_t} \sqrt{\alpha_l[m]}\, e^{j\theta_l[m]} h_{lk}[m] \right) x[m] + w_k[m]. \qquad (6.58)$$

In vector form, the scheme transmits $\mathbf{q}[m]x[m]$ at time m, where

$$\mathbf{q}[m] := \begin{bmatrix} \sqrt{\alpha_1[m]}\, e^{j\theta_1[m]} \\ \vdots \\ \sqrt{\alpha_{n_t}[m]}\, e^{j\theta_{n_t}[m]} \end{bmatrix} \qquad (6.59)$$

is a unit vector and

$$y_k[m] = (\mathbf{h}_k[m]^* \mathbf{q}[m])x[m] + w_k[m] \qquad (6.60)$$

where $\mathbf{h}_k[m]^* := (h_{1k}[m], \ldots, h_{n_t,k}[m])$ is the channel vector from the transmit antenna array to user k.

The overall channel gain seen by user k is now

$$\mathbf{h}_k[m]^* \mathbf{q}[m] = \sum_{l=1}^{n_t} \sqrt{\alpha_l[m]}\, e^{j\theta_l[m]} h_{lk}[m]. \qquad (6.61)$$

The $\alpha_l[m]$ denote the fractions of power allocated to each of the transmit antennas, and the $\theta_l[m]$ denote the phase shifts applied at each antenna to the

Figure 6.20 Pictorial
representation of the slow
fading channels of two users
before (left) and after (right)
applying opportunistic
beamforming.

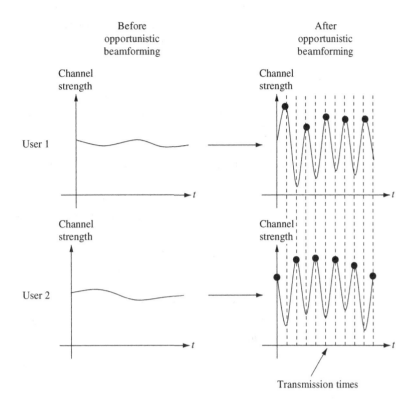

signal. By varying these quantities over time ($\alpha_l[m]$ from 0 to 1 and $\theta_l[m]$ from 0 to 2π), the antennas transmit signals in a time-varying direction, and fluctuations in the overall channel can be induced even if the physical channel gains $\{h_{lk}[m]\}$ have very little fluctuation (Figure 6.20).

As in the single transmit antenna system, each user k feeds back the overall received SNR of its own channel, $|\mathbf{h}_k[m]^*\mathbf{q}[m]|^2/N_0$, to the base-station (or equivalently the data rate that the channel can currently support) and the base-station schedules transmissions to users accordingly. There is no need to measure the individual channel gains $h_{lk}[m]$ (phase or magnitude); in fact, the existence of multiple transmit antennas is completely transparent to the users. Thus, only a single pilot signal is needed for channel measurement (as opposed to a pilot to measure each antenna gain). The pilot symbols are repeated at each transmit antenna, exactly like the data symbols.

The rate of variation of $\{\alpha_l[m]\}$ and $\{\theta_l[m]\}$ in time (or, equivalently, of the transmit direction $\mathbf{q}[m]$) is a design parameter of the system. We would like it to be as fast as possible to provide full channel fluctuations within the latency time-scale of interest. On the other hand, there is a practical limitation to how fast this can be. The variation should be slow enough and should happen at a time-scale that allows the channel to be reliably estimated by the users and the SNR fed back. Further, the variation should be slow enough to ensure that the channel seen by a user does not change abruptly and thus maintains stability of the channel tracking loop.

Slow fading: opportunistic beamforming

To get some insight into the performance of this scheme, consider the case of slow fading where the channel gain vector of each user k remains constant, i.e., $\mathbf{h}_k[m] = \mathbf{h}_k$, for all m. (In practice, this means for all m over the latency time-scale of interest.) The received SNR for this user would have remained constant if only one antenna were used. If all users in the system experience such slow fading, no multiuser diversity gain can be exploited. Under the proposed scheme, on the other hand, the overall channel gain $\mathbf{h}_k[m]^*\mathbf{q}[m]$ for each user k varies in time and provides opportunity for exploiting multiuser diversity.

Let us focus on a particular user k. Now if $\mathbf{q}[m]$ varies across all directions, the amplitude squared of the channel $|\mathbf{h}_k^*\mathbf{q}[m]|^2$ seen by user k varies from 0 to $||\mathbf{h}_k||^2$. The peak value occurs when the transmission is aligned along the direction of the channel of user k, i.e., $\mathbf{q}[m] = \mathbf{h}_k/||\mathbf{h}_k||$ (recall Example 5.2 in Section 5.3). The power and phase values are then in the *beamforming configuration*:

$$\alpha_l = \frac{|h_{lk}|^2}{||\mathbf{h}_k||^2}, \qquad l = 1, \ldots, n_t,$$

$$\theta_l = -\arg(h_{lk}), \qquad l = 1, \ldots, n_t.$$

To be able to beamform to a particular user, the base-station needs to know individual channel amplitude and phase responses from all the antennas, which requires much more information to feedback than just the overall SNR. However, if there are many users in the system, the proportional fair algorithm will schedule transmission to a user only when its overall channel SNR is near its peak. Thus, it is plausible that in a slow fading environment, the technique can approach the performance of coherent beamforming but with only overall SNR feedback (Figure 6.21). In this context, the technique can be interpreted as *opportunistic beamforming*: by varying the phases and powers allocated to the transmit antennas, a beam is randomly swept and at any time transmission is scheduled to the user currently closest to the beam. With many users, there is likely to be a user very close to the beam at any time. This intuition has been formally justified (see Exercise 6.29).

Fast fading: increasing channel fluctuations

We see that opportunistic beamforming can significantly improve performance in slow fading environments by adding fast time-scale fluctuations on the overall channel quality. The rate of channel fluctuation is artificially sped up. Can opportunistic beamforming help if the underlying channel variations are already fast (fast compared to the latency time-scale)?

Figure 6.21 Plot of spectral efficiency under opportunistic beamforming as a function of the total number of users in the system. The scenario is for slow Rayleigh faded channels for the users and the channels are fixed in time. The spectral efficiency plotted is the performance averaged over the Rayleigh distribution. As the number of users grows, the performance approaches the performance of true beamforming.

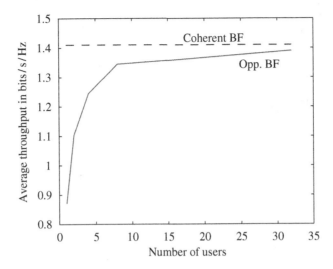

The long-term throughput under fast fading depends only on the stationary distribution of the channel gains. The impact of opportunistic beamforming in the fast fading scenario then depends on how the stationary distributions of the overall channel gains can be modified by power and phase randomization. Intuitively, better multiuser diversity gain can be exploited if the dynamic range of the distribution of h_k can be increased, so that the maximum SNRs can be larger. We consider two examples of common fading models.

- **Independent Rayleigh fading** In this model, appropriate for an environment where there is full scattering and the transmit antennas are spaced sufficiently, the channel gains $h_{1k}[m], \ldots, h_{n_t k}[m]$ are i.i.d. $\mathcal{CN}$ random variables. In this case, the channel vector $\mathbf{h}_k[m]$ is isotropically distributed, and $\mathbf{h}_k[m]^*\mathbf{q}[m]$ is circularly symmetric Gaussian for any choice of $\mathbf{q}[m]$; moreover the overall gains are independent across the users. Hence, the stationary statistics of the channel are *identical* to the original situation with one transmit antenna. Thus, in an independent fast Rayleigh fading environment, the opportunistic beamforming technique does not provide any performance gain.

- **Independent Rician fading** In contrast to the Rayleigh fading case, opportunistic beamforming has a significant impact in a Rician environment, particularly when the κ-factor is large. In this case, the scheme can significantly increase the dynamic range of the fluctuations. This is because the fluctuations in the underlying Rician fading process come from the diffused component, while with randomization of phase and powers, the fluctuations are from the coherent addition and cancellation of the direct path components in the signals from the different transmit antennas, in addition to the fluctuation of the diffused components. If the direct path

Figure 6.22 Total throughput as a function of the number of users under Rician fast fading, with and without opportunistic beamforming. The power allocations $\alpha_l[m]$ are uniformly distributed in $[0, 1]$ and the phases $\theta_l[m]$ uniform in $[0, 2\pi]$.

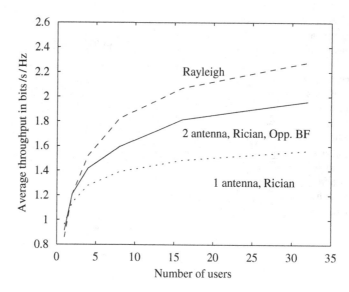

is much stronger than the diffused part (large κ values), then much larger fluctuations can be created with this technique.

This intuition is substantiated in Figure 6.22, which plots the total throughput with the proportional fair algorithm (large t_c, of the order of 100 time slots) for Rician fading with $\kappa = 10$. We see that there is a considerable improvement in performance going from the single transmit antenna case to dual transmit antennas with opportunistic beamforming. For comparison, we also plot the analogous curves for pure Rayleigh fading; as expected, there is no improvement in performance in this case. Figure 6.23 compares the stationary distributions of the overall channel gain $|\mathbf{h}_k[m]^*\mathbf{q}[m]|$ in the single-antenna and dual-antenna cases; one can see the increase in dynamic range due to opportunistic beamforming.

Antennas: dumb, smart and smarter

In this section so far, our discussion has focused on the use of multiple transmit antennas to induce larger and faster channel fluctuations for multiuser diversity benefits. It is insightful to compare this with the two other point-to-point transmit antenna techniques we have already discussed earlier in the book:

- **Space-time codes** like the Alamouti scheme (Section 3.3.2). They are primarily used to increase the diversity in slow fading point-to-point links.
- **Transmit beamforming** (Section 5.3.2). In addition to providing diversity, a power gain is also obtained through the coherent addition of signals at the users.

Figure 6.23 Comparison of the distribution of the overall channel gain with and without opportunistic beamforming using two transmit antennas, Rician fading. The Rayleigh distribution is also shown.

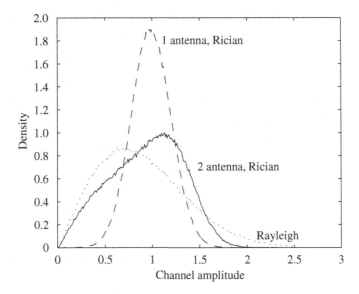

The three techniques have different system requirements. Coherent space-time codes like the Alamouti scheme require the users to track all the *individual* channel gains (amplitude and phase) from the transmit antennas. This requires separate pilot symbols on each of the transmit antennas. Transmit beamforming has an even stronger requirement that the channel should be known at the transmitter. In an FDD system, this means feedback of the individual channel gains (amplitude and phase). In contrast to these two techniques, the opportunistic beamforming scheme requires no knowledge of the individual channel gains, neither at the users nor at the transmitter. In fact, the users are *completely ignorant* of the fact that there are multiple transmit antennas and the receiver is identical to that in the single transmit antenna case. Thus, they can be termed *dumb antennas*. Opportunistic beamforming does rely on multiuser diversity scheduling, which requires the feedback of the overall SNR of each user. However, this only needs a *single* pilot to measure the overall channel.

What is the performance of these techniques when used in the downlink? In a slow fading environment, we have already remarked that opportunistic beamforming approaches the performance of transmit beamforming when there are many users in the system. On the other hand, space-time codes do not perform as well as transmit beamforming since they do not capture the array power gain. This means, for example, using the Alamouti scheme on dual transmit antennas in the downlink is 3 dB worse than using opportunistic beamforming combined with multiuser diversity scheduling when there are many users in the system. Thus, dumb antennas together with smart scheduling can surpass the performance of smart space-time codes and approach that of the even smarter transmit beamforming.

Table 6.1 A comparison between three methods of using transmit antennas.

	Dumb antennas (Opp. beamform)	Smart antennas (Space-time codes)	Smarter antennas (Transmit beamform)
Channel knowledge	Overall SNR	Entire CSI at Rx	Entire CSI at Rx, Tx
Slow fading performance gain	Diversity and power gains	Diversity gain only	Diversity and power gains
Fast fading performance gain	No impact	Multiuser diversity ↓	Multiuser diversity ↓ power ↑

How about in a fast Rayleigh fading environment? In this case, we have observed that dumb antennas have no effect on the overall channel as the full multiuser diversity gain has already been realized. Space-time codes, on the other hand, *increase* the diversity of the point-to-point links and consequently *decrease* the channel fluctuations and hence the multiuser diversity gain. (Exercise 6.31 makes this more precise.) Thus, the use of space-time codes as a point-to-point technology in a multiuser downlink with rate control and scheduling can actually be *harmful*, in the sense that even the naturally present multiuser diversity is removed. The performance impact of using transmit beamforming is not so clear: on the one hand it reduces the channel fluctuation and hence the multiuser diversity gain, but on the other hand it provides an array power gain. However, in an FDD system the fast fading channel may make it very difficult to feed back so much information to enable coherent beamforming.

The comparison between the three schemes is summarized in Table 6.1. All three techniques use the multiple antennas to transmit to only one user at a time. With full channel knowledge at the transmitter, an even smarter scheme can transmit to multiple users simultaneously, exploiting the multiple degrees of freedom existing inherently in the multiple antenna channel. We will discuss this in Chapter 10.

6.7.4 Multiuser diversity in multicell systems

So far we have considered a single-cell scenario, where the noise is assumed to be white Gaussian. For wideband cellular systems with full frequency reuse (such as the CDMA and OFDM based systems in Chapter 4), it is important to consider the effect of inter-cell interference on the performance of the system, particularly in interference-limited scenarios. In a cellular system, this effect is captured by measuring the channel quality of a user by the SINR, signal-to-interference-plus-noise ratio. In a fading environment, the energies in both the received signal and the received interference fluctuate over time. Since the multiuser diversity scheduling algorithm allocates resources based on the channel SINR (which depends on both the channel amplitude and the amplitude of the interference), it automatically exploits both the fluctuations in the energy of the received signal and those of the interference: the algorithm tries to schedule resource to a user whose instantaneous channel is good *and* the interference is weak. Thus, multiuser diversity naturally takes advantage of the time-varying interference to increase the spatial reuse of the network.

From this point of view, amplitude and phase randomization at the base-station transmit antennas plays an additional role: it increases not only the amount of fluctuations of the received signal to the intended users *within* the cells, it also increases the fluctuations of the interference that the base-station causes in *adjacent* cells. Hence, opportunistic beamforming has a dual benefit in an interference-limited cellular system. In fact, opportunistic beamforming performs *opportunistic nulling* simultaneously: while randomization of amplitude and phase in the transmitted signals from the antennas allows near coherent beamforming to some user within the cell, it will create near nulls at some other user in an adjacent cell. This in effect allows *interference avoidance* for that user if it is currently being scheduled.

Let us focus on the downlink and slow flat fading scenario to get some insight into the performance gain from opportunistic beamforming and nulling. Under amplitude and phase randomization at all base-stations, the received signal of a typical user that is interfered by J adjacent base-stations is given by

$$y[m] = (\mathbf{h}^*\mathbf{q}[m])x[m] + \sum_{j=1}^{J}(\mathbf{g}_j^*\mathbf{q}_j[m])u_j[m] + z[m]. \qquad (6.62)$$

Here, $x[m], \mathbf{h}, \mathbf{q}[m]$ are respectively the signal, channel vector and random transmit direction from the base-station of interest; $u_j[m], \mathbf{g}_j, \mathbf{q}_j[m]$ are respectively the interfering signal, channel vector and random transmit direction from the jth base-station. All base-stations have the same transmit power, P, and n_t transmit antennas and are performing amplitude and phase randomization independently.

By averaging over the signal $x[m]$ and the interference $u_j[m]$, the (time-varying) SINR of the user k can be computed to be

$$\mathsf{SINR}_k[m] = \frac{P|\mathbf{h}^*\mathbf{q}[m]|^2}{P\sum_{j=1}^{J}|\mathbf{g}_j^*\mathbf{q}_j[m]|^2 + N_0}. \qquad (6.63)$$

As the random transmit directions $\mathbf{q}[m], \mathbf{q}_j[m]$ vary, the overall SINR changes over time. This is due to the variations of the overall gain from the base-station of interest as well as those from the interfering base-stations. The SINR is high when $\mathbf{q}[m]$ is closely aligned to the channel vector $\mathbf{h}$, and/or for many j, $\mathbf{q}_j[m]$ is nearly orthogonal to $\mathbf{g}_j$, i.e., the user is near a null of the interference pattern from the jth base-station. In a system with many other users, the proportional fair scheduler will serve this user while its SINR is at its peak $P\|\mathbf{h}\|^2/N_0$, i.e., when the received signal is the strongest and the interference is completely nulled out. Thus, the opportunistic nulling and beamforming technique has the potential of shifting a user from a low SINR, interference-limited regime to a high SINR, noise-limited regime. An analysis of the tail of the distribution of SINR is conducted in Exercise 6.30.

6.7.5 A system view

A new design principle for wireless systems can now be seen through the lens of multiuser diversity. In the three systems in Chapter 4, many of the design techniques centered on making the *individual* point-to-point links as close to AWGN channels as possible, with a reliable channel quality that is constant over time. This is accomplished by *channel averaging*, and includes the use of diversity techniques such as multipath combining, time-interleaving and antenna diversity that attempt to keep the channel fading constant in time, as well as interference management techniques such as interference averaging by means of spreading.

However, if one shifts from the view of the wireless system as a set of point-to-point links to the view of a system with multiple users sharing the same resources (spectrum and time), then quite a different design objective suggests itself. Indeed, the results in this chapter suggest that one should instead try to *exploit* the channel fluctuations. This is done through an appropriate scheduling algorithm that "rides the peaks", i.e., each user is scheduled when it has a very strong channel, while taking into account real world traffic constraints such as delay and fairness. The technique of dumb antennas goes one step further by *creating* variations when there are none. This is accomplished by varying the strengths of *both* the signal and the interference that a user receives through opportunistic beamforming and nulling.

The viability of the opportunistic communication scheme depends on traffic that has some tolerance to scheduling delays. On the other hand, there are some forms of traffic that are not so flexible. The functioning of the wireless systems is supported by the overhead control channels, which are "circuit-switched" and hence have very tight latency requirements, unlike data, which have the flexibility to allow dynamic scheduling. From the perspective of these signals, it is preferable that the channel remain unfaded; a requirement that is contradictory to our scheduler-oriented observation that we would prefer the channel to have fast and large variations.

This issue suggests the following design perspective: separate very-low latency signals (such as control signals) from flexible latency data. One way to achieve this separation is to split the bandwidth into two parts. One part is made as flat as possible (by using the principles we saw in Chapter 4 such as spreading over this part of the bandwidth) and is used to transmit flows with very low latency requirements. The performance metric here is to make the channel as reliable as possible (equivalently keeping the probability of outage low) for some fixed data rate. The second part uses opportunistic beamforming to induce large and fast channel fluctuations and a scheduler to harness the multiuser diversity gains. The performance metric on this part is to maximize the multiuser diversity gain.

The gains of the opportunistic beamforming and nulling depend on the probability that the received signal is near beamformed *and* all the interference is near null. In the interference-limited regime and when $P/N_0 \gg 1$, the performance depends mainly on the probability of the latter event (see Exercise 6.30). In the downlink, this probability is large since there are only one or two base-stations contributing most of the interference. The uplink poses a contrasting picture: there is interference from many mobiles allowing interference averaging. Now the probability that the *total* interference is near null is much smaller. Interference averaging, which is one of the principle design features of the wideband full reuse systems (such as the ones we saw in Chapter 4 based on CDMA and OFDM), is actually unfavorable for the opportunistic scheme described here, since it reduces the likelihood of the nulling of the interference and hence the likelihood of the peaks of the SINR.

In a typical cell, there will be a distribution of users, some closer to the base-station and some closer to the cell boundaries. Users close to the base-station are at high SINR and are noise-limited; the contribution of the inter-cell interference is relatively small. These users benefit mainly from opportunistic beamforming. Users close to the cell boundaries, on the other hand, are at low SINR and are interference-limited; the average interference power can be much larger than the background noise. These users benefit both from opportunistic beamforming and from opportunistic nulling of inter-cell interference. Thus, the cell edge users benefit more in this system than users in the interior. This is rather desirable from a system fairness point-of-view, as the cell edge users tend to have poorer service. This feature is particularly important for a system without soft handoff (which is difficult to implement in a packet data scheduling system). To maximize the opportunistic nulling benefits, the transmit power at the base-station should be set as large as possible, subject to regulatory and hardware constraints. (See Exercise 6.30(5) where this is explored in more detail.)

We have seen the multiuser diversity as primarily a form of power gain. The opportunistic beamforming technique of using an array of multiple transmit antennas has approximately an n_t-fold improvement in received SNR to a user in a slow fading environment, as compared to the single-antenna case. With an array of n_r *receive* antennas at each mobile (and say a single transmit antenna at the base-station), the received SNR of any user gets an n_r-fold improvement as compared to a single receive antenna; this gain is realized by *receiver beamforming*. This operation is easy to accomplish since the mobile has full channel information at each of the antenna elements. Hence the gains of opportunistic beamforming are about the same order as that of installing a receive antenna array at *each* of the mobiles.

Thus, for a system designer, the opportunistic beamforming technique provides a compelling case for implementation, particularly in view of the constraints of space and cost of installing multiple antennas on *each* mobile device. Further, this technique needs neither any extra processing on the part of any user, nor any updates to an existing air-link interface standard. In other words, the mobile receiver can be completely ignorant of the use or non-use of this technique. This means that it does not have to be "designed in" (by appropriate inclusions in the air interface standard and the receiver design) and can be added/removed at any time. This is one of the important benefits of this technique from an overall system design point of view.

In the cellular wireless systems studied in Chapter 4, the cell is sectorized to allow better focusing of the power transmitted from the antennas and also to reduce the interference seen by mobile users from transmissions of the same base-station but intended for users in different sectors. This technique is particularly gainful in scenarios when the base-station is located at a fairly large height and thus there is limited scattering around the base-station. In contrast, in systems with far denser deployment of base-stations (a strategy that can be expected to be a good one for wireless systems aiming to provide mobile, broadband data services), it is unreasonable to stipulate that the base-stations be located high above the ground so that the local scattering (around the base-station) is minimal. In an urban environment, there is substantial local scattering around a base-station and the gains of sectorization are minimal; users in a sector also see interference from the same base-station (due to the local scattering) intended for another sector. The opportunistic beamforming scheme can be thought of as sweeping a random beam and scheduling transmissions to users when they are beamformed. Thus, the gains

Table 6.2 Contrast between conventional multiple access and opportunistic communication.

	Conventional multiple access	Opportunistic communication
Guiding principle	Averaging out fast channel fluctuations	Exploiting channel fluctuations
Knowledge at Tx	Track slow fluctuations No need to track fast ones	Track as many fluctuations as possible
Control	Power control the slow fluctuations	Rate control to all fluctuations
Delay requirement	Can support tight delay	Needs some laxity
Role of Tx antennas	Point-to-point diversity	Increase fluctuations
Power gain in downlink	Multiple Rx antennas	Opportunistic beamform via multiple Tx antennas
Interference management	Averaged	Opportunistically avoided

of sectorization are automatically realized. We conclude that the opportunistic beamforming technique is particularly suited to harness sectorization gains even in low-height base-stations with plenty of local scattering. In a cellular system, the opportunistic beamforming scheme also obtains the gains of nulling, a gain traditionally obtained by coordinated transmissions from neighboring base-stations in a full frequency reuse system or by appropriately designing the frequency reuse pattern.

The discussion is summarized in Table 6.2.

Chapter 6 The main plot

This chapter looked at the capacities of uplink and downlink channels. Two important sets of concepts emerged:
- successive interference cancellation (SIC) and superposition coding;
- multiuser opportunistic communication and multiuser diversity.

SIC and superposition coding
Uplink

Capacity is achieved by allowing users to simultaneously transmit on the full bandwidth and the use of SIC to decode the users.

SIC has a significant performance gain over conventional multiple access techniques in near–far situations. It takes advantage of the strong channel of the nearby user to give it high rate while providing the weak user with the best possible performance.

Downlink

Capacity is achieved by superimposing users' signals and the use of SIC at the receivers. The strong user decodes the weak user's signal first and then decodes its own.

Superposition coding/SIC has a significant gain over orthogonal techniques. Only a small amount of power has to be allocated to the strong user to give it a high rate, while delivering near-optimal performance to the weak user.

Opportunistic communication
Symmetric uplink fading channel:

$$y[m] = \sum_{k=1}^{K} h_k[m]x_k[m] + w[m]. \tag{6.64}$$

Sum capacity with CSI at receiver only:

$$C_{\text{sum}} = \mathbb{E}\left[\log\left(1 + \frac{\sum_{k=1}^{K}|h_k|^2 P}{N_0}\right)\right]. \tag{6.65}$$

Very close to AWGN capacity for large number of users. Orthogonal multiple access is strictly suboptimal.

Sum capacity with full CSI:

$$C_{\text{sum}} = \mathbb{E}\left[\log\left(1 + \frac{P_{k^*}(\mathbf{h})|h_{k^*}|^2}{N_0}\right)\right], \tag{6.66}$$

where k^* is the user with the strongest channel at joint channel state $\mathbf{h}$. This is achieved by transmitting only to the user with the best channel and a waterfilling power allocation $P_{k^*}(\mathbf{h})$ over the fading state.

Symmetric downlink fading channel:

$$y_k[m] = h_k[m]x[m] + w_k[m], \qquad k = 1, \ldots, K. \tag{6.67}$$

Sum capacity with CSI at receiver only:

$$C_{\text{sum}} = \mathbb{E}\left[\log\left(1 + \frac{|h_k|^2 P}{N_0}\right)\right]. \tag{6.68}$$

Can be achieved by orthogonal multiple access.
Sum capacity with full CSI: same as uplink.

Multiuser diversity

Multiuser diversity gain: under full CSI, capacity increases with the number of users: in a large system with high probability there is always a user with a very strong channel.

System issues in implementing multiuser diversity:

- **Fairness** Fair access to the channel when some users are statistically stronger than others.
- **Delay** Cannot wait too long for a good channel.
- **Channel tracking** Channel has to be measured and fed back fast enough.
- **Small and slow channel fluctuations** Multiuser diversity gain is limited when channel varies too slowly and/or has a small dynamic range.

The solutions discussed were:

- Proportional fair scheduler transmits to a user when its channel is near its peak within the delay constraint. Every user has access to the channel for roughly the same amount of time.
- Channel feedback delay can be reduced by having shorter time slots and feeding back more often. Aggregate feedback can be reduced by each user selectively feeding channel state back only when its channel is near its peak.

> • Channel fluctuations can be sped up and their dynamic range increased
>   by the use of multiple transmit antennas to perform opportunistic beam-
>   forming. The scheme sweeps a random beam and schedules transmis-
>   sions to users when they are beamformed.
>
> In a cellular system, multiuser diversity scheduling performs interference
> avoidance as well: a user is scheduled transmission when its channel is
> strong *and* the out-of-cell interference is weak.
>
> Multiple transmit antennas can perform opportunistic beamforming as well
> as nulling.

6.8 Bibliographical notes

Classical treatment of the general multiple access channel was initiated by Ahlswede [2] and Liao [73] who characterized the capacity region. The capacity region of the Gaussian multiple access channel is derived as a special case. A good survey of the literature on MACs was done by Gallager [45]. Hui [59] first observed that the sum capacity of the uplink channel with single-user decoding is bounded by 1.442 bits/s/Hz.

The general broadcast channel was introduced by Cover [25] and a complete characterization of its capacity is one of the famous open problems in information theory. Degraded broadcast channels, where the users can be "ordered" based on their channel quality, are fully understood with superposition coding being the optimal strategy; a textbook reference is Chapter 14.6 in Cover and Thomas [26]. The best inner and outer bounds are by Marton [81] and a good survey of the literature appears in [24].

The capacity region of the uplink fading channel with receiver CSI was derived by Gallager [44], where he also showed that orthogonal multiple access schemes are strictly suboptimal in fading channels. Knopp and Humblet [65] studied the sum capacity of the uplink fading channel with full CSI. They noted that transmitting to only one user is the optimal strategy. An analogous result was obtained earlier by Cheng and Verdú [20] in the context of the time-invariant uplink frequency-selective channels. Both these channels are instances of the parallel Gaussian multiple access channel, so the two results are mathematically equivalent. The latter authors also derived the capacity region in the two-user case. The solution for arbitrary number of users was obtained by Tse and Hanly [122], exploiting a basic polymatroid property of the region.

The study of downlink fading channels with full CSI was carried out by Tse [124] and Li and Goldsmith [74]. The key aspect of the study was to observe that the fading downlink is really a parallel degraded broadcast channel, the capacity of which has been fully understood (El Gamal [33]). There is an intriguing similarity between the downlink resource allocation solution and the uplink one. This connection is studied further in Chapter 10.

Multiuser diversity is a key distinguishing feature of the uplink and the downlink fading channel study as compared to our understanding of the point-to-point fading channel. The term multiuser diversity was coined by Knopp and Humblet [66]. The multiuser diversity concept was integrated into the downlink design of IS-856 (CDMA 2000 EV-DO) via the proportional fair scheduler by Tse [19]. In realistic scenarios, performance gains of 50% to 100% have been reported (Wu and Esteves [149]).

If the channels are slowly varying, then the multiuser diversity gains are limited. The opportunistic beamforming idea mitigates this defect by creating variations while maintaining the same average channel quality; this was proposed by Viswanath *et al.* [137], who also studied its impact on system design.

Several works have studied the design of schedulers that harness the multiuser diversity gain. A theoretical analysis of the proportional fair scheduler has appeared in several places including a work by Borst and Whiting [12].

6.9 Exercises

Exercise 6.1 The sum constraint in (6.6) applies because the two users send independent information and cannot cooperate in the encoding. If they could cooperate, what is the maximum sum rate they could achieve, still assuming individual power constraints P_1 and P_2 on the two users? In the case $P_1 = P_2$, quantify the cooperation gain at low and at high SNR. In which regime is the gain more significant?

Exercise 6.2 Consider the basic uplink AWGN channel in (6.1) with power constraints P_k on user k (for $k = 1, 2$). In Section 6.1.3, we stated that orthogonal multiple access is optimal when the degrees of freedom are split in direct proportion to the powers of the users. Verify this. Show also that any other split of degrees of freedom is strictly suboptimal, i.e., the corresponding rate pair lies strictly inside the capacity region given by the pentagon in Figure 6.2. *Hint*: Think of the sum rate as the performance of a point-to-point channel and apply the insight from Exercise 5.6.

Exercise 6.3 Calculate the symmetric capacity, (6.2), for the two-user uplink channel. Identify scenarios where there are definitely superior operating points.

Exercise 6.4 Consider the uplink of a single IS-95 cell where all the users are controlled to have the same received power P at the base-station.
1. In the IS-95 system, decoding is done by a conventional CDMA receiver which treats the interference of the other users as Gaussian noise. What is the maximum number of voice users that can be accommodated, assuming capacity-achieving point-to-point codes? You can assume a total bandwidth of 1.25 MHz and a data rate per user of 9.6 kbits/s. You can also assume that the background noise is negligible compared to the intra-cell interference.
2. Now suppose one of the users is a data user and it happens to be close to the base-station. By not controlling its power, its received power can be 20 dB above the rest. Propose a receiver that can give this user a higher rate while still delivering 9.6 kbits/s to the other (voice) users. What rate can it get?

Exercise 6.5 Consider the uplink of an IS-95 system.
1. A single cell is modeled as a disk of radius 1 km. If a mobile at the edge of the cell transmits at its maximum power limit, its received SNR at the base-station is 15 dB when no one else is transmitting. Estimate (via numerical simulations)

the average sum capacity of the uplink with 16 users that are independently and uniformly located in the disk. Compare this to the corresponding average total throughput in a system with conventional CDMA decoding and each user perfectly power controlled at the base-station. What is the potential percentage gain in spectral efficiency by using the more sophisticated receiver? You can assume that all mobiles have the same transmit power constraint and the path loss (power) attenuation is proportional to r^{-4}.

2. Part (1) ignores out-of-cell interference. With out-of-cell interference taken into consideration, the received SINR of the cell edge user is only -10 dB. Redo part (1). Is the potential gain from using a more sophisticated receiver still as impressive?

Exercise 6.6 Consider the downlink of the IS-856 system.

1. Suppose there are two users on the cell edge. Users are scheduled on a TDMA basis, with equal time for each user. The received SINR of each user is 0 dB when it is transmitted to. Find the rate that each user gets. The total bandwidth is 1.25 MHz and you can assume an AWGN channel and the use of capacity-achieving codes.

2. Now suppose there is an extra user which is near the base-station with a 20 dB SINR advantage over the other two users. Consider two ways to accommodate this user:
   • Give a fraction of the time slots to this user and divide the rest equally among the two cell edge users.
   • Give a fraction of the power to this user and superimpose its signal on top of the signals of both users. The two cell edge users are still scheduled on a TDMA basis with equal time, and the strong user uses a SIC decoder to extract its signal after decoding the other users' signals at each time slot.

   Since the two cell edge users have weak reception, it is important to maintain the best possible quality of service to them. So suppose the constraint is that we want each of them to have 95% of the rates they were getting before this strong user joined. Compare the performance that the strong user gets in the two schemes above.

Exercise 6.7 The capacity region of the two-user AWGN uplink channel is shown in Figure 6.2. The two corner points A and B can be achieved using successive cancellation. Points inside the line segment AB can be achieved by time sharing. In this exercise we will see another way to achieve every point (R_1, R_2) on the line segment AB using successive cancellation. By definition we must have

$$R_k < \log\left(1 + \frac{P_k}{N_0}\right) \quad k = 1, 2, \tag{6.69}$$

$$R_1 + R_2 = \log\left(1 + \frac{P_1 + P_2}{N_0}\right). \tag{6.70}$$

Define $\delta > 0$ by

$$R_2 = \log\left(1 + \frac{P_2}{\delta + N_0}\right). \tag{6.71}$$

Now consider the situation when user 1 splits itself into two users, say users 1a and 1b, with power constraints $P_1 - \delta$ and δ respectively. We decode the users with successive cancellation in the order user 1a, 2, 1b, i.e., user 1a is decoded first, user 2 is decoded next (with user 1a cancelled) and finally user 1b is decoded (seeing no interference from users 1a and 2).

1. Calculate the rates of reliable communication (r_{1a}, r_2, r_{1b}) for the users 1a, 2 and 1b using the successive cancellation just outlined.

2. Show that $r_2 = R_2$ and $r_{1a} + r_{1b} = R_1$. This means that the point (R_1, R_2) on the line segment AB can be achieved by successive cancellation of three users formed by one of the users "splitting" itself into two virtual users.

Exercise 6.8 In Exercise 6.7, we studied rate splitting multiple access for two users. A reading exercise is to study [101], where this result was introduced and generalized to the K-user uplink: $K - 1$ users can split themselves into two users each (with appropriate power splits) so that any rate vector on the boundary of the capacity region that meets the sum power constraint can be achieved via successive cancellation (with appropriate ordering of the $2K - 1$ users).

Exercise 6.9 Consider the K-user AWGN uplink channel with user power constraints $P_1, \ldots, P_K$. The capacity region is the set of rate vectors that lie in the intersection of the constraints (cf. (6.10)):

$$\sum_{k \in \mathcal{S}} R_k < \log \left(1 + \frac{\sum_{k \in \mathcal{S}} P_k}{N_0} \right), \tag{6.72}$$

for every subset $\mathcal{S}$ of the K users.

1. Fix an ordering of the users $\pi_1, \ldots, \pi_K$ (here π represents a permutation of set $\{1, \ldots, K\}$). Show that the rate vector $\left(R_1^{(\pi)}, \ldots, R_K^{(\pi)} \right)$:

$$R_{\pi_k}^{(\pi)} := \log \left(1 + \frac{P_{\pi_k}}{\sum_{i=k+1}^{K} P_{\pi_i} + N_0} \right) \quad k = 1, \ldots, K, \tag{6.73}$$

is in the capacity region. This rate vector can be interpreted using the successive cancellation viewpoint: the users are successively decoded in the order $\pi_1, \ldots, \pi_K$ with cancellation after each decoding step. So, user π_k has no interference from the previously decoded users $\pi_1, \ldots, \pi_{k-1}$, but experiences interference from the users following it (namely $\pi_{k+1}, \ldots, \pi_K$). In Figure 6.2, the point A corresponds to the permutation $\pi_1 = 2, \pi_2 = 1$ and the point B corresponds to the identity permutation $\pi_1 = 1, \pi_2 = 2$.

2. Consider maximizing the linear objective function $\sum_{k=1}^{K} a_k R_k$ with non-negative $a_1, \ldots, a_K$ over the rate vectors in the capacity region. (a_k can be interpreted as the *revenue* per unit rate for user k.) Show that the maximum occurs at the rate vector of the form in (6.73) with the permutation π defined by the property:

$$a_{\pi_1} \leq a_{\pi_2} \leq \cdots \leq a_{\pi_K}. \tag{6.74}$$

This means that optimizing linear objective functions on the capacity region can be done in a greedy way: we order the users based on their priority (a_k for user k). This ordering is denoted by the permutation π in (6.74). Next, the receiver decodes via successive cancellation using this order: the user with the least priority is decoded first (seeing full interference from all the other users) and the user with the highest priority decoded last (seeing no interference from the other users). *Hint*: Show that if the ordering is not according to (6.74), then one can always improve the objective function by changing the decoding order.

3. Since the capacity region is the intersection of hyperplanes, it is a convex polyhedron. An equivalent representation of a convex polyhedron is through enumerating its *vertices*: points which cannot be expressed as a strict convex combination of any subset of other points in the polyhedron. Show that $\left(R_1^{(\pi)}, \ldots, R_K^{(\pi)}\right)$ is a *vertex* of the capacity region. *Hint*: Consider the following fact: a linear object function is maximized on a convex polyhedron at one of the vertices. Further, every vertex must be optimal for some linear objective function.

4. Show that vertices of the form (6.73) (one for each permutation, so there are $K!$ of them) are the only *interesting* vertices of the capacity region. (This means that any other vertex of the capacity region is component-wise dominated by one of these $K!$ vertices.)

Exercise 6.10 Consider the K-user uplink AWGN channel. In the text, we focus on the capacity region $\mathcal{C}(\mathbf{P})$: the set of achievable rates for given power constraint vector $\mathbf{P} := (P_1, \ldots, P_K)^t$. A "dual" characterization is the *power region* $\mathcal{P}(\mathbf{R})$: set of all feasible received power vectors that can support a given target rate vector $\mathbf{R} := (R_1, \ldots, R_K)^t$.

1. Write down the constraints describing $\mathcal{P}(\mathbf{R})$. Sketch the region for $K = 2$.
2. What are the vertices of $\mathcal{P}(\mathbf{R})$?
3. Find a decoding strategy and a power allocation that minimizes $\sum_{k=1}^{K} b_k P_k$ while meeting the given target rates. Here, the constants b_k are positive and should be interpreted as "power prices". *Hint*: Exercise 6.9 may be useful.
4. Suppose users are at different distances from the base-station so that the transmit power of user k is attenuated by a factor of γ_i. Find a decoding strategy and a power allocation that minimizes the total transmit power of the users while meeting the target rates $\mathbf{R}$.
5. In IS-95, the code used by each user is not necessarily capacity-achieving but communication is considered reliable as long as a $\mathcal{E}_b/I_0$ requirement of 7 dB is met. Suppose these codes are used in conjunction with SIC. Find the optimal decoding order to minimize the total transmit power in the uplink.

Exercise 6.11 (Impact of using SIC on interference-limited capacity) Consider the two-cell system in Exercise 4.11. The interference-limited spectral efficiency in the many-user regime was calculated for both CDMA and OFDM. Now suppose SIC is used instead of the conventional receiver in the CDMA system. In the context of SIC, the interference I_0 in the target $\mathcal{E}_b/I_0$ requirement refers to the interference from the uncancelled users. Below you can always assume that interference cancellation is perfect.

1. Focus on a single cell first and assume a background noise power of N_0. Is the system interference-limited under the SIC receiver? Was it interference-limited under the conventional CDMA receiver?
2. Suppose there are K users with user k at a distance r_k from the base-station. Give an expression for the total transmit power saving (in dB) in using SIC with the optimal decoding order as compared to the conventional CDMA receiver (with an $\mathcal{E}_b/I_0$ requirement of β).
3. Give an expression for the power saving in the asymptotic regime with a large number of users and large bandwidth. The users are randomly located in the single cell as specified in Exercise 4.11. What is this value when $\beta = 7$ dB and the power decay is r^{-2} (i.e., $\alpha = 2$)?

4. Now consider the two-cell system. Explain why in this case the system is interference-limited even when using SIC.

5. Nevertheless, SIC increases the interference-limited capacity because of the reduction in transmit power, which translates into a reduction of out-of-cell interference. Give an expression for the asymptotic interference-limited spectral efficiency under SIC in terms of β and α. You can ignore the background noise and assume that users closer to the base-station are always decoded before the users further away.

6. For $\beta = 7\,\mathrm{dB}$ and $\alpha = 2$, compare the performance with the conventional CDMA system and the OFDM system.

7. Is the cancellation order in part 5 optimal? If not, find the optimal order and give an expression for the resulting asymptotic spectral efficiency. *Hint*: You might find Exercise 6.10 useful.

Exercise 6.12 Verify the bound (6.30) on the actual error probability of the kth user in the SIC, accounting for error propagation.

Exercise 6.13 Consider the two-user uplink fading channel,

$$y[m] = h_1[m]x_1[m] + h_2[m]x_2[m] + w[m]. \tag{6.75}$$

Here the user channels $\{h_1[m]\}, \{h_2[m]\}$ are statistically independent. Suppose that $h_1[m]$ and $h_2[m]$ are $\mathcal{CN}(0,1)$ and user k has power $P_k, k = 1, 2$, with $P_1 \gg P_2$. The background noise $w[m]$ is i.i.d. $\mathcal{CN}(0, N_0)$. An SIC receiver decodes user 1 first, removes its contribution from $\{y[m]\}$ and then decodes user 2. We would like to assess the effect of channel estimation error of h_1 on the performance of user 2.

1. Assuming that the channel coherence time is T_c seconds and user 1 spends 20% of its power on sending a training signal, what is the mean square estimation error of h_1? You can assume the same setup as in Section 3.5.2. You can ignore the effect of user 2 in this estimation stage, since $P_1 \gg P_2$.

2. The SIC receiver decodes the transmitted signal from user 1 and subtracts its contribution from $\{y[m]\}$. Assuming that the information is decoded correctly, the residual error is due to the channel estimation error of h_1. Quantify the degradation in SINR of user 2 due to this channel estimation error. Plot this degradation as a function of P_1/N_0 for $T_c = 10\,\mathrm{ms}$. Does the degradation worsen if the power P_1 of user 1 increases? Explain.

3. In part (2), user 2 still faced some interference due to the presence of user 1 despite decoding the information meant for user 1 accurately. This is due to the error in the channel estimate of user 1. In the calculation in part (2), we used the expression for the error of user 1's channel estimate as derived from the training symbol. However, conditioned on the event that the first user's information has been correctly decoded, the channel estimate of user 1 can be improved. Model this situation appropriately and arrive at an approximation of the error in user 1's channel estimate. Now redo part (2). Does your answer change qualitatively?

Exercise 6.14 Consider the probability of the outage event ($p_{\mathrm{out}}^{\mathrm{ul}}$, cf. (6.32)) in a symmetric slow Rayleigh fading uplink with the K users operating at the symmetric rate R bits/s/Hz.

1. Suppose $p_{\text{out}}^{\text{ul}}$ is fixed to be ϵ. Argue that at very high SNR (with SNR defined to be P/N_0), the dominating event is the one on the sum rate:

$$KR > \log\left(1 + \frac{\sum_{k=1}^K P|h_k|^2}{N_0}\right).$$

2. Show that the ϵ-outage symmetric capacity, C_ϵ^{sym}, can be approximated at very high SNR as

$$C_\epsilon^{\text{sym}} \approx \frac{1}{K} \log_2\left(1 + \frac{P\epsilon^{\frac{1}{K}}}{N_0}\right).$$

3. Argue that at very high SNR, the ratio of C_ϵ^{sym} to C_ϵ (the ϵ-outage capacity with just a single user in the uplink) is approximately $1/K$.

Exercise 6.15 In Section 6.3.3, we have discussed the optimal multiple access strategy for achieving the sum capacity of the uplink fading channel when users have identical channel statistics and power constraints.

1. Solve the problem for the general case when the channel statistics and the power constraints of the users are arbitrary. *Hint*: Construct a Lagrangian for the convex optimization problem (6.42) with a separate Lagrange multiplier for each of the individual power constraints (6.43).

2. Do you think the sum capacity is a reasonable performance measure in the asymmetric case?

Exercise 6.16 In Section 6.3.3, we have derived the optimal power allocation with full CSI in the symmetric uplink with the assumption that there is always a unique user with the strongest channel at any one time. This assumption holds with probability 1 when the fading distributions are continuous. Moreover, under this assumption, the solution is *unique*. This is in contrast to the uplink AWGN channel where there is a continuum of solutions that achieves the optimal sum rate, of which only one is orthogonal. We will see in this exercise that transmitting to only one user at a time is not necessarily the unique optimal solution even for fading channels, if the fading distribution is discrete (to model measurement realities, such as the feedback of a finite number of rate levels).

Consider the full CSI two-user uplink with identical, independent, stationary and ergodic flat fading processes for the two users. The stationary distribution of the flat fading for both of the users takes one of just two values: channel amplitude is either at 0 or at 1 (with equal probability). Both of the users are individually average power constrained (by $\bar{P}$). Calculate explicitly *all* the optimal joint power allocation and decoding policies to maximize the sum rate. Is the optimal solution unique? *Hint*: Clearly there is no benefit by allocating power to a user whose channel is fully faded (the zero amplitude state).

Exercise 6.17 In this exercise we further study the nature of the optimal power and rate control strategy that achieves the sum capacity of the symmetric uplink fading channel.

1. Show that the optimal power/rate allocation policy for achieving the sum capacity of the symmetric uplink fading channel can be obtained by solving for each fading state the optimization problem:

$$\max_{\mathbf{r},\mathbf{p}} \sum_{k=1}^{K} r_k - \lambda \sum_{k=1}^{K} p_k, \tag{6.76}$$

subject to the constraint that

$$\mathbf{r} \in \mathcal{C}(\mathbf{p},\mathbf{h}), \tag{6.77}$$

where $\mathcal{C}(\mathbf{p},\mathbf{h})$ is the uplink AWGN channel capacity region with received power $p_k|h_k|^2$. Here λ is chosen to meet the average power constraint of P for each user.

2. What happens when the channels are not symmetric but we are still interested in the sum rate?

Exercise 6.18 [122] In the text, we focused on computing the power/rate allocation policy that maximizes the *sum* rate. More generally, we can look for the policy that maximizes a weighted sum of rates $\sum_k \mu_k R_k$. Since the uplink fading channel capacity region is convex, solving this for all non-negative μ_i will enable us to characterize the entire capacity region (as opposed to just the sum capacity point).

In analogy with Exercise 6.17, it can be shown that the optimal power/rate allocation policy can be computed by solving for each fading state $\mathbf{h}$ the optimization problem:

$$\max_{\mathbf{r},\mathbf{p}} \sum_{k=1}^{K} \mu_k r_k - \sum_{k=1}^{K} \lambda_k p_k, \tag{6.78}$$

subject to the constraint that

$$\mathbf{r} \in \mathcal{C}(\mathbf{p},\mathbf{h}), \tag{6.79}$$

where the λ_k are chosen to meet the average power constraints P_k of the users (averaged over the fading distribution). If we define $q_k := p_k|h_k|^2$ as the received power, then we can rewrite the optimization problem as

$$\max_{\mathbf{r},\mathbf{q}} \sum_{k=1}^{K} \mu_k r_k - \sum_{k=1}^{K} \frac{\lambda_k}{|h_k|^2} p_k \tag{6.80}$$

subject to the constraint that

$$\mathbf{r} \in \mathcal{C}(\mathbf{q}), \tag{6.81}$$

where $\mathcal{C}(\mathbf{q})$ is the uplink AWGN channel capacity region. You are asked to solve this optimization problem in several steps below.

1. Verify that the capacity of a point-to-point AWGN channel can be written in the integral form:

$$C_{\text{awgn}} = \log\left(1 + \frac{P}{N_0}\right) = \int_0^P \frac{1}{N_0 + z}\, dz. \tag{6.82}$$

Give an interpretation in terms of splitting the single user into many infinitesimally small virtual users, each with power dz (cf. Exercise 6.7). What is the interpretation of the quantity $1/(N_0 + z)\,dz$?

2. Consider first $K = 1$ in the uplink fading channel above, i.e., the point-to-point scenario. Define the *utility function*:

$$u_1(z) = \frac{\mu_1}{N_0 + z} - \frac{\lambda_1}{|h_1|^2}, \qquad (6.83)$$

where N_0 is the background noise power. Express the optimal solution in terms of the graph of $u_1(z)$ against z. Interpret the solution as a greedy solution and also give an interpretation of $u_1(z)$. *Hint*: Make good use of the rate-splitting interpretation in part 1.

3. Now, for $K > 1$, define the utility function of user k to be

$$u_k(z) = \frac{\mu_k}{N_0 + z} - \frac{\lambda_k}{|h_k|^2}. \qquad (6.84)$$

Guess what the optimal solution should be in terms of the graphs of $u_k(z)$ against z for $k = 1, \ldots, K$.

4. Show that each pair of the utility functions intersects at most once for non-negative z.

5. Using the previous parts, verify your conjecture in part (3).

6. Can the optimal solution be achieved by successive cancellation?

7. Verify that your solution reduces to the known solution for the sum capacity problem (i.e., when $\mu_1 = \cdots = \mu_K$).

8. What does your solution look like when there are two groups of users such that within each group, users have the same μ_k and λ_k (but not necessarily the same h_k).

9. Using your solution to the optimization problem (6.78), compute numerically the boundary of the capacity region of the two-user Rayleigh uplink fading channel with average received SNR of $0\,\mathrm{dB}$ for each of the two users.

Exercise 6.19 [124] Consider the downlink fading channel.

1. Formulate and solve the downlink version of Exercise 6.18.

2. The total transmit power varies as a function of time in the optimal solution. But now suppose we fix the total transmit power to be P at all times (as in the IS-856 system). Re-derive the optimal solution.

Exercise 6.20 Within a cell in the IS-856 system there are eight users on the edge and one user near the base-station. Every user experiences independent Rayleigh fading, but the average SNR of the user near the base-station is γ times that of the users on the edge. Suppose the average SNR of a cell edge user is $0\,\mathrm{dB}$ when all the power of the base-station is allocated to it. A fixed transmit power of P is used at all times.

1. Simulate the proportional fair scheduling algorithm for t_c large and compute the performance of each user for a range of γ from 1 to 100. You can assume the use of capacity-achieving codes.

2. Fix γ. Show how you would compute the *optimal* achievable rate among *all strategies* for the user near the base-station, given a (equal) rate for all the users on the edge. *Hint*: Use the results in Exercise 6.19.

3. Plot the potential gain in rate for the strong user over what it gets under the proportional fair algorithm, for the same rate for the weak users.

Exercise 6.21 In Section 6.6, we have seen that the multiuser diversity gain comes about because the effective channel gain becomes the maximum of the channel gains of the K users:

$$|h|^2 := \max_{k=1\ldots K} |h_k|^2.$$

1. Let $h_1, \ldots, h_K$ be i.i.d. $\mathcal{CN}(0,1)$ random variables. Show that

$$\mathbb{E}[|h|^2] = \sum_{k=1}^{K} \frac{1}{k}. \tag{6.85}$$

Hint: You might find it easier to prove the following stronger result (using induction):

$$|h|^2 \text{ has the same distribution as } \sum_{k=1}^{K} \frac{|h_k|^2}{k}. \tag{6.86}$$

2. Using the previous part, or directly, show that

$$\frac{\mathbb{E}[|h|^2]}{\log_e K} \to 1 \qquad \text{as } K \to \infty; \tag{6.87}$$

thus the mean of the effective channel grows logarithmically with the number of users.

3. Now suppose $h_1, \ldots, h_K$ are i.i.d. $\mathcal{CN}(\sqrt{\kappa}/\sqrt{1+\kappa}, 1/(1+\kappa))$ (i.e., Rician random variables with the ratio of specular path power to diffuse path power equal to κ). Show that

$$\frac{\mathbb{E}[|h|^2]}{\log_e K} \to \frac{1}{1+\kappa} \qquad \text{as } K \to \infty; \tag{6.88}$$

i.e., the mean of the effective channel is now reduced by a factor $1 + \kappa$ compared to the Rayleigh fading case. Can you see this result intuitively as well? *Hint*: You might find the following limit theorem (p. 261 of [28]) useful for this exercise. Let $h_1, \ldots, h_K$ be i.i.d. real random variables with a common cdf $F(\cdot)$ and pdf $f(\cdot)$ satisfying $F(h)$ is less than 1 and is twice differentiable for all h, and is such that

$$\lim_{h \to \infty} \frac{d}{dh} \left[\frac{1 - F(h)}{f(h)} \right] = 0. \tag{6.89}$$

Then

$$\max_{1 \le k \le K} Kf(l_K)(h_k - l_K)$$

converges in distribution to a limiting random variable with cdf

$$\exp(-e^{-x}).$$

In the above, l_K is given by $F(l_K) = 1 - 1/K$. This result states that the maximum of K such i.i.d. random variables grows like l_K.

Exercise 6.22 (Selective feedback) The downlink of IS-856 has K users each experiencing i.i.d. Rayleigh fading with average SNR of $0\,\text{dB}$. Each user selectively feeds back the requested rate only if its channel is greater than a threshold γ. Suppose γ is chosen such that the probability that no one sends a requested rate is ϵ. Find the expected number of users that sends in a requested rate. Plot this number for $K = 2, 4, 8, 16, 32, 64$ and for $\epsilon = 0.1$ and $\epsilon = 0.01$. Is selective feedback effective?

Exercise 6.23 The discussions in Section 6.7.2 about channel measurement, prediction and feedback are based on an FDD system. Discuss the analogous issues for a TDD system, both in the uplink and in the downlink.

Exercise 6.24 Consider the two-user downlink AWGN channel (cf. (6.16)):

$$y_k[m] = h_k x[m] + z_k[m], \quad k = 1, 2. \tag{6.90}$$

Here $z_k[m]$ are i.i.d. $\mathcal{CN}(0, N_0)$ Gaussian processes marginally ($k = 1, 2$). Let us take $|h_1| > |h_2|$ for this problem.

1. Argue that the capacity region of this downlink channel does not depend on the correlation between the additive Gaussian noise processes $z_1[m]$ and $z_2[m]$. *Hint*: Since the two users cannot cooperate, it should be intuitive that the error probability for user k depends only on the marginal distribution of $z_k[m]$ (for both $k = 1, 2$).

2. Now consider the following specific correlation between the two additive noises of the users. The pair $(z_1[m], z_2[m])$ is i.i.d. with time m with the distribution $\mathcal{CN}(0, \mathbf{K}_z)$. To preserve the marginals, the diagonal entries of the covariance matrix $\mathbf{K}_z$ have to be both equal to N_0. The only parameter that is free to be chosen is the off-diagonal element (denoted by ρN_0 with $|\rho| \leq 1$):

$$\mathbf{K}_z = \begin{bmatrix} N_0 & \rho N_0 \\ \rho N_0 & N_0 \end{bmatrix}.$$

   Let us now allow the two users to cooperate, in essence creating a point-to-point AWGN channel with a single transmit but two receive antennas. Calculate the capacity ($C(\rho)$) of this channel as a function of ρ and show that if the rate pair (R_1, R_2) is within the capacity region of the downlink AWGN channel, then

$$R_1 + R_2 \leq C(\rho). \tag{6.91}$$

3. We can now choose the correlation ρ to minimize the upper bound in (6.91). Find the minimizing ρ (denoted by $\rho_{\min}$) and show that the corresponding (minimal) $C(\rho_{\min})$ is equal to $\log(1 + |h_1|^2 P/N_0)$.

4. The result of the calculation in the previous part is rather surprising: the rate $\log(1 + |h_1|^2 P/N_0)$ can be achieved by simply user 1 alone. This means that with a specific correlation ($\rho_{\min}$), cooperation among the users is not gainful. Show this formally by proving that for every time m with the correlation given by $\rho_{\min}$, the sequence of random variables $x[m], y_1[m], y_2[m]$ form a Markov chain (i.e., conditioned on $y_1[m]$, the random variables $x[m]$ and $y_2[m]$ are independent). This technique is useful in characterizing the capacity region of more involved downlinks, such as when there are multiple antennas at the base station.

Exercise 6.25 Consider the rate vectors in the downlink AWGN channel (cf. (6.16)) with superposition coding and orthogonal signaling as given in (6.22) and (6.23), respectively. Show that superposition coding is strictly better than the orthogonal schemes, i.e., for every non-zero rate pair achieved by an orthogonal scheme, there is a superposition coding scheme which allows *each* user to strictly increase its rate.

Exercise 6.26 A reading exercise is to study [8], where the sufficiency of superposition encoding and decoding for the downlink AWGN channel is shown.

Exercise 6.27 Consider the two-user symmetric downlink fading channel with receiver CSI alone (cf. (6.50)). We have seen that the capacity region of the downlink channel does not depend on the correlation between the additive noise processes $z_1[m]$ and $z_2[m]$ (cf. Exercise 6.24(1)). Consider the following specific correlation: $(z_1[m], z_2[m])$ are $\mathcal{CN}(0, \mathbf{K}[m])$ and independent in time m. To preserve the marginal variance, the diagonal entries of the covariance matrix $\mathbf{K}[m]$ must be N_0 each. Let us denote the off-diagonal term by $\rho[m]N_0$ (with $|\rho[m]| \leq 1$). Suppose now we let the two users cooperate.

1. Show that by a careful choice of $\rho[m]$ (as a function of $h_1[m]$ and $h_2[m]$), cooperation is not gainful: that is, for any reliable rates R_1, R_2 in the downlink fading channel,

$$R_1 + R_2 \leq \mathbb{E}\left[\log\left(1 + \frac{|h|^2 P}{N_0}\right)\right],\qquad(6.92)$$

   the same as can be achieved by a single user alone (cf. (6.51)). Here distribution of h is the symmetric stationary distribution of the fading processes $\{h_k[m]\}$ (for $k = 1, 2$). *Hint*: You will find Exercise 6.24(3) useful.

2. Conclude that the capacity region of the symmetric downlink fading channel is that given by (6.92).

Exercise 6.28 Show that the proportional fair algorithm with an infinite time-scale window maximizes (among all scheduling algorithms) the sum of the logarithms of the throughputs of the users. This justifies (6.57). This result has been derived in the literature at several places, including [12].

Exercise 6.29 Consider the opportunistic beamforming scheme in conjunction with a proportional fair scheduler operating in a slow fading environment. A reading exercise is to study Theorem 1 of [137], which shows that the rate available to each user is approximately equal to the instantaneous rate when it is being transmit beamformed, scaled down by the number of users.

Exercise 6.30 In a cellular system, the multiuser diversity gain in the downlink is expressed through the maximum SINR (cf. (6.63))

$$\text{SINR}_{\max} := \max_{k=1\ldots K} \text{SINR}_k = \frac{P|h_k|^2}{N_0 + P\sum_{j=1}^{J} |g_{kj}|^2},\qquad(6.93)$$

where we have denoted P by the average received power at a user. Let us denote the ratio P/N_0 by **SNR**. Let us suppose that $h_1, \ldots, h_K$ are i.i.d. $\mathcal{CN}(0, 1)$ random variables, and $\{g_{kj}, k = 1 \ldots K, j = 1 \ldots J\}$ are i.i.d. $\mathcal{CN}(0, 0.2)$ random variables independent of h. (A factor of 0.2 is used to model the average scenario of the mobile user being closer to the base-station it is communicating with as opposed to all the other base-stations it is hearing interference from, cf. Section 4.2.3.)

1. Show using the limit theorem in Exercise 6.21 that

$$\frac{\mathbb{E}[\text{SINR}_{\text{max}}]}{x_K} \to 1, \qquad \text{as } K \to \infty, \tag{6.94}$$

where x_K satisfies the non-linear equation:

$$\left(1 + \frac{x_K}{5}\right)^J = K \exp\left(-\frac{x_K}{\text{SNR}}\right). \tag{6.95}$$

2. Plot x_K for $K = 1, \ldots, 16$ for different values of SNR (ranging from 0 dB to 20 dB). Can you intuitively justify the observation from the plot that x_K increases with increasing SNR values? *Hint*: The probability that $|h_k|^2$ is less than or equal to a small positive number ϵ is approximately equal to ϵ itself, while the probability that $|h_k|^2$ is larger than a large number $1/\epsilon$ is $\exp(-1/\epsilon)$. Thus the likely way SINR becomes large is by the denominator being small as opposed to the numerator becoming large.

3. Show using part (1), or directly, that at small values of SNR the mean of the effective SINR grows like $\log K$. You can also see this directly from (6.93): at small values of SNR, the effective SINR is simply the maximum of K Rayleigh distributed random variables and from Exercise 6.21(2) we know that the mean value grows like $\log K$.

4. At very high values of SNR, we can approximate $\exp(-x_K/\text{SNR})$ in (6.95) by 1. With this approximation, show, using part (1), that the scaling x_K is approximately like $K^{1/J}$. This is a faster growth rate than the one at low SNR.

5. In a cellular system, typically the value of P is chosen such that the background noise N_0 and the interference term are of the same order. This makes sense for a system where there is no scheduling of users: since the system is interference plus noise limited, there is no point in making one of them (interference or background noise) much smaller than the other. In our notation here, this means that SNR is approximately 0 dB. From the calculations of this exercise what design setting of P can you infer for a system using the multiuser diversity harnessing scheduler? Thus, conventional transmit power settings will have to be revisited in this new system point of view.

Exercise 6.31 (Interaction between space-time codes and multiuser diversity scheduling) A design is proposed for the downlink IS-856 using dual transmit antennas at the base-station. It employs the Alamouti scheme when transmitting to a single user and among the users schedules the user with the best *effective* instantaneous SNR under the Alamouti scheme. We would like to compare the performance gain, if any, of using this scheme as opposed to using just a single transmit antenna and scheduling to the user with the best instantaneous SNR. Assume independent Rayleigh fading across the transmit antennas.

1. Plot the distribution of the instantaneous effective SNR under the Alamouti scheme, and compare that to the distribution of the SNR for a single antenna.

2. Suppose there is only a single user (i.e., $K = 1$). From your plot in part (1), do you think the dual transmit antennas provide any gain? Justify your answer. *Hint*: Use Jensen's inequality.

3. How about when $K > 1$? Plot the achievable throughput under both schemes at average SNR $= 0$ dB and for different values of K.

4. Is the proposed way of using dual transmit antennas smart?

7

MIMO I: spatial multiplexing and channel modeling

In this book, we have seen several different uses of multiple antennas in wireless communication. In Chapter 3, multiple antennas were used to provide *diversity gain* and increase the reliability of wireless links. Both receive and transmit diversity were considered. Moreover, receive antennas can also provide a *power gain*. In Chapter 5, we saw that with channel knowledge at the transmitter, multiple transmit antennas can also provide a power gain via transmit beamforming. In Chapter 6, multiple transmit antennas were used to induce channel variations, which can then be exploited by opportunistic communication techniques. The scheme can be interpreted as *opportunistic beamforming* and provides a *power gain* as well.

In this and the next few chapters, we will study a new way to use multiple antennas. We will see that under suitable channel fading conditions, having *both* multiple transmit and multiple receive antennas (i.e., a MIMO channel) provides an additional spatial dimension for communication and yields a *degree-of-freedom* gain. These additional degrees of freedom can be exploited by *spatially multiplexing* several data streams onto the MIMO channel, and lead to an increase in the capacity: the capacity of such a MIMO channel with n transmit and receive antennas is proportional to n.

Historically, it has been known for a while that a *multiple access* system with multiple antennas at the base-station allows several users to simultaneously communicate with the base-station. The multiple antennas allow spatial separation of the signals from the different users. It was observed in the mid 1990s that a similar effect can occur for a point-to-point channel with multiple transmit *and* receive antennas, i.e., even when the transmit antennas are not geographically far apart. This holds provided that the scattering environment is rich enough to allow the receive antennas to separate out the signals from the different transmit antennas. We have already seen how channel fading can be exploited by opportunistic communication techniques. Here, we see yet another example where channel fading is beneficial to communication.

It is insightful to compare and contrast the nature of the performance gains offered by opportunistic communication and by MIMO techniques.

Opportunistic communication techniques primarily provide a *power gain*. This power gain is very significant in the low SNR regime where systems are power-limited but less so in the high SNR regime where they are bandwidth-limited. As we will see, MIMO techniques can provide *both* a power gain and a degree-of-freedom gain. Thus, MIMO techniques become the primary tool to increase capacity significantly in the high SNR regime.

MIMO communication is a rich subject, and its study will span the remaining chapters of the book. The focus of the present chapter is to investigate the properties of the physical environment which enable spatial multiplexing and show how these properties can be succinctly captured in a statistical MIMO channel model. We proceed as follows. Through a capacity analysis, we first identify key parameters that determine the multiplexing capability of a deterministic MIMO channel. We then go through a sequence of physical MIMO channels to assess their spatial multiplexing capabilities. Building on the insights from these examples, we argue that it is most natural to model the MIMO channel in the *angular domain* and discuss a statistical model based on that approach. Our approach here parallels that in Chapter 2, where we started with a few idealized examples of multipath wireless channels to gain insights into the underlying physical phenomena, and proceeded to statistical fading models, which are more appropriate for the design and performance analysis of communication schemes. We will in fact see a lot of parallelism in the specific channel modeling technique as well.

Our focus throughout is on flat fading MIMO channels. The extensions to frequency-selective MIMO channels are straightforward and are developed in the exercises.

7.1 Multiplexing capability of deterministic MIMO channels

A narrowband time-invariant wireless channel with n_t transmit and n_r receive antennas is described by an n_r by n_t deterministic matrix $\mathbf{H}$. What are the key properties of $\mathbf{H}$ that determine how much spatial multiplexing it can support? We answer this question by looking at the capacity of the channel.

7.1.1 Capacity via singular value decomposition

The time-invariant channel is described by

$$\mathbf{y} = \mathbf{H}\mathbf{x} + \mathbf{w}, \tag{7.1}$$

where $\mathbf{x} \in \mathcal{C}^{n_t}$, $\mathbf{y} \in \mathcal{C}^{n_r}$ and $\mathbf{w} \sim \mathcal{CN}(0, N_0\mathbf{I}_{n_r})$ denote the transmitted signal, received signal and white Gaussian noise respectively at a symbol time (the time index is dropped for simplicity). The channel matrix $\mathbf{H} \in \mathcal{C}^{n_r \times n_t}$

is deterministic and assumed to be constant at all times and known to both the transmitter and the receiver. Here, h_{ij} is the channel gain from transmit antenna j to receive antenna i. There is a total power constraint, P, on the signals from the transmit antennas.

This is a *vector* Gaussian channel. The capacity can be computed by decomposing the vector channel into a set of parallel, independent scalar Gaussian sub-channels. From basic linear algebra, every linear transformation can be represented as a composition of three operations: a rotation operation, a scaling operation, and another rotation operation. In the notation of matrices, the matrix $\mathbf{H}$ has a *singular value decomposition* (SVD):

$$\mathbf{H} = \mathbf{U}\mathbf{\Lambda}\mathbf{V}^*, \tag{7.2}$$

where $\mathbf{U} \in \mathcal{C}^{n_r \times n_r}$ and $\mathbf{V} \in \mathcal{C}^{n_t \times n_t}$ are (rotation) unitary matrices[1] and $\mathbf{\Lambda} \in \Re^{n_r \times n_t}$ is a rectangular matrix whose diagonal elements are non-negative real numbers and whose off-diagonal elements are zero.[2] The diagonal elements $\lambda_1 \geq \lambda_2 \geq \cdots \geq \lambda_{n_{\min}}$ are the ordered *singular values* of the matrix $\mathbf{H}$, where $n_{\min} := \min(n_t, n_r)$. Since

$$\mathbf{H}\mathbf{H}^* = \mathbf{U}\mathbf{\Lambda}\mathbf{\Lambda}'\mathbf{U}^*, \tag{7.3}$$

the squared singular values λ_i^2 are the eigenvalues of the matrix $\mathbf{H}\mathbf{H}^*$ and also of $\mathbf{H}^*\mathbf{H}$. Note that there are $n_{\min}$ singular values. We can rewrite the SVD as

$$\mathbf{H} = \sum_{i=1}^{n_{\min}} \lambda_i \mathbf{u}_i \mathbf{v}_i^*, \tag{7.4}$$

i.e., the sum of rank-one matrices $\lambda_i \mathbf{u}_i \mathbf{v}_i^*$. It can be seen that the rank of $\mathbf{H}$ is precisely the number of non-zero singular values.

If we define

$$\tilde{\mathbf{x}} := \mathbf{V}^* \mathbf{x}, \tag{7.5}$$

$$\tilde{\mathbf{y}} := \mathbf{U}^* \mathbf{y}, \tag{7.6}$$

$$\tilde{\mathbf{w}} := \mathbf{U}^* \mathbf{w}, \tag{7.7}$$

then we can rewrite the channel (7.1) as

$$\tilde{\mathbf{y}} = \mathbf{\Lambda}\tilde{\mathbf{x}} + \tilde{\mathbf{w}}, \tag{7.8}$$

[1] Recall that a unitary matrix $\mathbf{U}$ satisfies $\mathbf{U}^*\mathbf{U} = \mathbf{U}\mathbf{U}^* = \mathbf{I}$.
[2] We will call this matrix diagonal even though it may not be square.

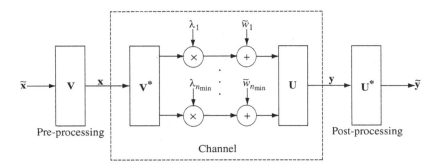

where $\tilde{\mathbf{w}} \sim \mathcal{CN}(0, N_0 \mathbf{I}_{n_r})$ has the same distribution as $\mathbf{w}$ (cf. (A.22) in Appendix A), and $\|\tilde{\mathbf{x}}\|^2 = \|\mathbf{x}\|^2$. Thus, the energy is preserved and we have an equivalent representation as a parallel Gaussian channel:

$$\tilde{y}_i = \lambda_i \tilde{x}_i + \tilde{w}_i, \qquad i = 1, 2, \ldots, n_{\min}. \tag{7.9}$$

The equivalence is summarized in Figure 7.1.

The SVD decomposition can be interpreted as two *coordinate transformations*: it says that if the input is expressed in terms of a coordinate system defined by the columns of $\mathbf{V}$ and the output is expressed in terms of a coordinate system defined by the columns of $\mathbf{U}$, then the input/output relationship is very simple. Equation (7.8) is a representation of the original channel (7.1) with the input and output expressed in terms of these new coordinates.

We have already seen examples of Gaussian parallel channels in Chapter 5, when we talked about capacities of time-invariant frequency-selective channels and about time-varying fading channels with full CSI. The time-invariant MIMO channel is yet another example. Here, the spatial dimension plays the same role as the time and frequency dimensions in those other problems. The capacity is by now familiar:

$$C = \sum_{i=1}^{n_{\min}} \log \left(1 + \frac{P_i^* \lambda_i^2}{N_0} \right) \text{ bits/s/Hz}, \tag{7.10}$$

where $P_1^*, \ldots, P_{n_{\min}}^*$ are the waterfilling power allocations:

$$P_i^* = \left(\mu - \frac{N_0}{\lambda_i^2} \right)^+, \tag{7.11}$$

with μ chosen to satisfy the total power constraint $\sum_i P_i^* = P$. Each λ_i corresponds to an *eigenmode* of the channel (also called an *eigenchannel*). Each non-zero eigenchannel can support a data stream; thus, the MIMO channel can support the spatial multiplexing of multiple streams. Figure 7.2 pictorially depicts the SVD-based architecture for reliable communication.

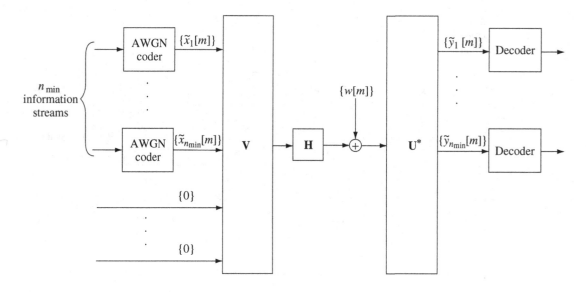

Figure 7.2 The SVD architecture for MIMO communication.

There is a clear analogy between this architecture and the OFDM system introduced in Chapter 3. In both cases, a transformation is applied to convert a matrix channel into a set of parallel independent sub-channels. In the OFDM setting, the matrix channel is given by the circulant matrix $\mathbf{C}$ in (3.139), defined by the ISI channel together with the cyclic prefix added onto the input symbols. In fact, the decomposition $\mathbf{C} = \mathbf{Q}^{-1}\Lambda\mathbf{Q}$ in (3.143) is the SVD decomposition of a circulant matrix $\mathbf{C}$, with $\mathbf{U} = \mathbf{Q}^{-1}$ and $\mathbf{V}^* = \mathbf{Q}$. The important difference between the ISI channel and the MIMO channel is that, for the former, the $\mathbf{U}$ and $\mathbf{V}$ matrices (DFTs) do not depend on the specific realization of the ISI channel, while for the latter, they do depend on the specific realization of the MIMO channel.

7.1.2 Rank and condition number

What are the key parameters that determine performance? It is simpler to focus separately on the high and the low SNR regimes. At high SNR, the water level is deep and the policy of allocating equal amounts of power on the non-zero eigenmodes is asymptotically optimal (cf. Figure 5.24(a)):

$$C \approx \sum_{i=1}^{k} \log\left(1 + \frac{P\lambda_i^2}{kN_0}\right) \approx k \log \mathsf{SNR} + \sum_{i=1}^{k} \log\left(\frac{\lambda_i^2}{k}\right) \text{ bits/s/Hz,} \qquad (7.12)$$

where k is the number of non-zero λ_i^2, i.e., the rank of $\mathbf{H}$, and $\mathsf{SNR} := P/N_0$. *The parameter k is the number of spatial degrees of freedom per second per Hertz.* It represents the dimension of the transmitted signal as modified by the MIMO channel, i.e., the dimension of the image of $\mathbf{H}$. This is equal to the rank of the matrix $\mathbf{H}$ and with full rank, we see that a MIMO channel provides $n_{\min}$ spatial degrees of freedom.

The rank is a first-order but crude measure of the capacity of the channel. To get a more refined picture, one needs to look at the non-zero singular values themselves. By Jensen's inequality,

$$\frac{1}{k}\sum_{i=1}^{k}\log\left(1+\frac{P}{kN_0}\lambda_i^2\right) \leq \log\left(1+\frac{P}{kN_0}\left(\frac{1}{k}\sum_{i=1}^{k}\lambda_i^2\right)\right) \qquad (7.13)$$

Now,

$$\sum_{i=1}^{k}\lambda_i^2 = \mathrm{Tr}[\mathbf{HH}^*] = \sum_{i,j}|h_{ij}|^2, \qquad (7.14)$$

which can be interpreted as the total power gain of the matrix channel if one spreads the energy equally between all the transmit antennas. Then, the above result says that among the channels with the same total power gain, the one that has the highest capacity is the one with all the singular values equal. More generally, the less spread out the singular values, the larger the capacity in the high SNR regime. In numerical analysis, $(\max_i \lambda_i / \min_i \lambda_i)$ is defined to be the *condition number* of the matrix $\mathbf{H}$. The matrix is said to be *well-conditioned* if the condition number is close to 1. From the above result, an important conclusion is:

> Well-conditioned channel matrices facilitate communication in the high SNR regime.

At low SNR, the optimal policy is to allocate power only to the strongest eigenmode (the bottom of the vessel to waterfill, cf. Figure 5.24(b)). The resulting capacity is

$$C \approx \frac{P}{N_0}\left(\max_i \lambda_i^2\right)\log_2 e \text{ bits/s/Hz}. \qquad (7.15)$$

The MIMO channel provides a *power gain* of $\max_i \lambda_i^2$. In this regime, the rank or condition number of the channel matrix is less relevant. What matters is how much energy gets transferred from the transmitter to the receiver.

7.2 Physical modeling of MIMO channels

In this section, we would like to gain some insight on how the spatial multiplexing capability of MIMO channels depends on the physical environment. We do so by looking at a sequence of idealized examples and analyzing the

rank and conditioning of their channel matrices. These deterministic examples
will also suggest a natural approach to statistical modeling of MIMO channels, which we discuss in Section 7.3. To be concrete, we restrict ourselves
to *uniform linear antenna arrays*, where the antennas are evenly spaced on a
straight line. The details of the analysis depend on the specific array structure
but the concepts we want to convey do not.

7.2.1 Line-of-sight SIMO channel

The simplest SIMO channel has a single line-of-sight (Figure 7.3(a)). Here,
there is only free space without any reflectors or scatterers, and only a
direct signal path between each antenna pair. The antenna separation is $\Delta_r \lambda_c$,
where λ_c is the carrier wavelength and Δ_r is the normalized receive antenna
separation, normalized to the unit of the carrier wavelength. The dimension
of the antenna array is much smaller than the distance between the transmitter
and the receiver.

The continuous-time impulse response $h_i(\tau)$ between the transmit antenna
and the ith receive antenna is given by

$$h_i(\tau) = a\delta(\tau - d_i/c), \qquad i = 1, \ldots, n_r, \qquad (7.16)$$

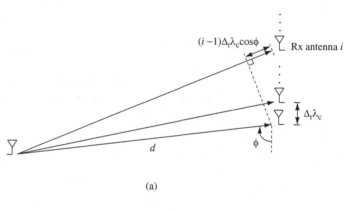

(a)

Figure 7.3 (a) Line-of-sight
channel with single transmit
antenna and multiple receive
antennas. The signals from the
transmit antenna arrive almost
in parallel at the receiving
antennas. (b) Line-of-sight
channel with multiple transmit
antennas and single receive
antenna.

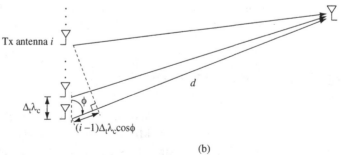

(b)

where d_i is the distance between the transmit antenna and ith receive antenna, c is the speed of light and a is the attenuation of the path, which we assume to be the same for all antenna pairs. Assuming $d_i/c \ll 1/W$, where W is the transmission bandwidth, the baseband channel gain is given by (2.34) and (2.27):

$$h_i = a \exp\left(-\frac{j2\pi f_c d_i}{c}\right) = a \exp\left(-\frac{j2\pi d_i}{\lambda_c}\right), \qquad (7.17)$$

where f_c is the carrier frequency. The SIMO channel can be written as

$$\mathbf{y} = \mathbf{h}x + \mathbf{w} \qquad (7.18)$$

where x is the transmitted symbol, $\mathbf{w} \sim \mathcal{CN}(0, N_0\mathbf{I})$ is the noise and $\mathbf{y}$ is the received vector. The vector of channel gains $\mathbf{h} = [h_1, \dots, h_{n_r}]^t$ is sometimes called the signal direction or the *spatial signature* induced on the receive antenna array by the transmitted signal.

Since the distance between the transmitter and the receiver is much larger than the size of the receive antenna array, the paths from the transmit antenna to each of the receive antennas are, to a first-order, parallel and

$$d_i \approx d + (i-1)\Delta_r\lambda_c \cos\phi, \quad i = 1, \dots, n_r, \qquad (7.19)$$

where d is the distance from the transmit antenna to the first receive antenna and ϕ is the angle of incidence of the line-of-sight onto the receive antenna array. (You are asked to verify this in Exercise 7.1.) The quantity $(i-1)\Delta_r\lambda_c \cos\phi$ is the displacement of receive antenna i from receive antenna 1 in the direction of the line-of-sight. The quantity

$$\Omega := \cos\phi$$

is often called the *directional cosine* with respect to the receive antenna array. The spatial signature $\mathbf{h} = [h_1, \dots, h_{n_r}]^t$ is therefore given by

$$\mathbf{h} = a \exp\left(-\frac{j2\pi d}{\lambda_c}\right) \begin{bmatrix} 1 \\ \exp(-j2\pi\Delta_r\Omega) \\ \exp(-j2\pi 2\Delta_r\Omega) \\ \vdots \\ \exp(-j2\pi(n_r-1)\Delta_r\Omega) \end{bmatrix}, \qquad (7.20)$$

i.e., the signals received at consecutive antennas differ in phase by $2\pi\Delta_r\Omega$ due to the relative delay. For notational convenience, we define

$$\mathbf{e}_r(\Omega) := \frac{1}{\sqrt{n_r}} \begin{bmatrix} 1 \\ \exp(-j2\pi\Delta_r\Omega) \\ \exp(-j2\pi 2\Delta_r\Omega) \\ \vdots \\ \exp(-j2\pi(n_r-1)\Delta_r\Omega) \end{bmatrix}, \qquad (7.21)$$

as the unit spatial signature in the directional cosine Ω.

The optimal receiver simply projects the noisy received signal onto the signal direction, i.e., maximal ratio combining or receive beamforming (cf. Section 5.3.1). It adjusts for the different delays so that the received signals at the antennas can be combined constructively, yielding an n_r-fold power gain. The resulting capacity is

$$C = \log\left(1 + \frac{P\|\mathbf{h}\|^2}{N_0}\right) = \log\left(1 + \frac{Pa^2 n_r}{N_0}\right) \quad \text{bits/s/Hz.} \qquad (7.22)$$

The SIMO channel thus provides a power gain but no degree-of-freedom gain.

In the context of a line-of-sight channel, the receive antenna array is sometimes called a *phased-array antenna*.

7.2.2 Line-of-sight MISO channel

The MISO channel with multiple transmit antennas and a single receive antenna is reciprocal to the SIMO channel (Figure 7.3(b)). If the transmit antennas are separated by $\Delta_t\lambda_c$ and there is a single line-of-sight with angle of departure of ϕ (directional cosine $\Omega := \cos\phi$), the MISO channel is given by

$$y = \mathbf{h}^*\mathbf{x} + w \qquad (7.23)$$

where

$$\mathbf{h} = a\exp\left(\frac{j2\pi d}{\lambda_c}\right) \begin{bmatrix} 1 \\ \exp(-j2\pi\Delta_t\Omega) \\ \exp(-j2\pi 2\Delta_t\Omega) \\ \vdots \\ \exp(-j2\pi(n_r-1)\Delta_t\Omega) \end{bmatrix}, \qquad (7.24)$$

The optimal transmission (transmit beamforming) is performed along the direction $\mathbf{e}_t(\Omega)$ of $\mathbf{h}$, where

$$\mathbf{e}_t(\Omega) := \frac{1}{\sqrt{n_t}} \begin{bmatrix} 1 \\ \exp(-\mathrm{j}2\pi\Delta_t\Omega) \\ \exp(-\mathrm{j}2\pi 2\Delta_t\Omega) \\ \vdots \\ \exp(-\mathrm{j}2\pi(n_t-1)\Delta_t\Omega) \end{bmatrix}, \qquad (7.25)$$

is the unit spatial signature in the transmit direction of Ω (cf. Section 5.3.2). The phase of the signal from each of the transmit antennas is adjusted so that they add constructively at the receiver, yielding an n_t-fold power gain. The capacity is the same as (7.22). Again there is no degree-of-freedom gain.

7.2.3 Antenna arrays with only a line-of-sight path

Let us now consider a MIMO channel with only direct line-of-sight paths between the antennas. Both the transmit and the receive antennas are in linear arrays. Suppose the normalized transmit antenna separation is Δ_t and the normalized receive antenna separation is Δ_r. The channel gain between the kth transmit antenna and the ith receive antenna is

$$h_{ik} = a \exp(-\mathrm{j}2\pi d_{ik}/\lambda_c), \qquad (7.26)$$

where d_{ik} is the distance between the antennas, and a is the attenuation along the line-of-sight path (assumed to be the same for all antenna pairs). Assuming again that the antenna array sizes are much smaller than the distance between the transmitter and the receiver, to a first-order:

$$d_{ik} = d + (i-1)\Delta_r\lambda_c \cos\phi_r - (k-1)\Delta_t\lambda_c \cos\phi_t, \qquad (7.27)$$

where d is the distance between transmit antenna 1 and receive antenna 1, and ϕ_t, ϕ_r are the angles of incidence of the line-of-sight path on the transmit and receive antenna arrays, respectively. Define $\Omega_t := \cos\phi_t$ and $\Omega_r := \cos\phi_r$. Substituting (7.27) into (7.26), we get

$$h_{ik} = a \exp\left(-\frac{\mathrm{j}2\pi d}{\lambda_c}\right) \cdot \exp(\mathrm{j}2\pi(k-1)\Delta_t\Omega_t) \cdot \exp(-\mathrm{j}2\pi(i-1)\Delta_r\Omega_r) \quad (7.28)$$

and we can write the channel matrix as

$$\mathbf{H} = a\sqrt{n_t n_r} \exp\left(-\frac{\mathrm{j}2\pi d}{\lambda_c}\right) \mathbf{e}_r(\Omega_r)\mathbf{e}_t(\Omega_t)^*, \qquad (7.29)$$

where $\mathbf{e_r}(\cdot)$ and $\mathbf{e_t}(\cdot)$ are defined in (7.21) and (7.25), respectively. Thus, $\mathbf{H}$ is a rank-one matrix with a unique non-zero singular value $\lambda_1 = a\sqrt{n_t n_r}$. The capacity of this channel follows from (7.10):

$$C = \log\left(1 + \frac{Pa^2 n_t n_r}{N_0}\right) \text{ bits/s/Hz.} \qquad (7.30)$$

Note that although there are multiple transmit and multiple receive antennas, the transmitted signals are all projected onto a single-dimensional space (the only non-zero eigenmode) and thus only one spatial degree of freedom is available. The receive spatial signatures at the receive antenna array from all the transmit antennas (i.e., the columns of $\mathbf{H}$) are along the same direction, $\mathbf{e_r}(\Omega_r)$. Thus, the number of available spatial degrees of freedom does not increase even though there are multiple transmit and multiple receive antennas.

The factor $n_t n_r$ is the *power gain* of the MIMO channel. If $n_t = 1$, the power gain is equal to the number of receive antennas and is obtained by maximal ratio combining at the receiver (receive beamforming). If $n_r = 1$, the power gain is equal to the number of transmit antennas and is obtained by transmit beamforming. For general numbers of transmit and receive antennas, one gets benefits from *both* transmit and receive beamforming: the transmitted signals are constructively added in-phase at each receive antenna, and the signal at each receive antenna is further constructively combined with each other.

In summary: in a line-of-sight only environment, a MIMO channel provides a power gain but no degree-of-freedom gain.

7.2.4 Geographically separated antennas

Geographically separated transmit antennas

How do we get a degree-of-freedom gain? Consider the thought experiment where the transmit antennas can now be placed very far apart, with a separation of the order of the distance between the transmitter and the receiver. For concreteness, suppose there are two transmit antennas (Figure 7.4). Each

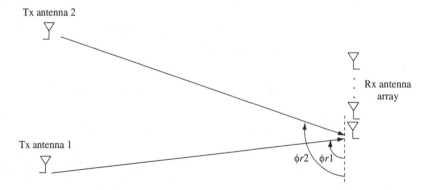

Figure 7.4 Two geographically separated transmit antennas each with line-of-sight to a receive antenna array.

transmit antenna has only a line-of-sight path to the receive antenna array, with attenuations a_1 and a_2 and angles of incidence ϕ_{r1} and ϕ_{r2}, respectively. Assume that the delay spread of the signals from the transmit antennas is much smaller than $1/W$ so that we can continue with the single-tap model. The spatial signature that transmit antenna k impinges on the receive antenna array is

$$\mathbf{h}_k = a_k \sqrt{n_r} \exp\left(\frac{-j2\pi d_{1k}}{\lambda_c}\right) \mathbf{e_r}(\Omega_{rk}), \quad k = 1, 2, \qquad (7.31)$$

where d_{1k} is the distance between transmit antenna k and receive antenna 1, $\Omega_{rk} := \cos \phi_{rk}$ and $\mathbf{e_r}(\cdot)$ is defined in (7.21).

It can be directly verified that the spatial signature $\mathbf{e_r}(\Omega)$ is a periodic function of Ω with period $1/\Delta_r$, and within one period it never repeats itself (Exercise 7.2). Thus, the channel matrix $\mathbf{H} = [\mathbf{h}_1, \mathbf{h}_2]$ has distinct and linearly independent columns as long as the separation in the directional cosines

$$\Omega_r := \Omega_{r2} - \Omega_{r1} \neq 0 \quad \mathrm{mod} \ \frac{1}{\Delta_r}. \qquad (7.32)$$

In this case, it has two non-zero singular values λ_1^2 and λ_2^2, yielding two degrees of freedom. Intuitively, the transmitted signal can now be received from two different directions that can be resolved by the receive antenna array. Contrast this with the example in Section 7.2.3, where the antennas are placed close together and the spatial signatures of the transmit antennas are all aligned with each other.

Note that since Ω_{r1}, Ω_{r2}, being directional cosines, lie in $[-1, 1]$ and cannot differ by more than 2, the condition (7.32) reduces to the simpler condition $\Omega_{r1} \neq \Omega_{r2}$ whenever the antenna spacing $\Delta_r \leq 1/2$.

Resolvability in the angular domain

The channel matrix $\mathbf{H}$ is full rank whenever the separation in the directional cosines $\Omega_r \neq 0 \mod 1/\Delta_r$. However, it can still be very ill-conditioned. We now give an order-of-magnitude estimate on how large the angular separation has to be so that $\mathbf{H}$ is well-conditioned and the two degrees of freedom can be effectively used to yield a high capacity.

The conditioning of $\mathbf{H}$ is determined by how aligned the spatial signatures of the two transmit antennas are: the less aligned the spatial signatures are, the better the conditioning of $\mathbf{H}$. The angle θ between the two spatial signatures satisfies

$$|\cos \theta| := |\mathbf{e_r}(\Omega_{r1})^* \mathbf{e_r}(\Omega_{r2})|. \qquad (7.33)$$

Note that $\mathbf{e_r}(\Omega_{r1})^* \mathbf{e_r}(\Omega_{r2})$ depends only on the difference $\Omega_r := \Omega_{r2} - \Omega_{r1}$. Define then

$$f_r(\Omega_{r2} - \Omega_{r1}) := \mathbf{e_r}(\Omega_{r1})^* \mathbf{e_r}(\Omega_{r2}). \qquad (7.34)$$

By direct computation (Exercise 7.3),

$$f_r(\Omega_r) = \frac{1}{n_r} \exp(j\pi\Delta_r\Omega_r(n_r-1))\frac{\sin(\pi L_r\Omega_r)}{\sin(\pi L_r\Omega_r/n_r)}, \qquad (7.35)$$

where $L_r := n_r\Delta_r$ is the normalized length of the receive antenna array. Hence,

$$|\cos\theta| = \left|\frac{\sin(\pi L_r\Omega_r)}{n_r\sin(\pi L_r\Omega_r/n_r)}\right|. \qquad (7.36)$$

The conditioning of the matrix $\mathbf{H}$ depends directly on this parameter. For simplicity, consider the case when the gains $a_1 = a_2 = a$. The squared singular values of $\mathbf{H}$ are

$$\lambda_1^2 = a^2 n_r(1+|\cos\theta|), \qquad \lambda_2^2 = a^2 n_r(1-|\cos\theta|) \qquad (7.37)$$

and the condition number of the matrix is

$$\frac{\lambda_1}{\lambda_2} = \sqrt{\frac{1+|\cos\theta|}{1-|\cos\theta|}}. \qquad (7.38)$$

The matrix is ill-conditioned whenever $|\cos\theta| \approx 1$, and is well-conditioned otherwise. In Figure 7.5, this quantity $|\cos\theta| = |f_r(\Omega_r)|$ is plotted as a function of Ω_r for a fixed array size and different values of n_r. The function $f_r(\cdot)$ has the following properties:

- $f_r(\Omega_r)$ is periodic with period $n_r/L_r = 1/\Delta_r$;
- $f_r(\Omega_r)$ peaks at $\Omega_r = 0$; $f(0) = 1$;
- $f_r(\Omega_r) = 0$ at $\Omega_r = k/L_r, k = 1, \ldots, n_r-1$.

The periodicity of $f_r(\cdot)$ follows from the periodicity of the spatial signature $\mathbf{e}_r(\cdot)$. It has a main lobe of width $2/L_r$ centered around integer multiples of $1/\Delta_r$. All the other lobes have significantly lower peaks. This means that the signatures are close to being aligned and the channel matrix is ill conditioned whenever

$$\left|\Omega_r - \frac{m}{\Delta_r}\right| \ll \frac{1}{L_r} \qquad (7.39)$$

for some integer m. Now, since Ω_r ranges from -2 to 2, this condition reduces to

$$|\Omega_r| \ll \frac{1}{L_r} \qquad (7.40)$$

whenever the antenna separation $\Delta_r \leq 1/2$.

Figure 7.5 The function $|f(\Omega_r)|$ plotted as a function of Ω_r for fixed $L_r = 8$ and different values of the number of receive antennas n_r.

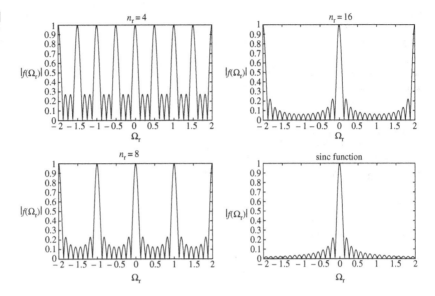

Increasing the number of antennas for a fixed antenna length L_r does not substantially change the qualitative picture above. In fact, as $n_r \to \infty$ and $\Delta_r \to 0$,

$$f_r(\Omega_r) \to e^{j\pi L_r \Omega_r} \mathrm{sinc}(L_r \Omega_r) \qquad (7.41)$$

and the dependency of $f_r(\cdot)$ on n_r vanishes. Equation (7.41) can be directly derived from (7.35), using the definition $\mathrm{sinc}(x) = \sin(\pi x)/\pi x$ (cf. (2.30)).

The parameter $1/L_r$ can be thought of as a measure of *resolvability* in the angular domain: if $\Omega_r \ll 1/L_r$, then the signals from the two transmit antennas cannot be resolved by the receive antenna array and there is effectively only one degree of freedom. Packing more and more antenna elements in a given amount of space does not increase the angular resolvability of the receive antenna array; it is intrinsically limited by the length of the array.

A common pictorial representation of the angular resolvability of an antenna array is the (receive) *beamforming pattern*. If the signal arrives from a single direction ϕ_0, then the optimal receiver projects the received signal onto the vector $\mathbf{e}_r(\cos\phi_0)$; recall that this is called the (receive) beamforming vector. A signal from any other direction ϕ is attenuated by a factor of

$$|\mathbf{e}_r(\cos\phi_0)^* \mathbf{e}_r(\cos\phi)| = |f_r(\cos\phi - \cos\phi_0)|. \qquad (7.42)$$

The beamforming pattern associated with the vector $\mathbf{e}_r(\cos\phi)$ is the polar plot

$$(\phi, |f_r(\cos\phi - \cos\phi_0)|) \qquad (7.43)$$

Figure 7.6 Receive beamforming patterns aimed at 90°, with antenna array length $L_r = 2$ and different numbers of receive antennas n_r. Note that the beamforming pattern is always symmetrical about the $0° - 180°$ axis, so lobes always appear in pairs. For $n_r = 4, 6, 32$, the antenna separation $\Delta_r \leq 1/2$, and there is a single main lobe around 90° (together with its mirror image). For $n_r = 2$, $\Delta_r = 1 > 1/2$ and there is an additional pair of main lobes.

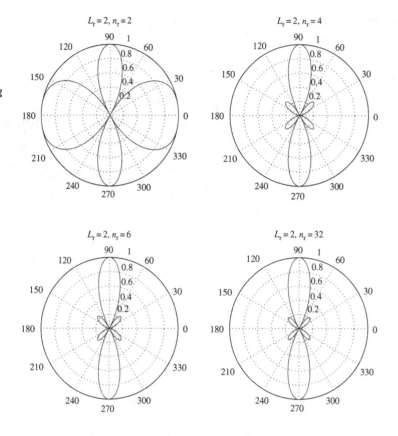

(Figures 7.6 and 7.7). Two important points to note about the beamforming pattern:

- It has main lobes around ϕ_0 and also around any angle ϕ for which

$$\cos \phi = \cos \phi_0 \quad \mod \frac{1}{\Delta_r}; \qquad (7.44)$$

this follows from the periodicity of $f_r(\cdot)$. If the antenna separation Δ_r is less than $1/2$, then there is only one main lobe at ϕ, together with its mirror image at $-\phi$. If the separation is greater than $1/2$, there can be several more pairs of main lobes (Figure 7.6).

- The main lobe has a directional cosine width of $2/L_r$; this is also called the *beam width*. The larger the array length L_r, the narrower the beam and the higher the angular resolution: the array filters out the signal from all directions except for a narrow range around the direction of interest (Figure 7.7). Signals that arrive along paths with angular seperation larger than $1/L_r$ can be discriminated by focusing different beams at them.

There is a clear analogy between the roles of the antenna array size L_r and the bandwidth W in a wireless channel. The parameter $1/W$ measures the

Figure 7.7 Beamforming patterns for different antenna array lengths. (Left) $L_r = 4$ and (right) $L_r = 8$. Antenna separation is fixed at half the carrier wavelength. The larger the length of the array, the narrower the beam.

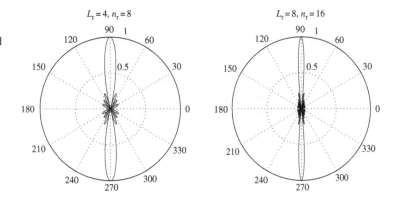

resolvability of signals in the time domain: multipaths arriving at time separation much less than $1/W$ cannot be resolved by the receiver. The parameter $1/L_r$ measures the resolvability of signals in the angular domain: signals that arrive within an angle much less than $1/L_r$ cannot be resolved by the receiver. Just as over-sampling cannot increase the time-domain resolvability beyond $1/W$, adding more antenna elements cannot increase the angular-domain resolvability beyond $1/L_r$. This analogy will be exploited in the statistical modeling of MIMO fading channels and explained more precisely in Section 7.3.

Geographically separated receive antennas

We have increased the number of degrees of freedom by placing the *transmit* antennas far apart and keeping the receive antennas close together, but we can achieve the same goal by placing the *receive* antennas far apart and keeping the transmit antennas close together (see Figure 7.8). The channel matrix is given by

$$\mathbf{H} = \begin{bmatrix} \mathbf{h}_1^* \\ \mathbf{h}_2^* \end{bmatrix}, \tag{7.45}$$

Figure 7.8 Two geographically separated receive antennas each with line-of-sight from a transmit antenna array.

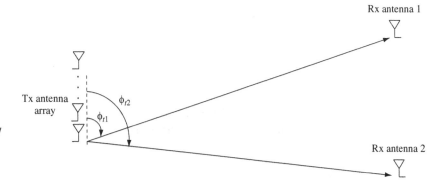

where

$$\mathbf{h}_i = a_i \exp\left(\frac{j2\pi d_{i1}}{\lambda_c}\right) \mathbf{e}_t(\Omega_{ti}), \tag{7.46}$$

and Ω_{ti} is the directional cosine of departure of the path from the transmit antenna array to receive antenna i and d_{i1} is the distance between transmit antenna 1 and receive antenna i. As long as

$$\Omega_t := \Omega_{t2} - \Omega_{t1} \neq 0 \quad \mathrm{mod} \ \frac{1}{\Delta_t}, \tag{7.47}$$

the two rows of $\mathbf{H}$ are linearly independent and the channel has rank 2, yielding 2 degrees of freedom. The output of the channel spans a two-dimensional space as we vary the transmitted signal at the transmit antenna array. In order to make $\mathbf{H}$ well-conditioned, the angular separation Ω_t of the two receive antennas should be of the order of or larger than $1/L_t$, where $L_t := n_t \Delta_t$ is the length of the transmit antenna array, normalized to the carrier wavelength.

Analogous to the receive beamforming pattern, one can also define a *transmit* beamforming pattern. This measures the amount of energy dissipated in other directions when the transmitter attempts to focus its signal along a direction ϕ_0. The beam width is $2/L_t$; the longer the antenna array, the sharper the transmitter can focus the energy along a desired direction and the better it can spatially multiplex information to the multiple receive antennas.

7.2.5 Line-of-sight plus one reflected path

Can we get a similar effect to that of the example in Section 7.2.4, without putting either the transmit antennas or the receive antennas far apart? Consider again the transmit and receive antenna arrays in that example, but now suppose in addition to a line-of-sight path there is another path reflected off a wall (see Figure 7.9(a)). Call the direct path, path 1 and the reflected path, path 2. Path i has an attenuation of a_i, makes an angle of ϕ_{ti} ($\Omega_{ti} := \cos\phi_{ti}$) with the transmit antenna array and an angle of $\phi_{ri}(\Omega_{ri} := \cos\phi_{ri})$ with the receive antenna array. The channel $\mathbf{H}$ is given by the principle of superposition:

$$\mathbf{H} = a_1^b \mathbf{e}_r(\Omega_{r1})\mathbf{e}_t(\Omega_{t1})^* + a_2^b \mathbf{e}_r(\Omega_{r2})\mathbf{e}_r(\Omega_{t2})^* \tag{7.48}$$

where for $i = 1, 2$,

$$a_i^b := a_i \sqrt{n_t n_r} \exp\left(-\frac{j2\pi d^{(i)}}{\lambda_c}\right), \tag{7.49}$$

and $d^{(i)}$ is the distance between transmit antenna 1 and receive antenna 1 along path i. We see that as long as

$$\Omega_{t1} \neq \Omega_{t2} \quad \mathrm{mod} \ \frac{1}{\Delta_t} \tag{7.50}$$

Figure 7.9 (a) A MIMO
channel with a direct path and
a reflected path. (b) Channel is
viewed as a concatenation of
two channels **H'** and **H''** with
intermediate (virtual) relays
A and B.

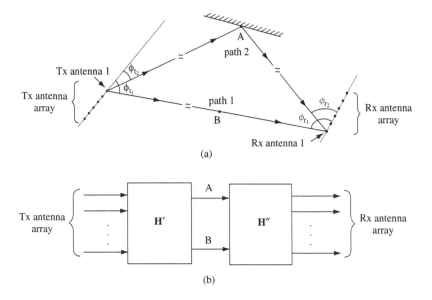

(a)

(b)

and

$$\Omega_{r1} \neq \Omega_{r2} \quad \mathrm{mod} \; \frac{1}{\Delta_r}, \tag{7.51}$$

the matrix **H** is of rank 2. In order to make **H** well-conditioned, the angular
separation $|\Omega_t|$ of the two paths at the transmit array should be of the same
order or larger than $1/L_t$ *and* the angular separation $|\Omega_r|$ at the receive array
should be of the same order as or larger than $1/L_r$, where

$$\Omega_t := \cos \phi_{t2} - \cos \phi_{t1}, \qquad L_t := n_t \Delta_t \tag{7.52}$$

and

$$\Omega_r := \cos \phi_{r2} - \cos \phi_{r1}, \qquad L_r := n_r \Delta_r. \tag{7.53}$$

To see clearly what the role of the multipath is, it is helpful to rewrite **H**
as $\mathbf{H} = \mathbf{H''H'}$, where

$$\mathbf{H''} = \left[a_1^b \mathbf{e_r}(\Omega_{r1}), \, a_2^b \mathbf{e_r}(\Omega_{r2}) \right], \qquad \mathbf{H'} = \begin{bmatrix} \mathbf{e_t^*}(\Omega_{t1}) \\ \mathbf{e_t^*}(\Omega_{t2}) \end{bmatrix}. \tag{7.54}$$

H' is a 2 by n_t matrix while **H''** is an n_r by 2 matrix. One can interpret **H'** as
the matrix for the channel from the transmit antenna array to two imaginary
receivers at point A and point B, as marked in Figure 7.9. Point A is the point
of incidence of the reflected path on the wall; point B is along the line-of-sight
path. Since points A and B are geographically widely separated, the matrix
H' has rank 2; its conditioning depends on the parameter $L_t \Omega_t$. Similarly,

one can interpret the second matrix $\mathbf{H}''$ as the matrix channel from two imaginary transmitters at A and B to the receive antenna array. This matrix has rank 2 as well; its conditioning depends on the parameter $L_r\Omega_r$. If both matrices are well-conditioned, then the overall channel matrix $\mathbf{H}$ is also well-conditioned.

The MIMO channel with two multipaths is essentially a concatenation of the n_t by 2 channel in Figure 7.8 and the 2 by n_r channel in Figure 7.4. Although both the transmit antennas and the receive antennas are close together, multipaths in effect provide virtual "relays", which are geographically far apart. The channel from the transmit array to the relays as well as the channel from the relays to the receive array both have two degrees of freedom, and so does the overall channel. Spatial multiplexing is now possible. In this context, *multipath fading can be viewed as providing an advantage that can be exploited.*

It is important to note in this example that significant angular separation of the two paths at *both* the transmit and the receive antenna arrays is crucial for the well-conditionedness of $\mathbf{H}$. This may not hold in some environments. For example, if the reflector is local around the receiver and is much closer to the receiver than to the transmitter, then the angular separation Ω_t at the transmitter is small. Similarly, if the reflector is local around the transmitter and is much closer to the transmitter than to the receiver, then the angular separation Ω_r at the receiver is small. In either case $\mathbf{H}$ would not be very well-conditioned (Figure 7.10). In a cellular system this suggests that if the base-station is high on top of a tower with most of the scatterers and reflectors locally around the mobile, then the size of the antenna array at the base-station

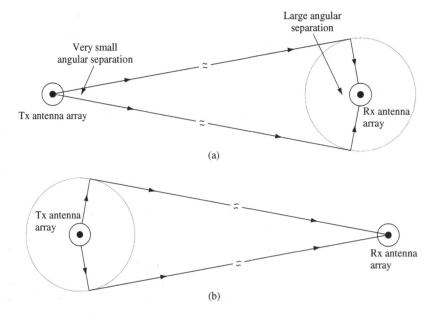

Figure 7.10 (a) The reflectors and scatterers are in a ring locally around the receiver; their angular separation at the transmitter is small. (b) The reflectors and scatterers are in a ring locally around the transmitter; their angular separation at the receiver is small.

will have to be many wavelengths to be able to exploit this spatial multiplexing effect.

Summary 7.1 Multiplexing capability of MIMO channels

SIMO and MISO channels provide a power gain but no degree-of-freedom gain.

Line-of-sight MIMO channels with co-located transmit antennas and co-located receive antennas also provide no degree-of-freedom gain.

MIMO channels with far-apart transmit antennas having angular separation greater than $1/L_r$ at the receive antenna array provide an effective degree-of-freedom gain. So do MIMO channels with far-apart receive antennas having angular separation greater than $1/L_t$ at the transmit antenna array.

Multipath MIMO channels with co-located transmit antennas and co-located receive antennas but with scatterers/reflectors far away also provide a degree-of-freedom gain.

7.3 Modeling of MIMO fading channels

The examples in the previous section are deterministic channels. Building on the insights obtained, we migrate towards statistical MIMO models which capture the key properties that enable spatial multiplexing.

7.3.1 Basic approach

In the previous section, we assessed the capacity of physical MIMO channels by first looking at the rank of the physical channel matrix $\mathbf{H}$ and then its condition number. In the example in Section 7.2.4, for instance, the rank of $\mathbf{H}$ is 2 but the condition number depends on how the angle between the two spatial signatures compares to the spatial resolution of the antenna array. The two-step analysis process is conceptually somewhat awkward. It suggests that physical models of the MIMO channel in terms of *individual* multipaths may not be at the right level of abstraction from the point of view of the design and analysis of communication systems. Rather, one may want to abstract the physical model into a higher-level model in terms of *spatially resolvable paths*.

We have in fact followed a similar strategy in the statistical modeling of frequency-selective fading channels in Chapter 2. There, the modeling is directly on the gains of the taps of the discrete-time sampled channel rather than on the gains of the individual physical paths. Each tap can be thought of as a (time-)resolvable path, consisting of an aggregation of individual physical paths. The bandwidth of the system dictates how finely or coarsely the physical paths are grouped into resolvable paths. From the point of view of communication, it is the behavior of the resolvable paths that matters, not that of the individual paths. Modeling the taps directly rather than the individual paths has the additional advantage that the aggregation makes statistical modeling more reliable.

Using the analogy between the finite time-resolution of a band-limited system and the finite angular-resolution of an array-size-limited system, we can follow the approach of Section 2.2.3 in modeling MIMO channels. The transmit and receive antenna array lengths L_t and L_r dictate the degree of resolvability in the angular domain: paths whose transmit directional cosines differ by less than $1/L_t$ and receive directional cosines by less than $1/L_r$ are not resolvable by the arrays. This suggests that we should "sample" the angular domain at fixed angular spacings of $1/L_t$ at the transmitter and at fixed angular spacings of $1/L_r$ at the receiver, and represent the channel in terms of these new input and output coordinates. The (k, l)th channel gain in these angular coordinates is then roughly the aggregation of all paths whose transmit directional cosine is within an angular window of width $1/L_t$ around l/L_t and whose receive directional cosine is within an angular window of width $1/L_r$ around k/L_r. See Figure 7.11 for an illustration of the linear transmit and receive antenna array with the corresponding angular windows. In the following subsections, we will develop this approach explicitly for uniform linear arrays.

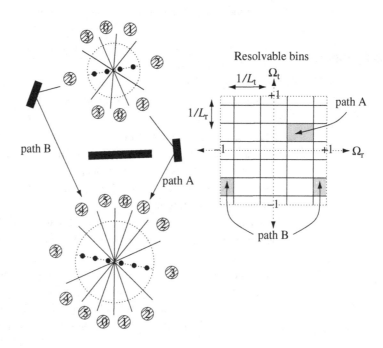

Figure 7.11 A representation of the MIMO channel in the angular domain. Due to the limited resolvability of the antenna arrays, the physical paths are partitioned into resolvable bins of angular widths $1/L_r$ by $1/L_t$. Here there are 6 receive antennas ($L_r = 3$) and 4 transmit antennas ($L_t = 2$).

7.3.2 MIMO multipath channel

Consider the narrowband MIMO channel:

$$\mathbf{y} = \mathbf{Hx} + \mathbf{w}. \tag{7.55}$$

The n_t transmit and n_r receive antennas are placed in uniform linear arrays of normalized lengths L_t and L_r, respectively. The normalized separation between the transmit antennas is $\Delta_t = L_t/n_t$ and the normalized separation between the receive antennas is $\Delta_r = L_r/n_r$. The normalization is by the wavelength λ_c of the passband transmitted signal. To simplify notation, we are now thinking of the channel $\mathbf{H}$ as fixed and it is easy to add the time-variation later on.

Suppose there is an arbitrary number of physical paths between the transmitter and the receiver; the ith path has an attenuation of a_i, makes an angle of ϕ_{ti} ($\Omega_{ti} := \cos \phi_{ti}$) with the transmit antenna array and an angle of ϕ_{ri} ($\Omega_{ri} := \cos \phi_{ri}$) with the receive antenna array. The channel matrix $\mathbf{H}$ is given by

$$\mathbf{H} = \sum_i a_i^b \mathbf{e_r}(\Omega_{ri}) \mathbf{e_t}(\Omega_{ti})^* \tag{7.56}$$

where, as in Section 7.2,

$$a_i^b := a_i \sqrt{n_t n_r} \exp\left(-\frac{\mathrm{j}2\pi d^{(i)}}{\lambda_c}\right),$$

$$\mathbf{e_r}(\Omega) := \frac{1}{\sqrt{n_r}} \begin{bmatrix} 1 \\ \exp(-\mathrm{j}2\pi\Delta_r\Omega) \\ \vdots \\ \exp(-\mathrm{j}2\pi(n_r-1)\Delta_r\Omega) \end{bmatrix}, \tag{7.57}$$

$$\mathbf{e_t}(\Omega) := \frac{1}{\sqrt{n_t}} \begin{bmatrix} 1 \\ \exp(-\mathrm{j}2\pi\Delta_t\Omega) \\ \vdots \\ \exp(-\mathrm{j}2\pi(n_t-1)\Delta_t\Omega) \end{bmatrix}. \tag{7.58}$$

Also, $d^{(i)}$ is the distance between transmit antenna 1 and receive antenna 1 along path i. The vectors $\mathbf{e_t}(\Omega)$ and $\mathbf{e_r}(\Omega)$ are, respectively, the transmitted and received unit spatial signatures along the direction Ω.

7.3.3 Angular domain representation of signals

The first step is to define precisely the angular domain representation of the transmitted and received signals. The signal arriving at a directional cosine Ω

onto the receive antenna array is along the unit spatial signature $\mathbf{e_r}(\Omega)$, given by (7.57). Recall (cf. (7.35))

$$f_{\mathrm{r}}(\Omega) := \mathbf{e_r}(0)^* \mathbf{e_r}(\Omega) = \frac{1}{n_{\mathrm{r}}} \exp(j\pi\Delta_{\mathrm{r}}\Omega(n_{\mathrm{r}} - 1)) \frac{\sin(\pi L_{\mathrm{r}}\Omega)}{\sin(\pi L_{\mathrm{r}}\Omega/n_{\mathrm{r}})}, \qquad (7.59)$$

analyzed in Section 7.2.4. In particular, we have

$$f_{\mathrm{r}}\left(\frac{k}{L_{\mathrm{r}}}\right) = 0, \text{ and } f_{\mathrm{r}}\left(\frac{-k}{L_{\mathrm{r}}}\right) = f_{\mathrm{r}}\left(\frac{n_{\mathrm{r}} - k}{L_{\mathrm{r}}}\right), \qquad k = 1, \dots, n_{\mathrm{r}} - 1 \quad (7.60)$$

(Figure 7.5). Hence, the n_{r} fixed vectors:

$$\mathcal{S}_{\mathrm{r}} := \left\{ \mathbf{e_r}(0), \mathbf{e_r}\left(\frac{1}{L_{\mathrm{r}}}\right), \dots, \mathbf{e_r}\left(\frac{n_{\mathrm{r}} - 1}{L_{\mathrm{r}}}\right) \right\} \qquad (7.61)$$

form an orthonormal basis for the received signal space $\mathcal{C}^{n_{\mathrm{r}}}$. This basis provides the representation of the received signals in the angular domain.

Why is this representation useful? Recall that associated with each vector $\mathbf{e_r}(\Omega)$ is its beamforming pattern (see Figures 7.6 and 7.7 for examples). It has one or more pairs of main lobes of width $2/L_{\mathrm{r}}$ and small side lobes. The different basis vectors $\mathbf{e_r}(k/L_{\mathrm{r}})$ have different main lobes. This implies that the received signal along any physical direction will have almost all of its energy along one particular $\mathbf{e_r}(k/L_{\mathrm{r}})$ vector and very little along all the others. Thus, this orthonormal basis provides a very simple (but approximate) decomposition of the total received signal into the multipaths received along the different physical directions, up to a resolution of $1/L_{\mathrm{r}}$.

We can similarly define the angular domain representation of the transmitted signal. The signal transmitted at a direction Ω is along the unit vector $\mathbf{e_t}(\Omega)$, defined in (7.58). The n_{t} fixed vectors:

$$\mathcal{S}_{\mathrm{t}} := \left\{ \mathbf{e_t}(0), \mathbf{e_t}\left(\frac{1}{L_{\mathrm{t}}}\right), \dots, \mathbf{e_t}\left(\frac{n_{\mathrm{t}} - 1}{L_{\mathrm{t}}}\right) \right\} \qquad (7.62)$$

form an orthonormal basis for the transmitted signal space $\mathcal{C}^{n_{\mathrm{t}}}$. This basis provides the representation of the transmitted signals in the angular domain. The transmitted signal along any physical direction will have almost all its energy along one particular $\mathbf{e_t}(k/L_{\mathrm{t}})$ vector and very little along all the others. Thus, this orthonormal basis provides a very simple (again, approximate)

Figure 7.12 Receive beamforming patterns of the angular basis vectors. Independent of the antenna spacing, the beamforming patterns all have the same beam widths for the main lobe, but the number of main lobes depends on the spacing. (a) Critically spaced case; (b) Sparsely spaced case. (c) Densely spaced case.

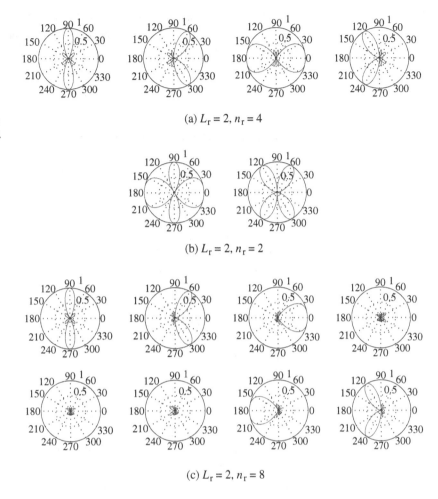

(a) $L_r = 2, n_r = 4$

(b) $L_r = 2, n_r = 2$

(c) $L_r = 2, n_r = 8$

decomposition of the overall transmitted signal into the components transmitted along the different physical directions, up to a resolution of $1/L_t$.

Examples of angular bases

Examples of angular bases, represented by their beamforming patterns, are shown in Figure 7.12. Three cases are distinguished:

- Antennas are *critically spaced* at half the wavelength ($\Delta_r = 1/2$). In this case, each basis vector $\mathbf{e}_r(k/L_r)$ has a single pair of main lobes around the angles $\pm \arccos(k/L_r)$.
- Antennas are *sparsely spaced* ($\Delta_r > 1/2$). In this case, some of the basis vectors have more than one pair of main lobes.
- Antennas are *densely spaced* ($\Delta_r < 1/2$). In this case, some of the basis vectors have no main lobes.

These statements can be understood from the fact that the function $f_r(\Omega_r)$ is periodic with period $1/\Delta_r$. The beamforming pattern of the vector $\mathbf{e}_r(k/L_r)$ is the polar plot

$$\left(\phi, \left|f_r\left(\cos\phi - \frac{k}{L_r}\right)\right|\right) \tag{7.63}$$

and the main lobes are at all angles ϕ for which

$$\cos\phi = \frac{k}{L_r} \bmod \frac{1}{\Delta_r} \tag{7.64}$$

In the critically spaced case, $1/\Delta_r = 2$ and k/L_r is between 0 and 2; there is a unique solution for $\cos\phi$ in (7.64). In the sparsely spaced case, $1/\Delta_r < 2$ and for some values of k there are multiple solutions: $\cos\phi = k/L_r + m/\Delta_r$ for integers m. In the densely spaced case, $1/\Delta_r > 2$, and for k satisfying $L_r < k < n_r - L_r$, there is no solution to (7.64). These angular basis vectors do not correspond to any physical directions. These cases are discussed further in Section 7.3.7.

Only in the critically spaced antennas is there a one-to-one correspondence between the angular windows and the angular basis vectors. This case is the simplest and *we will assume critically spaced antennas in the subsequent discussions*. The other cases are discussed further in Section 7.3.7.

Angular domain transformation as DFT

Actually the transformation between the spatial and angular domains is a familiar one! Let $\mathbf{U}_t$ be the $n_t \times n_t$ unitary matrix the columns of which are the basis vectors in $\mathcal{S}_t$. If $\mathbf{x}$ and $\mathbf{x}^a$ are the n_t-dimensional vector of transmitted signals from the antenna array and its angular domain representation respectively, then they are related by

$$\mathbf{x} = \mathbf{U}_t\mathbf{x}^a, \qquad \mathbf{x}^a = \mathbf{U}_t^*\mathbf{x}. \tag{7.65}$$

Now the (k, l)th entry of $\mathbf{U}_t$ is

$$\frac{1}{\sqrt{n_t}}\exp\left(\frac{-\mathrm{j}2\pi kl}{n_t}\right) \qquad k, l = 0, \dots, n_t - 1. \tag{7.66}$$

Hence, the angular domain representation $\mathbf{x}^a$ is nothing but the inverse discrete Fourier transform of $\mathbf{x}$ (cf. (3.142)). One should however note that the specific transformation for the angular domain representation is in fact a DFT because of the use of uniform linear arrays. On the other hand, the representation of signals in the angular domain is a more general concept and can be applied to other antenna array structures. Exercise 7.8 gives another example.

7.3.4 Angular domain representation of MIMO channels

We now represent the MIMO fading channel (7.55) in the angular domain. $\mathbf{U}_t$ and $\mathbf{U}_r$ are respectively the $n_t \times n_t$ and $n_r \times n_r$ unitary matrices the columns of which are the vectors in $\mathcal{S}_t$ and $\mathcal{S}_r$ respectively (IDFT matrices). The transformations

$$\mathbf{x}^a := \mathbf{U}_t^* \mathbf{x}, \tag{7.67}$$

$$\mathbf{y}^a := \mathbf{U}_r^* \mathbf{y} \tag{7.68}$$

are the changes of coordinates of the transmitted and received signals into the angular domain. (Superscript "a" denotes angular domain quantities.) Substituting this into (7.55), we have an equivalent representation of the channel in the angular domain:

$$\begin{aligned} \mathbf{y}^a &= \mathbf{U}_r^* \mathbf{H} \mathbf{U}_t \mathbf{x}^a + \mathbf{U}_r^* \mathbf{w} \\ &= \mathbf{H}^a \mathbf{x}^a + \mathbf{w}^a, \end{aligned} \tag{7.69}$$

where

$$\mathbf{H}^a := \mathbf{U}_r^* \mathbf{H} \mathbf{U}_t \tag{7.70}$$

is the channel matrix expressed in angular coordinates and

$$\mathbf{w}^a := \mathbf{U}_r^* \mathbf{w} \sim \mathcal{CN}(0, N_0 \mathbf{I}_{n_r}). \tag{7.71}$$

Now, recalling the representation of the channel matrix $\mathbf{H}$ in (7.56),

$$\begin{aligned} h_{kl}^a &= \mathbf{e}_r(k/L_r)^* \mathbf{H} \mathbf{e}_t(l/L_t) \\ &= \sum_i a_i^b [\mathbf{e}_r(k/L_r)^* \mathbf{e}_r(\Omega_{ri})] \cdot [\mathbf{e}_t(\Omega_{ti})^* \mathbf{e}_t(l/L_t)]. \end{aligned} \tag{7.72}$$

Recall from Section 7.3.3 that the beamforming pattern of the basis vector $\mathbf{e}_r(k/L_r)$ has a main lobe around k/L_r. The term $\mathbf{e}_r(k/L_r)^* \mathbf{e}_r(\Omega_{ri})$ is significant for the ith path if

$$\left| \Omega_{ri} - \frac{k}{L_r} \right| < \frac{1}{L_r}. \tag{7.73}$$

Define then $\mathcal{R}_k$ as the set of all paths whose receive directional cosine is within a window of width $1/L_r$ around k/L_r (Figure 7.13). The bin $\mathcal{R}_k$ can be interpreted as the set of all physical paths that have most of their energy along the receive angular basis vector $\mathbf{e}_r(k/L_r)$. Similarly, define $\mathcal{T}_l$ as the set of all paths whose transmit directional cosine is within a window of width $1/L_t$

Figure 7.13 The bin $\mathcal{R}_k$ is the set of all paths that arrive roughly in the direction of the main lobes of the beamforming pattern of $\mathbf{e}_r(k/L)$. Here $L_r = 2$ and $n_r = 4$.

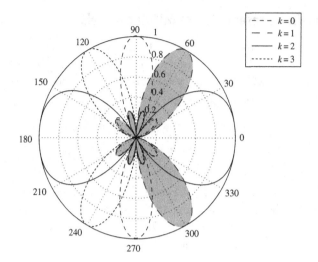

around l/L_t. The bin $\mathcal{T}_l$ can be interpreted as the set of all physical paths that have most of their energy along the transmit angular basis vector $\mathbf{e}_t(l/L_t)$. The entry h_{kl}^a is then mainly a function of the gains a_i^b of the physical paths that fall in $\mathcal{T}_l \cap \mathcal{R}_k$, and can be interpreted as the channel gain from the lth transmit angular bin to the kth receive angular bin.

The paths in $\mathcal{T}_l \cap \mathcal{R}_k$ are *unresolvable* in the angular domain. Due to the finite antenna aperture sizes (L_t and L_r), multiple unresolvable physical paths can be appropriately aggregated into one resolvable path with gain h_{kl}^a. Note that

$$\{\mathcal{T}_l \cap \mathcal{R}_k, l = 0, 1, \ldots, n_t - 1, k = 0, 1, \ldots, n_r - 1\}$$

forms a partition of the set of all physical paths. Hence, different physical paths (approximately) contribute to different entries in the angular representation $\mathbf{H}^a$ of the channel matrix.

The discussion in this section substantiates the intuitive picture in Figure 7.11. Note the similarity between (7.72) and (2.34); the latter quantifies how the underlying continuous-time channel is smoothed by the limited bandwidth of the system, while the former quantifies how the underlying continuous-space channel is smoothed by the limited antenna aperture. In the latter, the smoothing function is the sinc function, while in the former, the smoothing functions are f_r and f_t.

To simplify notations, we focus on a fixed channel as above. But time-variation can be easily incorporated: at time m, the ith time-varying path has attenuation $a_i[m]$, length $d^{(i)}[m]$, transmit angle $\phi_{t_i}[m]$ and receive angle $\phi_{r_i}[m]$. At time m, the resulting channel and its angular representation are time-varying: $\mathbf{H}[m]$ and $\mathbf{H}^a[m]$, respectively.

7.3.5 Statistical modeling in the angular domain

Figure 7.14 Some examples of $\mathbf{H}^{\mathrm{a}}$. (a) Small angular spread at the transmitter, such as the channel in Figure 7.10(a). (b) Small angular spread at the receiver, such as the channel in Figure 7.10(b). (c) Small angular spreads at both the transmitter and the receiver. (d) Full angular spreads at both the transmitter and the receiver.

The basis for the statistical modeling of MIMO fading channels is the approximation that the physical paths are partitioned into angularly resolvable bins and aggregated to form *resolvable paths* whose gains are $h_{kl}^{\mathrm{a}}[m]$. Assuming that the gains $a_i^{\mathrm{b}}[m]$ of the physical paths are independent, we can model the resolvable path gains $h_{kl}^{\mathrm{a}}[m]$ as independent. Moreover, the angles $\{\phi_{ri}[m]\}_m$ and $\{\phi_{ti}[m]\}_m$ typically evolve at a much slower time-scale than the gains $\{a_i^{\mathrm{b}}[m]\}_m$; therefore, within the time-scale of interest it is reasonable to assume that paths do not move from one angular bin to another, and the processes $\{h_{kl}^{\mathrm{a}}[m]\}_m$ can be modeled as independent across k and l (see Table 2.1 in Section 2.3 for the analogous situation for frequency-selective channels). In an angular bin (k, l), where there are many physical paths, one can invoke the Central Limit Theorem and approximate the aggregate gain $h_{kl}^{\mathrm{a}}[m]$ as a complex circular symmetric Gaussian process. On the other hand, in an angular bin (k, l) that contains no paths, the entries $h_{kl}^{\mathrm{a}}[m]$ can be approximated as 0. For a channel with limited angular spread at the receiver and/or the transmitter, many entries of $\mathbf{H}^{\mathrm{a}}[m]$ may be zero. Some examples are shown in Figures 7.14 and 7.15.

 (a) 60° spread at transmitter, 360° spread at receiver

 (b) 360° spread at transmitter, 60° spread at receiver

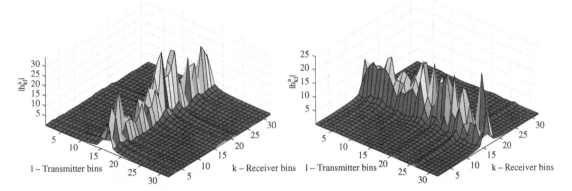

(c) 60° spread at transmitter, 60° spread at receiver

(d) 360° spread at transmitter, 360° spread at receiver

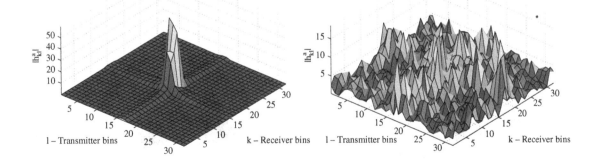

7.3.6 Degrees of freedom and diversity

Degrees of freedom

Given the statistical model, one can quantify the spatial multiplexing capability of a MIMO channel. With probability 1, the rank of the random matrix $\mathbf{H}^a$ is given by

$$\text{rank}(\mathbf{H}^a) = \min\{\text{number of non-zero rows, number of non-zero columns}\}$$

(7.74)

(Exercise 7.6). This yields the number of degrees of freedom available in the MIMO channel.

The number of non-zero rows and columns depends in turn on two separate factors:

Figure 7.15 Some examples of $\mathbf{H}^a$. (a) Two clusters of scatterers, with all paths going through a single bounce. (b) Paths scattered via multiple bounces.

- The *amount of scattering and reflection* in the multipath environment. The more scatterers and reflectors there are, the larger the number of non-zero entries in the random matrix $\mathbf{H}^a$, and the larger the number of degrees of freedom.

- The *lengths* L_t and L_r of the transmit and receive antenna arrays. With small antenna array lengths, many distinct multipaths may all be lumped into a single resolvable path. Increasing the array apertures allows the resolution

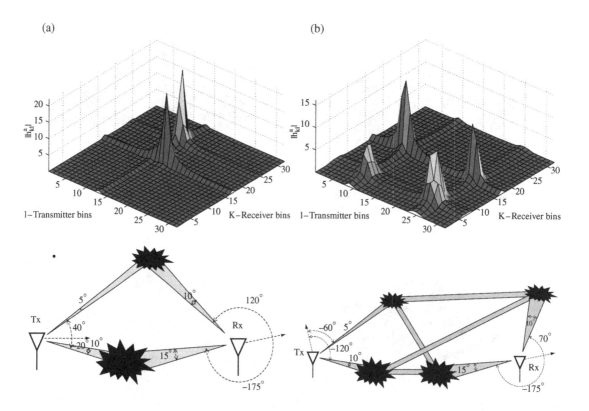

(a)   (b)

of more paths, resulting in more non-zero entries of $\mathbf{H}^a$ and an increased number of degrees of freedom.

The number of degrees of freedom is explicitly calculated in terms of the multipath environment and the array lengths in a *clustered response* model in Example 7.1.

Example 7.1 Degrees of freedom in clustered response models

Clarke's model

Let us start with Clarke's model, which was considered in Example 2.2. In this model, the signal arrives at the receiver along a continuum set of paths, uniformly from all directions. With a receive antenna array of length L_r, the number of receive angular bins is $2L_r$ and all of these bins are non-empty. Hence all of the $2L_r$ rows of $\mathbf{H}^a$ are non-zero. If the scatterers and reflectors are closer to the receiver than to the transmitter (Figures 7.10(a) and 7.14(a)), then at the transmitter the angular spread Ω_t (measured in terms of directional cosines) is less than the full span of 2. The number of non-empty rows in $\mathbf{H}^a$ is therefore $\lceil L_t \Omega_t \rceil$, such paths are resolved into bins of angular width $1/L_t$. Hence, the number of degrees of freedom in the MIMO channel is

$$\min\{\lceil L_t \Omega_t \rceil, 2L_r\}. \tag{7.75}$$

If the scatterers and reflectors are located at all directions from the transmitter as well, then $\Omega_t = 2$ and the number of degrees of freedom in the MIMO channel is

$$\min\{2L_t, 2L_r\}, \tag{7.76}$$

the maximum possible given the antenna array lengths. Since the antenna separation is assumed to be half the carrier wavelength, this formula can also be expressed as

$$\min\{n_t, n_r\},$$

the rank of the channel matrix $\mathbf{H}$

General clustered response model

In a more general model, scatterers and reflectors are not located at all directions from the transmitter or the receiver but are grouped into several *clusters* (Figure 7.16). Each cluster bounces off a continuum of paths. Table 7.1 summarizes several sets of indoor channel measurements that support such a *clustered response* model. In an indoor environment, clustering can be the result of reflections from walls and ceilings, scattering from furniture, diffraction from doorway openings and transmission through soft partitions. It is a reasonable model when the size of the channel objects is comparable to the distances from the transmitter and from the receiver.

Table 7.1 Examples of some indoor channel measurements. The Intel measurements span a very wide bandwidth and the number of clusters and angular spread measured are frequency dependent. This set of data is further elaborated in Figure 7.18.

	Frequency (GHz)	No. of clusters	Total angular spread (°)
USC UWB [27]	0–3	2–5	37
Intel UWB [91]	2–8	1–4	11–17
Spencer [112]	6.75–7.25	3–5	25.5
COST 259 [58]	24	3–5	18.5

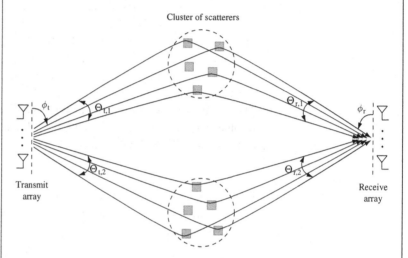

Figure 7.16 The clustered response model for the multipath environment. Each cluster bounces off a continuum of paths.

In such a model, the directional cosines Θ_r along which paths arrive are partitioned into several disjoint intervals: $\Theta_r = \cup_k \Theta_{rk}$. Similarly, on the transmit side, $\Theta_t = \cup_k \Theta_{tk}$. The number of degrees of freedom in the channel is

$$\min\left\{\sum_k \lceil L_t |\Theta_{tk}| \rceil, \sum_k \lceil L_r |\Theta_{tk}| \rceil\right\} \qquad (7.77)$$

For L_t and L_r large, the number of degrees of freedom is approximately

$$\min\{L_t \Omega_{t,\text{total}}, L_r \Omega_{r,\text{total}}\}, \qquad (7.78)$$

where

$$\Omega_{t,\text{total}} := \sum_k |\Theta_{tk}| \quad \text{and} \quad \Omega_{r,\text{total}} := \sum_k |\Theta_{rk}| \qquad (7.79)$$

are the total angular spreads of the clusters at the transmitter and at the receiver, respectively. This formula shows explicitly the *separate* effects of the antenna array and of the multipath environment on the number of degrees of freedom. The larger the angular spreads the more degrees of freedom there are. For fixed angular spreads, increasing the antenna array lengths allows zooming into and resolving the paths from each cluster, thus increasing the available degrees of freedom (Figure 7.17).

One can draw an analogy between the formula (7.78) and the classic fact that signals with bandwidth W and duration T have approximately $2WT$ degrees of freedom (cf. Discussion 2.1). Here, the antenna array lengths L_t and L_r play the role of the bandwidth W, and the total angular spreads $\Omega_{t,total}$ and $\Omega_{r,total}$ play the role of the signal duration T.

Effect of carrier frequency

As an application of the formula (7.78), consider the question of how the available number of degrees of freedom in a MIMO channel depends on the carrier frequency used. Recall that the array lengths L_t and L_r are quantities normalized to the carrier wavelength. Hence, for a fixed *physical* length of the antenna arrays, the normalized lengths L_t and L_r *increase* with the carrier frequency. Viewed in isolation, this fact would suggest an *increase* in the number of degrees of freedom with the carrier frequency; this is consistent with the intuition that, at higher carrier frequencies, one can pack more antenna elements in a given amount of area on the device. On the other hand, the angular spread of the environment

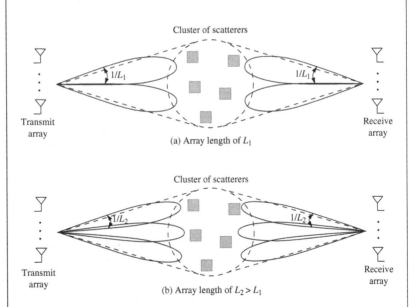

(a) Array length of L_1

(b) Array length of $L_2 > L_1$

Figure 7.17 Increasing the antenna array apertures increases path resolvability in the angular domain and the degrees of freedom.

typically *decreases* with the carrier frequency. The reasons are two-fold:

- signals at higher frequency attenuate more after passing through or bouncing off channel objects, thus reducing the number of effective clusters;
- at higher frequency the wavelength is small relative to the feature size of typical channel objects, so scattering appears to be more specular in nature and results in smaller angular spread.

These factors combine to reduce $\Omega_{t,\text{total}}$ and $\Omega_{r,\text{total}}$ as the carrier frequency increases. Thus the impact of carrier frequency on the overall degrees of freedom is not necessarily monotonic. A set of indoor measurements is shown in Figure 7.18. The number of degrees of freedom increases and then decreases with the carrier frequency, and there is in fact an optimal frequency at which the number of degrees of freedom is maximized. This example shows the importance of taking into account both the physical environment as well as the antenna arrays in determining the available degrees of freedom in a MIMO channel.

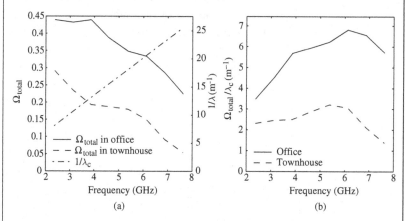

Figure 7.18 (a) The total angular spread Ω_{total} of the scattering environment (assumed equal at the transmitter side and at the receiver side) decreases with the carrier frequency; the normalized array length increases proportional to $1/\lambda_c$. (b) The number of degrees of freedom of the MIMO channel, proportional to $\Omega_{\text{total}}/\lambda_c$, first increases and then decreases with the carrier frequency. The data are taken from [91].

Diversity

In this chapter, we have focused on the phenomenon of spatial multiplexing and the key parameter is the number of degrees of freedom. In a slow fading environment, another important parameter is the amount of *diversity* in the channel. This is the number of independent channel gains that have to be in a deep fade for the entire channel to be in deep fade. In the angular domain MIMO model, *the amount of diversity is simply the number of non-zero*

Figure 7.19 Angular domain representation of three MIMO channels. They all have four degrees of freedom but they have diversity 4, 8 and 16 respectively. They model channels with increasing amounts of bounces in the paths (cf. Figure 7.15).

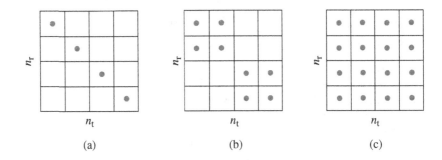

entries in $\mathbf{H}^a$. Some examples are shown in Figure 7.19. Note that channels that have the same degrees of freedom can have very different amounts of diversity. The number of degrees of freedom depends primarily on the angular spreads of the scatters/reflectors at the transmitter and at the receiver, while the amount of diversity depends also on the degree of connectivity between the transmit and receive angles. In a channel with multiple-bounced paths, signals sent along one transmit angle can arrive at several receive angles (see Figure 7.15). Such a channel would have more diversity than one with single-bounced paths with signal sent along one transmit angle received at a unique angle, even though the angular spreads may be the same.

7.3.7 Dependency on antenna spacing

So far we have been primarily focusing on the case of critically spaced antennas (i.e., antenna separations Δ_t and Δ_r are half the carrier wavelength). What is the impact of changing the antenna separation on the channel statistics and the key channel parameters such as the number of degrees of freedom?

To answer this question, we fix the antenna array lengths L_t and L_r and vary the antenna separation, or equivalently the number of antenna elements. Let us just focus on the receiver side; the transmitter side is analogous. Given the antenna array length L_r, the beamforming patterns associated with the basis vectors $\{\mathbf{e}_r(k/L_r)\}_k$ all have beam widths of $2/L_r$ (Figure 7.12). This dictates the maximum possible resolution of the antenna array: paths that arrive within an angular window of width $1/L_r$ cannot be resolved no matter how many antenna elements there are. There are $2L_r$ such angular windows, partitioning all the receive directions (Figure 7.20). Whether or not this maximum resolution can actually be achieved depends on the number of antenna elements.

Recall that the bins $\mathcal{R}_k$ can be interpreted as the set of all physical paths which have most of their energy along the basis vector $\mathbf{e}_t(k/L_r)$. The bins dictate the resolvability of the antenna array. In the critically spaced case $(\Delta_r = 1/2)$, the beamforming patterns of all the basis vectors have a single main lobe (together with its mirror image). There is a one-to-one correspondence between the angular windows and the resolvable bins $\mathcal{R}_k$, and paths arriving in different windows can be resolved by the array (Figure 7.21). In

Figure 7.20 An antenna array of length L_r partitions the receive directions into $2L_r$ angular windows. Here, $L_r = 3$ and there are six angular windows. Note that because of symmetry across the $0° - 180°$ axis, each angular window comes as a mirror image pair, and each pair is only counted as one angular window.

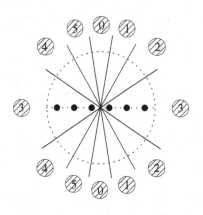

Figure 7.21 Antennas are critically spaced at half the wavelength. Each resolvable bin corresponds to exactly one angular window. Here, there are six angular windows and six bins.

$L_r = 3, n_r = 6$

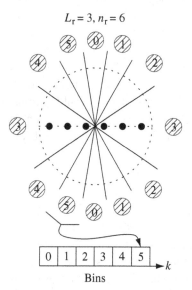

0	1	2	3	4	5

k

Bins

the sparsely spaced case ($\Delta_r > 1/2$), the beamforming patterns of some of the basis vectors have multiple main lobes. Thus, paths arriving in the different angular windows corresponding to these lobes are all lumped into one bin and cannot be resolved by the array (Figure 7.22). In the densely spaced case ($\Delta_r < 1/2$), the beamforming patterns of $2L_r$ of the basis vectors have a single main lobe; they can be used to resolve among the $2L_r$ angular windows. The beamforming patterns of the remaining $n_r - 2L_r$ basis vectors have no main lobe and do not correspond to any angular window. There is little received energy along these basis vectors and they do not participate significantly in the communication process. See Figure 7.23.

The key conclusion from the above analysis is that, given the antenna array lengths L_r and L_t, the maximum achievable angular resolution can be achieved by placing antenna elements half a wavelength apart. Placing antennas more sparsely reduces the resolution of the antenna array and can

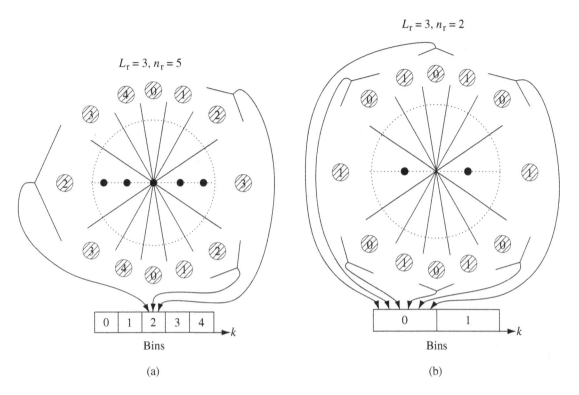

Figure 7.22 (a) Antennas are sparsely spaced. Some of the bins contain paths from multiple angular windows. (b) The antennas are very sparsely spaced. All bins contain several angular windows of paths.

reduce the number of degrees of freedom and the diversity of the channel. Placing the antennas more densely adds spurious basis vectors which do not correspond to any physical directions, and does not add resolvability. In terms of the angular channel matrix $\mathbf{H}^a$, this has the effect of adding zero rows and columns; in terms of the spatial channel matrix $\mathbf{H}$, this has the effect of making the entries more correlated. In fact, the angular domain representation makes it apparent that one can reduce the densely spaced system to an equivalent $2L_t \times 2L_r$ critically spaced system by just focusing on the basis vectors that do correspond to physical directions (Figure 7.24).

Increasing the antenna separation within a given *array length* L_r does not increase the number of degrees of freedom in the channel. What about increasing the antenna separation while keeping the number of *antenna elements* n_r the same? This question makes sense if the system is hardware-limited rather than limited by the amount of space to put the antenna array in. Increasing the antenna separation this way reduces the beam width of the n_r angular basis beamforming patterns but also increases the number of main lobes in each (Figure 7.25). If the scattering environment is rich enough such that the received signal arrives from all directions, the number of non-zero rows of the channel matrix $\mathbf{H}^a$ is already n_r, the largest possible, and increasing the spacing does not increase the number of degrees of freedom in the channel. On the other hand, if the scattering is clustered to within certain directions, increasing the separation makes it possible for the scattered signal to be

Figure 7.23 Antennas are densely spaced. Some bins contain no physical paths.

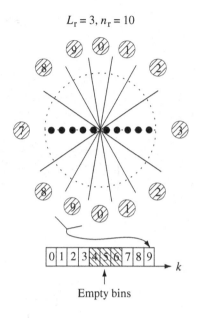

$L_r = 3, n_r = 10$

Empty bins

Figure 7.24 A typical $\mathbf{H}^a$ when the antennas are densely spaced.

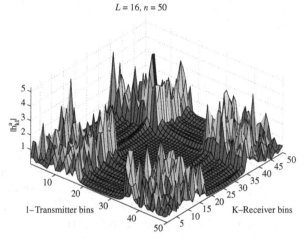

$L = 16, n = 50$

received in more bins, thus increasing the number of degrees of freedom (Figure 7.25). In terms of the spatial channel matrix $\mathbf{H}$, this has the effect of making the entries look more random and independent. At a base-station on a high tower with few local scatterers, the angular spread of the multipaths is small and therefore one has to put the antennas many wavelengths apart to decorrelate the channel gains.

Sampling interpretation

One can give a sampling interpretation to the above results. First, think of the discrete antenna array as a *sampling* of an underlying continuous array $[-L_r/2, L_r/2]$. On this array, the received signal $x(s)$ is a function of the

Figure 7.25 An example of a clustered response channel in which increasing the separation between a fixed number of antennas increases the number of degrees of freedom from 2 to 3.

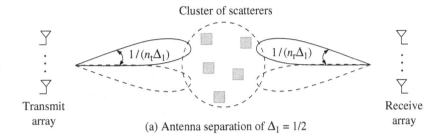

(a) Antenna separation of $\Delta_1 = 1/2$

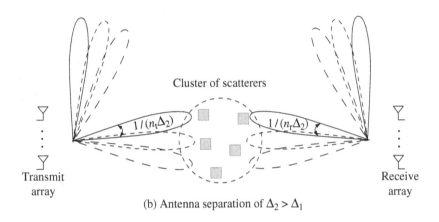

(b) Antenna separation of $\Delta_2 > \Delta_1$

continuous spatial location $s \in [-L_r/2, L_r/2]$. Just like in the discrete case (cf. Section 7.3.3), the spatial-domain signal $x(s)$ and its angular representation $x^a(\Omega)$ form a Fourier transform pair. However, since only $\Omega \in [-1, 1]$ corresponds to directional cosines of actual physical directions, the angular representation $x^a(\Omega)$ of the received signal is zero outside $[-1, 1]$. Hence, the spatial-domain signal $x(s)$ is "bandlimited" to $[-W, W]$, with "bandwidth" $W = 1$. By the sampling theorem, the signal $x(s)$ can be uniquely specified by samples spaced at distance $1/(2W) = 1/2$ apart, the Nyquist sampling rate. This is precise when $L_r \to \infty$ and approximate when L_r is finite. Hence, placing the antenna elements at the critical separation is sufficient to describe the received signal; a continuum of antenna elements is not needed. Antenna spacing greater than $1/2$ is not adequate: this is under-sampling and the loss of resolution mentioned above is analogous to the *aliasing* effect when one samples a bandlimited signal at below the Nyquist rate.

7.3.8 I.i.d. Rayleigh fading model

A very common MIMO fading model is the *i.i.d. Rayleigh fading model*: the entries of the channel gain matrix $\mathbf{H}[m]$ are independent, identically distributed and circular symmetric complex Gaussian. Since the matrix $\mathbf{H}[m]$ and its angular domain representation $\mathbf{H}^a[m]$ are related by

$$\mathbf{H}^a[m] := \mathbf{U}_r^* \mathbf{H}[m] \mathbf{U}_t, \qquad (7.80)$$

and $\mathbf{U}_r$ and $\mathbf{U}_t$ are fixed unitary matrices, this means that $\mathbf{H}^a$ should have the same i.i.d. Gaussian distribution as $\mathbf{H}$. Thus, using the modeling approach described here, we can see clearly the physical basis of the i.i.d Rayleigh fading model, in terms of both the multipath environment and the antenna arrays. There should be a significant number of multipaths in *each* of the resolvable angular bins, and the energy should be equally spread out across these bins. This is the so-called *richly scattered environment*. If there are very few or no paths in some of the angular directions, then the entries in $\mathbf{H}$ will be correlated. Moreover, the antennas should be either critically or sparsely spaced. If the antennas are densely spaced, then some entries of $\mathbf{H}^a$ are approximately zero and the entries in $\mathbf{H}$ itself are highly correlated. However, by a simple transformation, the channel can be reduced to an equivalent channel with fewer antennas which are critically spaced.

Compared to the critically spaced case, having sparser spacing makes it easier for the channel matrix to satisfy the i.i.d. Rayleigh assumption. This is because each bin now spans more distinct angular windows and thus contains more paths, from multiple transmit and receive directions. This substantiates the intuition that putting the antennas further apart makes the entries of $\mathbf{H}$ less dependent. On the other hand, if the physical environment already provides scattering in all directions, then having critical spacing of the antennas is enough to satisfy the i.i.d. Rayleigh assumption.

Due to the analytical tractability, we will use the i.i.d. Rayleigh fading model quite often to evaluate performance of MIMO communication schemes, but it is important to keep in mind the assumptions on both the physical environment and the antenna arrays for the model to be valid.

Chapter 7 The main plot

The angular domain provides a natural representation of the MIMO channel, highlighting the interaction between the antenna arrays and the physical environment.

The angular resolution of a linear antenna array is dictated by its length: an array of length L provides a resolution of $1/L$. Critical spacing of antenna elements at half the carrier wavelength captures the full angular resolution of $1/L$. Sparser spacing reduces the angular resolution due to aliasing. Denser spacing does not increase the resolution beyond $1/L$.

Transmit and receive antenna arrays of length L_t and L_r partition the angular domain into $2L_t \times 2L_r$ bins of unresolvable multipaths. Paths that fall within the same bin are aggregated to form one entry of the angular channel matrix $\mathbf{H}^a$.

A statistical model of $\mathbf{H}^a$ is obtained by assuming independent Gaussian distributed entries, of possibly different variances. Angular bins that contain no paths correspond to zero entries.

The number of degrees of freedom in the MIMO channel is the minimum of the number of non-zero rows and the number of non-zero columns of $\mathbf{H}^a$. The amount of diversity is the number of non-zero entries.

In a clustered-response model, the number of degrees of freedom is approximately:

$$\min\{L_t \Omega_{t,\text{total}}, L_r \Omega_{r,\text{total}}\} \qquad (7.81)$$

The multiplexing capability of a MIMO channel increases with the angular spreads $\Omega_{t,\text{total}}, \Omega_{r,\text{total}}$ of the scatterers/reflectors as well as with the antenna array lengths. This number of degrees of freedom can be achieved when the antennas are critically spaced at half the wavelength or closer. With a maximum angular spread of 2, the number of degrees of freedom is

$$\min\{2L_t, 2L_r\},$$

and this equals

$$\min\{n_t, n_r\}$$

when the antennas are critically spaced.

The i.i.d. Rayleigh fading model is reasonable in a richly scattering environment where the angular bins are fully populated with paths and there is roughly equal amount of energy in each bin. The antenna elements should be critically or sparsely spaced.

7.4 Bibliographical notes

The angular domain approach to MIMO channel modeling is based on works by Sayeed [105] and Poon *et al.* [90, 92]. [105] considered an array of discrete antenna elements, while [90, 92] considered a continuum of antenna elements to emphasize that spatial multiplexability is limited not by the number of antenna elements but by the size of the antenna array. We considered only linear arrays in this chapter, but [90] also treated other antenna array configurations such as circular rings and spherical surfaces. The degree-of-freedom formula (7.78) is derived in [90] for the clustered response model.

Other related approaches to MIMO channel modeling are by Raleigh and Cioffi [97], by Gesbert *et al.* [47] and by Shiu *et al.* [111]. The latter work used a Clarke-like model but with two rings of scatterers, one around the transmitter and one around the receiver, to derive the MIMO channel statistics.

7.5 Exercises

Exercise 7.1
1. For the SIMO channel with uniform linear array in Section 7.2.1, give an exact expression for the distance between the transmit antenna and the ith receive antenna. Make precise in what sense is (7.19) an approximation.
2. Repeat the analysis for the approximation (7.27) in the MIMO case.

Exercise 7.2 Verify that the unit vector $\mathbf{e}_r(\Omega_r)$, defined in (7.21), is periodic with period $1/\Delta_r$ and within one period never repeats itself.

Exercise 7.3 Verify (7.35).

Exercise 7.4 In an earlier work on MIMO communication [97], it is stated that the number of degrees of freedom in a MIMO channel with n_t transmit, n_r receive antennas and K multipaths is given by

$$\min\{n_t, n_r, K\} \tag{7.82}$$

and this is the key parameter that determines the multiplexing capability of the channel. What are the problems with this statement?

Exercise 7.5 In this question we study the role of antenna spacing in the angular representation of the MIMO channel.
1. Consider the critically spaced antenna array in Figure 7.21; there are six bins, each one corresponding to a specific physical angular window. All of these angular windows have the same width as measured in solid angle. Compute the angular window width in radians for each of the bins $\mathcal{T}_l$, with $l = 0, \ldots, 5$. Argue that the width in radians increases as we move from the line perpendicular to the antenna array to one that is parallel to it.
2. Now consider the sparsely spaced antenna arrays in Figure 7.22. Justify the depicted mapping from the angular windows to the bins $\mathcal{T}_l$ and evaluate the angular window width in radians for each of the bins $\mathcal{T}_l$ (for $l = 0, 1, \ldots, n_t - 1$). (The angular window width of a bin $\mathcal{T}_l$ is the sum of the widths of all the angular windows that correspond to the bin $\mathcal{T}_l$.)
3. Justify the depiction of the mapping from angular windows to the bins $\mathcal{T}_l$ in the densely spaced antenna array of Figure 7.23. Also evaluate the angular width of each bin in radians.

Exercise 7.6 The non-zero entries of the angular matrix $\mathbf{H}^a$ are distributed as independent complex Gaussian random variables. Show that with probability 1, the rank of the matrix is given by the formula (7.74).

Exercise 7.7 In Chapter 2, we introduced Clarke's flat fading model, where both the transmitter and the receiver have a single antenna. Suppose now that the receiver has n_r antennas, each spaced by half a wavelength. The transmitter still has one antenna (a SIMO channel). At time m

$$\mathbf{y}[m] = \mathbf{h}[m]x[m] + \mathbf{w}[m], \tag{7.83}$$

where $\mathbf{y}[m], \mathbf{h}[m]$ are the n_r-dimensional received vector and receive spatial signature (induced by the channel), respectively.

1. Consider first the case when the receiver is stationary. Compute approximately the joint statistics of the coefficients of **h** in the angular domain.

2. Now suppose the receiver is moving at a speed v. Compute the Doppler spread and the Doppler spectrum of each of the angular domain coefficients of the channel.

3. What happens to the Doppler spread as $n_r \to \infty$? What can you say about the difficulty of estimating and tracking the process $\{\mathbf{h}[m]\}$ as n grows? Easier, harder, or the same? Explain.

Exercise 7.8 [90] Consider a circular array of radius R normalized by the carrier wavelength with n elements uniformly spaced.

1. Compute the spatial signature in the direction ϕ.

2. Find the angle, $f(\phi_1, \phi_2)$, between the two spatial signatures in the direction ϕ_1 and ϕ_2.

3. Does $f(\phi_1, \phi_2)$ only depend on the difference $\phi_1 - \phi_2$? If not, explain why.

4. Plot $f(\phi_1, 0)$ for $R = 2$ and different values of n, from n equal to $\lceil \pi R/2 \rceil$, $\lceil \pi R \rceil$, $\lceil 2\pi R \rceil$, to $\lceil 4\pi R \rceil$. Observe the plot and describe your deductions.

5. Deduce the angular resolution.

6. Linear arrays of length L have a resolution of $1/L$ along the $\cos \phi$-domain, that is, they have non-uniform resolution along the ϕ-domain. Can you design a linear array with uniform resolution along the ϕ-domain?

Exercise 7.9 (Spatial sampling) Consider a MIMO system with $L_t = L_r = 2$ in a channel with $M = 10$ multipaths. The ith multipath makes an angle of $i\Delta\phi$ with the transmit array and an angle of $i\Delta\phi$ with the receive array where $\Delta\phi = \pi/M$.

1. Assuming there are n_t transmit and n_r receive antennas, compute the channel matrix.

2. Compute the channel eigenvalues for $n_t = n_r$ varying from 4 to 8.

3. Describe the distribution of the eigenvalues and contrast it with the binning interpretation in Section 7.3.4.

Exercise 7.10 In this exercise, we study the angular domain representation of frequency-selective MIMO channels.

1. Starting with the representation of the frequency-selective MIMO channel in time (cf. (8.112)) describe how you would arrive at the angular domain equivalent (cf. (7.69)):

$$\mathbf{y}^{\mathbf{a}}[m] = \sum_{\ell=0}^{L-1} \mathbf{H}_\ell^{\mathbf{a}}[m]\mathbf{x}^{\mathbf{a}}[m - \ell] + \mathbf{w}^{\mathbf{a}}[m]. \tag{7.84}$$

2. Consider the equivalent (except for the overhead in using the cyclic prefix) parallel MIMO channel as in (8.113).

(a) Discuss the role played by the density of the scatterers and the delay spread in the physical environment in arriving at an appropriate statistical model for $\tilde{\mathbf{H}}_n$ at the different OFDM tones n.

(b) Argue that the (marginal) distribution of the MIMO channel $\tilde{\mathbf{H}}_n$ is the same for each of the tones $n = 0, \ldots, N - 1$.

Exercise 7.11 A MIMO channel has a single cluster with the directional cosine ranges as $\Theta_t = \Theta_r = [0, 1]$. Compute the number of degrees of freedom of an $n \times n$ channel as a function of the antenna separation $\Delta_t = \Delta_r = \Delta$.

MIMO II: capacity and multiplexing architectures

In this chapter, we will look at the capacity of MIMO fading channels and discuss transceiver architectures that extract the promised multiplexing gains from the channel. We particularly focus on the scenario when the transmitter does not know the channel realization. In the fast fading MIMO channel, we show the following:

- At high SNR, the capacity of the i.i.d. Rayleigh fast fading channel scales like $n_{\min} \log \mathsf{SNR}$ bits/s/Hz, where $n_{\min}$ is the minimum of the number of transmit antennas n_t and the number of receive antennas n_r. This is a degree-of-freedom gain.
- At low SNR, the capacity is approximately $n_r \mathsf{SNR} \log_2 e$ bits/s/Hz. This is a receive beamforming power gain.
- At all SNR, the capacity scales linearly with $n_{\min}$. This is due to a combination of a power gain and a degree-of-freedom gain.

Furthermore, there is a transmit beamforming gain together with an opportunistic communication gain if the transmitter can track the channel as well.

Over a deterministic time-invariant MIMO channel, the capacity-achieving transceiver architecture is simple (cf. Section 7.1.1): independent data streams are multiplexed in an appropriate coordinate system (cf. Figure 7.2). The receiver transforms the received vector into another appropriate coordinate system to separately decode the different data streams. Without knowledge of the channel at the transmitter the choice of the coordinate system in which the independent data streams are multiplexed has to be fixed a priori. In conjunction with joint decoding, we will see that this transmitter architecture achieves the capacity of the fast fading channel. This architecture is also called V-BLAST[1] in the literature.

---

[1] Vertical Bell Labs Space-Time Architecture. There are several versions of V-BLAST with different receiver structures but they all share the same transmitting architecture of multiplexing independent streams, and we take this as its defining feature.

In Section 8.3, we discuss receiver architectures that are simpler than joint ML decoding of the independent streams. While there are several receiver architectures that can support the full degrees of freedom of the channel, a particular architecture, the MMSE-SIC, which uses a combination of minimum mean square error estimation (MMSE) and successive interference cancellation (SIC), achieves capacity.

The performance of the slow fading MIMO channel is characterized through the outage probability and the corresponding outage capacity. At low SNR, the outage capacity can be achieved, to a first order, by using one transmit antenna at a time, achieving a full diversity gain of $n_t\, n_r$ and a power gain of n_r. The outage capacity at high SNR, on the other hand, benefits from a degree-of-freedom gain as well; this is more difficult to characterize succinctly and its analysis is relegated until Chapter 9.

Although it achieves the capacity of the *fast fading* channel, the V-BLAST architecture is strictly suboptimal for the *slow fading* channel. In fact, it does not even achieve the full diversity gain promised by the MIMO channel. To see this, consider transmitting independent data streams directly over the transmit antennas. In this case, the diversity of each data stream is limited to just the receive diversity. To extract the full diversity from the channel, one needs to code *across* the transmit antennas. A modified architecture, D-BLAST[2], which combines transmit antenna coding with MMSE-SIC, not only extracts the full diversity from the channel but its performance also comes close to the outage capacity.

8.1 The V-BLAST architecture

We start with the time-invariant channel (cf. (7.1))

$$\mathbf{y}[m] = \mathbf{H}\mathbf{x}[m] + \mathbf{w}[m], \qquad m = 1, 2, \ldots \qquad (8.1)$$

When the channel matrix $\mathbf{H}$ is known to the transmitter, we have seen in Section 7.1.1 that the optimal strategy is to transmit independent streams in the directions of the eigenvectors of $\mathbf{H}^*\mathbf{H}$, i.e., in the coordinate system defined by the matrix $\mathbf{V}$, where $\mathbf{H} = \mathbf{U}\Lambda\mathbf{V}^*$ is the singular value decomposition of $\mathbf{H}$. This coordinate system is channel-dependent. With an eye towards dealing with the case of fading channels where the channel matrix is *unknown* to the transmitter, we generalize this to the architecture in Figure 8.1, where the independent data streams, n_t of them, are multiplexed in some arbitrary

---

[2] Diagonal Bell Labs Space-Time Architecture

Figure 8.1 The V-BLAST architecture for communicating over the MIMO channel.

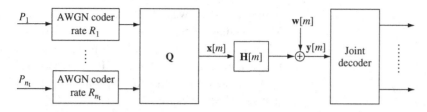

coordinate system given by a unitary matrix $\mathbf{Q}$, not necessarily dependent on the channel matrix $\mathbf{H}$. This is the V-BLAST architecture. The data streams are decoded jointly. The kth data stream is allocated a power P_k (such that the sum of the powers, $P_1 + \cdots + P_{n_t}$, is equal to P, the total transmit power constraint) and is encoded using a capacity-achieving Gaussian code with rate R_k. The total rate is $R = \sum_{k=1}^{n_t} R_k$.

As special cases:

- If $\mathbf{Q} = \mathbf{V}$ and the powers are given by the waterfilling allocations, then we have the capacity-achieving architecture in Figure 7.2.
- If $\mathbf{Q} = \mathbf{I}_{n_t}$, then independent data streams are sent on the different transmit antennas.

Using a sphere-packing argument analogous to the ones used in Chapter 5, we will argue an upper bound on the highest reliable rate of communication:

$$R < \log \det \left(\mathbf{I}_{n_r} + \frac{1}{N_0} \mathbf{H} \mathbf{K}_x \mathbf{H}^* \right) \text{ bits/s/Hz.} \tag{8.2}$$

Here $\mathbf{K}_x$ is the covariance matrix of the transmitted signal $\mathbf{x}$ and is a function of the multiplexing coordinate system and the power allocations:

$$\mathbf{K}_x := \mathbf{Q} \, \text{diag}\{P_1, \ldots, P_{n_t}\} \mathbf{Q}^*. \tag{8.3}$$

Considering communication over a block of time symbols of length N, the received vector, of length $n_r N$, lies with high probability in an ellipsoid of volume proportional to

$$\det(N_0 \mathbf{I}_{n_r} + \mathbf{H} \mathbf{K}_x \mathbf{H}^*)^N. \tag{8.4}$$

This formula is a direct generalization of the corresponding volume formula (5.50) for the parallel channel, and is justified in Exercise 8.2. Since we have to allow for non-overlapping noise spheres (of radius $\sqrt{N_0}$ and, hence, volume proportional to $N_0^{n_r N}$) around each codeword to ensure reliable communication, the maximum number of codewords that can be packed is the ratio

$$\frac{\det(N_0 \mathbf{I}_{n_r} + \mathbf{H} \mathbf{K}_x \mathbf{H}^*)^N}{N_0^{n_r N}}. \tag{8.5}$$

We can now conclude the upper bound on the rate of reliable communication in (8.2).

Is this upper bound actually achievable by the V-BLAST architecture? Observe that *independent* data streams are multiplexed in V-BLAST; perhaps coding across the streams is required to achieve the upper bound (8.2)? To get some insight on this question, consider the special case of a MISO channel ($n_r = 1$) and set $\mathbf{Q} = \mathbf{I}_{n_t}$ in the architecture, i.e., independent streams on each of the transmit antennas. This is precisely an uplink channel, as considered in Section 6.1, drawing an analogy between the transmit antennas and the users. We know from the development there that the sum capacity of this uplink channel is

$$\log \left(1 + \frac{\sum_{k=1}^{n_t} |h_k|^2 P_k}{N_0} \right). \tag{8.6}$$

This is precisely the upper bound (8.2) in this special case. Thus, the V-BLAST architecture, with independent data streams, is sufficient to achieve the upper bound (8.2). In the general case, an analogy can be drawn between the V-BLAST architecture and an uplink channel with n_r receive antennas and channel matrix $\mathbf{HQ}$; just as in the single receive antenna case, the upper bound (8.2) is the sum capacity of this uplink channel and therefore achievable using the V-BLAST architecture. This uplink channel is considered in greater detail in Chapter 10 and its information theoretic analysis is in Appendix B.9.

8.2 Fast fading MIMO channel

The fast fading MIMO channel is

$$\mathbf{y}[m] = \mathbf{H}[m]\mathbf{x}[m] + \mathbf{w}[m], \qquad m = 1, 2, \ldots, \tag{8.7}$$

where $\{\mathbf{H}[m]\}$ is a random fading process. To properly define a notion of capacity (achieved by averaging of the channel fading over time), we make the technical assumption (as in the earlier chapters) that $\{\mathbf{H}[m]\}$ is a stationary and ergodic process. As a normalization, let us suppose that $\mathbb{E}[|h_{ij}|^2] = 1$. As in our earlier study, we consider coherent communication: the receiver tracks the channel fading process exactly. We first start with the situation when the transmitter has only a statistical characterization of the fading channel. Finally, we look at the case when the transmitter also perfectly tracks the fading channel (full CSI); this situation is very similar to that of the time-invariant MIMO channel.

8.2.1 Capacity with CSI at receiver

Consider using the V-BLAST architecture (Figure 8.1) with a channel-independent multiplexing coordinate system $\mathbf{Q}$ and power allocations $P_1, \ldots, P_{n_t}$. The covariance matrix of the transmit signal is $\mathbf{K}_x$ and is not dependent on the channel realization. The rate achieved in a given channel state $\mathbf{H}$ is

$$\log \det \left(\mathbf{I}_{n_r} + \frac{1}{N_0} \mathbf{H} \mathbf{K}_x \mathbf{H}^* \right). \tag{8.8}$$

As usual, by coding over many coherence time intervals of the channel, a long-term rate of reliable communication equal to

$$\mathbb{E}_{\mathbf{H}} \left[\log \det \left(\mathbf{I}_{n_r} + \frac{1}{N_0} \mathbf{H} \mathbf{K}_x \mathbf{H}^* \right) \right] \tag{8.9}$$

is achieved. We can now choose the covariance $\mathbf{K}_x$ as a function of the channel *statistics* to achieve a reliable communication rate of

$$C = \max_{\mathbf{K}_x : \mathrm{Tr}[\mathbf{K}_x] \leq P} \mathbb{E} \left[\log \det \left(\mathbf{I}_{n_r} + \frac{1}{N_0} \mathbf{H} \mathbf{K}_x \mathbf{H}^* \right) \right]. \tag{8.10}$$

Here the trace constraint corresponds to the total transmit power constraint. This is indeed the capacity of the fast fading MIMO channel (a formal justification is in Appendix B.7.2). We emphasize that the input covariance is chosen to match the channel statistics rather than the channel realization, since the latter is not known at the transmitter.

The optimal $\mathbf{K}_x$ in (8.10) obviously depends on the stationary distribution of the channel process $\{\mathbf{H}[m]\}$. For example, if there are only a few dominant paths (no more than one in each of the angular bins) that are not time-varying, then we can view $\mathbf{H}$ as being deterministic. In this case, we know from Section 7.1.1 that the optimal coordinate system to multiplex the data streams is in the eigen-directions of $\mathbf{H}^*\mathbf{H}$ and, further, to allocate powers in a waterfilling manner across the eigenmodes of $\mathbf{H}$.

Let us now consider the other extreme: there are many paths (of approximately equal energy) in each of the angular bins. Some insight can be obtained by looking at the angular representation (cf. (7.80)): $\mathbf{H}^a := \mathbf{U}_r^* \mathbf{H} \mathbf{U}_t$. The key advantage of this viewpoint is in statistical modeling: the entries of $\mathbf{H}^a$ are generated by different physical paths and can be modeled as being statistically independent (cf. Section 7.3.5). Here we are interested in the case when the entries of $\mathbf{H}^a$ have zero mean (no single dominant path in any of the angular windows). Due to independence, it seems reasonable to separately send information in each of the transmit angular windows, with powers corresponding to the strength of the paths in the angular windows. That is, the multiplexing is done in the coordinate system given by $\mathbf{U}_t$ (so $\mathbf{Q} = \mathbf{U}_t$ in (8.3)). The covariance matrix now has the form

$$\mathbf{K}_x = \mathbf{U}_t \mathbf{\Lambda} \mathbf{U}_t^*, \tag{8.11}$$

where $\mathbf{\Lambda}$ is a diagonal matrix with non-negative entries, representing the powers transmitted in the angular windows, so that the sum of the entries is equal to P. This is shown formally in Exercise 8.3, where we see that this observation holds even if the entries of $\mathbf{H}^\mathbf{a}$ are only uncorrelated.

If there is additional symmetry among the transmit antennas, such as when the elements of $\mathbf{H}^\mathbf{a}$ are i.i.d. $\mathcal{CN}(0, 1)$ (the i.i.d. Rayleigh fading model), then one can further show that equal powers are allocated to each transmit angular window (see Exercises 8.4 and 8.6) and thus, in this case, the optimal covariance matrix is simply

$$\mathbf{K}_x = \left(\frac{P}{n_t}\right) \mathbf{I}_{n_t}. \tag{8.12}$$

More generally, the optimal powers (i.e., the diagonal entries of $\mathbf{\Lambda}$) are chosen to be the solution to the maximization problem (substituting the angular representation $\mathbf{H} = \mathbf{U}_r \mathbf{H}^\mathbf{a} \mathbf{U}_t^*$ and (8.11) in (8.10)):

$$C = \max_{\mathbf{\Lambda}:\text{Tr}[\mathbf{\Lambda}]\leq P} \mathbb{E}\left[\log\det\left(\mathbf{I}_{n_r} + \frac{1}{N_0}\mathbf{U}_r\mathbf{H}^\mathbf{a}\mathbf{\Lambda}\mathbf{H}^{\mathbf{a}*}\mathbf{U}_r^*\right)\right] \tag{8.13}$$

$$= \max_{\mathbf{\Lambda}:\text{Tr}[\mathbf{\Lambda}]\leq P} \mathbb{E}\left[\log\det\left(\mathbf{I}_{n_r} + \frac{1}{N_0}\mathbf{H}^\mathbf{a}\mathbf{\Lambda}\mathbf{H}^{\mathbf{a}*}\right)\right]. \tag{8.14}$$

With equal powers (i.e., the optimal $\mathbf{\Lambda}$ is equal to $(P/n_t)\mathbf{I}_{n_t}$), the resulting capacity is

$$\boxed{C = \mathbb{E}\left[\log\det\left(\mathbf{I}_{n_r} + \frac{\text{SNR}}{n_t}\mathbf{H}\mathbf{H}^*\right)\right],} \tag{8.15}$$

where $\text{SNR} := P/N_0$ is the common SNR at each receive antenna.

If $\lambda_1 \geq \lambda_2 \geq \cdots \geq \lambda_{n_{\min}}$ are the (random) ordered singular values of $\mathbf{H}$, then we can rewrite (8.15) as

$$C = \mathbb{E}\left[\sum_{i=1}^{n_{\min}} \log\left(1 + \frac{\text{SNR}}{n_t}\lambda_i^2\right)\right]$$

$$= \sum_{i=1}^{n_{\min}} \mathbb{E}\left[\log\left(1 + \frac{\text{SNR}}{n_t}\lambda_i^2\right)\right]. \tag{8.16}$$

Comparing this expression to the waterfilling capacity in (7.10), we see the contrast between the situation when the transmitter knows the channel and when it does not. When the transmitter knows the channel, it can allocate different amounts of power in the different eigenmodes depending on their strengths. When the transmitter does not know the channel but the channel is sufficiently random, the optimal covariance matrix is identity, resulting in equal amounts of power across the eigenmodes.

8.2.2 Performance gains

The capacity, (8.16), of the MIMO fading channel is a function of the distribution of the singular values, λ_i, of the random channel matrix $\mathbf{H}$. By Jensen's inequality, we know that

$$\sum_{i=1}^{n_{\min}} \log\left(1 + \frac{\mathsf{SNR}}{n_t}\lambda_i^2\right) \leq n_{\min} \log\left(1 + \frac{\mathsf{SNR}}{n_t}\left[\frac{1}{n_{\min}}\sum_{i=1}^{n_{\min}}\lambda_i^2\right]\right), \quad (8.17)$$

with equality if and only if the singular values are all equal. Hence, one would expect a high capacity if the channel matrix $\mathbf{H}$ is sufficiently random and statistically well conditioned, with the overall channel gain well distributed across the singular values. In particular, one would expect such a channel to attain the full degrees of freedom at high SNR.

We plot the capacity for the i.i.d. Rayleigh fading model in Figure 8.2 for different numbers of antennas. Indeed, we see that for such a random channel the capacity of a MIMO system can be very large. At moderate to high SNR, the capacity of an n by n channel is about n times the capacity of a 1 by 1 system. The asymptotic slope of capacity versus SNR in dB scale is proportional to n, which means that the capacity scales with SNR like $n \log \mathsf{SNR}$.

High SNR regime

The performance gain can be seen most clearly in the high SNR regime. At high SNR, the capacity for the i.i.d. Rayleigh channel is given by

$$C \approx n_{\min} \log \frac{\mathsf{SNR}}{n_t} + \sum_{i=1}^{n_{\min}} \mathbb{E}[\log \lambda_i^2], \quad (8.18)$$

and

$$\mathbb{E}[\log \lambda_i^2] > -\infty, \quad (8.19)$$

for all i. Hence, the full $n_{\min}$ degrees of freedom is attained. In fact, further analysis reveals that

$$\sum_{i=1}^{n_{\min}} \mathbb{E}[\log \lambda_i^2] = \sum_{i=|n_t - n_r|+1}^{\max\{n_t, n_r\}} \mathbb{E}[\log \chi_{2i}^2], \quad (8.20)$$

Figure 8.2 Capacity of an i.i.d. Rayleigh fading channel. Upper: 4 by 4 channel. Lower: 8 by 8 channel.

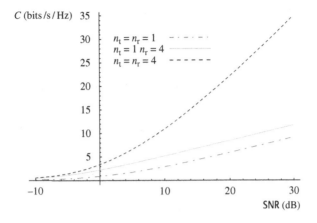

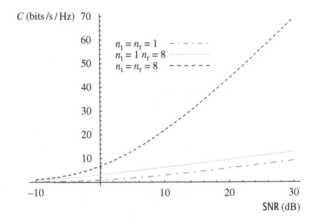

where χ_{2i}^2 is a χ-square distributed random variable with $2i$ degrees of freedom.

Note that the number of degrees of freedom is limited by the minimum of the number of transmit and the number of receive antennas, hence, to get a large capacity, we need multiple transmit *and* multiple receive antennas. To emphasize this fact, we also plot the capacity of a 1 by n_r channel in Figure 8.2. This capacity is given by

$$C = \mathbb{E}\left[\log\left(1 + \mathsf{SNR}\sum_{i=1}^{n_r}|h_i|^2\right)\right] \text{ bits/s/Hz.} \qquad (8.21)$$

We see that the capacity of such a channel is significantly less than that of an n_r by n_r system in the high SNR range, and this is due to the fact that there is only one degree of freedom in a 1 by n_r channel. The gain in going from a 1 by 1 system to a 1 by n_r system is a *power gain*, resulting in a parallel shift of the capacity versus SNR curves. At high SNR, a power gain is much less impressive than a degree-of-freedom gain.

Low SNR regime

Here we use the approximation $\log_2(1+x) \approx x \log_2 e$ for x small in (8.15) to get

$$
\begin{aligned}
C &= \sum_{i=1}^{n_{\min}} \mathbb{E}\left[\log\left(1+\frac{\text{SNR}}{n_t}\lambda_i^2\right)\right] \\
&\approx \sum_{i=1}^{n_{\min}} \frac{\text{SNR}}{n_t}\mathbb{E}\left[\lambda_i^2\right]\log_2 e \\
&= \frac{\text{SNR}}{n_t}\mathbb{E}[\text{Tr}[\mathbf{HH}^*]]\log_2 e \\
&= \frac{\text{SNR}}{n_t}\mathbb{E}\left[\sum_{i,j}|h_{ij}|^2\right]\log_2 e \\
&= n_r\text{SNR}\log_2 e \text{ bits/s/Hz}.
\end{aligned}
$$

Thus, at low SNR, an n_t by n_r system yields a *power gain* of n_r over a single antenna system. This is due to the fact that the multiple receive antennas can coherently combine their received signals to get a power boost. Note that increasing the number of transmit antennas does not increase the power gain since, unlike the case when the channel is known at the transmitter, *transmit beamforming* cannot be done to constructively add signals from the different antennas. Thus, at low SNR and without channel knowledge at the transmitter, multiple transmit antennas are not very useful: the performance of an n_t by n_r channel is comparable with that of a 1 by n_r channel. This is illustrated in Figure 8.3, which compares the capacity of an n by n channel with that of a 1 by n channel, as a fraction of the capacity of a 1 by 1 channel. We see that at an SNR of about -20 dB, the capacities of a 1 by 4 channel and a 4 by 4 channel are very similar.

Recall from Chapter 4 that the operating SINR of cellular systems with universal frequency reuse is typically very low. For example, an IS-95 CDMA system may have an SINR per chip of -15 to -17 dB. The above observation then suggests that just simply overlaying point-to-point MIMO technology on such systems to boost up per link capacity will not provide much additional benefit than just adding antennas at one end. On the other hand, the story is different if the multiple antennas are used to perform multiple access and interference management. This issue will be revisited in Chapter 10.

Another difference between the high and the low SNR regimes is that while channel randomness is crucial in yielding a large capacity gain in the high SNR regime, it plays little role in the low SNR regime. The low SNR result above does not depend on whether the channel gains, $\{h_{ij}\}$, are independent or correlated.

Figure 8.3 Low SNR capacities. Upper: a 1 by 4 and a 4 by 4 channel. Lower: a 1 by 8 an 8 by 8 channel. Capacity is a fraction of the 1 by 1 channel in each case.

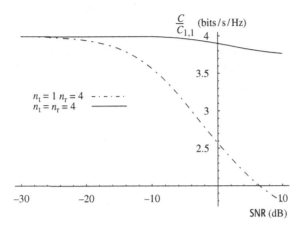

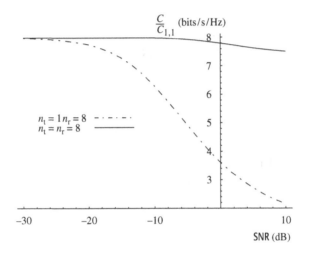

Large antenna array regime

We saw that in the high SNR regime, the capacity increases linearly with the minimum of the number of transmit and the number of receive antennas. This is a degree-of-freedom gain. In the low SNR regime, the capacity increases linearly with the number of receive antennas. This is a power gain. Will the combined effect of the two types of gain yield a linear growth in capacity at *any* SNR, as we scale up both n_t and n_r? Indeed, this turns out to be true. Let us focus on the square channel $n_t = n_r = n$ to demonstrate this.

With i.i.d. Rayleigh fading, the capacity of this channel is (cf. (8.15))

$$C_{nn}(\text{SNR}) = \mathbb{E}\left[\sum_{i=1}^{n} \log\left(1 + \text{SNR}\frac{\lambda_i^2}{n}\right)\right], \qquad (8.22)$$

where we emphasize the dependence on n and SNR in the notation. The $\lambda_1/\sqrt{n}$, $\ldots, \lambda_n/\sqrt{n}$ are the singular values of the random matrix $\mathbf{H}/\sqrt{n}$. By a random matrix result due to Marčenko and Pastur [78], the empirical distribution of the singular values of $\mathbf{H}/\sqrt{n}$ converges to a deterministic limiting distribution for almost all realizations of $\mathbf{H}$. Figure 8.4 demonstrates the convergence. The limiting distribution is the so-called *quarter circle law*.[3] The corresponding limiting density of the *squared* singular values is given by

$$f^*(x) = \begin{cases} \frac{1}{\pi}\sqrt{\frac{1}{x} - \frac{1}{4}} & 0 \leq x \leq 4, \\ 0 & \text{else.} \end{cases} \qquad (8.23)$$

Hence, we can conclude that, for increasing n,

$$\frac{1}{n}\sum_{i=1}^{n}\log\left(1 + \mathsf{SNR}\frac{\lambda_i^2}{n}\right) \to \int_0^4 \log(1 + \mathsf{SNR}x)f^*(x)\mathrm{d}x. \qquad (8.24)$$

If we denote

$$c^*(\mathsf{SNR}) := \int_0^4 \log(1 + \mathsf{SNR}x)f^*(x)\mathrm{d}x, \qquad (8.25)$$

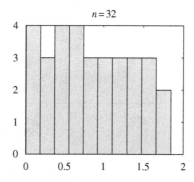

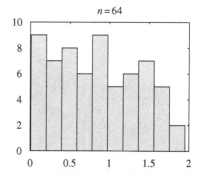

Figure 8.4 Convergence of the empirical singular value distribution of $\mathbf{H}/\sqrt{n}$. For each n, a single random realization of $\mathbf{H}/\sqrt{n}$ is generated and the empirical distribution (histogram) of the singular values is plotted. We see that as n grows, the histogram converges to the quarter circle law.

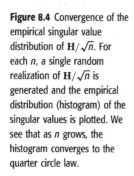

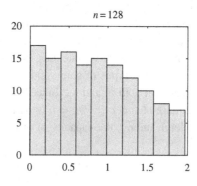

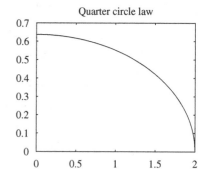

---

[3] Note that although the singular values are unbounded, in the limit they lie in the interval $[0, 2]$ with probability 1.

we can solve the integral for the density in (8.23) to arrive at (see Exercise 8.17)

$$c^*(\mathsf{SNR}) = 2\log\left(1 + \mathsf{SNR} - \frac{1}{4}F(\mathsf{SNR})\right) - \frac{\log e}{4\mathsf{SNR}}F(\mathsf{SNR}), \qquad (8.26)$$

where

$$F(\mathsf{SNR}) := \left(\sqrt{4\mathsf{SNR}+1} - 1\right)^2. \qquad (8.27)$$

The significance of $c^*(\mathsf{SNR})$ is that

$$\lim_{n\to\infty}\frac{C_{nn}(\mathsf{SNR})}{n} = c^*(\mathsf{SNR}). \qquad (8.28)$$

So capacity grows linearly in n at *any* SNR and the constant $c^*(\mathsf{SNR})$ is the rate of the growth.

We compare the large-n approximation

$$\boxed{C_{nn}(\mathsf{SNR}) \approx nc^*(\mathsf{SNR}),} \qquad (8.29)$$

with the actual value of the capacity for $n = 2, 4$ in Figure 8.5. We see the approximation is very good, even for such small values of n. In Exercise 8.7, we see statistical models other than i.i.d. Rayleigh, which also have a linear increase in capacity with an increase in n.

Linear scaling: a more in-depth look

To better understand why the capacity scales linearly with the number of antennas, it is useful to contrast the MIMO scenario here with three other scenarios:

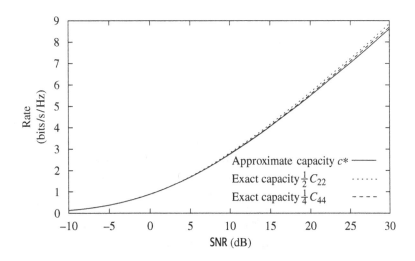

Figure 8.5 Comparison between the large-n approximation and the actual capacity for $n = 2, 4$.

- **MISO channel with a large transmit antenna array** Specializing (8.15) to the n by 1 MISO channel yields the capacity

$$C_{n1} = \mathbb{E}\left[\log\left(1 + \frac{\mathsf{SNR}}{n}\sum_{i=1}^{n}|h_i|^2\right)\right] \text{bits/s/Hz}. \quad (8.30)$$

As $n \to \infty$, by the law of large numbers,

$$C_{n1} \to \log(1 + \mathsf{SNR}) = C_{\text{awgn}}. \quad (8.31)$$

For $n = 1$, the 1 by 1 fading channel (with only receiver CSI) has lower capacity than the AWGN channel; this is due to the "Jensen's loss" (Section 5.4.5). But recall from Figure 5.20 that this loss is not large for the entire range of SNR. Increasing the number of transmit antennas has the effect of reducing the fluctuation of the instantaneous SNR

$$\frac{1}{n}\sum_{i=1}^{n}|h_i|^2 \cdot \mathsf{SNR}, \quad (8.32)$$

and hence reducing the Jensen's loss, but the loss was not big to start with, hence the gain is minimal. Since the total transmit power is fixed, the multiple transmit antennas provide neither a power gain nor a gain in spatial degrees of freedom. (In a *slow* fading channel, the multiple transmit antennas provide a diversity gain, but this is not relevant in the fast fading scenario considered here.)
- **SIMO channel with a large receive antenna array** A 1 by n SIMO channel has capacity

$$C_{1n} = \mathbb{E}\left[\log\left(1 + \mathsf{SNR}\sum_{i=1}^{n}|h_i|^2\right)\right]. \quad (8.33)$$

For large n

$$C_{1n} \approx \log(n\mathsf{SNR}) = \log n + \log \mathsf{SNR}, \quad (8.34)$$

i.e., the receive antennas provide a power gain (which increases linearly with the number of receive antennas) and the capacity increases logarithmically with the number of receive antennas. This is quite in contrast to the MISO case: the difference is due to the fact that now there is a linear increase in total *received* power due to a larger receive antenna array. However, the increase in capacity is only *logarithmic* in n; the increase in total received power is all accumulated in the single degree of freedom of the channel. There is power gain but no gain in the spatial degrees of freedom.

The capacities, as a function of n, are plotted for the SIMO, MISO and MIMO channels in Figure 8.6.

Figure 8.6 Capacities of the *n* by 1 MISO channel, 1 by *n* SIMO channel and the *n* by *n* MIMO channel as a function of *n*, for SNR = 0 dB

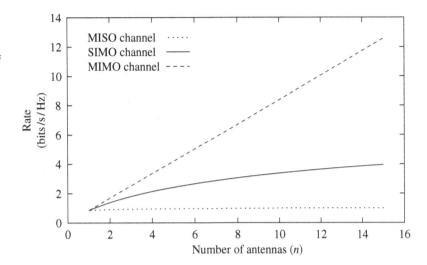

- **AWGN channel with infinite bandwidth** Given a power constraint of $\bar{P}$ and AWGN noise spectral density $N_0/2$, the infinite bandwidth limit is (cf. 5.18)

$$C_\infty = \lim_{W \to \infty} W \log\left(1 + \frac{\bar{P}}{N_0 W}\right) = \frac{\bar{P}}{N_0} \text{ bits/s.} \qquad (8.35)$$

Here, although the number of degrees of freedom increases, the capacity remains bounded. This is because the total received power is fixed and hence the SNR per degree of freedom vanishes. There is a gain in the degrees of freedom, but since there is no power gain the received power has to be spread across the many degrees of freedom.

In contrast to all of these scenarios, the capacity of an *n* by *n* MIMO channel increases linearly with *n*, because simultaneously:

- there is a linear increase in the total received power, and
- there is a linear increase in the degrees of freedom, due to the substantial randomness and consequent well-conditionedness of the channel matrix **H**.

Note that the well-conditionedness of the matrix depends on maintaining the uncorrelated nature of the channel gains, $\{h_{ij}\}$, while increasing the number of antennas. This can be achieved in a rich scattering environment by keeping the antenna spacing fixed at half the wavelength and increasing the aperture, L, of the antenna array. On the other hand, if we just pack more and more antenna elements in a *fixed* aperture, L, then the channel gains will become more and more correlated. In fact, we know from Section 7.3.7 that in the angular domain a MIMO channel with densely spaced antennas and aperture L can be reduced to an equivalent $2L$ by $2L$ channel with antennas spaced at half the wavelength. Thus, the number of degrees of freedom is ultimately limited by the antenna array aperture rather than the number of antenna elements.

8.2.3 Full CSI

We have considered the scenario when only the receiver can track the channel. This is the most interesting case in practice. In a TDD system or in an FDD system where the fading is very slow, it may be possible to track the channel matrix at the transmitter. We shall now discuss how channel capacity can be achieved in this scenario. Although channel knowledge at the transmitter does not help in extracting an additional degree-of-freedom gain, extra power gain is possible.

Capacity

The derivation of the channel capacity in the full CSI scenario is only a slight twist on the time-invariant case discussed in Section 7.1.1. At each time m, we decompose the channel matrix as $\mathbf{H}[m] = \mathbf{U}[m]\mathbf{\Lambda}[m]\mathbf{V}[m]^*$, so that the MIMO channel can be represented as a parallel channel

$$\tilde{y}_i[m] = \lambda_i[m]\tilde{x}_i[m] + \tilde{w}_i[m], \qquad i = 1, \ldots, n_{\min}, \tag{8.36}$$

where $\lambda_1[m] \geq \lambda_2[m] \geq \ldots \geq \lambda_{n_{\min}}[m]$ are the ordered singular values of $\mathbf{H}[m]$ and

$$\tilde{\mathbf{x}}[m] = \mathbf{V}^*[m]\mathbf{x}[m],$$
$$\tilde{\mathbf{y}}[m] = \mathbf{U}^*[m]\mathbf{y}[m],$$
$$\tilde{\mathbf{w}}[m] = \mathbf{U}^*[m]\mathbf{w}[m].$$

We have encountered the fast fading parallel channel in our study of the single antenna fast fading channel (cf. Section 5.4.6). We allocate powers to the sub-channels based on their strength according to the waterfilling policy

$$P^*(\lambda) = \left(\mu - \frac{N_0}{\lambda^2}\right)^+, \tag{8.37}$$

with μ chosen so that the total transmit power constraint is satisfied:

$$\sum_{i=1}^{n_{\min}} \mathbb{E}\left[\left(\mu - \frac{N_0}{\lambda_i^2}\right)^+\right] = P. \tag{8.38}$$

Note that this is waterfilling over time and space (the eigenmodes). The capacity is given by

$$C = \sum_{i=1}^{n_{\min}} \mathbb{E}\left[\log\left(1 + \frac{P^*(\lambda_i)\lambda_i^2}{N_0}\right)\right]. \tag{8.39}$$

Transceiver architecture

The transceiver architecture that achieves the capacity follows naturally from the SVD-based architecture depicted in Figure 7.2. Information bits are split into n_{min} parallel streams, each coded separately, and then augmented by $n_t - n_{min}$ streams of zeros. The symbols across the streams at time m form the vector $\tilde{\mathbf{x}}[m]$. This vector is pre-multiplied by the matrix $\mathbf{V}[m]$ before being sent through the channel, where $\mathbf{H}[m] = \mathbf{U}[m]\mathbf{\Lambda}[m]\mathbf{V}^*[m]$ is the singular value decomposition of the channel matrix at time m. The output is post-multiplied by the matrix $\mathbf{U}^*[m]$ to extract the independent streams, which are then separately decoded. The power allocated to each stream is time-dependent and is given by the waterfilling formula (8.37), and the rates are dynamically allocated accordingly. If an AWGN capacity-achieving code is used for each stream, then the entire system will be capacity-achieving for the MIMO channel.

Performance analysis

Let us focus on the i.i.d. Rayleigh fading model. Since with probability 1, the random matrix $\mathbf{HH}^*$ has full rank (Exercise 8.12), and is, in fact, well-conditioned (Exercise 8.14), it can be shown that at high SNR, the waterfilling strategy allocates an equal amount of power P/n_{min} to all the spatial modes, as well as an equal amount of power over time. Thus,

$$C \approx \sum_{i=1}^{n_{min}} \mathbb{E}\left[\log\left(1 + \frac{\mathsf{SNR}}{n_{min}}\lambda_i^2\right)\right], \qquad (8.40)$$

where $\mathsf{SNR} = P/N_0$. If we compare this to the capacity (8.16) with only receiver CSI, we see that the number of degrees of freedom is the same (n_{min}) but there is a power gain of a factor of n_t/n_{min} when the transmitter can track the channel. Thus, whenever there are more transmit antennas then receive antennas, there is a power boost of n_t/n_r from having transmitter CSI. The reason is simple. Without channel knowledge at the transmitter, the transmit energy is spread out equally across all directions in $\mathcal{C}^{n_t}$. With transmitter CSI, the energy can now be focused on only the n_r non-zero eigenmodes, which form a subspace of dimension n_r inside $\mathcal{C}^{n_t}$. For example, with $n_r = 1$, the capacity with only receiver CSI is

$$\mathbb{E}\left[\log\left(1 + \mathsf{SNR}/n_t \sum_{i=1}^{n_t} |h_i|^2\right)\right],$$

while the high SNR capacity when there is full CSI is

$$\mathbb{E}\left[\log\left(1 + \mathsf{SNR} \sum_{i=1}^{n_t} |h_i|^2\right)\right].$$

Thus a power gain of a factor of n_t is achieved by transmit beamforming. With dual transmit antennas, this is a gain of 3 dB.

At low SNR, there is a further gain from transmitter CSI due to dynamic allocation of power across the eigenmodes: at any given time, more power is given to stronger eigenmodes. This gain is of the same nature as the one from opportunistic communication discussed in Chapter 6.

What happens in the large antenna array regime? Applying the random matrix result of Marčenko and Pastur from Section 8.2.2, we conclude that the random singular values $\lambda_i[m]/\sqrt{n}$ of the channel matrix $\mathbf{H}[m]/\sqrt{n}$ converge to the same deterministic limiting distribution f^* across all times m. This means that in the waterfilling strategy, there is no dynamic power allocation over time, only over space. This is sometimes known as a *channel hardening* effect.

Summary 8.1 Performance gains in a MIMO channel

The capacity of an $n_t \times n_r$ i.i.d. Rayleigh fading MIMO channel $\mathbf{H}$ with receiver CSI is

$$C_{nn}(\text{SNR}) = \mathbb{E}\left[\log \det \left(\mathbf{I}_{n_r} + \frac{\text{SNR}}{n_t}\mathbf{H}\mathbf{H}^*\right)\right]. \qquad (8.41)$$

At high SNR, the capacity is approximately equal (up to an additive constant) to $n_{\min} \log \frac{\text{SNR}}{n_t}$ bits/s/Hz.

At low SNR, the capacity is approximately equal to $n_r \text{SNR} \log_2 e$ bits/s/Hz, so only a receive beamforming gain is realized.

With $n_t = n_r = n$, the capacity can be approximated by $nc^*(\text{SNR})$ where $c^*(\text{SNR})$ is the constant in (8.26).

Conclusion: In an $n \times n$ MIMO channel, the capacity increases *linearly* with n over the entire SNR range.

With channel knowledge at the transmitter, an additional n_t/n_r-fold transmit beamforming gain can be realized with an additional power gain from temporal–spatial waterfilling at low SNR.

8.3 Receiver architectures

The transceiver architecture of Figure 8.1 achieves the capacity of the fast fading MIMO channel with receiver CSI. The capacity is achieved by joint ML decoding of the data streams at the receiver, but the complexity grows exponentially with the number of data streams. Simpler decoding rules that provide soft information to feed to the decoders of the individual data streams is an active area of research; some of the approaches are reviewed in Exercise 8.15. In this section, we consider receiver architectures that use *linear* operations to convert the problem of joint decoding of the data streams into one of individual decoding of the data streams. These architectures extract the spatial degree of freedom gains characterized in the previous section. In conjunction with successive cancellation of data streams, we can achieve the capacity of the fast fading MIMO channel. To be able to focus on the receiver design, we start with transmitting the independent data streams directly over the antenna array (i.e., $\mathbf{Q} = \mathbf{I}_{n_t}$ in Figure 8.1).

8.3.1 Linear decorrelator

Geometric derivation

Is it surprising that the full degrees of freedom of $\mathbf{H}$ can be attained even when the transmitter does not track the channel matrix? When the transmitter does know the channel, the SVD architecture enables the transmitter to send parallel data streams through the channel so that they arrive orthogonally at the receiver without interference between the streams. This is achieved by pre-rotating the data so that the parallel streams can be sent along the eigenmodes of the channel. When the transmitter does not know the channel, this is not possible. Indeed, after passing through the MIMO channel of (7.1), the independent data streams sent on the transmit antennas all arrive cross-coupled at the receiver. It is not clear a priori that the receiver can separate the data streams efficiently enough so that the resulting performance has full degrees of freedom. But in fact we have already seen such a receiver: the channel inversion receiver in the 2×2 example discussed in Section 3.3.3. We develop the structure of this receiver in full generality here.

To simplify notations, let us first focus on the time-invariant case, where the channel matrix is fixed. We can write the received vector at symbol time m as

$$\mathbf{y}[m] = \sum_{i=1}^{n_t} \mathbf{h}_i x_i[m] + \mathbf{w}[m], \qquad (8.42)$$

where $\mathbf{h}_1, \ldots, \mathbf{h}_{n_t}$ are the columns of $\mathbf{H}$ and the data streams transmitted on the antennas, $\{x_i[m]\}$ on the ith antenna, are all independent. Focusing on the kth data stream, we can rewrite (8.42):

$$\mathbf{y}[m] = \mathbf{h}_k x_k[m] + \sum_{i \neq k} \mathbf{h}_i x_i[m] + \mathbf{w}. \qquad (8.43)$$

Compared to the SIMO point-to-point channel from Section 7.2.1, we see that the kth data stream faces an extra source of interference, that from the other data streams. One idea that can be used to remove this inter-stream interference is to project the received signal $\mathbf{y}$ onto the subspace orthogonal to the one spanned by the vectors $\mathbf{h}_1, \ldots, \mathbf{h}_{k-1}, \mathbf{h}_{k+1}, \ldots, \mathbf{h}_{n_t}$

(denoted henceforth by V_k). Suppose that the dimension of V_k is d_k. Projection is a linear operation and we can represent it by a d_k by n_r matrix $\mathbf{Q}_k$, the rows of which form an orthonormal basis of V_k; they are all orthogonal to $\mathbf{h}_1, \ldots, \mathbf{h}_{k-1}, \mathbf{h}_{k+1}, \ldots, \mathbf{h}_{n_t}$. The vector $\mathbf{Q}_k \mathbf{v}$ should be interpreted as the projection of the vector $\mathbf{v}$ onto V_k, but expressed in terms of the coordinates defined by the basis of V_k formed by the rows of $\mathbf{Q}_k$. A pictorial depiction of this projection operation is in Figure 8.7.

Now, the inter-stream interference "nulling" is successful (that is, the resulting projection of $\mathbf{h}_k$ is a non-zero vector) if the kth data stream "spatial signature" $\mathbf{h}_k$ is not a linear combination of the spatial signatures of the other data streams. In other words, if there are more data streams than the dimension of the received signal (i.e., $n_t > n_r$), then the nulling operation will not be successful, even for a full rank $\mathbf{H}$. Hence, we should choose the number of data streams to be no more than n_r. Physically, this corresponds to using only a subset of the transmit antennas and for notational convenience we will count only the transmit antennas that are used, by just making the assumption $n_t \le n_r$ in the decorrelator discussion henceforth.

After the projection operation,

$$\tilde{\mathbf{y}}[m] := \mathbf{Q}_k \mathbf{y}[m] = \mathbf{Q}_k \mathbf{h}_k x_k[m] + \tilde{\mathbf{w}}[m]$$

where $\tilde{\mathbf{w}}[m] := \mathbf{Q}_k \mathbf{w}[m]$ is the noise, still white, after the projection. Optimal demodulation of the kth stream can now be performed by match filtering to the vector $\mathbf{Q}_k \mathbf{h}_k$. The output of this matched filter (or maximal ratio combiner) has SNR

$$\frac{P_k \|\mathbf{Q}_k \mathbf{h}_k\|^2}{N_0}, \tag{8.44}$$

where P_k is the power allocated to stream k.

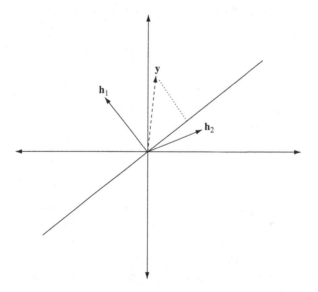

Figure 8.7 A schematic representation of the projection operation: $\mathbf{y}$ is projected onto the subspace orthogonal to $\mathbf{h}_1$ to demodulate stream 2.

The combination of the projection operation followed by the matched filter is called the *decorrelator* (also known as *interference nulling* or *zero-forcing* receiver). Since projection and matched filtering are both linear operations, the decorrelator is a linear filter. The filter $\mathbf{c}_k$ is given by

$$\mathbf{c}_k^* = (\mathbf{Q}_k \mathbf{h}_k)^* \mathbf{Q}_k, \qquad (8.45)$$

or

$$\mathbf{c}_k = (\mathbf{Q}_k^* \mathbf{Q}_k) \mathbf{h}_k, \qquad (8.46)$$

which is the projection of $\mathbf{h}_k$ onto the subspace V_k, expressed in terms of the original coordinates. Since the matched filter maximizes the output SNR, the decorrelator can also be interpreted as the linear filter that maximizes the output SNR subject to the constraint that the filter nulls out the interference from all other streams. Intuitively, we are projecting the received signal in the direction within V_k that is closest to $\mathbf{h}_k$.

Only the kth stream has been in focus so far. We can now decorrelate each of the streams separately, as illustrated in Figure 8.8. We have described the decorrelator geometrically; however, there is a simple explicit formula for the entire bank of decorrelators: the decorrelator for the kth stream is the kth row of the pseudoinverse $\mathbf{H}^\dagger$ of the matrix $\mathbf{H}$, defined by

$$\mathbf{H}^\dagger := (\mathbf{H}^* \mathbf{H})^{-1} \mathbf{H}^*. \qquad (8.47)$$

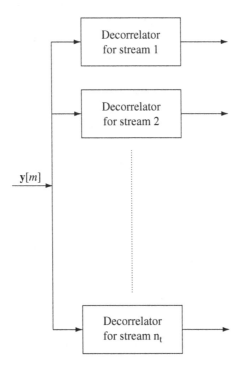

Figure 8.8 A bank of decorrelators, each estimating the parallel data streams.

The validity of this formula is verified in Exercise 8.11. In the special case when $\mathbf{H}$ is square and invertible, $\mathbf{H}^\dagger = \mathbf{H}^{-1}$ and the decorrelator is precisely the channel inversion receiver we already discussed in Section 3.3.3.

Performance for a deterministic $\mathbf{H}$

The channel from the kth stream to the output of the corresponding decorrelator is a Gaussian channel with SNR given by (8.44). A Gaussian code achieves the maximum data rate, given by

$$C_k := \log\left(1 + \frac{P_k \|\mathbf{Q}_k \mathbf{h}_k\|^2}{N_0}\right). \tag{8.48}$$

To get a better feel for this performance, let us compare it with the ideal situation of no inter-stream interference in (8.43). As we observed above, if there were no inter-stream interference in (8.43), the situation is exactly the SIMO channel of Section 7.2.1; the filter would be matched to $\mathbf{h}_k$ and the achieved SNR would be

$$\frac{P_k \|\mathbf{h}_k\|^2}{N_0}. \tag{8.49}$$

Since the inter-stream interference only hampers the recovery of the kth stream, the performance of the decorrelator (in terms of the SNR in (8.44)) must in general be less than that achieved by a matched filter with no inter-stream interference. We can also see this explicitly: the projection operation cannot increase the length of a vector and hence $\|\mathbf{Q}_k \mathbf{h}_k\| \leq \|\mathbf{h}_k\|$. We can further say that the projection operation always reduces the length of $\mathbf{h}_k$ unless $\mathbf{h}_k$ is already orthogonal to the spatial signatures of the other data streams.

Let us return to the bank of decorrelators in Figure 8.8. The total rate of communication supported here with efficient coding in each of the data streams is the sum of the individual rates in (8.48) and is given by

$$\sum_{k=1}^{n_t} C_k.$$

Performance in fading channels

So far our analysis has focused on a deterministic channel $\mathbf{H}$. As usual, in the time-varying fast fading scenario, coding should be done over time across the different fades, usually in combination with interleaving. The maximum achievable rate can be computed by simply averaging over the stationary distribution of the channel process $\{\mathbf{H}[m]\}_m$, yielding

$$R_{\text{decorr}} = \sum_{k=1}^{n_t} \bar{C}_k, \tag{8.50}$$

where

$$\bar{C}_k = \mathbb{E}\left[\log\left(1 + \frac{P_k \|\mathbf{Q}_k\mathbf{h}_k\|^2}{N_0}\right)\right]. \tag{8.51}$$

The achievable rate in (8.50) is in general less than or equal to the capacity of the MIMO fading channel with CSI at the receiver (cf. (8.10)) since transmission using independent data streams and receiving using the bank of decorrelators is only one of several possible communication strategies. To get some further insight, let us look at a specific statistical model, that of i.i.d. Rayleigh fading. Motivated by the fact that the optimal covariance matrix is of the form of scaled identity (cf. (8.12)), let us choose equal powers for each of the data streams (i.e., $P_k = P/n_t$). Continuing from (8.50), the decorrelator bank performance specialized to i.i.d. Rayleigh fading is (recall that for successful decorrelation $n_{\min} = n_t$)

$$R_{\text{decorr}} = \mathbb{E}\left[\sum_{k=1}^{n_{\min}} \log\left(1 + \frac{\text{SNR}}{n_t}\|\mathbf{Q}_k\mathbf{h}_k\|^2\right)\right]. \tag{8.52}$$

Since $\mathbf{h}_k \sim \mathcal{CN}(0, \mathbf{I}_{n_r})$, we know that $\|\mathbf{h}_k\|^2 \sim \chi^2_{2n_r}$, where χ^2_{2i} is a χ-squared random variable with $2i$ degrees of freedom (cf. (3.36)). Here $\mathbf{Q}_k\mathbf{h}_k \sim \mathcal{CN}(0, \mathbf{I}_{\dim V_k})$ (since $\mathbf{Q}_k\mathbf{Q}_k^* = \mathbf{I}_{\dim V_k}$). It can be shown that the channel $\mathbf{H}$ is full rank with probability 1 (see Exercise 8.12), and this means that $\dim V_k = n_r - n_t + 1$ (see Exercise 8.13). Thus $\|\mathbf{Q}_k\mathbf{h}_k\|^2 \sim \chi^2_{2(n_r-n_t+1)}$. This provides us with an explicit example for our earlier observation that the projection operation reduces the length. In the special case of a square system, $\dim V_k = 1$, and $\mathbf{Q}_k\mathbf{h}_k$ is a scalar distributed as circular symmetric Gaussian; we have already seen this in the 2×2 example of Section 3.3.3.

R_{decorr} is plotted in Figure 8.9 for different numbers of antennas. We see that the asymptotic slope of the rate obtained by the decorrelator bank as a

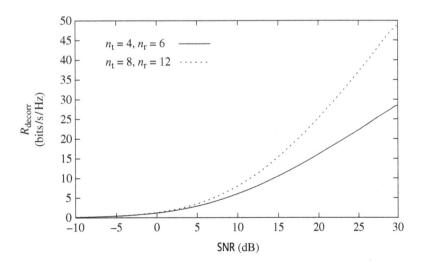

Figure 8.9 Rate achieved (in bits/s/Hz) by the decorrelator bank.

function of SNR in dB is proportional to $n_{\min}$; the same slope in the capacity of the MIMO channel. More specifically, we can approximate the rate in (8.52) at high SNR as

$$R_{\text{decorr}} \approx n_{\min} \log \frac{\text{SNR}}{n_t} + \mathbb{E}\left[\sum_{k=1}^{n_t} \log \left(\| \mathbf{Q}_k \mathbf{h}_k \|^2 \right) \right], \tag{8.53}$$

$$= n_{\min} \log \left(\frac{\text{SNR}}{n_t} \right) + n_t \mathbb{E}\left[\log \chi^2_{2(n_r - n_t + 1)} \right]. \tag{8.54}$$

Comparing (8.53) and (8.54) with the corresponding high SNR expansion of the capacity of this MIMO channel (cf. (8.18) and (8.20)), we can make the following observations:

- The first-order term (in the high SNR expansion) is the same for both the rate achieved by the decorrelator bank and the capacity of the MIMO channel. Thus, the decorrelator bank is able to fully harness the spatial degrees of freedom of the MIMO channel.
- The next term in the high SNR expansion (constant term) shows the performance degradation, in rate, of using the decorrelator bank as compared to the capacity of the channel. Figure 8.10 highlights this difference in the special case of $n_t = n_r = n$.

The above analysis is for the high SNR regime. At any *fixed* SNR, it is also straightforward to show that, just like the capacity, the total rate achievable by the bank of decorrelators scales linearly with the number of antennas (see Exercise 8.21).

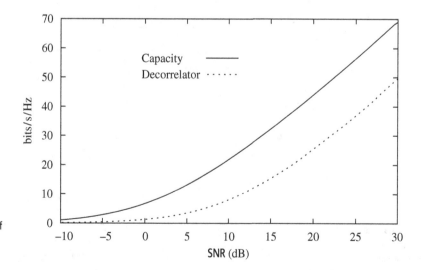

Figure 8.10 Plot of rate achievable with the decorrelator bank for the $n_t = n_r = 8$ i.i.d. Rayleigh fading channel. The capacity of the channel is also plotted for comparison.

8.3.2 Successive cancellation

We have just considered a bank of *separate* filters to estimate the data streams. However, the result of one of the filters could be used to aid the operation of the others. Indeed, we can use the *successive cancellation* strategy described in the uplink capacity analysis (in Section 6.1): once a data stream is successfully recovered, we can subtract it off from the received vector and reduce the burden on the receivers of the remaining data streams. With this motivation, consider the following modification to the bank of separate receiver structures in Figure 8.8. We use the first decorrelator to decode the data stream $x_1[m]$ and then subtract off this decoded stream from the received vector. If the first stream is successfully decoded, then the second decorrelator has to deal only with streams $x_3, \ldots, x_{n_t}$ as interference, since x_1 has been correctly subtracted off. Thus, the second decorrelator projects onto the subspace orthogonal to that spanned by $\mathbf{h}_3, \ldots, \mathbf{h}_{n_t}$. This process is continued until the final decorrelator does not have to deal with any interference from the other data streams (assuming successful subtraction in each preceding stage). This decorrelator–SIC (decorrelator with successive interference cancellation) architecture is illustrated in Figure 8.11.

One problem with this receiver structure is *error propagation*: an error in decoding the kth data stream means that the subtracted signal is incorrect and this error propagates to all the streams further down, $k+1, \ldots, n_t$. A careful analysis of the performance of this scheme is complicated, but can be made easier if we take the data streams to be well coded and the block length to be very large, so that streams are successfully cancelled with very high probability. With this assumption the kth data stream sees only down-stream interference, i.e., from the streams $k+1, \ldots, n_t$. Thus,

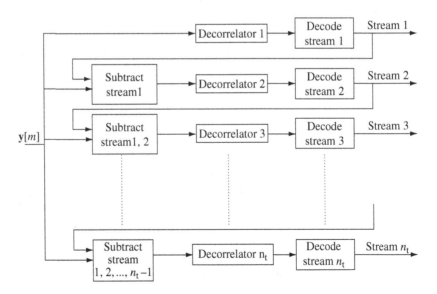

Figure 8.11 Decorrelator–SIC: A bank of decorrelators with successive cancellation of streams.

the corresponding projection operation (denoted by $\tilde{\mathbf{Q}}_k$) is onto a higher dimensional subspace (one orthogonal to that spanned by $\mathbf{h}_{k+1}, \ldots, \mathbf{h}_{n_t}$, as opposed to being orthogonal to the span of $\mathbf{h}_1, \ldots, \mathbf{h}_{k-1}, \mathbf{h}_{k+1}, \ldots, \mathbf{h}_{n_t}$). As in the calculation of the previous section, the SNR of the kth data stream is (cf. (8.44))

$$\frac{P_k \|\tilde{\mathbf{Q}}_k \mathbf{h}_k\|^2}{N_0}. \tag{8.55}$$

While we clearly expect this to be an improvement over the simple bank of decorrelators, let us again turn to the i.i.d. Rayleigh fading model to see this concretely. Analogous to the high SNR expansion of (8.52) in (8.53) for the simple decorrelator bank, with SIC and equal power allocation to each stream, we have

$$R_{\text{dec}-\text{sic}} \approx n_{\min} \log \frac{\text{SNR}}{n_t} + \mathbb{E}\left[\sum_{k=1}^{n_t} \log(\|\tilde{\mathbf{Q}}_k \mathbf{h}_k\|^2)\right]. \tag{8.56}$$

Similar to our analysis of the basic decorrelator bank, we can argue that $\|\tilde{\mathbf{Q}}_k \mathbf{h}_k\|^2 \sim \chi^2_{2(n_r - n_t + k)}$ with probability 1 (cf. Exercise 8.13), thus arriving at

$$\mathbb{E}\left[\log(\|\tilde{\mathbf{Q}}_k \mathbf{h}_k\|^2)\right] = \mathbb{E}[\log \chi^2_{2(n_r - n_t + k)}]. \tag{8.57}$$

Comparing this rate at high SNR with both the simple decorrelator bank and the capacity of the channel (cf. (8.53) and (8.18)), we observe the following

- The first-order term in the high SNR expansion is the same as that in the rate of the decorrelator bank and in the capacity: successive cancellation does not provide additional degrees of freedom.
- Moving to the next (constant) term, we see the performance boost in using the decorrelator–SIC over the simple decorrelator bank: the improved constant term is now *equal* to that in the capacity expansion. This boost in performance can be viewed as a *power gain*: by *decoding and subtracting* instead of *linear nulling*, the effective SNR at each stage is improved.

8.3.3 Linear MMSE receiver

Limitation of the decorrelator

We have seen the performance of the basic decorrelator bank and the decorrelator–SIC. At high SNR, for i.i.d. Rayleigh fading, the basic decorrelator bank achieves the full degrees of freedom in the channel. With SIC even the constant term in the high SNR capacity expansion is achieved. What about low SNR? The performance of the decorrelator bank (both with and without the modification of successive cancellation) as compared to the capacity of the MIMO channel is plotted in Figure 8.12.

Figure 8.12 Performance of the decorrelator bank, with and without successive cancellation at low SNR. Here $n_t = n_r = 8$.

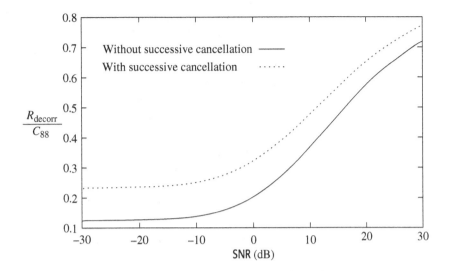

The main observation is that while the decorrelator bank performs well at high SNR, it is really far away from the capacity at low SNR. What is going on here?

To get more insight, let us plot the performance of a bank of matched filters, the kth filter being matched to the spatial signature $\mathbf{h}_k$ of transmit antenna k. From Figure 8.13 we see that the performance of the bank of matched filters is far superior to the decorrelator bank at low SNR (although far inferior at high SNR).

Derivation of the MMSE receiver

The decorrelator was motivated by the fact that it completely nulls out inter-stream interference; in fact it maximizes the output SNR among all linear

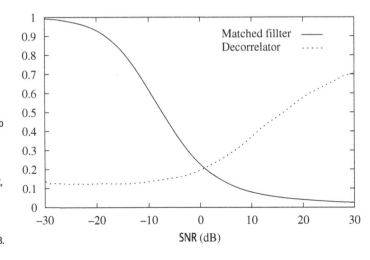

Figure 8.13 Performance (ratio of the rate to the capacity) of the matched filter bank as compared to that of the decorrelator bank. At low SNR, the matched filter is superior. The opposite is true for the decorrelator. The channel is i.i.d. Rayleigh with $n_t = n_r = 8$.

receivers that completely null out the interference. On the other hand, matched filtering (maximal ratio combining) is the optimal strategy for SIMO channels without any inter-stream interference. We called this *receive beamforming* in Example 1 in Section 7.2.1. Thus, we see a tradeoff between completely eliminating inter-stream interference (without any regard to how much energy of the stream of interest is lost in this process) and preserving as much energy content of the stream of interest as possible (at the cost of possibly facing high inter-stream interference). The decorrelator and the matched filter operate at two extreme ends of this tradeoff. At high SNR, the inter-stream interference is dominant over the additive Gaussian noise and the decorrelator performs well. On the other hand, at low SNR the inter-stream interference is not as much of an issue and receive beamforming (matched filter) is the superior strategy. In fact, the bank of matched filters achieves capacity at low SNR (Exercise 8.20).

We can ask for a linear receiver that *optimally* trades off fighting inter-stream interference and the background Gaussian noise, i.e., the receiver that maximizes the output signal-to-interference-plus-noise ratio (SINR) for any value of SNR. Such a receiver looks like the decorrelator when the inter-stream interference is large (i.e., when SNR is large) and like the matched filter when the interference is small (i.e., when SNR is small) (Figure 8.14). This can be thought of as the natural generalization of receive beamforming to the case when there is interference as well as noise.

To formulate this tradeoff precisely, let us first look at the following generic vector channel:

$$\mathbf{y} = \mathbf{h}x + \mathbf{z}, \tag{8.58}$$

where $\mathbf{z}$ is complex circular symmetric colored noise with an invertible covariance matrix $\mathbf{K}_z$, $\mathbf{h}$ is a deterministic vector and x is the unknown scalar symbol

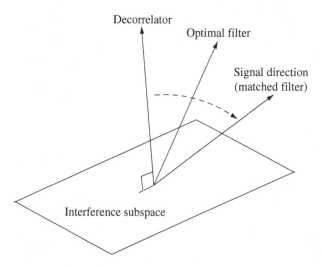

Figure 8.14 The optimal filter goes from being the decorrelator at high SNR to being the matched filter at low SNR.

to be estimated. $\mathbf{z}$ and x are assumed to be uncorrelated. We would like to choose a filter with maximum output SNR. If the noise is white, we know that it is optimal to project $\mathbf{y}$ onto the direction along $\mathbf{h}$. This observation suggests a natural strategy for the colored noise situation: first whiten the noise, and then follow the strategy used with white additive noise. That is, we first pass $\mathbf{y}$ through the invertible[4] linear transformation $\mathbf{K}_z^{-\frac{1}{2}}$ such that the noise $\tilde{\mathbf{z}} := \mathbf{K}_z^{-\frac{1}{2}}\mathbf{z}$ becomes white:

$$\mathbf{K}_z^{-\frac{1}{2}}\mathbf{y} = \mathbf{K}_z^{-\frac{1}{2}}\mathbf{h}x + \tilde{\mathbf{z}}. \tag{8.59}$$

Next, we project the output in the direction of $\mathbf{K}_z^{-\frac{1}{2}}\mathbf{h}$ to get an effective scalar channel

$$(\mathbf{K}_z^{-\frac{1}{2}}\mathbf{h})^*\mathbf{K}_z^{-\frac{1}{2}}\mathbf{y} = \mathbf{h}^*\mathbf{K}_z^{-1}\mathbf{y} = \mathbf{h}^*\mathbf{K}_z^{-1}\mathbf{h}x + \mathbf{h}^*\mathbf{K}_z^{-1}\mathbf{z}. \tag{8.60}$$

Thus the linear receiver in (8.60), represented by the vector

$$\boxed{\mathbf{v}_{\mathrm{mmse}} := \mathbf{K}_z^{-1}\mathbf{h},} \tag{8.61}$$

maximizes the SNR. It can also be shown that this receiver, with an appropriate scaling, minimizes the mean square error in estimating x (see Exercise 8.18), and hence it is also called the linear MMSE (minimum mean squared error) receiver. The corresponding SINR achieved is

$$\sigma_x^2\mathbf{h}^*\mathbf{K}_z^{-1}\mathbf{h}. \tag{8.62}$$

We can now upgrade the receiver structure in Section 8.3.1 by replacing the decorrelator for each stream by the linear MMSE receiver. Again, let us first consider the case where the channel $\mathbf{H}$ is fixed. The effective channel for the kth stream is

$$\mathbf{y}[m] = \mathbf{h}_k x_k[m] + \mathbf{z}_k[m], \tag{8.63}$$

where $\mathbf{z}_k$ represents the noise plus interference faced by data stream k:

$$\mathbf{z}_k[m] := \sum_{i\neq k}\mathbf{h}_i x_i[m] + \mathbf{w}[m]. \tag{8.64}$$

---

[4] $\mathbf{K}_z$ is an invertible covariance matrix and so it can be written as $\mathbf{U}\Lambda\mathbf{U}^*$ for rotation matrix $\mathbf{U}$ and diagonal matrix Λ with positive diagonal elements. Now $\mathbf{K}_z^{\frac{1}{2}}$ is defined as $\mathbf{U}\Lambda^{\frac{1}{2}}\mathbf{U}^*$, with $\Lambda^{\frac{1}{2}}$ defined as a diagonal matrix with diagonal elements equal to the square root of the diagonal elements of Λ.

With power P_i associated with the data stream i, we can explicitly calculate the covariance of $\mathbf{z}_k$

$$\mathbf{K}_{z_k} := N_0\mathbf{I}_{n_r} + \sum_{i \neq k}^{n_t} P_i\mathbf{h}_i\mathbf{h}_i^*, \tag{8.65}$$

and also note that the covariance is invertible. Substituting this expression for the covariance matrix into (8.61) and (8.62), we see that the linear receiver in the kth stage is given by

$$\left(N_0\mathbf{I}_{n_r} + \sum_{i \neq k}^{n_t} P_i\mathbf{h}_i\mathbf{h}_i^*\right)^{-1}\mathbf{h}_k, \tag{8.66}$$

and the corresponding output SINR is

$$P_k\mathbf{h}_k^*\left(N_0\mathbf{I}_{n_r} + \sum_{i \neq k}^{n_t} P_i\mathbf{h}_i\mathbf{h}_i^*\right)^{-1}\mathbf{h}_k. \tag{8.67}$$

Performance

We motivated the design of the linear MMSE receiver as something in between the decorrelator and receiver beamforming. Let us now see this explicitly. At very low SNR (i.e., $P_1, \ldots, P_{n_t}$ are very small compared to N_0) we see that

$$\mathbf{K}_{z_k} \approx N_0\mathbf{I}_{n_r}, \tag{8.68}$$

and the linear MMSE receiver in (8.66) reduces to the matched filter. On the other hand, at high SNR, the $\mathbf{K}_{z_k}^{-\frac{1}{2}}$ operation reduces to the projection of $\mathbf{y}$ onto the subspace orthogonal to that spanned by $\mathbf{h}_1, \ldots, \mathbf{h}_{k-1}, \mathbf{h}_{k+1}, \ldots, \mathbf{h}_{n_t}$ and the linear MMSE receiver reduces to the decorrelator.

Assuming the use of capacity-achieving codes for each stream, the maximum data rate that stream k can reliably carry is

$$C_k = \log\left(1 + P_k\mathbf{h}_k^*\mathbf{K}_{z_k}^{-1}\mathbf{h}_k\right). \tag{8.69}$$

As usual, the analysis directly carries over to the time-varying fading scenario, with data rate of the kth stream being

$$\bar{C}_k = \mathbb{E}[\log(1 + P_k\mathbf{h}_k^*\mathbf{K}_{z_k}^{-1}\mathbf{h}_k)], \tag{8.70}$$

where the average is over the stationary distribution of $\mathbf{H}$.

The performance of a bank of MMSE filters with equal power allocation over an i.i.d. Rayleigh fading channel is plotted in Figure 8.15. We see that the MMSE receiver performs strictly better than both the decorrelator and the matched filter over the entire range of SNRs.

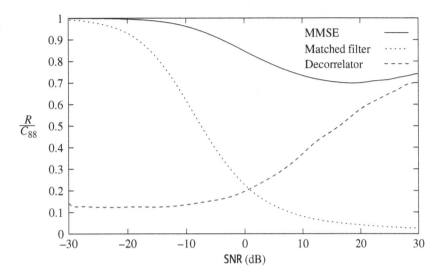

Figure 8.15 Performance (the ratio of rate to the capacity) of a basic bank of MMSE receivers as compared to the matched filter bank and to the decorrelator bank. MMSE performs better than both, over the entire range of SNR. The channel is i.i.d. Rayleigh with $n_t = n_r = 8$.

MMSE-SIC

Analogous to what we did in Section 8.3.2 for the decorrelator, we can now upgrade the basic bank of linear MMSE receivers by allowing successive cancellation of streams as well, as depicted in Figure 8.16. What is the performance improvement in using the MMSE–SIC receiver? Figure 8.17 plots the performance as compared to the capacity of the channel (with $n_t = n_r = 8$) for i.i.d. Rayleigh fading. We observe a startling fact: the bank of linear MMSE receivers with successive cancellation and equal power allocation *achieves* the capacity of the i.i.d. Rayleigh fading channel.

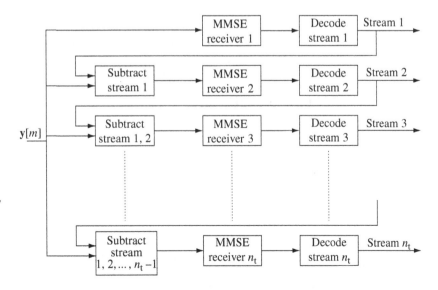

Figure 8.16 MMSE–SIC: a bank of linear MMSE receivers, each estimating one of the parallel data streams, with streams successively cancelled from the received vector at each stage.

Figure 8.17 The MMSE–SIC receiver achieves the capacity of the MIMO channel when fading is i.i.d. Rayleigh.

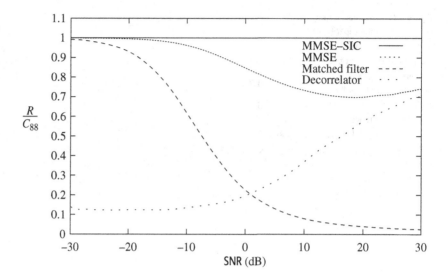

In fact, the MMSE–SIC receiver is optimal in a much stronger sense: it achieves the best possible sum rate (8.2) of the transceiver architecture in Section 8.1 for any given **H**. That is, if the MMSE–SIC receiver is used for demodulating the streams and the SINR and rate for stream k are SINR_k and $\log(1 + \mathsf{SINR}_k)$ respectively, then the rates sum up to

$$\sum_{k=1}^{n_t} \log(1 + \mathsf{SINR}_k) = \log \det \left(\mathbf{I}_{n_r} + \frac{1}{N_0} \mathbf{H} \mathbf{K}_x \mathbf{H}^* \right), \qquad (8.71)$$

which is the best possible sum rate. While this result can be verified directly by matrix manipulations (Exercise 8.22), the following section gives a deeper explanation in terms of the underlying information theory (the background of which is covered in Appendix B). Understanding at this level will be very useful as we adapt the MMSE–SIC architecture to the analysis of the uplink with multiple antennas in Chapter 10.

8.3.4 Information theoretic optimality*

MMSE is information lossless

As a key step to understanding why the MMSE–SIC receiver is optimal, let us go back to the generic vector channel with additive colored noise (8.58):

$$\mathbf{y} = \mathbf{h}x + \mathbf{z}, \qquad (8.72)$$

---

* This section can be skipped on a first reading. It requires knowledge of material in Appendix B and is not essential for understanding the rest of the book, except for the analysis of the MIMO uplink in Chapter 10.

but now with the further assumption that x and $\mathbf{z}$ are Gaussian. In this case, it can be seen that the linear MMSE filter ($\mathbf{v}_{\text{mmse}} := \mathbf{K}_z^{-1}\mathbf{h}$, cf. (8.61)) not only maximizes the SNR, but also provides a *sufficient statistic* to detect x, i.e., it is information lossless. Thus,

$$I(x; \mathbf{y}) = I(x; \mathbf{v}_{\text{mmse}}^*\mathbf{y}). \tag{8.73}$$

The justification for this step is carried out in Exercise 8.19.

A time-invariant channel

Consider again the MIMO channel with a time-invariant channel matrix $\mathbf{H}$:

$$\mathbf{y}[m] = \mathbf{H}\mathbf{x}[m] + \mathbf{w}[m].$$

We choose the input $\mathbf{x}$ to be $\mathcal{CN}(0, \text{diag}\{P_1, \ldots, P_{n_t}\})$. We can rewrite the mutual information between the input and the output as

$$\begin{aligned} I(\mathbf{x}; \mathbf{y}) &= I(x_1, x_2, \ldots, x_{n_t}; \mathbf{y}) \\ &= I(x_1; \mathbf{y}) + I(x_2; \mathbf{y}|x_1) + \cdots + I(x_{n_t}; \mathbf{y}|x_1, \ldots, x_{n_t-1}), \end{aligned} \tag{8.74}$$

where the last equality is a consequence of the chain rule of mutual information (see (B.18) in Appendix B). Let us look at the kth term in the chain rule expansion: $I(x_k; \mathbf{y}|x_1, \ldots, x_{k-1})$. Conditional on $x_1, \ldots, x_{k-1}$, we can subtract their effect from the output and obtain

$$\mathbf{y}' := \mathbf{y} - \sum_{i=1}^{k-1} \mathbf{h}_i x_i = \mathbf{h}_k x_k + \sum_{i>k} \mathbf{h}_i x_i + \mathbf{w}.$$

Thus,

$$I(x_k; \mathbf{y}|x_1, \ldots, x_{k-1}) = I(x_k; \mathbf{y}') = I(x_k; \mathbf{v}_{\text{mmse}}^*\mathbf{y}'), \tag{8.75}$$

where $\mathbf{v}_{\text{mmse}}$ is the MMSE filter for estimating x_k from $\mathbf{y}'$ and the last equality follows directly from the fact that the MMSE receiver is information-lossless. Hence, the rate achieved in kth stage of the MMSE–SIC receiver is precisely $I(x_k; \mathbf{y}|x_1, \ldots, x_{k-1})$, and the total rate achieved by this receiver is precisely the overall mutual information between the input $\mathbf{x}$ and the output $\mathbf{y}$ of the MIMO channel.

We now see why the MMSE filter is special: its scalar output preserves the information in the received vector about x_k. This property does not hold for other filters such as the decorrelator or the matched filter.

In the special case of a MISO channel with a scalar output

$$y[m] = \sum_{k=1}^{n_t} h_k x_k[m] + w[m], \tag{8.76}$$

the MMSE receiver at the kth stage is reduced to simple scalar multiplication followed by decoding; thus it is equivalent to decoding x_k while treating signals from antennas $k+1, k+2, \ldots, n_t$ as Gaussian interference. If we interpret (8.76) as an uplink channel with n_t users, the MMSE–SIC receiver thus reduces to the SIC receiver introduced in Section 6.1. Here we see another explanation why the SIC receiver is optimal in the sense of achieving the sum rate $I(x_1, x_2, \ldots, x_K; y)$ of the K-user uplink channel: it "implements" the chain rule of mutual information.

Fading channel

Now consider communicating using the transceiver architecture in Figure 8.1 but with the MMSE–SIC receiver on a time-varying fading MIMO channel with receiver CSI. If $\mathbf{Q} = \mathbf{I}_{n_t}$, the MMSE–SIC receiver allows reliable communication at a sum of the rates of the data streams equal to the mutual information of the channel under inputs of the form

$$\mathcal{CN}(0, \operatorname{diag}\{P_1, \ldots, P_{n_t}\}). \tag{8.77}$$

In the case of i.i.d. Rayleigh fading, the optimal input is precisely $\mathcal{CN}(0, \mathbf{I}_{n_t})$, and so the MMSE–SIC receiver achieves the capacity as well.

More generally, we have seen that if a MIMO channel, viewed in the angular domain, can be modeled by a matrix $\mathbf{H}$ having zero mean, uncorrelated entries, then the optimal input distribution is always of the form in (8.77) (cf. Section 8.2.1 and Exercise 8.3). Independent data streams decoded using the MMSE–SIC receiver still achieve the capacity of such MIMO channels, but the data streams are now transmitted over the transmit angular windows (instead of directly on the antennas themselves). This means that the transceiver architecture of Figure 8.1 with $\mathbf{Q} = \mathbf{U}_t$ and the MMSE-SIC receiver, achieves the capacity of the fast fading MIMO channel.

Discussion 8.1 Connections with CDMA multiuser detection and ISI equalization

Consider the situation where independent data streams are sent out from each antenna (cf. (8.42)). Here the received vector is a combination of the streams arriving in different receive spatial signatures, with stream k having a receive spatial signature of $\mathbf{h}_k$. If we make the analogy between space and bandwidth, then (8.42) serves as a model for the uplink of a CDMA system: the streams are replaced by the users (since the users cannot cooperate, the independence between them is justified naturally) and $\mathbf{h}_k$ now represents the *received* signature sequence of user k. The number of receive antennas is replaced by the number of chips in the CDMA signal. The base-station has access to the received signal and decodes the information simultaneously communicated by the different users. The base-station could use a bank of linear filters with or without successive cancellation. The study of the receiver design at the base-station, its complexities and performance, is called *multiuser detection*. The progress of multiuser detection is well chronicled in [131].

Another connection can be drawn to point-to-point communication over frequency-selective channels. In our study of the OFDM approach to communicating over frequency-selective channels in Section 3.4.4, we expressed the effect of the ISI in a matrix form (see (3.139)). This representation suggests the following interpretation: communicating over a block length of N_c on the L-tap time-invariant frequency-selective channel (see (3.129)) is equivalent to communicating over an $N_c \times N_c$ MIMO channel. The equivalent MIMO channel **H** is related to the taps of the frequency-selective channel, with the ℓth tap denoted by h_ℓ (for $\ell \geq L$, the tap $h_\ell = 0$), is

$$
H_{ij} = \begin{cases} h_{i-j} & \text{for } i \geq j, \\ 0 & \text{otherwise.} \end{cases} \tag{8.78}
$$

Due to the nature of the frequency-selective channel, previously transmitted symbols act as interference to the current symbol. The study of appropriate techniques to recover the transmit symbols in a frequency-selective channel is part of classical communication theory under the rubric of *equalization*. In our analogy, the transmitted symbols at different times in the frequency-selective channel correspond to the ones sent over the transmit antennas. Thus, there is a natural analogy between equalization for frequency-selective channels and transceiver design for MIMO channels (Table 8.1).

Table 8.1 Analogies between ISI equalization and MIMO communication techniques. We have covered all of these except the last one, which will be discussed in Chapter 10.

ISI equalization	MIMO communication
OFDM	SVD
Linear zero-forcing equalizer	Decorrelator/interference nuller
Linear MMSE equalizer	Linear MMSE receiver
Decision feedback equalizer (DFE)	Successive interference cancellation (SIC)
ISI precoding	Costa precoding

8.4 Slow fading MIMO channel

We now turn our attention to the slow fading MIMO channel,

$$\mathbf{y}[m] = \mathbf{Hx}[m] + \mathbf{w}[m], \tag{8.79}$$

where $\mathbf{H}$ is fixed over time but random. The receiver is aware of the channel realization but the transmitter only has access to its statistical characterization. As usual, there is a total transmit power constraint P. Suppose we want to communicate at a target rate R bits/s/Hz. If the transmitter were aware of the channel realization, then we could use the transceiver architecture in Figure 8.1 with an appropriate allocation of rates to the data streams to achieve reliable communication as long as

$$\log \det \left(\mathbf{I}_{n_r} + \frac{1}{N_0} \mathbf{H} \mathbf{K}_x \mathbf{H}^* \right) > R, \tag{8.80}$$

where the total transmit power constraint implies a condition on the covariance matrix: $\mathrm{Tr}[\mathbf{K}_x] \leq P$. However, remarkably, information theory guarantees the existence of a *channel-state independent* coding scheme that achieves reliable communication whenever the condition in (8.80) is met. Such a code is *universal*, in the sense that it achieves reliable communication on every MIMO channel satisfying (8.80). This is similar to the universality of the code achieving the outage performance on the slow fading parallel channel (cf. Section 5.4.4). When the MIMO channel does not satisfy the condition in (8.80), then we are in outage. We can choose the transmit strategy (parameterized by the covariance) to minimize the probability of the outage event:

$$p_{\mathrm{out}}^{\mathrm{mimo}}(R) = \min_{\mathbf{K}_x : \mathrm{Tr}[\mathbf{K}_x] \leq P} \mathbb{P} \left\{ \log \det \left(\mathbf{I}_{n_r} + \frac{1}{N_0} \mathbf{H} \mathbf{K}_x \mathbf{H}^* \right) < R \right\}. \tag{8.81}$$

Section 8.5 describes a transceiver architecture which achieves this outage performance.

The solution to this optimization problem depends, of course, on the statistics of channel $\mathbf{H}$. For example, if $\mathbf{H}$ is deterministic, the optimal solution is to perform a singular value decomposition of $\mathbf{H}$ and waterfill over the eigenmodes. When $\mathbf{H}$ is random, then one cannot tailor the covariance matrix to one particular channel realization but should instead seek a covariance matrix that works well statistically over the ensemble of the channel realizations.

It is instructive to compare the outage optimization problem (8.81) with that of computing the fast fading capacity with receiver CSI (cf. (8.10)). If we think of

$$f(\mathbf{K}_x, \mathbf{H}) := \log \det \left(\mathbf{I}_{n_r} + \frac{1}{N_0} \mathbf{H} \mathbf{K}_x \mathbf{H}^* \right), \tag{8.82}$$

as the rate of information flow over the channel $\mathbf{H}$ when using a coding strategy parameterized by the covariance matrix $\mathbf{K}_x$, then the fast fading capacity is

$$C = \max_{\mathbf{K}_x : \mathrm{Tr}[\mathbf{K}_x] \leq P} \mathbb{E}_\mathbf{H}[f(\mathbf{K}_x, \mathbf{H})], \qquad (8.83)$$

while the outage probability is

$$p_{\mathrm{out}}(R) = \min_{\mathbf{K}_x : \mathrm{Tr}[\mathbf{K}_x] \leq P} \mathbb{P}\{f(\mathbf{K}_x, \mathbf{H}) < R\}. \qquad (8.84)$$

In the fast fading scenario, one codes over the fades through time and the relevant performance metric is the long-term average rate of information flow that is permissible through the channel. In the slow fading scenario, one is only provided with a single realization of the channel and the objective is to minimize the probability that the rate of information flow falls below the target rate. Thus, the former is concerned with maximizing the expected value of the random variable $f(\mathbf{K}_x, \mathbf{H})$ and the latter with minimizing the tail probability that the same random variable is less than the target rate. While maximizing the expected value typically helps to reduce this tail probability, in general there is no one-to-one correspondence between these two quantities: the tail probability depends on higher-order moments such as the variance.

We can consider the i.i.d. Rayleigh fading model to get more insight into the nature of the optimizing covariance matrix. The optimal covariance matrix over the fast fading i.i.d. Rayleigh MIMO channel is $\mathbf{K}_x^* = P/n_t \cdot \mathbf{I}_{n_t}$. This covariance matrix transmits isotropically (in all directions), and thus one would expect that it is also good in terms of reducing the variance of the information rate $f(\mathbf{K}_x, \mathbf{H})$ and, indirectly, the tail probability. Indeed, we have seen (cf. Section 5.4.3 and Exercise 5.16) that this is the optimal covariance in terms of outage performance for the MISO channel, i.e., $n_r = 1$, at high SNR. In general, [119] conjectures that this is the optimal covariance matrix for the i.i.d. Rayleigh slow fading MIMO channel at high SNR. Hence, the resulting outage probability

$$p_{\mathrm{out}}^{\mathrm{iid}}(R) = \mathbb{P}\left\{\log\det\left(\mathbf{I}_{n_r} + \frac{\mathsf{SNR}}{n_t}\mathbf{HH}^*\right) < R\right\}, \qquad (8.85)$$

is often taken as a good upper bound to the actual outage probability at high SNR.

More generally, the conjecture is that it is optimal to restrict to a subset of the antennas and then transmit isotropically among the antennas used. The number of antennas used depends on the SNR level: the lower the SNR level relative to the target rate, the smaller the number of antennas used. In particular, at very low SNR relative to the target rate, it is optimal to use just one transmit antenna. We have already seen the validity of this conjecture in the context of a single receive antenna (cf. Section 5.4.3) and we are considering a natural extension to the MIMO situation. However, at typical outage probability levels, the SNR is high relative to the target rate and it is expected that using all the antennas is a good strategy.

High SNR

What outage performance can we expect at high SNR? First, we see that the MIMO channel provides increased diversity. We know that with $n_r = 1$ (the MISO channel) and i.i.d. Rayleigh fading, we get a diversity gain equal to n_t. On the other hand, we also know that with $n_t = 1$ (the SIMO channel) and i.i.d. Rayleigh fading, the diversity gain is equal to n_r. In the i.i.d. Rayleigh fading MIMO channel, we can achieve a diversity gain of $n_t \cdot n_r$, which is the number of independent random variables in the channel. A simple repetition scheme of using one transmit antenna at a time to send the same symbol x successively on the different n_t antennas over n_t consecutive symbol periods, yields an equivalent scalar channel

$$\tilde{y} = \sum_{i=1}^{n_r} \sum_{j=1}^{n_t} |h_{ij}|^2 x + w, \qquad (8.86)$$

whose outage probability decays like $1/\mathsf{SNR}^{n_t n_r}$. Exercise 8.23 shows the unsurprising fact that the outage probability of the i.i.d. Rayleigh fading MIMO channel decays no faster than this.

Thus, a MIMO channel yields a diversity gain of exactly $n_t \cdot n_r$. The corresponding ϵ-outage capacity of the MIMO channel benefits from both the diversity gain and the spatial degrees of freedom. We will explore the high SNR characterization of the combined effect of these two gains in Chapter 9.

8.5 D-BLAST: an outage-optimal architecture

We have mentioned that information theory guarantees the existence of coding schemes (parameterized by the covariance matrix) that ensure reliable communication at rate R on every MIMO channel that satisfies the condition (8.80). In this section, we will derive a transceiver architecture that achieves the outage performance. We begin with considering the performance of the V-BLAST architecture in Figure 8.1 on the slow fading MIMO channel.

8.5.1 Suboptimality of V-BLAST

Consider the V-BLAST architecture in Figure 8.1 with the MMSE–SIC receiver structure (cf. Figure 8.16) that we have shown to achieve the capacity of the *fast* fading MIMO channel. This architecture has two main features:

- Independently coded data streams are multiplexed in an appropriate coordinate system $\mathbf{Q}$ and transmitted over the antenna array. Stream k is allocated an appropriate power P_k and an appropriate rate R_k.
- A bank of linear MMSE receivers, in conjunction with successive cancellation, is used to demodulate the streams (the MMSE–SIC receiver).

The MMSE–SIC receiver demodulates the stream from transmit antenna 1 using an MMSE filter, decodes the data, subtracts its contribution from the stream, and proceeds to stream 2, and so on. Each stream is thought of as a layer.

Can this same architecture achieve the optimal outage performance in the *slow fading* channel? In general, the answer is no. To see this concretely, consider the i.i.d. Rayleigh fading model. Here the data streams are transmitted over separate antennas and it is easy to see that each stream has a diversity of *at most* n_r: if the channel gains from the kth transmit antenna to *all* the n_r receive antennas are in deep fade, then the data in the kth stream will be lost. On the other hand, the MIMO channel itself provides a diversity gain of $n_t \cdot n_r$. Thus, V-BLAST does not exploit the full diversity available in the channel and therefore cannot be outage-optimal. The basic problem is that there is no coding across the streams so that if the channel gains from one transmit antenna are bad, the corresponding stream will be decoded in error.

We have said that, under the i.i.d. Rayleigh fading model, the diversity of each stream in V-BLAST is *at most* n_r. The diversity would be *exactly* n_r if it were the only stream being transmitted; with simultaneous transmission of streams, the diversity could be even lower depending on the receiver. This can be seen most clearly if we replace the bank of linear MMSE receivers in V-BLAST with a bank of decorrelators and consider the case $n_t \leq n_r$. In this case, the distribution of the output SNR at each stage can be explicitly computed; this was actually done in Section 8.3.2:

$$\mathsf{SINR}_k \sim \frac{P_k}{N_0} \cdot \chi^2_{2[n_r - (n_t - k)]}. \tag{8.87}$$

The diversity of the kth stream is therefore $n_r - (n_t - k)$. Since $n_t - k$ is the number of uncancelled interfering streams at the kth stage, one can interpret this as saying that the loss of diversity due to interference is precisely the number of interferers needed to be nulled out. The first stream has the worst diversity of $n_r - n_t + 1$; this is also the bottleneck of the whole system because the correct decoding of subsequent streams depends on the correct decoding and cancellation of this stream. In the case of a square system, the first stream has a diversity of only 1, i.e., no diversity gain. We have already seen this result in the special case of the 2×2 example in Section 3.3.3. Though this analysis is for the decorrelator, it turns out that the MMSE receiver yields exactly the same diversity gain (see Exercise 8.24). Using joint ML detection of the streams, on the other hand, a diversity of n_r can be recovered (as in the 2×2 example in Section 3.3.3). However, this is still far away from the full diversity gain $n_t n_r$ of the channel.

There are proposed improvements to the basic V-BLAST architecture. For instance, adapting the cancellation order as a function of the channel, and allocating different rates to different streams depending on their position in the cancellation order. However, none of these variations can provide a diversity larger than n_r, as long as we are sending independently coded streams on the transmit antennas.

A more careful look

Here is a more precise understanding of why V-BLAST is suboptimal, which will suggest how V-BLAST can be improved. For a given $\mathbf{H}$, (8.71) yields the following decomposition:

$$\log \det(\mathbf{I}_{n_r} + \frac{1}{N_0} \mathbf{H} \mathbf{K}_x \mathbf{H}^*) = \sum_{k=1}^{n_t} \log(1 + \mathsf{SINR}_k). \tag{8.88}$$

SINR_k is the output signal-to-interference-plus-noise ratio of the MMSE demodulator at the kth stage of the cancellation. The output SINRs are random since they are a function of the channel matrix $\mathbf{H}$. Suppose we have a target rate of R and we split this into rates $R_1, \ldots, R_{n_t}$ allocated to the individual streams. Suppose that the transmit strategy (parameterized by the covariance matrix $\mathbf{K}_x := \mathbf{Q} \, \mathrm{diag}\{P_1, \ldots, P_{n_t}\} \mathbf{Q}^*$, cf. (8.3)) is chosen to be the one that yields the outage probability in (8.81). Now we note that the *channel* is in outage if

$$\log \det(\mathbf{I}_{n_r} + \frac{1}{N_0} \mathbf{H} \mathbf{K}_x \mathbf{H}^*) < R, \tag{8.89}$$

or equivalently,

$$\sum_{k=1}^{n_t} \log(1 + \mathsf{SINR}_k) < \sum_{k=1}^{n_t} R_k. \tag{8.90}$$

However, V-BLAST is in outage as long as the random SINR in *any* stream cannot support the rate allocated to that stream, i.e.,

$$\log(1 + \mathsf{SINR}_k) < R_k, \tag{8.91}$$

for any k. Clearly, this can occur even when the channel is not in outage. Hence, V-BLAST cannot be universal and is not outage-optimal. This problem did not appear in the fast fading channel because there we code over the temporal channel *variations* and thus kth stream gets a *deterministic* rate of

$$\mathbb{E}[\log(1 + \mathsf{SINR}_k)] \text{ bits/s/Hz.} \qquad (8.92)$$

8.5.2 Coding across transmit antennas: D-BLAST

Significant improvement of V-BLAST has to come from coding *across* the transmit antennas. How do we improve the architecture to allow that? To see more clearly how to proceed, one can draw an analogy between V-BLAST and the *parallel fading channel*. In V-BLAST, the kth stream effectively sees a channel with a (random) signal-to-noise ratio SINR_k; this can therefore be viewed as a parallel channel with n_t sub-channels. In V-BLAST, there is no coding across these sub-channels: outage therefore occurs whenever one of these sub-channels is in a deep fade and cannot support the rate of the stream using that sub-channel. On the other hand, by coding across the sub-channels, we can average over the randomness of the individual sub-channels and get better outage performance. From our discussion on parallel channels in Section 5.4.4, we know reliable communication is possible whenever

$$\sum_{k=1}^{n_t} \log(1 + \mathsf{SINR}_k) > R. \qquad (8.93)$$

From the decomposition (8.88), we see that this is exactly the no-outage condition of the original MIMO channel as well. Therefore, it seems that universal codes for the parallel channel can be transformed directly into universal codes for the original MIMO channel.

However, there is a problem here. To obtain the second sub-channel (with SINR_2), we are assuming that the first stream is already decoded and its received signal is cancelled off. However, to code across the sub-channels, the two streams should be *jointly* decoded. There seems to be a chicken-and-egg problem: without decoding the first stream, one cannot cancel its signal and get the second stream in the first place. The key idea to solve this problem is to *stagger* multiple codewords so that each codeword spans multiple transmit antennas but the symbols sent simultaneously by the different transmit antennas belong to different codewords.

Let us go through a simple example with two transmit antennas (Figure 8.18). The ith codeword $\mathbf{x}^{(i)}$ is made up of two blocks, $\mathbf{x}_A^{(i)}$ and $\mathbf{x}_B^{(i)}$, each of length N. In the first N symbol times, the first antenna sends nothing. The second antenna sends $\mathbf{x}_A^{(1)}$, block A of the first codeword. The receiver performs maximal ratio combining of the signals at the receive antennas to estimate $\mathbf{x}_A^{(1)}$; this yields an equivalent sub-channel with signal-to-noise ratio SINR_2, since the other antenna is sending nothing.

In the second N symbol times, the first antenna sends $\mathbf{x}_B^{(1)}$ (block B of the first codeword), while the second antenna sends $\mathbf{x}_A^{(2)}$ (block A of the second

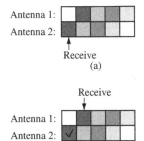

Antenna 1:
Antenna 2:

Receive
(a)

Receive

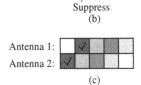

Antenna 1:
Antenna 2:

Suppress
(b)

Antenna 1:
Antenna 2:

(c)

Cancel

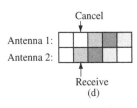

Antenna 1:
Antenna 2:

Receive
(d)

Figure 8.18 How D-BLAST works. (a) A soft estimate of block A of the first codeword (layer) obtained without interference. (b) A soft MMSE estimate of block B is obtained by suppressing the interference from antenna 2. (c) The soft estimates are combined to decode the first codeword (layer). (d) The first codeword is cancelled and the process restarts with the second codeword (layer).

codeword). The receiver does a linear MMSE estimation of $\mathbf{x}_B^{(1)}$, treating $\mathbf{x}_A^{(2)}$ as interference to be suppressed. This produces an equivalent sub-channel of signal-to-noise ratio SINR_1. Thus, the first codeword as a whole now sees the parallel channel described above (Exercise 8.25), and, assuming the use of a universal parallel channel code, can be decoded provided that

$$\log(1 + \mathsf{SINR}_1) + \log(1 + \mathsf{SINR}_2) > R. \tag{8.94}$$

Once codeword 1 is decoded, $\mathbf{x}_B^{(1)}$ can be subtracted off the received signal in the second N symbol times. This leaves $\mathbf{x}_A^{(2)}$ alone in the received signal, and the process can be repeated. Exercise 8.26 generalizes this architecture to arbitrary number of transmit antennas.

In V-BLAST, each coded stream, or layer, extends horizontally in the space-time grid and is placed vertically above another. In the improved architecture above, each layer is striped *diagonally* across the space-time grid (Figure 8.18). This architecture is naturally called Diagonal BLAST, or D-BLAST for short.

The D-BLAST scheme suffers from a rate loss because in the initialization phase some of the antennas have to be kept silent. For example, in the two transmit antenna architecture illustrated in Figure 8.18 (with $N = 1$ and 5 layers), two symbols are set to zero among the total of 10; this reduces the rate by a factor of 4/5 (Exercise 8.27 generalizes this calculation). So for a finite number of layers, D-BLAST does not achieve the outage performance of the MIMO channel. As the number of layers grows, the rate loss gets amortized and the MIMO outage performance is approached. In practice, D-BLAST suffers from *error propagation*: if one layer is decoded incorrectly, all subsequent layers are affected. This puts a practical limit on the number of layers which can be transmitted consecutively before re-initialization. In this case, the rate loss due to initialization and termination is not negligible.

8.5.3 Discussion

D-BLAST should really be viewed as a transceiver architecture rather than a space-time code: through signal processing and interleaving of the codewords across the antennas, it converts the MIMO channel into a parallel channel. As such, it allows the leveraging of any good parallel-channel code for the MIMO channel. In particular, a universal code for the parallel channel, when used in conjunction with D-BLAST, is a universal space-time code for the MIMO channel.

It is interesting to compare D-BLAST with the Alamouti scheme discussed in Chapters 3 and 5. The Alamouti scheme can also be considered as a transceiver architecture: it converts the 2×1 MISO slow fading channel into a SISO slow fading channel. Any universal code for the SISO channel when used in conjunction with the Alamouti scheme yields a universal code for the MISO channel. Compared to D-BLAST, the signal processing is much simpler, and there are no rate loss or error propagation issues. On the other hand, D-BLAST works for an arbitrary number of transmit and receive antennas. As we have seen, the Alamouti scheme does not generalize to arbitrary numbers of transmit antennas (cf. Exercise 3.16). Further, we will see in Chapter 9 that the Alamouti scheme is strictly suboptimal in MIMO channels with multiple transmit *and* receive antennas. This is because, unlike D-BLAST, the Alamouti scheme does not exploit all the available degrees of freedom in the channel.

Chapter 8 The main plot

Capacity of fast fading MIMO channels

In a rich scattering environment with receiver CSI, the capacity is approximately

- $\min(n_t, n_r) \log \mathsf{SNR}$ at high SNR: a gain in spatial degrees of freedom;
- $n_r \mathsf{SNR} \log_2 e$ at low SNR: a receive beamforming gain.

With $n_t = n_r = n$, the capacity is approximately $nc^*(\mathsf{SNR})$ for all SNR. Here $c^*(\mathsf{SNR})$ is a constant.

Transceiver architectures

- **With full CSI** convert the MIMO channel into $n_{\min}$ parallel channels by an appropriate change in the basis of the transmit and receive signals. This transceiver structure is motivated by the *singular value decomposition* of any linear transformation: a composition of a rotation, a scaling operation, followed by another rotation.
- **With receiver CSI** send independent data streams over each of the transmit antennas. The ML receiver decodes the streams jointly and achieves capacity. This is called the V-BLAST architecture.

Reciever structures

- **Simple receiver structure** Decode the data streams *separately*. Three main structures:
  - *matched filter*: use the receive antenna array to beamform to the receive spatial signature of the stream. Performance close to capacity at low SNR.
  - *decorrelator*: project the received signal onto the subspace orthogonal to the receive spatial signatures of all the other streams.
    - to be able to do the projection operation, need $n_r \geq n_t$.
    - For $n_r \geq n_t$, the decorrelator bank captures all the spatial degrees of freedom at high SNR.
  - *MMSE*: linear receiver that optimally trades off capturing the energy of the data stream of interest and nulling the inter-stream interference. Close to optimal performance at both low and high SNR.

- **Successive cancellation** Decode the data streams sequentially, using the results of the decoding operation to cancel the effect of the decoded data streams on the received signal.

Bank of linear MMSE receivers with successive cancellation achieves the capacity of the fast fading MIMO channel at all SNR.

Outage performance of slow fading MIMO channels
The i.i.d. Rayleigh slow fading MIMO channel provides a diversity gain equal to the product of n_t and n_r. Since the V-BLAST architecture does not code across the transmit antennas, it can achieve a diversity gain of at most n_r. Staggered interleaving of the streams of V-BLAST among the transmit antennas achieves the full outage performance of the MIMO channel. This is the D-BLAST architecture.

8.6 Bibliographical notes

The interest in MIMO communications was sparked by the capacity analysis of Foschini [40], Foschini and Gans [41] and Telatar [119]. Foschini and Gans focused on analyzing the outage capacity of the slow fading MIMO channel, while Telatar studied the capacity of fixed MIMO channels under optimal waterfilling, ergodic capacity of fast fading channels under receiver CSI, as well as outage capacity of slow fading channels. The D-BLAST architecture was introduced by Foschini [40], while the V-BLAST architecture was considered by Wolniansky *et al.* [147] in the context of point-to-point MIMO communication.

The study of the linear receivers, decorrelator and MMSE, was initiated in the context of multiuser detection of CDMA signals. The research in multiuser detection is very well exposed and summarized in a book by Verdú [131], who was the pioneer in this field. In particular, decorrelators were introduced by Lupas and Verdú [77] and the MMSE receiver by Madhow and Honig [79]. The optimality of the MMSE receiver in conjunction with successive cancellation was shown by Varanasi and Guess [129].

The literature on random matrices as applied in communication theory is summarized by Tulino and Verdú [127]. The key result on the asymptotic distribution of the singular values of large random matrices used in this chapter is by Marčenko and Pastur [78].

8.7 Exercises

Exercise 8.1 (reciprocity) Show that the capacity of a time-invariant MIMO channel with n_t transmit, n_r receive antennas and channel matrix $\mathbf{H}$ is the same as that of the channel with n_r transmit, n_t receive antennas, matrix $\mathbf{H}^*$, and same total power constraint.

Exercise 8.2 Consider coding over a block of length N on the data streams in the transceiver architecture in Figure 8.1 to communicate over the time-invariant MIMO channel in (8.1).

1. Fix $\epsilon > 0$ and consider the ellipsoid $E^{(\epsilon)}$ defined as

$$\{\mathbf{a} : \mathbf{a}^* (\mathbf{HK}_x\mathbf{H}^* \otimes \mathbf{I}_N + N_0\mathbf{I}_{n_r N})^{-1}\mathbf{a} \leq N(n_r + \epsilon)\}. \tag{8.95}$$

Here we have denoted the tensor product (or Kronecker product) between matrices by the symbol $\otimes$. In particular, $\mathbf{HK}_x\mathbf{H}^* \otimes \mathbf{I}_N$ is a $n_r N \times n_r N$ block diagonal matrix:

$$\mathbf{HK}_x\mathbf{H}^* \otimes \mathbf{I}_N = \begin{bmatrix} \mathbf{HK}_x\mathbf{H}^* & & & 0 \\ & \mathbf{HK}_x\mathbf{H}^* & & \\ & & \ddots & \\ 0 & & & \mathbf{HK}_x\mathbf{H} \end{bmatrix}.$$

Show that, for every ϵ, the received vector $\mathbf{y}^N$ (of length $n_r N$) lies with high probability in the ellipsoid $E^{(\epsilon)}$, i.e.,

$$\mathbb{P}\{\mathbf{y}^N \in E^{(\epsilon)}\} \to 1, \quad \text{as } N \to \infty. \tag{8.96}$$

2. Show that the volume of the ellipsoid $E^{(0)}$ is equal to

$$\det(N_0\mathbf{I}_{n_r} + \mathbf{HK}_x\mathbf{H}^*)^N \tag{8.97}$$

times the volume of a $2n_r N$-dimensional real sphere with radius $\sqrt{n_r N}$. This justifies the expression in (8.4).

3. Show that the noise vector $\mathbf{w}^N$ of length $n_r N$ satisfies

$$\mathbb{P}\{\|\mathbf{w}^N\|^2 \leq N_0 N(n_r + \epsilon)\} \to 1, \quad \text{as } N \to \infty. \tag{8.98}$$

Thus $\mathbf{w}^N$ lives, with high probability, in a $2n_r N$-dimensional real sphere of radius $\sqrt{N_0 n_r N}$. Compare the volume of this sphere to the volume of the ellipsoid in (8.97) to justify the expression in (8.5).

Exercise 8.3 [130, 126] Consider the angular representation $\mathbf{H}^a$ of the MIMO channel $\mathbf{H}$. We statistically model the entries of $\mathbf{H}^a$ as zero mean and jointly uncorrelated.

1. Starting with the expression in (8.10) for the capacity of the MIMO channel with receiver CSI and substituting $\mathbf{H} := \mathbf{U}_r\mathbf{H}^a\mathbf{U}_t^*$, show that

$$C = \max_{\mathbf{K}_x:Tr\mathbf{K}_x \leq P} \mathbb{E}\left[\log \det\left(\mathbf{I}_{n_r} + \frac{1}{N_0}\mathbf{H}^a\mathbf{U}_t^*\mathbf{K}_x\mathbf{U}_t\mathbf{H}^{a*}\right)\right]. \tag{8.99}$$

2. Show that we can restrict the input covariance in (8.99), without changing the maximal value, to be of the following special structure:

$$\mathbf{K}_x = \mathbf{U}_t\mathbf{\Lambda}\mathbf{U}_t^*, \tag{8.100}$$

where Λ is a diagonal matrix with non-negative entries that sum to P. *Hint*: We can always consider a covariance matrix of the form

$$\mathbf{K}_x = \mathbf{U}_t \tilde{\mathbf{K}}_x \mathbf{U}_t^*, \tag{8.101}$$

with $\tilde{\mathbf{K}}$ also a covariance matrix satisfying the total power constraint. To show that $\tilde{\mathbf{K}}$ can be restricted to be diagonal, consider the following decomposition:

$$\tilde{\mathbf{K}}_x = \Lambda + \mathbf{K}_{\text{off}}, \tag{8.102}$$

where Λ is a diagonal matrix and $\mathbf{K}_{\text{off}}$ has zero diagonal elements (and thus contains all the off-diagonal elements of $\tilde{\mathbf{K}}$). Validate the following sequence of inequalities:

$$\mathbb{E}\left[\log\det\left(\mathbf{I}_{n_r} + \frac{1}{N_0}\mathbf{H}^{\mathbf{a}}\mathbf{K}_{\text{off}}\mathbf{H}^{\mathbf{a}*}\right)\right] \le \log\mathbb{E}\left[\det\left(\mathbf{I}_{n_r} + \frac{1}{N_0}\mathbf{H}^{\mathbf{a}}\mathbf{K}_{\text{off}}\mathbf{H}^{\mathbf{a}*}\right)\right], \tag{8.103}$$

$$= \log\det\left(\mathbb{E}\left[\mathbf{I}_{n_r} + \frac{1}{N_0}\mathbf{H}^{\mathbf{a}}\mathbf{K}_{\text{off}}\mathbf{H}^{\mathbf{a}*}\right]\right), \tag{8.104}$$

$$= 0. \tag{8.105}$$

You can use Jensen's inequality (cf. Exercise B.2) to get (8.103). In (8.104), we have denoted $\mathbb{E}[\mathbf{X}]$ to be the matrix with (i, j)th entry equal to $\mathbb{E}[\mathbf{X}_{ij}]$. Now use the property that the elements of $\mathbf{H}^{\mathbf{a}}$ are uncorrelated in arriving at (8.104) and (8.105). Finally, using the decomposition in (8.102), conclude (8.100), i.e., it suffices to consider covariance matrices $\tilde{\mathbf{K}}_x$ in (8.101) to be diagonal.

Exercise 8.4 [119] Consider i.i.d. Rayleigh fading, i.e., the entries of $\mathbf{H}$ are i.i.d. $\mathcal{CN}(0, 1)$, and the capacity of the fast fading channel with only receiver CSI (cf. (8.10)).

1. For i.i.d. Rayleigh fading, show that the distribution of $\mathbf{H}$ and that of $\mathbf{HU}$ are identical for every unitary matrix $\mathbf{U}$. This is a generalization of the rotational invariance of an i.i.d. complex Gaussian vector (cf. (A.22) in Appendix A).
2. Show directly for i.i.d. Rayleigh fading that the input covariance $\mathbf{K}_x$ in (8.10) can be restricted to be diagonal (without resorting to Exercise 8.3(2)).
3. Show further that among the diagonal matrices, the optimal input covariance is $(P/n_t)\mathbf{I}_{n_t}$. *Hint*: Show that the map

$$(p_1, \ldots, p_{n_t}) \mapsto \mathbb{E}\left[\log\det\left(\mathbf{I}_{n_r} + \frac{1}{N_0}\mathbf{H}\,\text{diag}\{p_1, \ldots, p_{n_t}\}\mathbf{H}^*\right)\right] \tag{8.106}$$

is jointly concave. Further show that the map is *symmetric*, i.e., reordering the argument $p_1, \ldots, p_{n_t}$ does not change the value. Observe that a jointly concave, symmetric function is maximized, subject to a sum constraint, exactly when all the function arguments are the same and conclude the desired result.

Exercise 8.5 Consider the uplink of the cellular systems studied in Chapter 4: the narrowband system (GSM), the wideband CDMA system (IS-95), and the wideband OFDM system (Flash-OFDM).

1. Suppose that the base-station is equipped with an array of multiple receive antennas. Discuss the impact of the receive antenna array on the performance of the three systems discussed in Chapter 4. Which system benefits the most?

2. Now consider the MIMO uplink, i.e., the mobiles are also equipped with multiple (transmit) antennas. Discuss the impact on the performance of the three cellular systems. Which system benefits the most?

Exercise 8.6 In Exercise 8.3 we have seen that the optimal input covariance is of the form $\mathbf{K}_x = \mathbf{U}_t \mathbf{\Lambda} \mathbf{U}_t^*$ with $\mathbf{\Lambda}$ a diagonal matrix. In this exercise, we study the situations under which $\mathbf{\Lambda}$ is $(P/n_t)\mathbf{I}_{n_t}$, making the optimal input covariance also equal to $(P/n_t)\mathbf{I}_{n_t}$. (We have already seen one instance when this is true in Exercise 8.4: the i.i.d. Rayleigh fading scenario.) Intuitively, this should be true whenever there is *complete* symmetry among the transmit angular windows. This heuristic idea is made precise below.

1. The symmetry condition formally corresponds to the following assumption on the columns (there are n_t of them, one for each of the transmit angular windows) of the angular representation $\mathbf{H}^a = \mathbf{U}_t \mathbf{H} \mathbf{U}_r^*$: the n_t column vectors are independent and, further, the vectors are identically distributed. We do not specify the joint distribution of the entries within any of the columns other than requiring that they have zero mean. With this symmetry condition, show that the optimal input covariance is $(P/n_t)\mathbf{I}_{n_t}$.

2. Using the previous part, or directly, strengthen the result of Exercise 8.4 by showing that the optimal input covariance is $(P/n_t)\mathbf{I}_{n_t}$ whenever

$$\mathbf{H} := [\mathbf{h}_1 \ldots \mathbf{h}_{n_t}], \qquad (8.107)$$

where $\mathbf{h}_1, \ldots, \mathbf{h}_{n_t}$ are i.i.d. $\mathcal{CN}(0, \mathbf{K}_h)$ for some covariance matrix $\mathbf{K}_h$.

Exercise 8.7 In Section 8.2.2, we showed that with receiver CSI the capacity of the i.i.d. Rayleigh fading $n \times n$ MIMO channel grows linearly with n at all SNR. In this reading exercise, we consider other statistical channel models which also lead to a linear increase of the capacity with n.

1. The capacity of the MIMO channel with i.i.d. entries (not necessarily Rayleigh), grows linearly with n. This result is derived in [21].

2. In [21], the authors also consider a *correlated* channel model: the entries of the MIMO channel are jointly complex Gaussian (with invertible covariance matrix). The authors show that the capacity still increases linearly with the number of antennas.

3. In [75], the authors show a linear increase in capacity for a MIMO channel with the number of i.i.d. entries growing quadratically in n (i.e., the number of i.i.d. entries is proportional to n^2, with the rest of the entries equal to zero).

Exercise 8.8 Consider the *block fading* MIMO channel (an extension of the single antenna model in Exercise 5.28):

$$\mathbf{y}[m + nT_c] = \mathbf{H}[n]\mathbf{x}[m + nT_c] + \mathbf{w}[m + nT_c], \qquad m = 1, \ldots, T_c, n \geq 1, \quad (8.108)$$

where T_c is the coherence time of the channel (measured in terms of the number of samples). The channel variations across the blocks $\mathbf{H}[n]$ are i.i.d. Rayleigh. A pilot based communication scheme transmits known symbols for k time samples at the beginning of each coherence time interval: each known symbol is sent over a different transmit antenna, with the other transmit antennas silent. At high SNR, the k pilot symbols allow the receiver to *partially* estimate the channel: over the nth block, k of the n_t columns of $\mathbf{H}[n]$ are estimated with a high degree of accuracy. This allows us to reliably communicate on the $k \times n_r$ MIMO channel *with* receiver CSI.

1. Argue that the rate of reliable communication using this scheme at high SNR is approximately at least

$$\left(\frac{T_c - k}{T_c}\right) \min(k, n_r) \log \mathsf{SNR} \text{ bits/s/Hz.} \tag{8.109}$$

   Hint: An information theory fact says that replacing the effect of channel uncertainty as Gaussian noise (with the same covariance) can only make the reliable communication rate smaller.

2. Show that the optimal training time (and the corresponding number of transmit antennas to use) is

$$k^* := \min\left(n_t, n_r, \frac{T_c}{2}\right). \tag{8.110}$$

   Substituting this in (8.109) we see that the number of spatial degrees of freedom using the pilot scheme is equal to

$$\left(\frac{T_c - k^*}{T_c}\right) k^*. \tag{8.111}$$

3. A reading exercise is to study [155], which shows that the *capacity* of the non-coherent block fading channel at high SNR also has the same number of spatial degrees freedom as in (8.111).

Exercise 8.9 Consider the time-invariant frequency-selective MIMO channel:

$$\mathbf{y}[m] = \sum_{\ell=0}^{L-1} \mathbf{H}_\ell \mathbf{x}[m - \ell] + \mathbf{w}[m]. \tag{8.112}$$

Construct an appropriate OFDM transmission and reception scheme to transform the original channel to the following *parallel* MIMO channel:

$$\tilde{\mathbf{y}}_n = \tilde{\mathbf{H}}_n \tilde{\mathbf{x}}_n + \tilde{\mathbf{w}}_n, \qquad n = 0, \dots, N_c - 1. \tag{8.113}$$

Here N_c is the number of OFDM tones. Identify $\tilde{\mathbf{H}}_n$, $n = 0, \dots, N_c - 1$ in terms of $\mathbf{H}_\ell$, $\ell = 0, \dots, L - 1$.

Exercise 8.10 Consider a fixed physical environment and a corresponding flat fading MIMO channel. Now suppose we double the transmit power constraint and the bandwidth. Argue that the capacity of the MIMO channel with receiver CSI exactly doubles. This scaling is consistent with that in the single antenna AWGN channel.

Exercise 8.11 Consider (8.42) where independent data streams $\{x_i[m]\}$ are transmitted on the transmit antennas ($i = 1 \dots, n_t$):

$$\mathbf{y}[m] = \sum_{i=1}^{n_t} \mathbf{h}_i x_i[m] + \mathbf{w}[m]. \tag{8.114}$$

Assume $n_t \le n_r$.

1. We would like to study the operation of the decorrelator in some detail here. So we make the assumption that $\mathbf{h}_i$ is not a linear combination of the other vectors $\mathbf{h}_1, \ldots, \mathbf{h}_{i-1}, \mathbf{h}_{i+1}, \ldots, \mathbf{h}_{n_t}$ for every $i = 1, \ldots, n_t$. Denoting $\mathbf{H} = [\mathbf{h}_1 \cdots \mathbf{h}_{n_t}]$, show that this assumption is equivalent to the fact that $\mathbf{H}^*\mathbf{H}$ is invertible.

2. Consider the following operation on the received vector in (8.114):

$$\hat{\mathbf{x}}[m] := (\mathbf{H}^*\mathbf{H})^{-1}\mathbf{H}^*\mathbf{y}[m] \tag{8.115}$$

$$= \mathbf{x}[m] + (\mathbf{H}^*\mathbf{H})^{-1}\mathbf{H}^*\mathbf{w}[m]. \tag{8.116}$$

Thus $\hat{x}_i[m] = x_i[m] + \tilde{w}_i[m]$ where $\tilde{\mathbf{w}}[m] := (\mathbf{H}^*\mathbf{H})^{-1}\mathbf{H}^*\mathbf{w}[m]$ is colored Gaussian noise. This means that the ith data stream sees no interference from any of the other streams in the received signal $\hat{x}_i[m]$. Show that $\hat{x}_i[m]$ must be the output of the decorrelator (up to a scaling constant) for the ith data stream and hence conclude the validity of (8.47). This property, and many more, about the decorrelator can be learnt from Chapter 5 of [131]. The special case of $n_t = n_r = 2$ can be verified by explicit calculations.

Exercise 8.12 Suppose $\mathbf{H}$ (with $n_t < n_r$) has i.i.d. $\mathcal{CN}(0, 1)$ entries and denote $\mathbf{h}_1, \ldots, \mathbf{h}_{n_t}$ as the columns of $\mathbf{H}$. Show that the probability that the columns are linearly dependent is zero. Hence, conclude that the probability that the rank of $\mathbf{H}$ is strictly smaller than n_t is zero.

Exercise 8.13 Suppose $\mathbf{H}$ (with $n_t < n_r$) has i.i.d. $\mathcal{CN}(0, 1)$ entries and denote the columns of $\mathbf{H}$ as $\mathbf{h}_1, \ldots, \mathbf{h}_{n_t}$. Use the result of Exercise 8.12 to show that the dimension of the subspace spanned by the vectors $\mathbf{h}_1, \ldots, \mathbf{h}_{k-1}, \mathbf{h}_{k+1}, \ldots, \mathbf{h}_{n_t}$ is $n_t - 1$ with probability 1. Hence conclude that the dimension of the subspace V_k, orthogonal to this one, has dimension $n_r - n_t + 1$ with probability 1.

Exercise 8.14 Consider the Rayleigh fading $n \times n$ MIMO channel $\mathbf{H}$ with i.i.d. $\mathcal{CN}(0, 1)$ entries. In the text we have discussed a random matrix result about the convergence of the empirical distribution of the singular values of $\mathbf{H}/\sqrt{n}$. It turns out that the condition number of $\mathbf{H}/\sqrt{n}$ converges to a deterministic limiting distribution. This means that the random matrix $\mathbf{H}$ is well-conditioned. The corresponding limiting density is given by

$$f(x) := \frac{4}{x^3}\,e^{-2/x^2}. \tag{8.117}$$

A reading exercise is to study the derivation of this result proved in Theorem 7.2 of [32].

Exercise 8.15 Consider communicating over the time-invariant $n_t \times n_r$ MIMO channel:

$$\mathbf{y}[m] = \mathbf{H}\mathbf{x}[m] + \mathbf{w}[m]. \tag{8.118}$$

The information bits are encoded using, say, a capacity-achieving Gaussian code such as an LDPC code. The encoded bits are then modulated into the transmit signal $\mathbf{x}[m]$; typically the components of the transmit vector belong to a regular constellation such as QAM. The receiver, typically, operates in two stages. The first stage is demodulation: at each time, soft information (a posteriori probabilities of the bits that modulated the transmit vector) about the transmitted QAM symbol is evaluated. In the second stage, the soft information about the bits is fed to a channel decoder.

In this reading exercise, we study the first stage of the receiver. At time m, the demodulation problem is to find the QAM points composing the vector $\mathbf{x}[m]$ such that $\|\mathbf{y}[m] - \mathbf{H}\mathbf{x}[m]\|^2$ is the smallest possible. This problem is one of classical "least squares", but with the domain restricted to a finite set of points. When the modulation is QAM, the domain is a finite subset of the integer lattice. Integer least squares is known to be a computationally hard problem and several heuristic solutions, with less complexity, have been proposed. One among them is the *sphere decoding* algorithm. A reading exercise is to use [133] to understand the algorithm and an analysis of the average (over the fading channel) complexity of decoding.

Exercise 8.16 In Section 8.2.2 we showed two facts for the i.i.d. Rayleigh fading channel: (i) for fixed n and at low SNR, the capacity of a 1 by n channel approaches that of an n by n channel; (ii) for fixed SNR but large n, the capacity of a 1 by n channel grows only logarithmically with n while that of an n by n channel grows linearly with n. Resolve the apparent paradox.

Exercise 8.17 Verify (8.26). This result is derived in [132].

Exercise 8.18 Consider the channel (8.58):

$$\mathbf{y} = \mathbf{h}x + \mathbf{z}, \tag{8.119}$$

where $\mathbf{z}$ is $\mathcal{CN}(0, \mathbf{K}_z)$, $\mathbf{h}$ is a (complex) deterministic vector and x is the zero mean unknown (complex) random variable to be estimated. The noise $\mathbf{z}$ and the data symbol x are assumed to be uncorrelated.

1. Consider the following estimate of x from $\mathbf{y}$ using the vector $\mathbf{c}$ (normalized so that $\|\mathbf{c}\| = 1$):

$$\hat{x} := a\mathbf{c}^*\mathbf{y} = a\mathbf{c}^*\mathbf{h}x + a\mathbf{c}^*\mathbf{z}. \tag{8.120}$$

Show that the constant a that minimizes the mean square error ($\mathbb{E}[|x - \hat{x}|^2]$) is equal to

$$\frac{\mathbb{E}[|x|^2]|\mathbf{c}^*\mathbf{h}|^2}{\mathbb{E}[|x|^2]|\mathbf{c}^*\mathbf{h}|^2 + \mathbf{c}^*\mathbf{K}_z\mathbf{c}} \frac{\mathbf{h}^*\mathbf{c}}{|\mathbf{h}^*\mathbf{c}|}. \tag{8.121}$$

2. Calculate the minimal mean square error (denoted by MMSE) of the linear estimate in (8.120) (by using the value of a in (8.121)). Show that

$$\frac{\mathbb{E}[|x|^2]}{\text{MMSE}} = 1 + \text{SNR} := 1 + \frac{\mathbb{E}[|x|^2]|\mathbf{c}^*\mathbf{h}|^2}{\mathbf{c}^*\mathbf{K}_z\mathbf{c}}. \tag{8.122}$$

3. Since we have shown that $\mathbf{c} = \mathbf{K}_z^{-1}\mathbf{h}$ maximizes the SNR (cf. (8.61)) among all linear estimators, conclude that this linear estimate (along with an appropriate choice of the scaling a, as in (8.121)), minimizes the mean square error in the linear estimation of x from (8.119).

Exercise 8.19 Consider detection on the generic vector channel with additive colored Gaussian noise (cf. (8.72)).

1. Show that the output of the linear MMSE receiver,

$$\mathbf{v}_{\text{mmse}}^* \mathbf{y}, \tag{8.123}$$

is a sufficient statistic to detect x from $\mathbf{y}$. This is a generalization of the scalar sufficient statistic extracted from the vector detection problem in Appendix A (cf. (A.55)).

2. From the previous part, we know that the random variables $\mathbf{y}$ and x are independent conditioned on $\mathbf{v}_{\text{mmse}}^* \mathbf{y}$. Use this to verify (8.73).

Exercise 8.20 We have seen in Figure 8.13 that, at low SNR, the bank of linear matched filter achieves capacity of the 8 by 8 i.i.d. Rayleigh fading channel, in the sense that the ratio of the total achievable rate to the capacity approaches 1. Show that this is true for general n_t and n_r.

Exercise 8.21 Consider the n by n i.i.d. flat Rayleigh fading channel. Show that the total achievable rate of the following receiver architectures scales linearly with n: (a) bank of linear decorrelators; (b) bank of matched filters; (c) bank of linear MMSE receivers. You can assume that independent information streams are coded and sent out of each of the transmit antennas and the power allocation across antennas is uniform. *Hint*: The calculation involving the linear MMSE receivers is tricky. You have to show that the linear MMSE receiver performance, asymptotically for large n, depends on the covariance matrix of the interference faced by each stream only through its empirical eigenvalue distribution, and then apply the large-n random matrix result used in Section 8.2.2. To show the first step, compute the mean and variance of the output SINR, conditional on the spatial signatures of the interfering streams. This calculation is done in [132, 123]

Exercise 8.22 Verify (8.71) by direct matrix manipulations.
Hint: You might find useful the following matrix inversion lemma (for invertible $\mathbf{A}$),

$$(\mathbf{A} + \mathbf{x}\mathbf{x}^*)^{-1} = \mathbf{A}^{-1} - \frac{\mathbf{A}^{-1}\mathbf{x}\mathbf{x}^*\mathbf{A}^{-1}}{1 + \mathbf{x}^*\mathbf{A}^{-1}\mathbf{x}}. \tag{8.124}$$

Exercise 8.23 Consider the outage probability of an i.i.d. Rayleigh MIMO channel (cf. (8.81)). Show that its decay rate in SNR (equal to P/N_0) is no faster than $n_t \cdot n_r$ by justifying each of the following steps.

$$p_{\text{out}}(R) \geq \mathbb{P}\{\log \det(\mathbf{I}_{n_r} + \mathsf{SNR}\,\mathbf{HH}^*) < R\} \tag{8.125}$$

$$\geq \mathbb{P}\{\mathsf{SNR}\,\text{Tr}[\mathbf{HH}^*] < R\} \tag{8.126}$$

$$\geq (\mathbb{P}\{\mathsf{SNR}\,|h_{11}|^2 < R\})^{n_t n_r} \tag{8.127}$$

$$= \left(1 - e^{-\frac{R}{\mathsf{SNR}}}\right)^{n_t n_r} \tag{8.128}$$

$$\approx \frac{R^{n_t n_r}}{\mathsf{SNR}^{n_t n_r}}. \tag{8.129}$$

Exercise 8.24 Calculate the maximum diversity gains for each of the streams in the V-BLAST architecture using the MMSE–SIC receiver. *Hint*: At high SNR, interference seen by each stream is very high and the SINR of the linear MMSE receiver is very close to that of the decorrelator in this regime.

Exercise 8.25 Consider communicating over a 2×2 MIMO channel using the D-BLAST architecture with $N = 1$ and equal power allocation $P_1 = P_2 = P$ for both the layers. In this exercise, we will derive some properties of the parallel channel (with $L = 2$ diversity branches) created by the MMSE–SIC operation. We denote the MIMO channel by $\mathbf{H} = [\mathbf{h}_1, \mathbf{h}_2]$ and the projections

$$\mathbf{h}_{1\|2} := \frac{\mathbf{h}_1^* \mathbf{h}_2}{\|\mathbf{h}_2\|^2} \, \mathbf{h}_2, \qquad \mathbf{h}_{1\perp 2} := \mathbf{h}_1 - \mathbf{h}_{1\|2}. \tag{8.130}$$

Let us denote the induced parallel channel as

$$y_\ell = g_\ell x_\ell + w_\ell, \qquad \ell = 1, 2. \tag{8.131}$$

1. Show that

$$|g_1|^2 = \|\mathbf{h}_{1\perp 2}\|^2 + \frac{\|\mathbf{h}_{1\|2}\|^2}{\mathsf{SNR}\|\mathbf{h}_2\|^2 + 1}, \qquad |g_2|^2 = \|\mathbf{h}_2\|^2, \tag{8.132}$$

where $\mathsf{SNR} = P/N_0$.

2. What is the marginal distribution of $|g_1|^2$ at high SNR? Are $|g_1|^2$ and $|g_2|^2$ positively correlated or negatively correlated?

3. What is the maximum diversity gain offered by this parallel channel?

4. Now suppose $|g_1|^2$ and $|g_2|^2$ in the parallel channel in (8.131) are independent, while still having the same marginal distribution as before. What is the maximum diversity gain offered by this parallel channel?

Exercise 8.26 Generalize the staggered stream structure (discussed in the context of a $2 \times n_r$ MIMO channel in Section 8.5) of the D-BLAST architecture to a MIMO channel with $n_t > 2$ transmit antennas.

Exercise 8.27 Consider a block length N D-BLAST architecture on a MIMO channel with n_t transmit antennas. Determine the rate loss due to the initialization phase as a function of N and n_t.

9 MIMO III: diversity–multiplexing tradeoff and universal space-time codes

In the previous chapter, we analyzed the performance benefits of MIMO communication and discussed architectures that are designed to reap those benefits. The focus was on the *fast fading* scenario. The story on *slow fading* MIMO channels is more complex. While the communication capability of a fast fading channel can be described by a single *number*, its capacity, that of a slow fading channel has to be described by the outage probability curve $p_{out}(\cdot)$, as a function of the target rate. This curve is in essence a tradeoff between the data rate and error probability. Moreover, in addition to the power and degree-of-freedom gains in the fast fading scenario, multiple antennas provide a diversity gain in the slow fading scenario as well. A clear characterization of the performance benefits of multiple antennas in slow fading channels and the design of good space-time coding schemes that reap those benefits are the subjects of this chapter.

The outage probability curve $p_{out}(\cdot)$ is the natural benchmark for evaluating the performance of space-time codes. However, it is difficult to characterize analytically the outage probability curves for MIMO channels. We develop an approximation that captures the dual benefits of MIMO communication in the high SNR regime: increased data rate (via an increase in the spatial degrees of freedom or, equivalently, the multiplexing gain) and increased reliability (via an increase in the diversity gain). The dual benefits are captured as a fundamental *tradeoff* between these two types of gains.[1] We use the optimal *diversity–multiplexing tradeoff* as a benchmark to compare the various space-time schemes discussed previously in the book. The tradeoff curve also suggests how optimal space-time coding schemes should look. A powerful idea for the design of tradeoff-optimal schemes is *universality*, which we discuss in the second part of the chapter.

We have studied an approach to space-time code design in Chapter 3. Codes designed using that approach have small error probabilities, averaged over

---

[1] The careful reader will note that we saw an inkling of the tension between these two types of gains in our study of the 2×2 MIMO Rayleigh fading channel in Chapter 3.

the distribution of the fading channel gains. The drawback of the approach is that the performance of the designed codes may be sensitive to the supposed fading distribution. This is problematic, since, as we mentioned in Chapter 2, accurate statistical modeling of wireless channels is difficult. The outage formulation, however, suggests a different approach. The operational interpretation of the outage performance is based on the existence of *universal codes*: codes that *simultaneously* achieve reliable communication over *every* MIMO channel that is not in outage. Such codes are robust from an engineering point of view: they achieve the best possible outage performance for every fading distribution. This result motivates a universal code design criterion: instead of using the pairwise error probability averaged over the fading distribution of the channel, we consider the *worst-case* pairwise error probability over all channels that are not in outage. Somewhat surprisingly, the universal code-design criterion is closely related to the product distance, which is obtained by averaging over the Rayleigh distribution. Thus, the product distance criterion, while seemingly tailored for the Rayleigh distribution, is actually more fundamental. Using universal code design ideas, we construct codes that achieve the optimal diversity–multiplexing tradeoff.

Throughout this chapter, the receiver is assumed to have perfect knowledge of the channel matrix while the transmitter has none.

9.1 Diversity–multiplexing tradeoff

In this section, we use the outage formulation to characterize the performance capability of slow fading MIMO channels in terms of a tradeoff between diversity and multiplexing gains. This tradeoff is then used as a unified framework to compare the various space-time coding schemes described in this book.

9.1.1 Formulation

When we analyzed the performance of communication schemes in the slow fading scenario in Chapters 3 and 5, the emphasis was on the *diversity gain*. In this light, a key measure of the performance capability of a slow fading channel is the *maximum diversity gain* that can be extracted from it. For example, a slow i.i.d. Rayleigh faded MIMO channel with n_t transmit and n_r receive antennas has a maximum diversity gain of $n_t \cdot n_r$: i.e., for a fixed target rate R, the outage probability $p_{out}(R)$ decays like $1/\text{SNR}^{n_t n_r}$ at high SNR.

On the other hand, we know from Chapter 7 that the key performance benefit of a *fast fading* MIMO channel is the spatial multiplexing capability it provides through the additional degrees of freedom. For example, the capacity of an i.i.d. Rayleigh fading channel scales like $n_{\min} \log \mathsf{SNR}$, where $n_{\min} := \min(n_t, n_r)$ is the number of spatial degrees of freedom in the channel. This fast fading (ergodic) capacity is achieved by averaging over the variation of the channel over time. In the slow fading scenario, no such averaging is possible and one cannot communicate at this rate reliably. Instead, the information rate allowed through the channel is a random variable fluctuating around the fast fading capacity. Nevertheless, one would still expect to be able to benefit from the increased degrees of freedom even in the slow fading scenario. Yet the maximum diversity gain provides no such indication; for example, both an $n_t \times n_r$ channel and an $n_t n_r \times 1$ channel have the same maximum diversity gain and yet one would expect the former to allow better spatial multiplexing than the latter. One needs something more than the maximum diversity gain to capture the spatial multiplexing benefit.

Observe that to achieve the maximum diversity gain, one needs to communicate at a *fixed* rate R, which becomes vanishingly small compared to the fast fading capacity at high SNR (which grows like $n_{\min} \log \mathsf{SNR}$). Thus, one is actually sacrificing all the spatial multiplexing benefit of the MIMO channel to maximize the reliability. To reclaim some of that benefit, one would instead want to communicate at a rate $R = r \log \mathsf{SNR}$, which is a *fraction* of the fast fading capacity. Thus, it makes sense to formulate the following *diversity–multiplexing tradeoff* for a slow fading channel.

A diversity gain $d^*(r)$ is achieved at multiplexing gain r if

$$R = r \log \mathsf{SNR} \tag{9.1}$$

and

$$p_{\text{out}}(R) \approx \mathsf{SNR}^{-d^*(r)}, \tag{9.2}$$

or more precisely,

$$\lim_{\mathsf{SNR} \to \infty} \frac{\log p_{\text{out}}(r \log \mathsf{SNR})}{\log \mathsf{SNR}} = -d^*(r). \tag{9.3}$$

The curve $d^*(\cdot)$ is the diversity–multiplexing tradeoff of the slow fading channel.

The above tradeoff characterizes the slow fading performance limit of the *channel*. Similarly, we can formulate a diversity–multiplexing tradeoff for any space-time coding *scheme*, with outage probabilities replaced by *error probabilities*.

> A space-time coding scheme is a family of codes, indexed by the signal-to-noise ratio SNR. It attains a multiplexing gain r and a diversity gain d if the data rate scales as
>
> $$R = r \log \mathsf{SNR} \tag{9.4}$$
>
> and the error probability scales as
>
> $$p_e \approx \mathsf{SNR}^{-d}, \tag{9.5}$$
>
> i.e.,
>
> $$\lim_{\mathsf{SNR} \to \infty} \frac{\log p_e}{\log \mathsf{SNR}} = -d. \tag{9.6}$$

The diversity–multiplexing tradeoff formulation may seem abstract at first sight. We will now go through a few examples to develop a more concrete feel. The tradeoff performance of specific coding schemes will be analyzed and we will see how they perform compared to each other and to the optimal diversity–multiplexing tradeoff of the channel. For concreteness, we use the i.i.d. Rayleigh fading model. In Section 9.2, we will describe a general approach to tradeoff-optimal space-time code based on universal coding ideas.

9.1.2 Scalar Rayleigh channel

PAM and QAM

Consider the scalar slow fading Rayleigh channel,

$$y[m] = hx[m] + w[m], \tag{9.7}$$

with the additive noise i.i.d. $\mathcal{CN}(0, 1)$ and the power constraint equal to SNR. Suppose h is $\mathcal{CN}(0, 1)$ and consider uncoded communication using PAM with a data rate of R bits/s/Hz. We have done the error probability analysis in Section 3.1.2 for $R = 1$; for general R, the analysis is similar. The average error probability is governed by the minimum distance between the PAM points. The constellation ranges from approximately $-\sqrt{\mathsf{SNR}}$ to $+\sqrt{\mathsf{SNR}}$, and since there are 2^R constellation points, the minimum distance is approximately

$$D_{\min} \approx \frac{\sqrt{\mathsf{SNR}}}{2^R}, \tag{9.8}$$

and the error probability at high SNR is approximately (cf. (3.28)),

$$p_e \approx \frac{1}{2}\left(1 - \sqrt{\frac{D_{min}^2}{4 + D_{min}^2}}\right) \approx \frac{1}{D_{min}^2} \approx \frac{2^{2R}}{\mathsf{SNR}}. \tag{9.9}$$

By setting the data rate $R = r \log \mathsf{SNR}$, we get

$$p_e \approx \frac{1}{\mathsf{SNR}^{1-2r}}, \tag{9.10}$$

yielding a diversity–multiplexing tradeoff of

$$\boxed{d_{pam}(r) = 1 - 2r, \qquad r \in \left[0, \frac{1}{2}\right].} \tag{9.11}$$

Note that in the approximate analysis of the error probability above, we focus on the scaling of the error probability with the SNR and the data rate but are somewhat careless with constant multipliers: they do not matter as far as the diversity–multiplexing tradeoff is concerned.

We can repeat the analysis for QAM with data rate R. There are now $2^{R/2}$ constellation points in each of the real and imaginary dimensions, and hence the minimum distance is approximately

$$D_{min} \approx \frac{\sqrt{\mathsf{SNR}}}{2^{R/2}}, \tag{9.12}$$

and the error probability at high SNR is approximately

$$p_e \approx \frac{2^R}{\mathsf{SNR}}, \tag{9.13}$$

yielding a diversity–multiplexing tradeoff of

$$\boxed{d_{qam}(r) = 1 - r, \qquad r \in [0, 1].} \tag{9.14}$$

The tradeoff curves are plotted in Figure 9.1.

Let us relate the two endpoints of a tradeoff curve to notions that we already know. The value $d_{max} := d(0)$ can be interpreted as the SNR exponent that describes how fast the error probability can be decreased with the SNR for a *fixed* data rate; this is the *classical diversity gain* of a scheme. It is 1 for both PAM and QAM. The decrease in error probability is due to an increase in D_{min}. This is illustrated in Figure 9.2.

In a dual way, the value r_{max} for which $d(r_{max}) = 0$ describes how fast the data rate can be increased with the SNR for a *fixed* error probability. This number can be interpreted as the number of (complex) degrees of freedom that are exploited by the scheme. It is 1 for QAM but only 1/2 for PAM.

Figure 9.1 Tradeoff curves for the single antenna slow fading Rayleigh channel.

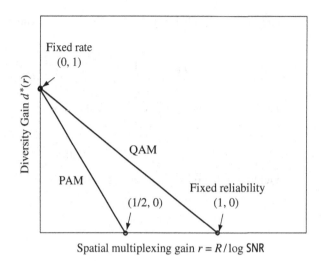

Figure 9.2 Increasing the SNR by 6 dB decreases the error probability by 1/4 for both PAM and QAM due to a doubling of the minimum distance.

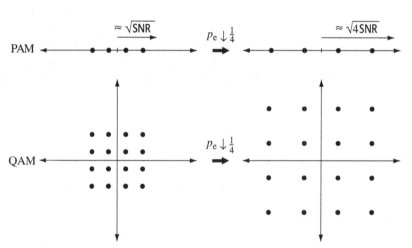

This is consistent with our observation in Section 3.1.3 that PAM uses only half the degrees of freedom of QAM. The increase in data rate is due to the packing of more constellation points for a given D_{min}. This is illustrated in Figure 9.3.

The two endpoints represent two extreme ways of using the increase in the resource (SNR): increasing the reliability for a fixed data rate, or increasing the data rate for a fixed reliability. More generally, we can simultaneously increase the data rate (positive multiplexing gain r) and increase the reliability (positive diversity gain $d > 0$) but there is a tradeoff between how much of each type of gain we can get. The diversity–multiplexing curve describes this tradeoff. Note that the classical diversity gain only describes the rate of decay of the error probability for a fixed data rate, but does not provide any information on how well a scheme exploits the available degrees of freedom. For example, PAM and QAM have the same classical diversity

Figure 9.3 Increasing the SNR by 6 dB increases the data rate for QAM by 2 bits/s/Hz but only increases the data rate of PAM by 1 bit/s/Hz.

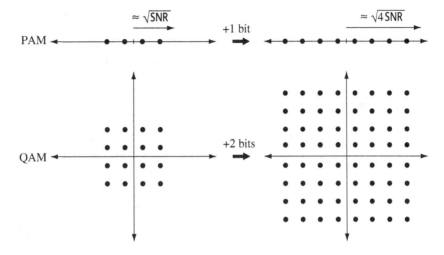

Figure 9.3 Increasing the SNR by 6 dB increases the data rate for QAM by 2 bits/s/Hz but only increases the data rate of PAM by 1 bit/s/Hz.

gain, even though clearly QAM is more efficient in exploiting the available degrees of freedom. The tradeoff curve, by treating error probability and data rate in a symmetrical manner, provides a more complete picture. We see that in terms of their tradeoff curves, QAM is indeed superior to PAM (see Figure 9.1).

Optimal tradeoff

So far, we have considered the tradeoff between diversity and multiplexing in the context of two specific schemes: uncoded PAM and QAM. What is the fundamental diversity–multiplexing tradeoff of the scalar channel itself? For the slow fading Rayleigh channel, the outage probability at a target data rate $R = r \log \mathsf{SNR}$ is

$$
\begin{aligned}
p_{\text{out}} &= \mathbb{P}\{\log(1 + |h|^2 \mathsf{SNR}) < r \log \mathsf{SNR}\} \\
&= \mathbb{P}\left\{|h|^2 < \frac{\mathsf{SNR}^r - 1}{\mathsf{SNR}}\right\} \\
&\approx \frac{1}{\mathsf{SNR}^{1-r}},
\end{aligned}
\tag{9.15}
$$

at high SNR. In the last step, we used the fact that for Rayleigh fading, $\mathbb{P}\{|h|^2 < \epsilon\} \approx \epsilon$ for small ϵ. Thus

$$
\boxed{d^*(r) = 1 - r, \qquad r \in [0, 1].}
\tag{9.16}
$$

Hence, the uncoded QAM scheme trades off diversity and multiplexing gains optimally.

The tradeoff between diversity and multiplexing gains can be viewed as a *coarser* way of capturing the fundamental tradeoff between error probability and data rate over a fading channel at high SNR. Even very simple, low-complexity schemes can trade off optimally in this coarser context (the uncoded QAM achieved the tradeoff for the Rayleigh slow fading channel). To achieve the *exact* tradeoff between outage probability and data rate, we need to code over long block lengths, at the expense of higher complexity.

9.1.3 Parallel Rayleigh channel

Consider the slow fading parallel channel with i.i.d. Rayleigh fading on each sub-channel:

$$y_\ell[m] = h_\ell x_\ell[m] + w_\ell[m], \qquad \ell = 1, \ldots, L. \tag{9.17}$$

Here, the w_ℓ are i.i.d. $\mathcal{CN}(0, 1)$ additive noise and the transmit power per sub-channel is constrained by SNR. We have seen that L Rayleigh faded sub-channels provide a (classical) diversity gain equal to L (cf. Section 3.2 and Section 5.4.4): this is an L-fold improvement over the basic single antenna slow fading channel. In the parlance we introduced in the previous section, this says that $d^*(0) = L$. How about the diversity gain at any positive multiplexing rate?

Suppose the target data rate is $R = r \log \mathsf{SNR}$ bits/s/Hz per sub-channel. The optimal diversity $d^*(r)$ can be calculated from the rate of decay of the outage probability with increasing SNR. For the i.i.d. Rayleigh fading parallel channel, the outage probability at rate per sub-channel $R = r \log \mathsf{SNR}$ is (cf. (5.83))

$$p_{\text{out}} = \mathbb{P}\left\{\sum_{\ell=1}^{L} \log(1 + |h_\ell|^2 \mathsf{SNR}) < Lr \log \mathsf{SNR}\right\}. \tag{9.18}$$

Outage typically occurs when each of the sub-channels cannot support the rate R (Exercise 9.1): so we can write

$$p_{\text{out}} \approx (\mathbb{P}\{\log(1 + |h_1|^2 \mathsf{SNR}) < r \log \mathsf{SNR}\})^L \approx \frac{1}{\mathsf{SNR}^{L(1-r)}}. \tag{9.19}$$

So, the optimal diversity–multiplexing tradeoff for the parallel channel with L diversity branches is

$$\boxed{d^*(r) = L(1 - r), \qquad r \in [0, 1],} \tag{9.20}$$

an L-fold gain over the scalar single antenna performance (cf. (9.16)) at *every* multiplexing gain r; this performance is illustrated in Figure 9.4.

One particular scheme is to transmit the same QAM symbol over the L sub-channels; the repetition converts the parallel channel into a scalar channel with squared amplitude $\sum_\ell |h_\ell|^2$, but with the rate reduced by a factor of $1/L$.

Figure 9.4 The diversity-
multiplexing tradeoff of the
i.i.d. Rayleigh fading parallel
channel with *L* sub-channels
together with that of the
repetition scheme.

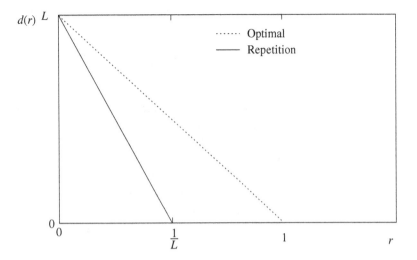

The diversity–multiplexing tradeoff achieved by this scheme can be computed to be

$$d_{\mathrm{rep}}(r) = L(1 - Lr), \qquad r \in \left[0, \frac{1}{L}\right], \tag{9.21}$$

(Exercise 9.2). The classical diversity gain $d_{\mathrm{rep}}(0)$ is L, the full diversity of the parallel channel, but the number of degrees of freedom per sub-channel is only $1/L$, due to the repetition.

9.1.4 MISO Rayleigh channel

Consider the n_{t} transmit and single receive antenna MISO channel with i.i.d. Rayleigh coefficients:

$$y[m] = \mathbf{h}^* \mathbf{x}[m] + w[m]. \tag{9.22}$$

As usual, the additive noise $w[m]$ is i.i.d. $\mathcal{CN}(0, 1)$ and there is an overall transmit power constraint of SNR. We have seen that the Rayleigh fading MISO channel with n_{t} transmit antennas provides the (classical) diversity gain of n_{t} (cf. Section 3.3.2 and Section 5.4.3). By how much is the diversity gain increased at a positive multiplexing rate of r?

We can answer this question by looking at the outage probability at target data rate $R = r \log \mathsf{SNR}$ bits/s/Hz:

$$p_{\mathrm{out}} = \mathbb{P}\left\{ \log\left(1 + \|\mathbf{h}\|^2 \frac{\mathsf{SNR}}{n_{\mathrm{t}}}\right) < r \log \mathsf{SNR} \right\}. \tag{9.23}$$

Now $\|\mathbf{h}\|^2$ is a χ^2 random variable with $2n_{\mathrm{t}}$ degrees of freedom and we have seen that $\mathbb{P}\{\|\mathbf{h}\|^2 < \epsilon\} \approx \epsilon^{n_{\mathrm{t}}}$ (cf. (3.44)). Thus, p_{out} decays as $\mathsf{SNR}^{-n_{\mathrm{t}}(1-r)}$ with increasing SNR and the optimal diversity–multiplexing tradeoff for the i.i.d. Rayleigh fading MISO channel is

$$d^*(r) = n_t(1-r), \qquad r \in [0,1].$$

(9.24)

Thus the MISO channel provides an n_t-fold increase in diversity at all multiplexing gains.

In the case of $n_t = 2$, we know that the Alamouti scheme converts the MISO channel into a scalar channel with the same outage behavior as the original MISO channel. Hence, if we use QAM symbols in conjunction with the Alamouti scheme, we achieve the diversity–multiplexing tradeoff of the MISO channel. In contrast, the repetition scheme that transmits the same QAM symbol from each of the two transmit antennas one at a time achieves a diversity–multiplexing tradeoff curve of

$$d_{\text{rep}}(r) = 2(1-2r), \qquad r \in \left[0, \frac{1}{2}\right].$$

(9.25)

The tradeoff curves of these schemes as well as that of the 2×1 MISO channel are shown in Figure 9.5.

9.1.5 2 × 2 MIMO Rayleigh channel

Four schemes revisited

In Section 3.3.3, we analyzed the (classical) diversity gains and degrees of freedom utilized by four schemes for the 2×2 i.i.d. Rayleigh fading

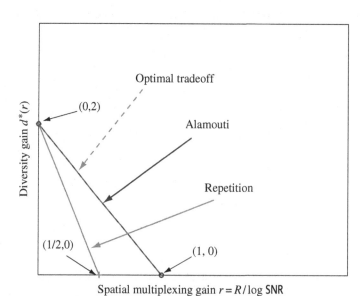

Figure 9.5 The diversity-multiplexing tradeoff of the 2 x 1 i.i.d. Rayleigh fading MISO channel along with those of two schemes.

Table 9.1 A summary of the performance of the four schemes for the 2 × 2 channel.

	Classical diversity gain	Degrees of freedom utilized	D–M tradeoff
Repetition	4	1/2	$4 - 8r, \ r \in [0, 1/2]$
Alamouti	4	1	$4 - 4r, \ r \in [0, 1]$
V-BLAST (ML)	2	2	$2 - r, \ r \in [0, 2]$
V-BLAST (nulling)	1	2	$1 - r/2, \ r \in [0, 2]$
Channel itself	4	2	$4 - 3r, \ r \in [0, 1]$ $2 - r, \ r \in [1, 2]$

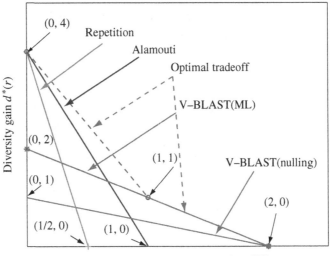

Figure 9.6 The diversity–multiplexing tradeoff of the 2 × 2 i.i.d. Rayleigh fading MIMO channel along with those of four schemes.

MIMO channel (with the results summarized in Summary 3.2). The diversity–multiplexing tradeoffs of these schemes when used in conjunction with uncoded QAM can be computed as well; they are summarized in Table 9.1 and plotted in Figure 9.6. The classical diversity gains and degrees of freedom utilized correspond to the endpoints of these curves.

The repetition, Alamouti and V-BLAST with nulling schemes all convert the MIMO channel into scalar channels for which the diversity–multiplexing tradeoffs can be computed in a straightforward manner (Exercises 9.3, 9.4 and 9.5). The diversity–multiplexing tradeoff of V-BLAST with ML decoding can be analyzed starting from the pairwise error probability between two codewords $\mathbf{x}_A$ and $\mathbf{x}_B$ (with average transmit energy normalized to 1):

$$\mathbb{P}\{\mathbf{x}_A \to \mathbf{x}_B | \mathbf{H}\} \leq \frac{16}{\mathsf{SNR}^2 \|\mathbf{x}_A - \mathbf{x}_B\|^4}, \tag{9.26}$$

(cf. 3.92). Each codeword is a pair of QAM symbols transmitted on the two antennas, and hence the distance between the two closest codewords is that between two adjacent constellation points in one of the QAM constellation, i.e., $\mathbf{x}_A$ and $\mathbf{x}_B$ differ only in one of the two QAM symbols. With a total data rate of R bits/s/Hz, each QAM symbol carries $R/2$ bits, and hence each of the I and Q channels carries $R/4$ bits. The distance between two adjacent constellation points is of the order of $1/2^{R/4}$. Thus, the worst-case pairwise error probability is of the order

$$\frac{16 \cdot 2^R}{\mathsf{SNR}^2} = 16 \cdot \mathsf{SNR}^{-(2-r)}, \tag{9.27}$$

where the data rate $R = r \log \mathsf{SNR}$. This is the worst-case pairwise error probability, but Exercise 9.6 shows that the overall error probability is also of the same order. Hence, the diversity–multiplexing tradeoff of V-BLAST with ML decoding is

$$d(r) = 2 - r \qquad r \in [0, 2]. \tag{9.28}$$

As already remarked in Section 3.3.3, the (classical) diversity gain and the degrees of freedom utilized are not always sufficient to say which scheme is best. For example, the Alamouti scheme has a higher (classical) diversity gain than V-BLAST but utilizes fewer degrees of freedom. The tradeoff curves, in contrast, provide a clear basis for the comparison. We see that which scheme is better depends on the target diversity gain (error probability) of the operating point: for smaller target diversity gains, V-BLAST is better than the Alamouti scheme, while the situation reverses for higher target diversity gains.

Optimal tradeoff

Do any of the four schemes actually achieve the optimal tradeoff of the 2×2 channel? The tradeoff curve of the 2×2 i.i.d. Rayleigh fading MIMO channel turns out to be piecewise linear joining the points $(0, 4)$, $(1, 1)$ and $(2, 0)$ (also shown in Figure 9.6). Thus, all of the schemes are tradeoff-suboptimal, except for V-BLAST with ML, which is optimal but only for $r > 1$.

The endpoints of the optimal tradeoff curve are $(0, 4)$ and $(2, 0)$, consistent with the fact that the 2×2 MIMO channel has a maximum diversity gain of 4 and 2 degrees of freedom. More interestingly, unlike all the tradeoff curves we have computed before, this curve is not a line but piecewise linear, consisting of two linear segments. V-BLAST with ML decoding sends two symbols per symbol time with (classical) diversity of 2 for each symbol, and achieves the gentle part, $2 - r$, of this curve. But what about the steep part, $4 - 3r$? Intuitively, there should be a scheme that sends 4 symbols over 3 symbol times (with a rate of $4/3$ symbols/s/Hz) and achieves the full diversity gain of 4. We will see such a scheme in Section 9.2.4.

9.1.6 $n_t \times n_r$ MIMO i.i.d. Rayleigh channel

Optimal tradeoff

Consider the $n_t \times n_r$ MIMO channel with i.i.d. Rayleigh faded gains. The optimal diversity gain at a data rate $r \log \mathsf{SNR}$ bits/s/Hz is the rate at which the outage probability (cf. (8.81)) decays with SNR:

$$p_{\text{out}}^{\text{mimo}}(r \log \mathsf{SNR}) = \min_{\mathbf{K}_x : \text{Tr}[\mathbf{K}_x] \leq \mathsf{SNR}} \mathbb{P}\{\log \det(\mathbf{I}_{n_r} + \mathbf{HK}_x\mathbf{H}^*) < r \log \mathsf{SNR}\}. \quad (9.29)$$

While the optimal covariance matrix $\mathbf{K}_x$ depends on the SNR and the data rate, we argued in Section 8.4 that the choice of $\mathbf{K}_x = \mathsf{SNR}/n_t \mathbf{I}_{n_t}$ is often used as a good approximation to the actual outage probability. In the coarser scaling of the tradeoff curve formulation, that argument can be made precise: the decay rate of the outage probability in (9.29) is the same as when the covariance matrix is the scaled identity. (See Exercise 9.8.) Thus, for the purpose of identifying the optimal diversity gain at a multiplexing rate r it suffices to consider the expression in (8.85):

$$p_{\text{out}}^{\text{iid}}(r \log \mathsf{SNR}) = \mathbb{P}\left\{\log \det\left(\mathbf{I}_{n_r} + \frac{\mathsf{SNR}}{n_t}\mathbf{HH}^*\right) < r \log \mathsf{SNR}\right\}. \quad (9.30)$$

By analyzing this expression, the diversity–multiplexing tradeoff of the $n_t \times n_r$ i.i.d. Rayleigh fading channel can be computed. It is the piecewise linear curve joining the points

$$(k, (n_t - k)(n_r - k)), k = 0, \ldots, n_{\min}, \quad (9.31)$$

as shown in Figure 9.7.

The tradeoff curve summarizes succinctly the performance capability of the slow fading MIMO channel. At one extreme where $r \to 0$, the maximal diversity gain $n_t \cdot n_r$ is achieved, at the expense of very low multiplexing gain. At the other extreme where $r \to n_{\min}$, the full degrees of freedom are attained. However, the system is now operating very close to the fast fading capacity and there is little protection against the randomness of the slow fading channel; the diversity gain is approaching 0. The tradeoff curve bridges between the two extremes and provides a more complete picture of the slow fading performance capability than the two extreme points. For example, adding one transmit and one receive antenna to the system increases the degrees of freedom $\min(n_t, n_r)$ by 1; this corresponds to increasing the maximum possible multiplexing gain by 1. The tradeoff curve gives a more refined picture of the system benefit: for *any* diversity requirement d, the supported multiplexing gain is increased by 1.

Figure 9.7
Diversity–multiplexing tradeoff, $d^*(r)$ for the i.i.d. Rayleigh fading channel.

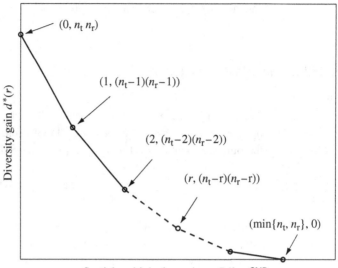

Figure 9.8 Adding one transmit and one receive antenna increases spatial multiplexing gain by 1 at each diversity level.

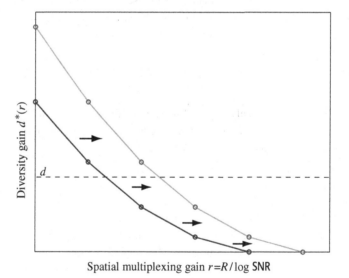

This is because the entire tradeoff curve is shifted by 1 to the right; see Figure 9.8.

The optimal tradeoff curve is based on the outage probability, so in principle arbitrarily large block lengths are required to achieve the optimal tradeoff curve. However, it has been shown that, in fact, space-time codes of block length $l = n_t + n_r - 1$ achieve the curve. In Section 9.2.4, we will see a scheme that achieves the tradeoff curve but requires arbitrarily large block lengths.

Geometric interpretation

To provide more intuition let us consider the geometric picture behind the optimal tradeoff for integer values of r. The outage probability is given by

$$p_{\text{out}}(r \log \text{SNR}) = \mathbb{P}\left\{\log \det\left(\mathbf{I}_{n_r} + \frac{\text{SNR}}{n_t}\mathbf{H}\mathbf{H}^*\right) < r \log \text{SNR}\right\}$$

$$= \mathbb{P}\left\{\sum_{i=1}^{n_{\min}} \log\left(1 + \frac{\text{SNR}}{n_t}\lambda_i^2\right) < r \log \text{SNR}\right\}, \qquad (9.32)$$

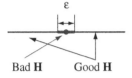

Figure 9.9 Geometric picture for the 1 × 1 channel. Outage occurs when $|h|$ is close to 0.

where λ_i are the (random) singular values of the matrix $\mathbf{H}$. There are $n_{\min}$ possible modes for communication but the effectiveness of mode i depends on how large the received signal strength $\text{SNR}\lambda_i^2/n_t$ is for that mode; we can think of a mode as *fully effective* if $\text{SNR}\lambda_i^2/n_t$ is of order SNR and not effective at all when $\text{SNR}\lambda_i^2/n_t$ is of order 1 or smaller.

At low multiplexing gains ($r \to 0$), outage occurs when none of the modes are effective at all; i.e., *all* the squared singular values are small, of the order of $1/\text{SNR}$. Geometrically, this event happens when the channel matrix $\mathbf{H}$ is close to the zero matrix; see Figure 9.9 and 9.10. Since $\sum_i \lambda_i^2 = \sum_{i,j} |h_{ij}|^2$, this event occurs only when *all* of the $n_t n_r$ squared magnitude channel gains, $|h_{ij}|^2$, are small, each on the order of $1/\text{SNR}$. As the channel gains are independent and $\mathbb{P}\{|h_{ij}|^2 < 1/\text{SNR}\} \approx 1/\text{SNR}$, the probability of this event is on the order of $1/\text{SNR}^{n_t n_r}$.

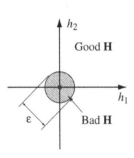

Figure 9.10 Geometric picture for the 1 × 2 channel. Outage occurs when $|h_1|^2 + |h_2|^2$ is close to 0.

Now consider the case when r is a positive integer. The situation is more complicated. For the outage event in (9.32) to occur, there are now many possible combinations of values that the singular values, λ_i, can take on, with modes taking on different shades of effectiveness. However, at high SNR, it can be shown that the *typical* way for outage to occur is when precisely r of the modes are fully effective and the rest completely ineffective. This means the largest r singular values of $\mathbf{H}$ are of order 1, while the rest are of the order $1/\text{SNR}$ or smaller; geometrically, $\mathbf{H}$ is close to a rank r matrix. What is the probability of this event?

In the case of $r = 0$, the outage event is when the channel matrix $\mathbf{H}$ is close to a rank 0 matrix. The channel matrix lies in the $n_t n_r$-dimensional space $\mathcal{C}^{n_r \times n_t}$, so for this to occur, there is a collapse in all $n_t n_r$ dimensions. This leads to an outage probability of $1/\text{SNR}^{n_t n_r}$. At general multiplexing gain r (r positive integer), outage occurs when $\mathbf{H}$ is close to $\mathcal{V}_r$, the space of all rank r matrices. This requires a collapse in the component of $\mathbf{H}$ "orthogonal" to $\mathcal{V}_r$. Thus, one would expect the probability of this event to be approximately $1/\text{SNR}^d$, where d is the number of such dimensions.[2] See Figure 9.11. It is

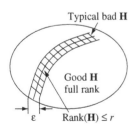

Figure 9.11 Geometric picture for the $n_t \times n_r$ channel at multiplexing gain r (r integer). Outage occurs when the channel matrix $\mathbf{H}$ is close to a rank r matrix.

---

[2] $\mathcal{V}_r$ is not a linear space. So, strictly speaking, we cannot talk about the concept of orthogonal dimensions. However, $\mathcal{V}_r$ is a *manifold*, which means that the neighborhood of every point looks like a Euclidean space of the same dimension. So the notion of orthogonal dimensions (called the "co-dimension" of $\mathcal{V}_r$) still makes sense.

easy to compute d. A $n_r \times n_t$ matrix $\mathbf{H}$ of rank r is described by $rn_t + (n_r - r)r$ parameters: rn_t parameters to specify r linearly independent row vectors of $\mathbf{H}$ and $(n_r - r)r$ parameters to specify the remaining $n_r - r$ rows in terms of linear combinations of the first r row vectors. Hence $\mathcal{V}_r$ is $n_t r + (n_r - r)r$-dimensional and the number of dimensions orthogonal to $\mathcal{V}_r$ in $\mathcal{C}^{n_t n_r}$ is simply

$$n_t n_r - (n_t r + (n_r - r)r) = (n_t - r)(n_r - r).$$

This is precisely the SNR exponent of the outage probability in (9.32).

9.2 Universal code design for optimal diversity–multiplexing tradeoff

The operational interpretation of the outage formulation is based on the existence of universal codes that can achieve arbitrarily small error whenever the channel is not in outage. To achieve such performance, arbitrarily long block lengths and powerful codes are required. In the high SNR regime, we have seen in Chapter 3 that the typical error event is the event that the channel is in a deep fade, where the deep-fade event depends on the channel as well as the scheme. This leads to a natural high SNR relaxation of the universality concept:

> A scheme is *approximately universal* if it is in deep fade only when the channel itself is in outage.

Being approximately universal is sufficient for a scheme to achieve the diversity–multiplexing tradeoff of the channel. Moreover, one can explicitly construct approximately universal schemes of short block lengths. We describe this approach towards optimal diversity–multiplexing tradeoff code design in this section. We start with the scalar channel and progress towards more complex models, culminating in the general $n_t \times n_r$ MIMO channel.

9.2.1 QAM is approximately universal for scalar channels

In Section 9.1.2 we have seen that uncoded QAM achieves the optimal diversity–multiplexing tradeoff of the scalar Rayleigh fading channel. One can obtain a deeper understanding of why this is so via a typical error event analysis. Conditional on the channel gain h, the probability of error of uncoded QAM at data rate R is approximately

$$Q\left(\sqrt{\frac{\mathsf{SNR}}{2}|h|^2 d_{\min}^2}\right), \tag{9.33}$$

where $d_{\min}$ is the minimum distance between two normalized constellation points, given by

$$d_{\min} \approx \frac{1}{2^{R/2}}. \tag{9.34}$$

When $\sqrt{\text{SNR}}|h|d_{\min} \gg 1$, i.e. the separation of the constellation points at the receiver is much larger than the standard deviation of the additive Gaussian noise, errors occur very rarely due to the very rapid drop off of the Gaussian tail probability. Thus, as an order-of-magnitude approximation, errors typically occur due to:

$$\boxed{\text{Deep-fade event}: |h|^2 < \frac{2^R}{\text{SNR}}.} \tag{9.35}$$

This deep-fade event is analogous to that of BPSK in Section 3.1.2. On the other hand, the *channel outage* condition is given by

$$\log\left(1 + |h|^2 \text{SNR}\right) < R, \tag{9.36}$$

or equivalently

$$|h|^2 < \frac{2^R - 1}{\text{SNR}}. \tag{9.37}$$

At high SNR and high rate, the channel outage condition (9.37) and the deep-fade event of QAM (9.35) coincide. Thus, *typically errors occur for QAM only when the channel is in outage.* Since the optimal diversity–multiplexing tradeoff is determined by the outage probability of the channel, this explains why QAM achieves the optimal tradeoff. (A rigorous proof of the tradeoff optimality of QAM based solely on this typical error event view is carried out in Exercise 9.9, which is the generalization of Exercise 3.3 where we used the typical error event to analyze classical diversity gain.)

In Section 9.1.2, the diversity–multiplexing tradeoff of QAM is computed by averaging the error probability over the Rayleigh fading. It happens to be equal to the optimal tradeoff. The present explanation based on relating the deep-fade event of QAM and the outage condition is more insightful. For one thing, this explanation is in terms of conditions on the channel gain h and has nothing to do with the distribution of h. This means that QAM achieves the optimal diversity–multiplexing tradeoff not only under Rayleigh fading but in fact under *any* channel statistics. This is the true meaning of universality. For example, for a channel with the near-zero behavior of $\mathbb{P}\{|h|^2 < \epsilon\} \approx \epsilon^k$, the optimal diversity–multiplexing tradeoff curve follows directly from (9.15): $d^*(r) = k(1 - r)$. Uncoded QAM on this channel can achieve this tradeoff as well.

Note that the approximate universality of QAM depends only on a condition on its normalized minimum distance:

$$\boxed{d^2_{\min} > \frac{1}{2^R}.}$$
(9.38)

Any other constellation with this property is also approximately universal (Exercise 9.9).

Summary 9.1 Approximate universality

A scheme is *approximately universal* if it is in deep fade only when the channel itself is in outage.

Being approximately universal is sufficient for a scheme to achieve the diversity–multiplexing tradeoff of the channel.

9.2.2 Universal code design for parallel channels

In Section 3.2.2 we derived design criteria for codes that have a good coding gain while extracting the maximum diversity from the parallel channel. The criterion was derived based on averaging the error probability over the statistics of the fading channel. For example, the i.i.d. Rayleigh fading parallel channel yielded the product distance criterion (cf. Summary 3.1). In this section, we consider instead a *universal* design criterion based on considering the performance of the code over the *worst-case* channel that is not in outage. Somewhat surprisingly, this universal code design criterion reduces to the product distance criterion at high SNR. Using this universal design criterion, we can characterize codes that are approximately universal using the idea of typical error event used in the last section.

Universal code design criterion

We begin with the parallel channel with L diversity branches, focusing on just one time symbol (and dropping the time index):

$$y_\ell = h_\ell x_\ell + w_\ell$$
(9.39)

for $\ell = 1, \ldots, L$. Here, as before, the w_ℓ are i.i.d. $\mathcal{CN}(0, 1)$ noise. Suppose the rate of communication is R bits/s/Hz per sub-channel. Each codeword is a vector of length L. The ℓth component of any codeword is transmitted over the ℓth sub-channel in (9.39). Here, a codeword consists of one symbol for each of the L sub-channels; more generally, we can consider coding over multiple symbols for each of the sub-channels as well as coding across the different sub-channels. The derivation of a code design criterion for the more general case is done in Exercise 9.10.

The channels that are not in outage are those whose gains satisfy

$$\sum_{\ell=1}^{L} \log(1 + |h_\ell|^2 \mathsf{SNR}) \geq LR. \tag{9.40}$$

As before, SNR is the transmit power constraint per sub-channel.

For a fixed pair of codewords $\mathbf{x}_A, \mathbf{x}_B$, the probability that $\mathbf{x}_B$ is more likely than $\mathbf{x}_A$ when $\mathbf{x}_A$ is transmitted, conditional on the channel gains $\mathbf{h}$, is (cf. (3.51))

$$\mathbb{P}\{\mathbf{x}_A \to \mathbf{x}_B | \mathbf{h}\} = Q\left(\sqrt{\frac{\mathsf{SNR}}{2} \sum_{\ell=1}^{L} |h_\ell|^2 |d_\ell|^2} \right), \tag{9.41}$$

where d_ℓ is the ℓth component of the normalized codeword difference (cf. (3.52)):

$$d_\ell := \frac{1}{\sqrt{\mathsf{SNR}}}(x_{A\ell} - x_{B\ell}). \tag{9.42}$$

The *worst-case* pairwise error probability over the channels that are not in outage is the $Q(\sqrt{\cdot})$ function evaluated at the solution to the optimization problem

$$\min_{h_1,\ldots,h_L} \frac{\mathsf{SNR}}{2} \sum_{\ell=1}^{L} |h_\ell|^2 |d_\ell|^2, \tag{9.43}$$

subject to the constraint (9.40). If we define $Q_\ell := \mathsf{SNR} \cdot |h_\ell|^2 |d_\ell|^2$, then the optimization problem can be rewritten as

$$\min_{Q_1 \geq 0,\ldots,Q_L \geq 0} \frac{1}{2} \sum_{\ell=1}^{L} Q_\ell \tag{9.44}$$

subject to the constraint

$$\sum_{\ell=1}^{L} \log\left(1 + \frac{Q_\ell}{|d_\ell|^2}\right) \geq LR. \tag{9.45}$$

This is analogous to the problem of minimizing the total power required to support a target rate R bits/s/Hz per sub-channel over a parallel Gaussian channel; the solution is just standard waterfilling, and the worst-case channel is

$$|h_\ell|^2 = \frac{1}{\mathsf{SNR}} \cdot \left(\frac{1}{\lambda |d_\ell|^2} - 1\right)^+, \qquad \ell = 1,\ldots,L. \tag{9.46}$$

Here λ is the Lagrange multiplier chosen such that the channel in (9.46) satisfies (9.40) with equality. The worst-case pairwise error probability is

$$Q\left(\sqrt{\frac{1}{2}\sum_{\ell=1}^{L}\left(\frac{1}{\lambda}-|d_\ell|^2\right)^+}\right),\qquad(9.47)$$

where λ satisfies

$$\sum_{\ell=1}^{L}\left[\log\left(\frac{1}{\lambda|d_\ell|^2}\right)\right]^+=LR.\qquad(9.48)$$

Examples

We look at some simple coding schemes to better understand the universal design criterion, the argument of the $Q\left(\sqrt{\cdot/2}\right)$ function in (9.47):

$$\sum_{\ell=1}^{L}\left(\frac{1}{\lambda}-|d_\ell|^2\right)^+,\qquad(9.49)$$

where λ satisfies the constraint in (9.48).

1. **No coding** Here symbols from L independent constellations (say, QAM), with 2^R points each, are transmitted separately on each of the sub-channels. This has very poor performance since all but one of the $|d_\ell|^2$ can be simultaneously zero. Thus the design criterion in (9.49) evaluates to zero.
2. **Repetition coding** Suppose the symbol is drawn from a QAM constellation (with 2^{RL} points) but the same symbol is repeated over each of the sub-channels. For the 2-parallel channel with $R = 2$ bits/s/Hz per sub-channel, the repetition code is illustrated in Figure 9.12. The smallest value of $|d_\ell|^2$ is 4/9. Due to the repetition, for any pair of codewords, the differences in the sub-channels are equal. With the choice of the worst pairwise differences, the universal criterion in (9.49) evaluates to 8/3 (see Exercise 9.12).
3. **Permutation coding** Consider the 2-parallel channel where the symbol on each of the sub-channels is drawn from a separate QAM constellation. This

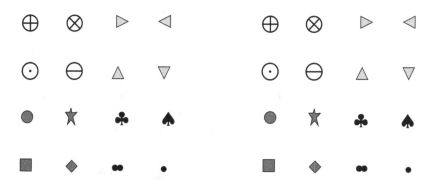

Figure 9.12 A repetition code for the 2-parallel channel with rate $R = 2$ bits/s/Hz per sub-channel.

Figure 9.13 A permutation code for the 2-parallel channel with rate $R = 2$ bits/s/Hz per sub-channel.

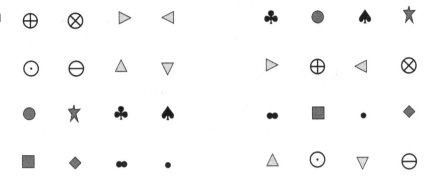

is similar to the repetition code (Figure 9.12), but we consider different mappings of the QAM points in the sub-channels. In particular, we map the points such that if two points are close to each other in one QAM constellation, their images in the other QAM constellation are far apart. One such choice is illustrated in Figure 9.13, for $R = 2$ bits/s/Hz per sub-channel where two points that are nearest neighbors in one QAM constellation have their images in the other QAM constellation separated by at least double the minimum distance. With the choice of the worst pairwise differences for this code, the universal design criterion in (9.49) can be explicitly evaluated to be 44/9 (see Exercise 9.13).

This code involves a one-to-one map between the two QAM constellations and can be parameterized by a *permutation* of the QAM points. The repetition code is a special case of this class of codes: it corresponds to the *identity* permutation.

Universal code design criterion at high SNR

Although the universal criterion (9.49) can be computed given the codewords, the expression is quite complicated (Exercise 9.11) and is not amenable to use as a criterion for code design. We can however find a simple bound by relaxing the non-negativity constraint in the optimization problem (9.44). This allows the water depth to go negative, resulting in the following lower bound on (9.49):

$$L2^R|d_1 d_2 \cdots d_L|^{2/L} - \sum_{\ell=1}^{L} |d_\ell|^2. \tag{9.50}$$

When the rate of communication per sub-channel R is large, the water level in the waterfilling problem (9.44) is deep at every sub-channel for good codes, and this lower bound is tight. Moreover, for good codes the second term is small compared to the first term, and so in this regime the universal criterion is approximately

$$L2^R|d_1 d_2 \cdots d_L|^{2/L}. \tag{9.51}$$

Thus, the universal code design problem is to choose the codewords maximizing the pairwise product distance; in this regime, the criterion coincides with that of the i.i.d. Rayleigh parallel fading channel (cf. Section 3.2.2).

Property of an approximately universal code

We can use the universal code design criterion developed above to characterize the property of a code that makes it approximately universal over the parallel channel at high SNR. Following the approach in Section 9.2.1, we first define a pairwise typical error event: this is when the argument of the $Q(\sqrt{\cdot/2})$ in (9.41) is less than 1:

$$\text{SNR} \cdot \sum_{\ell=1}^{L} |h_\ell|^2 |d_\ell|^2 < 1. \qquad (9.52)$$

For a code to be approximately universal, we want this event to occur only when the channel is in outage; equivalently, this event should not occur whenever the channel is not in outage. This translates to saying that the worst-case code design criterion derived above should be greater than 1. At high SNR, using (9.51), the condition becomes

$$\boxed{|d_1 d_2 \cdots d_L|^{2/L} > \frac{1}{L 2^R}.} \qquad (9.53)$$

Moreover, this condition should hold for any pair of codewords. It is verified in Exercise 9.14 that this is sufficient to guarantee that a coding scheme achieves the optimal diversity–multiplexing tradeoff of the parallel channel.

We saw the permutation code in Figure 9.13 as an example of a code with good universal design criterion value. This class of codes contains approximately universal codes. To see this, we first need to generalize the essential structure in the permutation code example in Figure 9.13 to higher rates and to more than two sub-channels. We consider codes of just a single block length to carry out the following generalization.

We fix the constellation from which the codeword is chosen in each sub-channel to be a QAM. Each of these QAM constellations contains the entire information to be transmitted: so, the total number of points in the QAM constellation is 2^{LR} if R is the data rate per sub-channel. The overall code is specified by the maps between the QAM points for each of the sub-channels. Since the maps are one-to-one, they can be represented by permutations of the QAM points. In particular, the code is specified by $L - 1$ permutations $\pi_2, \ldots, \pi_L$: for each message, say m, we identify one of the QAM points, say q, in the QAM constellation for the first sub-channel. Then, to convey the message m, the transmit codeword is

$$(q, \pi_2(q), \ldots, \pi_L(q)),$$

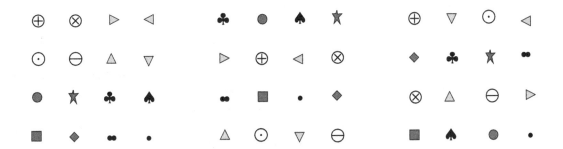

Figure 9.14 A permutation code for a parallel channel with three sub-channels. The entire information (4 bits) is contained in each of the QAM constellations.

i.e., the QAM point transmitted over the ℓth sub-channel is $\pi_\ell(q)$ with π_1 defined to be the identity permutation. An example of a permutation code with a rate of 4/3 bits/s/Hz per sub-channel for $L = 3$ (so the QAM constellation has 2^4 points) is illustrated in Figure 9.14.

Given the physical constraints (the operating SNR, the data rate, and the number of sub-channels), the engineer can now choose appropriate permutations to maximize the universal code design criterion. Thus permutation codes provide a framework within which specific codes can be designed based on the requirements. This framework is quite rich: Exercise 9.15 shows that even randomly chosen permutations are approximately universal with high probability.

Bit-reversal scheme: an operational interpretation of the outage condition

We can use the concept of approximately universal codes to give an operational interpretation of the outage condition for the parallel channel. To be able to focus on the essential issues, we restrict our attention to just two sub-channels, so $L = 2$. If we communicate at a total rate $2R$ bits/s/Hz over the parallel channel, the no-outage condition is

$$\log(1 + |h_1|^2 \text{SNR}) + \log(1 + |h_2|^2 \text{SNR}) > 2R. \qquad (9.54)$$

One way of interpreting this condition is as though the first sub-channel provides $\log(1 + |h_1|^2 \text{SNR})$ bits of information and the second sub-channel provides $\log(1 + |h_2|^2 \text{SNR})$ bits of information, and as long as the total number of bits provided exceeds the target rate, then reliable communication is possible. In the high SNR regime, we exhibit below a permutation code that makes the outage condition concrete.

Suppose we independently code over the I and Q channels of the two sub-channels. So we can focus on only one of them, say, the I channel. We wish to communicate R bits over two uses of the I-channel. Analogous to the typical event analysis for the scalar channel, we can exactly recover all the R information bits from the first I sub-channel alone if

$$|h_1|^2 > \frac{2^{2R}}{\text{SNR}}, \qquad (9.55)$$

or

$$|h_1|^2 \text{SNR} > 2^{2R}. \tag{9.56}$$

However, we do not need to use just the first I sub-channel to recover all the information bits: the second I sub-channel also contains the same information and can be used in the recovery process. Indeed, if we create x_1^I by treating the ordered R bits as the binary representation of the points x_1^I, then one would intuitively expect that if

$$|h_1|^2 \text{SNR} > 2^{2R_1}, \tag{9.57}$$

then one should be able to recover at least R_1 of the most significant bits of information. Now, if we create x_2^I by treating the *reversal* of the R bits as its binary representation, then one should be able to recover at least R_2 of the most significant bits, if

$$|h_2|^2 \text{SNR} > 2^{2R_2}. \tag{9.58}$$

But due to the reversal, the most significant bits in the representation in the second I sub-channel are the least significant bits in the representation in the first I sub-channel. Hence, as long as $R_1 + R_2 \geq R$, then we can recover *all R* bits. This translates to the condition

$$\log(|h_1|^2 \text{SNR}) + \log(|h_2|^2 \text{SNR}) > 2R, \tag{9.59}$$

which is precisely the no-outage condition (9.54) at high SNR.

The bit-reversal scheme described here with some slight modifications can be shown to be approximately universal (Exercise 9.16). A simple variant of this scheme is also approximately universal (Exercise 9.17).

Summary 9.2 Universal codes for the parallel channel

A universal code design criterion between two codewords can be computed by finding the channel not in outage that yields the worst-case pairwise error probability.

At high SNR and high rate, the universal code design criterion becomes proportional to the product distance:

$$|d_1 \ldots d_L|^{2/L} \tag{9.60}$$

where L is the number of sub-channels and d_ℓ is the difference between the ℓth components of the codewords.

> A code is approximately universal for the parallel channel if its product distance is large enough: for a code at a data rate of R bits/s/Hz per sub-channel, we require
>
> $$|d_1 d_2 \cdots d_L|^2 > \frac{1}{(L2^R)^L}. \tag{9.61}$$
>
> Simple bit-reversal schemes are approximately universal for the 2-parallel channel. Random permutation codes are approximately universal for the L-parallel channel with high probability.

9.2.3 Universal code design for MISO channels

The outage event for the $n_t \times 1$ MISO channel (9.22) is

$$\log\left(1 + \|\mathbf{h}\|^2 \frac{\mathrm{SNR}}{n_t}\right) < R. \tag{9.62}$$

In the case when $n_t = 2$, the Alamouti scheme converts the MISO channel to a scalar channel with gain $\|\mathbf{h}\|$ and SNR reduced by a factor of 2. Hence, the outage behavior is exactly the same as in the original MISO channel, and the Alamouti scheme provides a *universal* conversion of the 2×1 MISO channel to a scalar channel. Any approximately universal scheme for the scalar channel, such as QAM, when used in conjunction with the Alamouti scheme is also approximately optimal for the MISO channel and achieves its diversity–multiplexing tradeoff.

In the general case when the number of transmit antennas is greater than two, there is no equivalence of the Alamouti scheme. Here we explore two approaches to constructing universal schemes for the general MISO channel.

MISO channel viewed as a parallel channel

Using one transmit antenna at a time converts the MISO channel into a parallel channel. We have used this conversion in conjunction with repetition coding to argue the classical diversity gain of the MISO channel (cf. Section 3.3.2). Replacing the repetition code with an appropriate parallel channel code (such as the bit-reversal scheme from Section 9.2.2), we will see that converting the MISO channel into a parallel channel is actually tradeoff-optimal for the i.i.d. Rayleigh fading channel.

Suppose we want to communicate at rate $R = r \log \mathrm{SNR}$ bits/s/Hz on the MISO channel. Using one transmit antenna at a time yields a parallel channel with n_t diversity branches and the data rate of communication is R bits/s/Hz per sub-channel. The optimal diversity gain for the i.i.d. Rayleigh parallel fading channel is $n_t(1 - r)$ (cf. (9.20)); thus, using one antenna at a

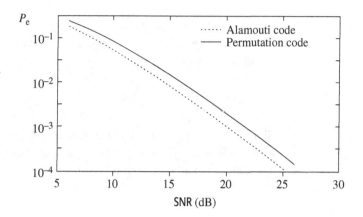

Figure 9.15 The error probability of uncoded QAM with the Alamouti scheme and that of a permutation code over one antenna at a time for the Rayleigh fading MISO channel with two transmit antennas: the permutation code is about 1.5 dB worse than the Alamouti scheme over the plotted error probability range.

time in conjunction with a tradeoff-optimal parallel channel code achieves the largest diversity gain over the i.i.d. Rayleigh fading MISO channel (cf. (9.24)).

To understand how much loss the conversion of the MISO channel into a parallel channel entails with respect to the optimal outage performance, we plot the error probabilities of two schemes with the same rate ($R = 2$ bits/s/Hz): uncoded QAM over the Alamouti scheme and the permutation code in Figure 9.13. This performance is plotted in Figure 9.15 where we see that the conversion of the MISO channel into a parallel channel entails a loss of about 1.5 dB in SNR for the same error probability performance.

Universality of conversion to parallel channel

We have seen that the conversion of the MISO channel into a parallel channel is tradeoff-optimal for the i.i.d. Rayleigh fading channel. Is this conversion universal? In other words, will a tradeoff-optimal scheme for the parallel channel also be tradeoff-optimal for the MISO channel, under *any* channel statistics? In general, the answer is no. To see this, consider the following MISO channel model: suppose the channels from all but the first transmit antenna are very poor. To make this example concrete, suppose $h_\ell = 0$, $\ell = 2, \ldots, n_t$. The tradeoff curve depends on the outage probability (which depends only on the statistics of the first channel)

$$p_{\text{out}} = \mathbb{P}\left\{\log\left(1 + \text{SNR}|h_1|^2\right) < R\right\}. \tag{9.63}$$

Using one transmit antenna at a time is a waste of degrees of freedom: since the channels from all but the first antenna are zero, there is no point in transmitting any signal on them. This loss in degrees of freedom is explicit in the outage probability of the parallel channel formed by transmitting from one antenna at a time:

$$p_{\text{out}}^{\text{parallel}} = \mathbb{P}\left\{\log\left(1 + \text{SNR}|h_1|^2\right) < n_t R\right\}. \tag{9.64}$$

Comparing (9.64) with (9.63), we see clearly that the conversion to the parallel channel is not tradeoff-optimal for this channel model.

Essentially, using one antenna at a time equates temporal degrees of freedom with spatial ones. All temporal degrees of freedom are the same, but the spatial ones need not be the same: in the extreme example above, the spatial channels from all but the first transmit antenna are zero. Thus, it seems reasonable that when all the spatial channels are *symmetric* then the parallel channel conversion of the MIMO channel is justified. This sentiment is justified in Exercise 9.18, which shows that the parallel channel conversion is approximately *universal* over a restricted class of MISO channels: those with i.i.d. spatial channel coefficients.

Universal code design criterion

Instead of converting to a parallel channel, one can design universal schemes directly for the MISO channel. What is an appropriate code design criterion? In the context of the i.i.d. Rayleigh fading channel, we derived the determinant criterion for the codeword difference matrices in Section 3.3.2. What is the corresponding criterion for universal MISO schemes? We can answer this question by considering the worst-case pairwise error probability over all MISO channels that are not in outage.

The pairwise error probability (of confusing the transmit codeword matrix $\mathbf{X}_A$ with $\mathbf{X}_B$) conditioned on a specific MISO channel realization is (cf. (3.82))

$$\mathbb{P}\{\mathbf{X}_A \to \mathbf{X}_B | \mathbf{h}\} = Q\left(\frac{\|\mathbf{h}^*(\mathbf{X}_A - \mathbf{X}_B)\|}{\sqrt{2}}\right). \tag{9.65}$$

In Section 3.3.2 we averaged this quantity over the statistics of the MISO channel (cf. (3.83)). Here we consider the worst-case over all channels not in outage:

$$\max_{\mathbf{h}: \|\mathbf{h}\|^2 > \frac{n_t(2^R - 1)}{\mathsf{SNR}}} Q\left(\frac{\|\mathbf{h}^*(\mathbf{X}_A - \mathbf{X}_B)\|}{\sqrt{2}}\right). \tag{9.66}$$

From a basic result in linear algebra, the worst-case pairwise error probability in (9.66) can be explicitly written as (Exercise 9.19)

$$Q\left(\sqrt{\frac{1}{2}\lambda_1^2 n_t(2^R - 1)}\right), \tag{9.67}$$

where λ_1 is the *smallest* singular value of the normalized codeword difference matrix

$$\frac{1}{\sqrt{\mathsf{SNR}}}(\mathbf{X}_A - \mathbf{X}_B). \tag{9.68}$$

Essentially, the worst-case channel aligns itself in the direction of the weakest singular value of the codeword difference matrix. So, the universal code design criterion for the MISO channel is to ensure that no singular value is too small; equivalently

> *maximize the minimum singular value of the codeword difference matrices.*

$$(9.69)$$

There is an intuitive explanation for this design criterion: a universal code has to protect itself against the worst channel that is not in outage. The condition of no-outage only puts a constraint on the *norm* of the channel vector $\mathbf{h}$ but not on its direction. So, the worst channel aligns itself to the "weakest direction" of the codeword difference matrix to create the most havoc. The corresponding worst-case pairwise error probability will be governed by the smallest singular value of the codeword difference matrix. On the other hand, the i.i.d. Rayleigh channel does not prefer any specific direction: thus the design criterion tailored to its statistics requires that the *average* direction be well protected and this translates to the determinant criterion. While the two criteria are different, codes with large determinant tend to also have a large value for the smallest singular value; the two criteria (based on worst-case and average-case) are related in this aspect.

We can use the universal code design criterion to derive a property that makes a code universally achieve the tradeoff curve (as we did for the parallel channel in the previous section). We want the typical error event to occur only when the channel is in outage. This corresponds to the argument of $Q(\sqrt{(\cdot)/2})$ in the worst-case error probability (9.67) to be greater than 1, i.e.,

$$\lambda_1^2 > \frac{1}{n_t(2^R - 1)} \approx \frac{1}{n_t 2^R}.$$

$$(9.70)$$

for every pair of codewords. We can explicitly verify that the Alamouti scheme with independent uncoded QAMs on the two data streams satisfies the approximate universality property in (9.70). This is done in Exercise 9.20.

Summary 9.3 Universal codes for the MISO channel

The MISO channel can be converted into a parallel channel by using one transmit antenna at a time. This conversion is approximately universal for the class of MISO channels with i.i.d. fading coefficients.

The universal code design criterion is to maximize the minimum singular value of the codeword difference matrices.

9.2.4 Universal code design for MIMO channels

We finally arrive at the multiple transmit *and* multiple receive antenna slow fading channel:

$$\mathbf{y}[m] = \mathbf{H}\mathbf{x}[m] + \mathbf{w}[m]. \tag{9.71}$$

The outage event of this channel is

$$\log \det(\mathbf{I}_{n_r} + \mathbf{H}\mathbf{K}_x\mathbf{H}^*) < R, \tag{9.72}$$

where $\mathbf{K}_x$ is the optimizing covariance in (9.29).

Universality of D-BLAST

In Section 8.5, we have seen that the D-BLAST architecture with the MMSE–SIC receiver converts the MIMO channel into a parallel channel with n_t sub-channels. Suppose we pick the transmit strategy $\mathbf{K}_x$ in the D-BLAST architecture (the covariance matrix represents the combination of the power allocated to the streams and coordinate system under which they are mixed before transmitting, cf. (8.3)) to be the one in (9.72). The important property of this conversion is the conservation expressed in (8.88): denoting the effective SNR of the kth sub-channel of the parallel channel by SINR_k,

$$\log \det \left(\mathbf{I}_{n_r} + \mathbf{H}\mathbf{K}_x\mathbf{H}^*\right) = \sum_{k=1}^{n_t} \log(1 + \mathsf{SINR}_k). \tag{9.73}$$

However, $\mathsf{SINR}_1, \ldots, \mathsf{SINR}_{n_t}$, across the sub-channels are *correlated*. On the other hand, we saw codes (with just block length 1) that universally achieve the tradeoff curve for any parallel channel (in Section 9.2.2). This means that, using approximately universal parallel channel codes for each of the interleaved streams, the D-BLAST architecture with the MMSE–SIC receiver at a rate of $R = r \log \mathsf{SNR}$ bits/s/Hz per stream has a diversity gain determined by the decay rate of

$$\mathbb{P}\left\{\sum_{k=1}^{n_t} \log(1 + \mathsf{SINR}_k) < R\right\}, \tag{9.74}$$

with increasing SNR. With n interleaved streams, each having block length 1 (i.e., $N = 1$ in the notation of Section 8.5.2), the initialization loss in D-BLAST reduces a data rate of R bits/s/Hz per stream into a data rate of $nR/(n+n_t-1)$ bits/s/Hz on the MIMO channel (Exercise 8.27). Suppose we use the D-BLAST architecture in conjunction with a block length 1 universal parallel channel code for each of n interleaved streams. If this code operates at a multiplexing gain of r on the MIMO channel, the diversity gain obtained is, substituting for the rate in (9.74) and comparing with (9.73), the decay rate of

$$\mathbb{P} \left\{ \log \det \left(\mathbf{I}_{n_r} + \mathbf{H} \mathbf{K}_x \mathbf{H}^* \right) < \frac{r(n + n_t - 1)}{n} \log \mathsf{SNR} \right\}. \tag{9.75}$$

Now comparing this with the actual decay behavior of the outage probability (cf. (9.29)), we see that the D-BLAST/MMSE–SIC architecture with n interleaved streams used to operate at a multiplexing gain of r over the MIMO channel has a diversity gain equal to the decay rate of

$$p_{\text{out}}^{\text{mimo}} \left(\frac{r(n + n_t - 1)}{n} \log \mathsf{SNR} \right). \tag{9.76}$$

Thus, with a large number, n, of interleaved streams, the D-BLAST/MMSE–SIC architecture achieves universally the tradeoff curve of the MIMO channel. With a finite number of streams, it is strictly tradeoff-suboptimal. In fact, the tradeoff performance can be improved by replacing the MMSE–SIC receiver by joint ML decoding of all the streams. To see this concretely, let us consider the 2×2 MIMO Rayleigh fading channel (so $n_t = n_r = 2$) with just two interleaved streams (so $n = 2$). The transmit signal lasts 3 time symbols:

$$\begin{bmatrix} 0 & x_B^{(1)} & x_B^{(2)} \\ x_A^{(1)} & x_A^{(2)} & 0 \end{bmatrix}. \tag{9.77}$$

With the MMSE–SIC receiver, the diversity gain obtained at the multiplexing rate of r is the optimal diversity gain at the multiplexing rate of $3r/2$. This scaled version of the optimal tradeoff curve is depicted in Figure 9.16. On the other hand, with the ML receiver the performance is significantly improved, also depicted in Figure 9.16. This achieves the optimal diversity performance for multiplexing rates between 0 and 1, and in fact is the scheme that sends 4 symbols over 3 symbol times that we were seeking in Section 9.1.5! The performance analysis of the D-BLAST architecture with the joint ML receiver is rather intricate and is carried out in Exercise 9.21. Basically, MMSE–SIC is suboptimal because it favors stream 1 over stream 2 while ML treats them equally. This asymmetry is only a small edge effect when there are many interleaved streams but does impact performance when there are only a small number of streams.

Universal code design criterion

We have seen that the D-BLAST architecture is a universal one, but how do we recognize when another space-time code also has good outage performance universally? To answer this question, we can derive a code design criterion based on the worst-case MIMO channel that is not in outage. Consider space-time code matrices with block length n_t. The worst-case channel aligns itself in the "weakest directions" afforded by a codeword pair difference matrix. With just

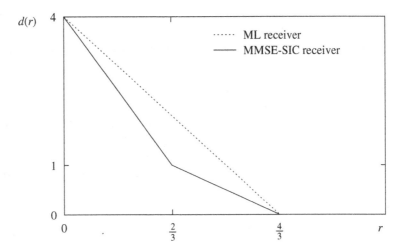

Figure 9.16 Tradeoff performance for the D-BLAST architecture with the ML receiver and with the MMSE–SIC receiver.

one receive antenna, the MISO channel is simply a row vector and it aligns itself in the direction of the smallest singular value of the codeword difference matrix (cf. Section 9.2.3). Here, there are $n_{\min}$ directions for the MIMO channel and the corresponding design criterion is an extension of that for the MISO channel: the universal code design criterion at high SNR is to maximize

$$\lambda_1 \lambda_2 \cdots \lambda_{n_{\min}}, \tag{9.78}$$

where $\lambda_1, \ldots, \lambda_{n_{\min}}$ are the smallest $n_{\min}$ singular values of the normalized codeword difference matrices (cf. (9.68)). The derivation is carried out in Exercise 9.22. With $n_t \leq n_r$, this is just the *determinant* criterion, derived in Chapter 3 by averaging the code performance over the i.i.d. Rayleigh statistics.

The exact code design criterion at an intermediate value of SNR is similar to the expression for the universal code design for the parallel channel (cf. (9.49)).

Property of an approximately universal code

Using exactly the same arguments as in Section 9.2.2, we can use the universal code design criterion developed above to characterize the property of a code that makes it approximately universal over the MIMO channel (see Exercise 9.23):

$$\boxed{\left|\lambda_1 \lambda_2 \cdots \lambda_{n_{\min}}\right|^{2/n_{\min}} > \frac{1}{n_{\min} 2^{R/n_{\min}}}.} \tag{9.79}$$

As in the parallel channel (cf. Exercise 9.14), this condition is only an order-of-magnitude one. A relaxed condition

$$\left|\lambda_1 \lambda_2 \cdots \lambda_{n_{\min}}\right|^{2/n_{\min}} > c \cdot \frac{1}{n_{\min} 2^{R/n_{\min}}}, \quad \text{for some constant } c > 0, \tag{9.80}$$

can also be used for approximate universality: it is sufficient to guarantee that the code achieves the optimal diversity–multiplexing tradeoff. We can make a couple of interesting observations immediately from this result.

- If a code satisfies the condition for approximate universality in (9.80) for an $n_t \times n_r$ MIMO channel with $n_r \geq n_t$, i.e., the number of receive antennas is equal to or larger than the number of transmit antennas, then it is also approximately universal for an $n_t \times l$ MIMO channel with $l \geq n_r$.
- The singular values of the normalized codeword matrices are upper bounded by $2\sqrt{n_t}$ (Exercise 9.24). Thus, a code that satisfies (9.80) for an $n_t \times n_r$ MIMO channel also satisfies the criterion in (9.80) for an $n_t \times l$ MIMO channel with $l \leq n_r$. Thus is it also approximately universal for the $n_t \times l$ MIMO channel with $l \leq n_r$.

We can conclude the following from the above two observations:

> A code that satisfies (9.80) for an $n_t \times n_t$ MIMO channel is approximately universal for an $n_t \times n_r$ MIMO channel for *every* value of the number of receive antennas n_r.

Exercise 9.25 shows a rotation code that satisfies (9.80) for the 2×2 MIMO channel; so this code is approximately universal for every $2 \times n_r$ MIMO channel.

We have already observed that the D-BLAST architecture with approximately universal parallel channel codes for the interleaved streams is approximately universal for the MIMO channel. Alternatively, we can see its approximate universality by explicitly verifying that it satisfies the condition in (9.80) with $n_t = n_r$. Here, we will see this for the 2×2 channel with two interleaved streams in the D-BLAST transmit codeword matrix (cf. (9.77)). The normalized codeword difference matrix can be written as

$$\mathbf{D} = \begin{bmatrix} 0 & d_B^{(1)} & d_B^{(2)} \\ d_A^{(1)} & d_A^{(2)} & 0 \end{bmatrix}, \tag{9.81}$$

where $\left(d_B^{(\ell)}, d_A^{(\ell)} \right)$ is the normalized pairwise difference codeword for an approximately universal parallel channel code and satisfies the condition in (9.53):

$$|d_B^{(\ell)} d_A^{(\ell)}| > \frac{1}{4 \cdot 2^R}, \qquad \ell = 1, 2. \tag{9.82}$$

Here R is the rate in bits/s/Hz in each of the streams. The product of the two singular values of $\mathbf{D}$ is

$$\begin{aligned} \lambda_1^2 \lambda_2^2 &= \det(\mathbf{D}\mathbf{D}^*) \\ &= |d_B^{(1)} d_A^{(1)}|^2 + |d_B^{(2)} d_A^{(2)}|^2 + |d_B^{(2)} d_A^{(1)}|^2 \\ &> \frac{1}{4 \cdot 2^R}, \end{aligned} \tag{9.83}$$

where the last inequality follows from (9.82). A rate of R bits/s/Hz on each of the streams corresponds to a rate of $2R/3$ bits/s/Hz on the MIMO channel. Thus, comparing (9.83) with (9.79), we have verified the approximate universality of D-BLAST at a reduced rate due to the initialization loss. In other words, the diversity gain obtained by the D-BLAST architecture in (9.77) at a multiplexing rate of r over the MIMO channel is $d^*(3r/2)$.

Discussion 9.1 Universal codes in the downlink

Consider the downlink of a cellular system where the base-stations are equipped with multiple transmit antennas. Suppose we want to broadcast the same information to all the users in the cell in the downlink. We would like our transmission scheme to not depend on the number of receive antennas at the users: each user could have a different number of receive antennas, depending on the model, age, and type of the mobile device.

Universal MIMO codes provide an attractive solution to this problem. Suppose we broadcast the common information at rate R using a space-time code that satisfies (9.79) for an $n_t \times n_t$ MIMO channel. Since this code is approximately universal for every $n_t \times n_r$ MIMO channel, the diversity seen by each user is *simultaneously* the best possible at rate R. To summarize: the diversity gain obtained by each user is the best possible with respect to both

- the number of receive antennas it has, and
- the statistics of the fading channel the user is currently experiencing.

Chapter 9 The main plot

For a slow fading channel at high SNR, the tradeoff between data rate and error probability is captured by the tradeoff between multiplexing and diversity gains. The optimal diversity gain $d^*(r)$ is the rate at which outage probability decays with increasing SNR when the data rate is increasing as $r \log \mathsf{SNR}$. The classical diversity gain is the diversity gain at a fixed rate, i.e., the multiplexing gain $r = 0$.

The optimal diversity gain $d^*(r)$ is determined by the outage probability of the channel at a data rate of $r \log \mathsf{SNR}$ bits/s/Hz. The operational interpretation is via the existence of a universal code that achieves reliable communication simultaneously over all channels that are not in outage.

The universal code viewpoint provides a new code design criterion. Instead of averaging over the channel statistics, we consider the performance of a code over the worst-case channel that is not in outage.

- For the parallel channel, the universal criterion is to maximize the product of the codeword differences. Somewhat surprisingly, this is the same as the criterion arrived at by averaging over the Rayleigh channel statistics.
- For the MISO channel, the universal criterion is to maximize the smallest singular value of the codeword difference matrices.
- For the $n_t \times n_r$ MIMO channel, the universal criterion is to maximize the product of the n_{min} smallest singular values of the codeword difference matrices. With $n_r \geq n_t$, this criterion is the same as that arrived at by averaging over the i.i.d. Rayleigh statistics.

The MIMO channel can be transformed into a parallel channel via D-BLAST. This transformation is universal: universal parallel channel codes for each of the interleaved streams in D-BLAST serve as a universal code for the MIMO channel. The rate loss due to initialization in D-BLAST can be reduced by increasing the number of interleaved streams. For the MISO channel, however, the D-BLAST transformation with only one stream, i.e., using the transmit antennas one at a time, is approximately universal within the class of channels that have i.i.d. fading coefficients.

9.3 Bibliographical notes

The design of space-time codes has been a fertile area of research. There are books that provide a comprehensive view of the subject: for example, see the books by Larsson, Stoica and Ganesan [72], and Paulraj *et al.* [89]. Several works have recognized the tradeoff between diversity and multiplexing gains. The formulation of the coarser scaling of error probability and data rate and the corresponding characterization of their fundamental tradeoff for the i.i.d. Rayleigh fading channel is the work of Zheng and Tse [156].

The notion of universal communication, i.e., communicating reliably over a class of channel, was first formulated in the context of discrete memoryless channels by Blackwell *et al.* [10], Dobrushin [31] and Wolfowitz [146]. They showed the existence of universal codes. The results were later extended to Gaussian channels by Root and Varaiya [103]. Motivated by these information theoretic results, Wesel and his coauthors have studied the problem of universal code design in a sequence of works, starting with his Ph.D. thesis [142]. The worst-case code design metric for the parallel channel and a heuristic derivation of the product distance criterion were obtained in [143]. This was extended to MIMO channels in [67]. The general concept of approximate universality in the high SNR regime was formulated by Tavildar and Viswanath [118]; earlier, in the special case of the 2×2 MIMO channel, Yao and Wornell [152] used the determinant condition (9.80) to show the tradeoff-optimality of their rotation-based codes. The conditions derived for approximate universality, (cf. (9.38), (9.53), (9.70) and (9.80)) are also necessary; this is derived in Tavildar and Viswanath [118].

The design of tradeoff-optimal space-time codes is an active area of research, and several approaches have been presented recently. They include: rotation-based codes for the 2×2 channel, by Yao and Wornell [152] and Dayal and Varanasi [29]; lattice space-time (LAST) codes, by El Gamal *et al.* [34]; permutation codes for the parallel channel derived from D-BLAST, by Tavildar and Viswanath [118]; Golden code, by Belfiore *et al.* [5] for the 2×2 channel; codes based on cyclic divisional algebras, by Elia *et al.* [35]. The tradeoff-optimality of most of these codes is demonstrated by verifying the approximate universality conditions.

9.4 Exercises

Exercise 9.1 Consider the L-parallel channel with i.i.d. Rayleigh coefficients. Show that the optimal diversity gain at a multiplexing rate of r per sub-channel is $L - Lr$.

Exercise 9.2 Consider the repetition scheme where the same codeword is transmitted over the L i.i.d. Rayleigh sub-channels of a parallel channel. Show that the largest diversity gain this scheme can achieve at a multiplexing rate of r per sub-channel is $L(1 - Lr)$.

Exercise 9.3 Consider the repetition scheme of transmitting the same codeword over the n_t transmit antennas, one at a time, of an i.i.d. Rayleigh fading $n_t \times n_r$ MIMO channel. Show that the maximum diversity gain this scheme can achieve, at a multiplexing rate of r, is $n_t n_r (1 - n_t r)$.

Exercise 9.4 Consider using the Alamouti scheme over a $2 \times n_r$ i.i.d. Rayleigh fading MIMO channel. The transmit codeword matrix spans two symbol times $m = 1, 2$ (cf. Section 3.3.2):

$$\begin{bmatrix} u_1 & -u_2^* \\ u_2 & u_1^* \end{bmatrix}. \tag{9.84}$$

1. With this input to the MIMO channel in (9.71), show that we can write the output over the two time symbols as (cf. (3.75))

$$\begin{bmatrix} \mathbf{y}[1] \\ (\mathbf{y}[2]^*)^t \end{bmatrix} = \begin{bmatrix} \mathbf{h}_1 & \mathbf{h}_2 \\ (\mathbf{h}_2^*)^t & -(\mathbf{h}_1^*)^t \end{bmatrix} \begin{bmatrix} u_1 \\ u_2 \end{bmatrix} + \begin{bmatrix} \mathbf{w}[1] \\ (\mathbf{w}[2]^*)^t \end{bmatrix}. \tag{9.85}$$

Here we have denoted the two columns of $\mathbf{H}$ by $\mathbf{h}_1$ and $\mathbf{h}_2$.

2. Observing that the two columns of the effective channel matrix in (9.85) are orthogonal, show that we can extract simple sufficient statistics for the data symbols u_1, u_2 (cf. (3.76)):

$$r_i = \|\mathbf{H}\| u_i + w_i, \qquad i = 1, 2. \tag{9.86}$$

Here $\|\mathbf{H}\|^2$ denotes $\|\mathbf{h}_1\|^2 + \|\mathbf{h}_2\|^2$ and the additive noises w_1 and w_2 are i.i.d. $\mathcal{CN}(0, 1)$.

3. Conclude that the maximum diversity gain seen by either stream (u_1 or u_2) at a multiplexing rate of r per stream is $2n_r(1 - r)$.

Exercise 9.5 Consider the V-BLAST architecture with a bank of decorrelators for the $n_t \times n_r$ i.i.d. Rayleigh fading MIMO channel with $n_r \geq n_t$. Show that the effective channel seen by each stream is a scalar fading channel with distribution $\chi^2_{2(n_r - n_t + 1)}$. Conclude that the diversity gain with a multiplexing gain of r is $(n_r - n_t + 1)(1 - r/n_t)$.

Exercise 9.6 Verify the claim in (9.28) by showing that the sum of the pairwise error probabilities in (9.26), with $\mathbf{x}_A, \mathbf{x}_B$ each a pair of QAM symbols (the union bound on the error probability) has a decay rate of $2 - r$ with increasing SNR.

Exercise 9.7 The result in Exercise 9.6 can be generalized. Show that the diversity gain of transmitting uncoded QAMs (each at a rate of $R = r/n \log \text{SNR}$ bits/s/Hz) on the n transmit antennas of an i.i.d. Rayleigh fading MIMO channel with n receive antennas is $n - r$.

Exercise 9.8 Consider the expression for p_out^mimo in (9.29) and for p_out^iid in (9.30). Suppose that the entries of the MIMO channel $\mathbf{H}$ have some joint distribution and are not necessarily i.i.d. Rayleigh.
1. Show that

$$p_\text{out}^\text{iid}(r \log \text{SNR}) \geq p_\text{out}^\text{mimo}(r \log \text{SNR}) \geq \mathbb{P}\{\log \det(\mathbf{I}_{n_r} + \text{SNR} \, \mathbf{H}\mathbf{H}^*) < r \log \text{SNR}\}. \tag{9.87}$$

2. Show that the lower bound above decays at the same polynomial rate as p_out^iid with increasing SNR.
3. Conclude that the polynomial decay rates of both p_out^mimo and p_out^iid with increasing SNR are the same.

Exercise 9.9 Consider a scalar slow fading channel

$$y[m] = hx[m] + w[m], \tag{9.88}$$

with an optimal diversity–multiplexing tradeoff $d^*(\cdot)$, i.e.,

$$\lim_{\text{SNR}\to\infty} \frac{\log p_\text{out}(r \log \text{SNR})}{\log \text{SNR}} = -d^*(r). \tag{9.89}$$

Let $\epsilon > 0$ and consider the following event on the channel gain h:

$$\mathbb{E}_\epsilon := \{h : \log(1 + |h|^2 \text{SNR}^{1-\epsilon}) < R\}. \tag{9.90}$$

1. Show, by conditioning on the event $\mathbb{E}_\epsilon$ or otherwise, that the probability of error $p_\text{e}(\text{SNR})$ of QAM with rate $R = r \log \text{SNR}$ bits/symbol satisfies

$$\lim_{\text{SNR}\to\infty} \frac{\log p_\text{e}(\text{SNR})}{\log \text{SNR}} \leq -d^*(r)(1-\epsilon). \tag{9.91}$$

Hint: you should show that conditional on the $\mathbb{E}_\epsilon$ not happening, the probability of error decays very fast and is negligible compared to the probability of error conditional on $\mathbb{E}_\epsilon$ happening.
2. Hence, conclude that QAM achieves the diversity–multiplexing tradeoff of any scalar channel.
3. More generally, show that any constellation that satisfies the condition (9.38) achieves the diversity–multiplexing tradeoff curve of the channel.
4. Even more generally, show that any constellation that satisfies the condition

$$d_\text{min}^2 > c \cdot \frac{1}{2^R} \qquad \text{for any constant } c > 0 \tag{9.92}$$

achieves the diversity–multiplexing tradeoff curve of the channel. This shows that the condition (9.38) is really only an order-of-magnitude condition. A slightly weaker version of this condition is also necessary for a code to be approximately universal; see [118].

Exercise 9.10 Consider coding over a block length N for communication over the parallel channel in (9.17). Derive the universal code design criterion, generalizing the derivation in Section 9.2.2 over a block length of 1.

Exercise 9.11 In this exercise we will try to explicitly calculate the universal code design criterion for the parallel fading channel; for given differences between a pair of normalized codewords, the criterion is to maximize the expression in (9.49).

1. Suppose the codeword differences on all the sub-channels have the same magnitude, i.e., $|d_1| = \cdots = |d_L|$. Show that in this case the worst case channel is the same over all the sub-channels and the universal criterion in (9.49) simplifies considerably to

$$L(2^R - 1)|d_1|^2. \tag{9.93}$$

2. Suppose the codeword differences are ordered: $|d_1| \leq \cdots \leq |d_L|$.
   (a) Argue that if the worst case channel h_ℓ on the ℓth sub-channel is non-zero, then it is also non-zero on all the sub-channels $1, \ldots, \ell - 1$.
   (b) Consider the largest k such that

$$|d_k|^{2k} \leq 2^{RL}|d_1 \cdots d_k|^2 \leq |d_{k+1}|^{2k}, \tag{9.94}$$

   with $|d_{L+1}|$ defined as $+\infty$. Argue that the worst-case channel is zero on all the sub-channels $k+1, \ldots, L$. Observe that $k = L$ when all the codeword differences have the same magnitude; this is in agreement with the result in part (1).

3. Use the results of the previous part (and the notation of k from (9.94)) to derive an explicit expression for λ in (9.49):

$$\lambda^k |d_1 \cdots d_k|^2 = 2^{-RL}. \tag{9.95}$$

Conclude that the universal code design criterion is to maximize

$$\left(k(2^{RL}|d_1 d_2 \cdots d_k|^2)^{1/k} - \sum_{\ell=1}^{k} |d_\ell|^2 \right). \tag{9.96}$$

Exercise 9.12 Consider the repetition code illustrated in Figure 9.12. This code is for the 2-parallel channel with $R = 2$ bits/s/Hz per sub-channel. We would like to evaluate the value of the universal design criterion, minimized over all pairs of codewords. Show that this value is equal to 8/3. *Hint*: The smallest value is yielded by choosing the pair of codewords as nearest neighbors in the QAM constellation. Since this is a repetition code, the codeword differences are the same for both the channels; now use (9.93) to evaluate the universal design criterion.

Exercise 9.13 Consider the permutation code illustrated in Figure 9.13 (with $R = 2$ bits/s/Hz per sub-channel). Show that the smallest value of the universal design criterion, minimized over all choices of codeword pairs, is equal to 44/9.

Exercise 9.14 In this exercise we will explore the implications of the condition for approximate universality in (9.53).

1. Show that if a parallel channel scheme satisfies the condition (9.53), then it achieves the diversity–multiplexing tradeoff of the parallel channel. *Hint*: Do Exercise 9.9 first.

2. Show that the diversity–multiplexing tradeoff can still be achieved even when the scheme satisfies a more relaxed condition:

$$|d_1 d_2 \cdots d_L|^{2/L} > c \cdot \frac{1}{L 2^R}, \qquad \text{for some constant } c > 0. \tag{9.97}$$

Exercise 9.15 Consider the class of permutatation codes for the L-parallel channel described in Section 9.2.2. The codeword is described as $(q, \pi_2(q), \ldots, \pi_L(q))$ where q belongs to a normalized QAM (so that each of the I and Q channels are peak constrained by ± 1) with 2^{LR} points; so, the rate of the code is R bits/s/Hz per sub-channel. In this exercise we will see that this class contains approximately universal codes.

1. Consider *random* permutations with the uniform measure; since there are $2^{LR}!$ of them, each of the permutations occurs with probability $1/2^{LR}!$. Show that the average inverse product of the pairwise codeword differences, averaged over both the codeword pairs and the random permutations, is upper bounded as follows:

$$\mathbb{E}_{\pi_2, \ldots, \pi_L} \left[\frac{1}{2^{LR}(2^{LR} - 1)} \right.$$

$$\left. \times \sum_{q_1 \neq q_2} \frac{1}{|q_1 - q_2|^2 |\pi_2(q_1) - \pi_2(q_2)|^2 \cdots |\pi_L(q_1) - \pi_L(q_2)|^2} \right] \leq L^L R^L. \tag{9.98}$$

2. Conclude from the previous part that there exist permutations $\pi_2, \ldots, \pi_L$ such that

$$\frac{1}{2^{LR}} \sum_{q_1} \left(\sum_{q_2 \neq q_1} \frac{1}{|q_1 - q_2|^2 |\pi_2(q_1) - \pi_2(q_2)|^2 \cdots |\pi_L(q_1) - \pi_L(q_2)|^2} \right)$$

$$\leq L^L R^L 2^{LR}. \tag{9.99}$$

3. Now suppose we fix q_1 and consider the sum of the inverse product of all the possible pairwise codeword differences:

$$f(q_1) := \sum_{q_2 \neq q_1} \frac{1}{|q_1 - q_2|^2 |\pi_2(q_1) - \pi_2(q_2)|^2 \cdots |\pi_L(q_1) - \pi_L(q_2)|^2}. \tag{9.100}$$

Since $f(q_1) \geq 0$, argue from (9.99) that at least half the QAM points q_1 must have the property that

$$f(q_1) \leq 2 L^L R^L 2^{LR}. \tag{9.101}$$

Further, conclude that for such q_1 (they make up at least half of the total QAM points) we must have for every $q_2 \neq q_1$ that

$$|q_1 - q_2|^2 |\pi_2(q_1) - \pi_2(q_2)|^2 \cdots |\pi_L(q_1) - \pi_L(q_2)|^2 \geq \frac{1}{2 L^L R^L 2^{LR}}. \tag{9.102}$$

4. Finally, conclude that there exists a permutation code that is approximately universal for the parallel channel by arguing the following:
   - Expurgating no more than half the number of QAM points only reduces the total rate LR by no more than 1 bit/s/Hz and thus does not affect the multiplexing gain.
   - The product distance condition on the permutation codeword differences in (9.102) does not quite satisfy the condition for approximate universality in (9.97). Relax the condition in (9.97) to

$$|d_1 d_2 \cdots d_L|^{2/L} > c \cdot \frac{1}{R 2^R}, \qquad \text{for some constant } c > 0, \qquad (9.103)$$

   and show that this is sufficient for a code to achieve the optimal diversity–multiplexing tradeoff curve.

Exercise 9.16 Consider the bit-reversal scheme for the parallel channel described in Section 9.2.2. Strictly speaking, the condition in (9.57) is not true for every integer between 0 and $2^R - 1$. However, the set of integers for which this is not true is small (i.e., expurgating them will not change the multiplexing rate of the scheme). Thus the bit-reversal scheme with an appropriate expurgation of codewords is approximately universal for the 2-parallel channel. A reading exercise is to study [118] where the expurgated bit-reversal scheme is described in detail.

Exercise 9.17 Consider the bit-reversal scheme described in Section 9.2.2 but with every alternate bit *flipped* after the reversal. Then for every pair of normalized codeword differences, it can be shown that

$$|d_1 d_2|^2 > \frac{1}{64 \cdot 2^{2R}}, \qquad (9.104)$$

where the data rate is R bits/s/Hz per sub-channel. Argue now that the bit-reversal scheme with alternate bit flipping is approximately universal for the 2-parallel channel. A reading exercise is to study the proof of (9.104) in [118]. *Hint*: Compare (9.104) with (9.53) and use the result derived in Exercise 9.14.

Exercise 9.18 Consider a MISO channel with the fading channels from the n_t transmit antennas, $h_1, \ldots, h_{n_t}$, i.i.d.
1. Show that

$$\mathbb{P}\left\{ \log\left(1 + \frac{\mathsf{SNR}}{n_t} \sum_{\ell=1}^{n_t} |h_\ell|^2\right) < r \log \mathsf{SNR} \right\} \qquad (9.105)$$

   and

$$\mathbb{P}\left\{ \sum_{\ell=1}^{n_t} \log(1 + \mathsf{SNR}|h_\ell|^2) < n_t r \log \mathsf{SNR} \right\} \qquad (9.106)$$

   have the same decay rate with increasing SNR.
2. Interpret (9.105) and (9.106) with the outage probabilities of the MISO channel and that of a parallel channel obtained through an appropriate transformation of the MISO channel, respectively. Argue that the conversion of the MISO channel into a parallel channel discussed in Section 9.2.3 is approximately universal for the class of i.i.d. fading coefficients.

Exercise 9.19 Consider an $n_t \times n_t$ matrix $\mathbf{D}$. Show that

$$\min_{\mathbf{h}: \|\mathbf{h}\|=1} \mathbf{h}^*\mathbf{D}\mathbf{D}^*\mathbf{h} = \lambda_1^2, \tag{9.107}$$

where λ_1 is the smallest singular value of $\mathbf{D}$.

Exercise 9.20 Consider the Alamouti transmit codeword (cf. (9.84)) with u_1, u_2 independent uncoded QAMs with 2^R points in each.

1. For every codeword difference matrix

$$\begin{bmatrix} d_1 & -d_2^* \\ d_2 & d_1^* \end{bmatrix}, \tag{9.108}$$

show that the two singular values are the same and equal to $\sqrt{|d_1|^2 + |d_2|^2}$.

2. With the codeword difference matrix normalized as in (9.68) and each of the QAM symbols u_1, u_2 constrained in power of $\mathsf{SNR}/2$ (i.e., both the I and Q channels are peak constrained by $\pm\sqrt{\mathsf{SNR}/2}$), show that if the codeword difference d_ℓ is not zero, then it is

$$|d_\ell|^2 \geq \frac{2}{2^R}, \qquad \ell = 1, 2.$$

3. Conclude from the previous steps that the square of the smallest singular value of the codeword difference matrix is lower bounded by $2/2^R$. Since the condition for approximate universality in (9.70) is an order-of-magnitude one (the constant factor next to the 2^R term does not matter, see Exercises 9.9 and 9.14), we have explicitly shown that the Alamouti scheme with uncoded QAMs on the two streams is approximately universal for the two transmit antenna MISO channel.

Exercise 9.21 Consider the D-BLAST architecture in (9.77) with just two interleaved streams for the 2×2 i.i.d. Rayleigh fading MIMO channel. The two streams are independently coded at rate $R = r \log \mathsf{SNR}$ bits/s/Hz each and composed of the pair of codewords $\left(x_A^{(\ell)}, x_B^{(\ell)}\right)$ for $\ell = 1, 2$. The two streams are coded using an approximately universal parallel channel code (say, the bit-reversal scheme described in Section 9.2.2).

A union bound averaged over the Rayleigh MIMO channel can be used to show that the diversity gain obtained by each stream with joint ML decoding is $4 - 2r$. A reading exercise is to study the proof of this result in [118].

Exercise 9.22 [67] Consider transmitting codeword matrices of length at least n_t on the $n_t \times n_r$ MIMO slow fading channel at rate R bits/s/Hz (cf. (9.71)).

1. Show that the pairwise error probability between two codeword matrices $\mathbf{X}_A$ and $\mathbf{X}_B$, conditioned on a specific realization of the MIMO channel $\mathbf{H}$, is

$$Q\left(\sqrt{\frac{\mathsf{SNR}}{2}\|\mathbf{HD}\|^2}\right), \tag{9.109}$$

where $\mathbf{D}$ is the normalized codeword difference matrix (cf. (9.68)).

2. Writing the SVDs $\mathbf{H} := \mathbf{U}_1 \mathbf{\Psi} \mathbf{V}_1^*$ and $\mathbf{D} := \mathbf{U}_2 \mathbf{\Lambda} \mathbf{V}_2^*$, show that the pairwise error probability in (9.109) can be written as

$$Q\left(\sqrt{\frac{\mathsf{SNR}}{2} \|\mathbf{\Psi} \mathbf{V}_1^* \mathbf{U}_2 \mathbf{\Lambda}\|^2}\right). \tag{9.110}$$

3. Suppose the singular values are increasingly ordered in $\mathbf{\Lambda}$ and decreasingly ordered in $\mathbf{\Psi}$. For fixed $\mathbf{\Psi}, \mathbf{\Lambda}, \mathbf{U}_2$, show that the channel eigendirections $\mathbf{V}_1^*$ that minimize the pairwise error probability in (9.110) are

$$\mathbf{V}_1 = \mathbf{U}_2. \tag{9.111}$$

4. Observe that the channel outage condition depends only on the singular values $\mathbf{\Psi}$ of $\mathbf{H}$ (cf. Exercise 9.8). Use the previous parts to conclude that the calculation of the worst-case pairwise error probability for the MIMO channel reduces to the optimization problem

$$\min_{\psi_1, \ldots, \psi_{n_{\min}}} \frac{\mathsf{SNR}}{2} \sum_{\ell=1}^{L} |\psi_\ell|^2 |\lambda_\ell|^2, \tag{9.112}$$

subject to the constraint

$$\sum_{\ell=1}^{n_{\min}} \log\left(1 + \frac{\mathsf{SNR}}{n_{\mathsf{t}}} |\psi_\ell|^2\right) \geq R. \tag{9.113}$$

Here we have written

$$\mathbf{\Psi} := \mathrm{diag}\{\psi_1, \ldots, \psi_{n_{\min}}\}, \quad \text{and} \quad \mathbf{\Lambda} := \mathrm{diag}\{\lambda_1, \ldots, \lambda_{n_{\mathsf{t}}}\}.$$

5. Observe that the optimization problem in (9.112) and the constraint (9.113) are very similar to the corresponding ones in the *parallel* channel (cf. (9.43) and (9.40), respectively). Thus the universal code design criterion for the MIMO channel is the same as that of a parallel channel (cf. (9.47)) with the following parameters:
   - there are $n_{\min}$ sub-channels,
   - the rate per sub-channel is $R/n_{\min}$ bits/s/Hz,
   - the parallel channel coefficients are $\psi_1, \ldots, \psi_{n_{\min}}$, the singular values of the MIMO channel, and
   - the codeword differences are the smallest singular values, $\lambda_1, \ldots, \lambda_{n_{\min}}$, of the codeword difference matrix.

Exercise 9.23 Using the analogy between the worst-case pairwise error probability of a MIMO channel and that of an appropriately defined parallel channel (cf. Exercise 9.22), justify the condition for approximate universality for the MIMO channel in (9.79).

Exercise 9.24 Consider transmitting codeword matrices of length $l \geq n_{\mathsf{t}}$ on the $n_{\mathsf{t}} \times n_{\mathsf{r}}$ MIMO slow fading channel. The total power constraint is SNR, so for any transmit codeword matrix $\mathbf{X}$, we have $\|\mathbf{X}\|^2 \leq l\mathsf{SNR}$. For a pair of codeword matrices $\mathbf{X}_A$ and $\mathbf{X}_B$, let the normalized codeword difference matrix be $\mathbf{D}$ (normalized as in (9.68)).

1. Show that **D** satisfies

$$\|\mathbf{D}\|^2 \le \frac{2}{\mathsf{SNR}}(\|\mathbf{X}_A\|^2 + \|\mathbf{X}_B\|^2) \le 4l. \tag{9.114}$$

2. Writing the singular values of **D** as $\lambda_1, \ldots, \lambda_{n_t}$, show that

$$\sum_{\ell=1}^{n_t} \lambda_\ell^2 \le 4l. \tag{9.115}$$

Thus, each of the singular values is upper bounded by $2\sqrt{l}$, a constant that does not increase with SNR.

Exercise 9.25 [152] Consider the following transmission scheme (spanning two symbols) for the two transmit antenna MIMO channel. The entries of the transmit codeword matrix $\mathbf{X} := [x_{ij}]$ are defined as

$$\begin{bmatrix} x_{11} \\ x_{22} \end{bmatrix} := \mathbf{R}(\theta_1)\begin{bmatrix} u_1 \\ u_2 \end{bmatrix}, \quad \text{and} \quad \begin{bmatrix} x_{21} \\ x_{12} \end{bmatrix} := \mathbf{R}(\theta_2)\begin{bmatrix} u_3 \\ u_4 \end{bmatrix}. \tag{9.116}$$

Here u_1, u_2, u_3, u_4 are independent QAMs of size $2^{R/2}$ each (so the data rate of this scheme is R bits/s/Hz). The rotation matrix $\mathbf{R}(\theta)$ is (cf. (3.46))

$$\mathbf{R}(\theta) := \begin{bmatrix} \cos\theta & -\sin\theta \\ \sin\theta & \cos\theta \end{bmatrix}. \tag{9.117}$$

With the choice of the angles θ_1, θ_2 equal to $1/2\tan^{-1}2$ and $1/2\tan^{-1}(1/2)$ radians respectively, Theorem 2 of [152] shows that the determinant of every normalized codeword difference matrix **D** satisfies

$$|\det \mathbf{D}|^2 \ge \frac{1}{10 \cdot 2^R}. \tag{9.118}$$

Conclude that the code described in (9.116), with the appropriate choice of the angles θ_1, θ_2 above, is approximately universal for every MIMO channel with two transmit antennas.

10 MIMO IV: multiuser communication

In Chapters 8 and 9, we have studied the role of multiple transmit and receive antennas in the context of point-to-point channels. In this chapter, we shift the focus to multiuser channels and study the role of multiple antennas in both the uplink (many-to-one) and the downlink (one-to-many). In addition to allowing spatial multiplexing and providing diversity to each user, multiple antennas allow the base-station to simultaneously transmit or receive data from multiple users. Again, this is a consequence of the increase in degrees of freedom from having multiple antennas.

We have considered several MIMO transceiver architectures for the point-to-point channel in Chapter 8. In some of these, such as linear receivers with or without successive cancellation, the complexity is mainly at the receiver. Independent data streams are sent at the different transmit antennas, and no cooperation across transmit antennas is needed. Equating the transmit antennas with users, these receiver structures can be directly used in the uplink where the users have a single transmit antenna each but the base-station has multiple receive antennas; this is a common configuration in cellular wireless systems.

It is less apparent how to come up with good strategies for the *downlink*, where the *receive* antennas are at the different users; thus the receiver structure has to be separate, one for each user. However, as will see, there is an interesting duality between the uplink and the downlink, and by exploiting this duality, one can map each receive architecture for the uplink to a corresponding transmit architecture for the downlink. In particular, there is an interesting *precoding* strategy, which is the "transmit dual" to the receiver-based successive cancellation strategy. We will spend some time discussing this.

The chapter is structured as follows. In Section 10.1, we first focus on the uplink with a single transmit antenna for each user and multiple receive antennas at the base-station. We then, in Section 10.2, extend our study to the MIMO uplink where there are multiple transmit antennas for each user. In Sections 10.3 and 10.4, we turn our attention to the use of multiple antennas in the downlink. We study precoding strategies that achieve the capacity of the downlink. We conclude in Section 10.5 with a discussion of the system implications of using MIMO in cellular networks; this will link up the new insights obtained here with those in Chapters 4 and 6.

10.1 Uplink with multiple receive antennas

We begin with the narrowband time-invariant uplink with each user having a single transmit antenna and the base-station equipped with an array of antennas (Figure 10.1). The channels from the users to the base-station are time-invariant. The baseband model is

$$\mathbf{y}[m] = \sum_{k=1}^{K} \mathbf{h}_k x_k[m] + \mathbf{w}[m], \tag{10.1}$$

with $\mathbf{y}[m]$ being the received vector (of dimension n_r, the number of receive antennas) at time m, and $\mathbf{h}_k$ the spatial signature of user k impinged on the receive antenna array at the base-station. User k's scalar transmit symbol at time m is denoted by $x_k[m]$ and $\mathbf{w}[m]$ is i.i.d. $\mathcal{CN}(0, N_0\mathbf{I}_{n_r})$ noise.

10.1.1 Space-division multiple access

In the literature, the use of multiple receive antennas in the uplink is often called *space-division multiple access* (SDMA): we can discriminate amongst the users by exploiting the fact that different users impinge different spatial signatures on the receive antenna array.

An easy observation we can make is that this uplink is very similar to the MIMO point-to-point channel in Chapter 7 except that the signals sent out on the transmit antennas cannot be coordinated. We studied precisely such a signaling scheme using separate data streams on each of the transmit antennas in Section 8.3. We can form an analogy between users and transmit antennas (so n_t, the number of transmit antennas in the MIMO point-to-point channel in Section 8.3, is equal to the number of users K). Further, the equivalent MIMO point-to-point channel $\mathbf{H}$ is $[\mathbf{h}_1, \ldots, \mathbf{h}_K]$, constructed from the SIMO channels of the users.

Thus, the transceiver architecture in Figure 8.1 in conjunction with the receiver structures in Section 8.3 can be used as an SDMA strategy. For example, each of the user's signal can be demodulated using a linear decorrelator or an MMSE receiver. The MMSE receiver is the optimal compromise between maximizing the signal strength from the user of interest and suppressing the interference from the other users. To get better performance, one can also augment the linear receiver structure with successive cancellation to yield the MMSE–SIC receiver (Figure 10.2). With successive cancellation, there is also a further choice of cancellation ordering. By choosing a

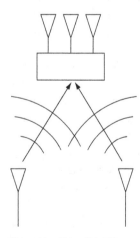

Figure 10.1 The uplink with single transmit antenna at each user and multiple receive antennas at the base-station.

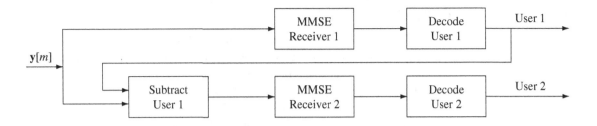

Figure 10.2 The MMSE–SIC receiver: user 1's data is first decoded and then the corresponding transmit signal is subtracted off before the next stage. This receiver structure, by changing the ordering of cancellation, achieves the two corner points in the capacity region.

different order, users are prioritized differently in the sharing of the common resource of the uplink channel, in the sense that users canceled later are treated better.

Provided that the overall channel matrix $\mathbf{H}$ is well-conditioned, all of these SDMA schemes can fully exploit the total number of degrees of freedom $\min\{K, n_r\}$ of the uplink channel (although, as we have seen, different schemes have different power gains). This translates to being able to simultaneously support multiple users, each with a data rate that is not limited by interference. Since the users are geographically separated, their transmit signals arrive in different directions at the receive array even when there is limited scattering in the environment, and the assumption of a well-conditioned $\mathbf{H}$ is usually valid. (Recall Example 7.4 in Section 7.2.4.) Contrast this to the point-to-point case when the transmit antennas are co-located, and a rich scattering environment is needed to provide a well-conditioned channel matrix $\mathbf{H}$.

Given the power levels of the users, the achieved SINR of each user can be computed for the different SDMA schemes using the formulas derived in Section 8.3 (Exercise 10.1). Within the class of linear receiver architecture, we can also formulate a power control problem: given target SINR requirements for the users, how does one optimally choose the powers and linear filters to meet the requirements? This is similar to the uplink CDMA power control problem described in Section 4.3.1, except that there is a further flexibility in the choice of the receive filters as well as the transmit powers. The first observation is that for *any* choice of transmit powers, one always wants to use the MMSE filter for each user, since that choice maximizes the SINR for every user. Second, the power control problem shares the basic *monotonicity* property of the CDMA problem: when a user lowers its transmit power, it creates less interference and benefits all other users in the system. As a consequence, there is a component-wise optimal solution for the powers, where every user is using the minimum possible power to support the SINR requirements. (See Exercise 10.2.) A simple distributed power control algorithm will converge to the optimal solution: at each step, each user first updates its MMSE filter as a function of the current power levels of the other users, and then updates its own transmit power so that its SINR requirement is just met. (See Exercise 10.3.)

10.1.2 SDMA capacity region

In Section 8.3.4, we have seen that the MMSE–SIC receiver achieves the best total rate among all the receiver structures. The performance limit of the uplink channel is characterized by the notion of a *capacity region*, introduced in Chapter 6. How does the performance achieved by MMSE–SIC compare to this limit?

With a *single* receive antenna at the base-station, the capacity region of the two-user uplink channel was presented in Chapter 6; it is the pentagon in Figure 6.2:

$$R_1 < \log\left(1 + \frac{P_1}{N_0}\right),$$

$$R_2 < \log\left(1 + \frac{P_2}{N_0}\right),$$

$$R_1 + R_2 < \log\left(1 + \frac{P_1 + P_2}{N_0}\right),$$

where P_1 and P_2 are the average power constraints on users 1 and 2 respectively. The individual rate constraints correspond to the maximum rate that each user can get if it has the entire channel to itself; the sum rate constraint is the total rate of a point-to-point channel with the two users acting as two transmit antennas of a single user, but sending independent signals.

The SDMA capacity region, for the *multiple* receive antenna case, is the natural extension (Appendix B.9 provides a formal justification):

$$R_1 < \log\left(1 + \frac{\|\mathbf{h}_1\|^2 P_1}{N_0}\right), \tag{10.2}$$

$$R_2 < \log\left(1 + \frac{\|\mathbf{h}_2\|^2 P_2}{N_0}\right), \tag{10.3}$$

$$R_1 + R_2 < \log \det\left(\mathbf{I}_{n_r} + \frac{1}{N_0}\mathbf{H}\mathbf{K}_x\mathbf{H}^*\right), \tag{10.4}$$

where $\mathbf{K}_x = \mathrm{diag}(P_1, P_2)$. The capacity region is plotted in Figure 10.3.

The capacities of the point-to-point SIMO channels from each user to the base-station serve as the maximum rate each user can reliably communicate at if it has the entire channel to itself. These yield the constraints (10.2) and (10.3). The point-to-point capacity for user $k(k = 1, 2)$ is achieved by receive beamforming (projecting the received vector $\mathbf{y}$ in the direction of $\mathbf{h}_k$), converting the effective channel into a SISO one, and then decoding the data of the user.

Inequality (10.4) is a constraint on the sum of the rates that the users can communicate at. The right hand side is the total rate achieved in a point-to-point channel with the two users acting as two transmit antennas of one user with independent inputs at the antennas (cf. (8.2)).

Figure 10.3 Capacity region of the two-user SDMA uplink.

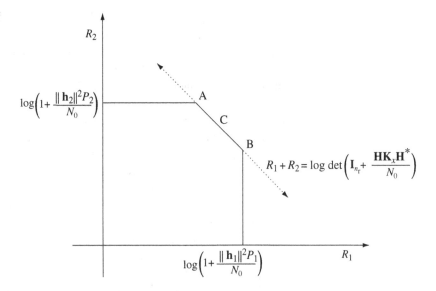

Since MMSE–SIC receivers (in Figure 10.2) are optimal with respect to achieving the total rate of the point-to-point channel with the two users acting as two transmit antennas of one user, it follows that the rates for the two users that this architecture can achieve in the uplink meets inequality (10.4) with equality. Moreover, if we cancel user 1 first, user 2 only has to contend with the background Gaussian noise and its performance meets the single-user bound (10.2). Hence, we achieve the corner point A in Figure 10.3. By reversing the cancellation order, we achieve the corner point B. Thus, MMSE–SIC receivers are information theoretically optimal for SDMA in the sense of achieving rate pairs corresponding to the two corner points A and B. Explicitly, the rate point A is given by the rate tuple (R_1, R_2):

$$R_2 = \log\left(1 + \frac{P_2 \|\mathbf{h}_2\|^2}{N_0}\right),$$
$$R_1 = \log(1 + P_1 \mathbf{h}_1^* (N_0 \mathbf{I}_{n_r} + P_2 \mathbf{h}_2 \mathbf{h}_2^*)^{-1} \mathbf{h}_1), \tag{10.5}$$

where $P_1 \mathbf{h}_1^* (N_0 \mathbf{I}_{n_r} + P_2 \mathbf{h}_2^* \mathbf{h}_2^*)^{-1} \mathbf{h}_1$ is the output SIR of the MMSE receiver for user 1 treating user 2's signal as colored Gaussian interference (cf. (8.62)).

For the single receive antenna (scalar) uplink channel, we have already seen in Section 6.1 that the corner points are also achievable by the SIC receiver, where at each stage a user is decoded treating all the uncanceled users as Gaussian noise. In the vector case with multiple receive antennas, the uncanceled users are also treated as Gaussian noise, but now this is a colored vector Gaussian noise. The MMSE filter is the optimal demodulator for a user in the face of such colored noise (cf. Section 8.3.3). Thus, we see that successive cancellation with MMSE filtering at each stage is the natural generalization of the SIC receiver we developed for the single antenna channel. Indeed, as explained in

Section 8.3.4, the SIC receiver is really just a special case of the MMSE–SIC receiver when there is only one receive antenna, and they are optimal for the same reason: they "implement" the chain rule of mutual information.

A comparison between the capacity regions of the uplink with and without multiple receive antennas (Figure 6.2 and Figure 10.3, respectively) highlights the importance of having multiple receive antennas in allowing SDMA. Let us focus on the high SNR scenario when N_0 is very small as compared with P_1 and P_2. With a single receive antenna at the base-station, we see from Figure 6.2 that there is a total of only one spatial degree of freedom, shared between the users. In contrast, with multiple receive antennas we see from Figure 10.3 that while the individual rates of the users have no more than one spatial degree of freedom, the sum rate has *two* spatial degrees of freedom. This means that both users can simultaneously enjoy one spatial degree of freedom, a scenario made possible by SDMA and not possible with a single receive antenna. The intuition behind this is clear when we look back at our discussion of the decorrelator (cf. Section 8.3.1). The received signal space has more dimensions than that spanned by the transmit signals of the users. Thus in decoding user 1's signal we can project the received signal in a direction orthogonal to the transmit signal of user 2, completely eliminating the inter-user interference (the analogy between streams and users carries forth here as well). This allows two effective parallel channels at high SNR. Improving the simple decorrelator by using the MMSE–SIC receiver allows us to *exactly* achieve the information theoretic limit.

In the light of this observation, we can take a closer look at the two corner points in the boundary of the capacity region (points A and B in Figure 10.3). If we are operating at point A we see that both users 1 and 2 have one spatial degree of freedom each. The point C, which corresponds to the symmetric capacity of the uplink (cf. (6.2)), also allows both users to have unit spatial degree of freedom. (In general, the symmetric capacity point C need not lie on the line segment joining points A and B; however it will be the center of this line segment when the channels are symmetric, i.e., $\|\mathbf{h}_1\| = \|\mathbf{h}_2\|$.) While the point C cannot be achieved directly using the receiver structure in Figure 10.2, we can achieve that rate pair by time-sharing between the operating points A and B (these two latter points can be achieved by the MMSE–SIC receiver).

Our discussion has been restricted to the two-user uplink. The extension to K users is completely natural. The capacity region is now a K-dimensional polyhedron: the set of rates $(R_1, \ldots, R_K)$ such that

$$\sum_{k \in S} R_k < \log \det \left(\mathbf{I}_{n_r} + \frac{1}{N_0} \sum_{k \in S} P_k \mathbf{h}_k \mathbf{h}_k^* \right), \quad \text{for each } S \subset \{1, \ldots, K\}. \quad (10.6)$$

There are $K!$ corner points on the boundary of the capacity region and each corner point is specified by an ordering of the K users and the corresponding rates are achieved by an MMSE–SIC receiver with that ordering of cancelling users.

10.1.3 System implications

What are the practical ways of exploiting multiple receive antennas in the uplink, and how does their performance compare to capacity? Let us first consider the narrowband system from Chapter 4 where the allocation of resources among the users is orthogonal. In Section 6.1 we studied orthogonal multiple access for the uplink with a single receive antenna at the base-station. Analogous to (6.8) and (6.9), the rates achieved by two users, when the base-station has multiple receive antennas and a fraction α of the degrees of freedom is allocated to user 1, are

$$\left(\alpha \log \left(1 + \frac{P_1 \|\mathbf{h}_1\|^2}{\alpha N_0} \right), (1 - \alpha) \log \left(1 + \frac{P_2 \|\mathbf{h}_2\|^2}{(1 - \alpha) N_0} \right) \right). \tag{10.7}$$

It is instructive to compare this pair of rates with the one obtained with orthogonal multiple access in the single receive antenna setting (cf. (6.8) and (6.9)). The difference is that the received SNR of user k is boosted by a factor $\|\mathbf{h}_k\|^2$; this is the receive beamforming power gain. There is however no gain in the degrees of freedom: the total is still one. The power gain allows the users to reduce their transmit power for the same received SNR level. However, due to orthogonal resource allocation and sparse reuse of the bandwidth, narrowband systems already operate at high SNR and in this situation a power gain is not much of a system benefit. A degree-of-freedom gain would have made a larger impact.

At high SNR, we have already seen that the two-user SDMA sum capacity has two spatial degrees of freedom as opposed to the single one with only one receive antenna at the base-station. Thus, orthogonal multiple access makes very poor use of the available spatial degrees of freedom when there are multiple receive antennas. Indeed, this can be seen clearly from a comparison of the orthogonal multiple access rates with the capacity region. With a single receive antenna, we have found that we can get to exactly one point on the boundary of the uplink capacity region (see Figure 6.4); the gap is not too large unless there is a significant power disparity. With multiple receive antennas, Figure 10.4 shows that the orthogonal multiple access rates are strictly suboptimal at all points[1] and the gap is also larger.

Intuitively, to exploit the available degrees of freedom *both* users must access the channel simultaneously and their signals should be separable at the base-station (in the sense that $\mathbf{h}_1$ and $\mathbf{h}_2$, the receive spatial signatures of the users at the base-station, are linearly independent). To get this benefit, more complex signal processing is required at the receiver to extract the signal of each user from the aggregate. The complexity of SDMA grows with the number of users K when there are more users in the system. On the

---

[1] Except for the degenerate case when $\mathbf{h}_1$ and $\mathbf{h}_2$ are multiples of each other; see Exercise 10.4.

Figure 10.4 The two-user uplink with multiple receive antennas at the base-station: performance of orthogonal multiple access is strictly inferior to the capacity.

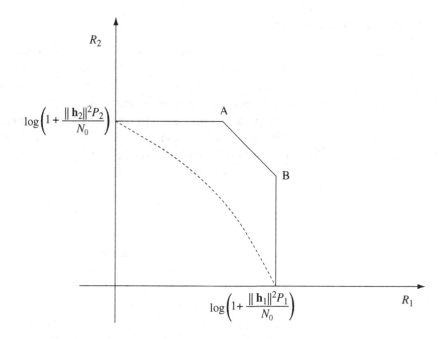

other hand, the available degrees of freedom are limited by the number of receive antennas, n_r, and so there is no further degree-of-freedom gain beyond having n_r users performing SDMA simultaneously. This suggests a nearly optimal multiple access strategy where the users are divided into groups of n_r users with SDMA *within* each group and orthogonal multiple access *between* the groups. Exercise 10.5 studies the performance of this scheme in greater detail.

On the other hand, at low SNR, the channel is power-limited rather than degrees-of-freedom-limited and SDMA provides little performance gain over orthogonal multiple access. This can be observed by an analysis as in the characterization of the capacity of MIMO channels at low SNR, cf. Section 8.2.2, and is elaborated in Exercise 10.6.

In general, multiple receive antennas can be used to provide beamforming gain for the users. While this power gain is not of much benefit to the narrowband systems, both the wideband CDMA and wideband OFDM uplink operate at low SNR and the power gain is more beneficial.

Summary 10.1 SDMA and orthogonal multiple access

The MMSE–SIC receiver is optimal for achieving SDMA capacity.

SDMA with n_r receive antennas and K users provides $\min(n_r, K)$ spatial degrees of freedom.

> Orthogonal multiple access with n_r receive antennas provides only one spatial degree of freedom but n_r-fold power gain.
>
> Orthogonal multiple access provides comparable performance to SDMA at low SNR but is far inferior at high SNR.

10.1.4 Slow fading

We introduce fading first in the scenario when the delay constraint is small relative to the coherence time of all the users: the slow fading scenario. The uplink fading channel can be written as an extension of (10.1), as

$$\mathbf{y}[m] = \sum_{k=1}^{K} \mathbf{h}_k[m] x_k[m] + \mathbf{w}[m]. \tag{10.8}$$

In the slow fading model, for every user k, $\mathbf{h}_k[m] = \mathbf{h}_k$ for all time m. As in the uplink with a single antenna (cf. Section 6.3.1), we will analyze only the symmetric uplink: the users have the same transmit power constraint, P, and further, the channels of the users are statistically independent and identical. In this situation, symmetric capacity is a natural performance measure and we suppose the users are transmitting at the same rate R bits/s/Hz.

Conditioned on a realization of the received spatial signatures $\mathbf{h}_1, \ldots, \mathbf{h}_K$, we have the time-invariant uplink studied in Section 10.1.2. When the symmetric capacity of this channel is less than R, an outage results. The probability of the outage event is, from (10.6),

$$p_{\text{out}}^{\text{ul-mimo}} := \mathbb{P}\left\{ \log \det \left(\mathbf{I}_{n_r} + \mathsf{SNR} \sum_{k \in \mathcal{S}} \mathbf{h}_k \mathbf{h}_k^* \right) < |\mathcal{S}| R, \right.$$
$$\left. \text{for some } \mathcal{S} \subset \{1, \ldots, K\} \right\}. \tag{10.9}$$

Here we have written $\mathsf{SNR} := P/N_0$. The corresponding largest rate R such that $p_{\text{out}}^{\text{ul-mimo}}$ is less than or equal to ϵ is the ϵ-outage symmetric capacity C_ϵ^{sym}. With a single user in the system, C_ϵ^{sym} is simply the ϵ-outage capacity, $C_\epsilon(\mathsf{SNR})$, of the point-to-point channel with receive diversity studied in Section 5.4.2. More generally, with $K > 1$, C_ϵ^{sym} is upper bounded by this quantity: with more users, inter-user interference is another source of error.

Orthogonal multiple access completely eliminates inter-user interference and the corresponding largest symmetric outage rate is, as in (6.33),

$$\frac{C_{\epsilon/K}(K\mathsf{SNR})}{K}. \tag{10.10}$$

We can see, just as in the situation when the base-station has a single receive antenna (cf. Section 6.3.1), that orthogonal multiple access at low SNR is close to optimal. At low SNR, we can approximate $p_{\text{out}}^{\text{ul-mimo}}$ (with $n_r = 1$, a similar approximation is in (6.34)):

$$p_{\text{out}}^{\text{ul-mimo}} \approx K p_{\text{out}}^{\text{rx}}, \qquad (10.11)$$

where $p_{\text{out}}^{\text{rx}}$ is the outage probability of the point-to-point channel with receive diversity (cf. (5.62)). Thus C_ϵ^{sym} is approximately $C_{\epsilon/K}(\text{SNR})$. On the other hand, the rate in (10.10) is also approximately equal to $C_{\epsilon/K}(\text{SNR})$ at low SNR.

At high SNR, we have seen that orthogonal multiple access is suboptimal, both in the context of outage performance with a single receive antenna and the capacity region of SDMA. A better baseline performance can be obtained by considering the outage performance of the bank of decorrelators: this receiver structure performed well in terms of the capacity of the point-to-point MIMO channel, cf. Figure 8.9. With the decorrelator bank, the inter-user interference is completely nulled out (assuming $n_r \geq K$). Further, with i.i.d. Rayleigh fading, each user sees an effective point-to-point channel with $n_r - K + 1$ receive diversity branches (cf. Section 8.3.1). Thus, the largest symmetric outage rate is exactly the ϵ-outage capacity of the point-to-point channel with $n_r - K + 1$ receive diversity branches, leading to the following interpretation:

> Using the bank of decorrelators, increasing the number of receive antennas, n_r, by 1 allows us to *either* admit one extra user with the same outage performance for each user, *or* increase the effective number of diversity branches seen by each user by 1.

How does the outage performance improve if we replace the bank of decorrelators with the joint ML receiver? The direct analysis of C_ϵ^{sym} at high SNR is quite involved, so we resort to the use of the coarser diversity–multiplexing tradeoff introduced in Chapter 9 to answer this question. For the bank of decorrelators, the diversity gain seen by each user is $(n_r - K + 1)(1 - r)$ where r is the multiplexing gain of each user (cf. Exercise 9.5). This provides a lower bound to the diversity–multiplexing performance of the joint ML receiver. On the other hand, the outage performance of the uplink cannot be better than the situation when there is no inter-user interference, i.e., each user sees a point-to-point channel with receiver diversity of n_r branches. This is the *single-user* upper bound. The corresponding single-user tradeoff curve is $n_r(1 - r)$. These upper and lower bounds to the outage performance are plotted in Figure 10.5.

The tradeoff curve with the joint ML receiver in the uplink can be evaluated: with more receive antennas than the number of users (i.e., $n_r \geq K$), the tradeoff curve is the *same* as the upper bound derived with each user seeing no inter-user interference. In other words, the tradeoff curve is $n_r(1 - r)$ and single-user performance is achieved even though there are other users in

Figure 10.5 The diversity–multiplexing tradeoff curves for the uplink with a bank of decorrelators (equal to $(n_r - K + 1)(1 - r)$, a lower bound to the outage performance with the joint ML receiver) and that when there is no inter-user interference (equal to $n_r(1 - r)$, the single-user upper bound to the outage performance of the uplink). The latter is actually achievable.

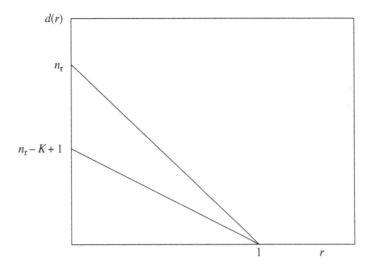

the system. This allows the following interpretation of the performance of the joint ML receiver, in contrast to the decorrelator bank:

> Using the joint ML receiver, increasing the number of receive antennas, n_r, by 1 allows us to *both* admit one extra user *and* simultaneously increase the effective number of diversity branches seen by each user by 1.

With $n_r < K$, the optimal uplink tradeoff curve is more involved. We can observe that the total spatial degrees of freedom in the uplink is now limited by n_r and thus the largest multiplexing rate *per user* can be no more than n_r/K. On the other hand, with no inter-user interference, each user can have a multiplexing gain up to 1; thus, this upper bound can never be attained for large enough multiplexing rates. It turns out that for slightly smaller multiplexing rates $r \le n_r/(K + 1)$ per user, the diversity gain obtained is still equal to the single-user bound of $n_r(1 - r)$. For r larger than this threshold (but still smaller than n_r/K), the diversity gain is that of a $K \times n_r$ MIMO channel at a total multiplexing rate of Kr; this is as if the K users pooled their total rate together. The overall optimal uplink tradeoff curve is plotted in Figure 10.6: it has two line segments joining the points

$$(0, n_r), \qquad \left(\frac{n_r}{K+1}, \frac{n_r(K - n_r + 1)}{K+1}\right), \qquad \text{and} \left(\frac{n_r}{K}, 0\right).$$

Exercise 10.7 provides the justification to the calculation of this tradeoff curve.

In Section 6.3.1, we plotted the ratio of C_ϵ^{sym} for a single receive antenna uplink to $C_\epsilon(\text{SNR})$, the outage capacity of a point-to-point channel with no inter-user interference. For a fixed outage probability ϵ, increasing the SNR

Figure 10.6 The diversity–multiplexing tradeoff curve for the uplink with the joint ML receiver for $n_r < K$. The multiplexing rate r is measured per user. Up to a multiplexing gain of $n_r/(K+1)$, single-user tradeoff performance of $n_r(1-r)$ is achieved. The maximum number of degrees of freedom per user is n_r/K, limited by the number of receive antennas.

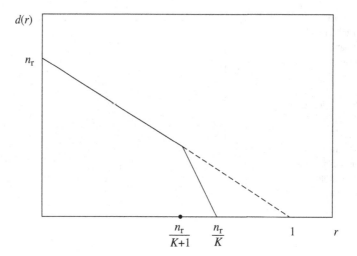

corresponds to decreasing the required diversity gain. Substituting $n_r = 1$ and $K = 2$, in Figure 10.6, we see that as long as the required diversity gain is larger than 2/3, the corresponding multiplexing gain is as if there is no inter-user interference. This explains the behavior in Figure 6.10, where the ratio of C_ϵ^{sym} to $C_\epsilon(\text{SNR})$ increases initially with SNR. With a further increase in SNR, the corresponding desired diversity gain drops below 2/3 and now there is a penalty in the achievable multiplexing rate due to the inter-user interference. This penalty corresponds to the drop of the ratio in Figure 6.10 as SNR increases further.

10.1.5 Fast fading

Here we focus on the case when communication is over several coherence intervals of the user channels; this way most channel fade levels are experienced. This is the *fast fading* assumption studied for the single antenna uplink in Section 6.3 and the point-to-point MIMO channel in Section 8.2. As usual, to simplify the analysis we assume that the base-station can perfectly track the channels of all the users.

Receiver CSI

Let us first consider the case when the users have only a statistical model of the channel (taken to be stationary and ergodic, as in the earlier chapters). In our notation, this is the case of receiver CSI. For notational simplicity, let us consider only two users in the uplink (i.e., $K = 2$). Each user's rate cannot be larger than when it is the only user transmitting (an extension of (5.91) with multiple receive antennas):

$$R_k \leq \mathbb{E}\left[\log\left(1 + \frac{\|\mathbf{h}_k\|^2 P_k}{N_0}\right)\right], \qquad k = 1, 2. \qquad (10.12)$$

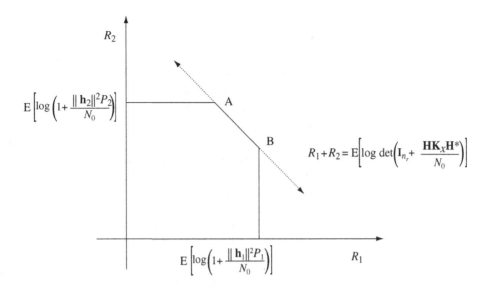

Figure 10.7 Capacity region of the two-user SIMO uplink with receiver CSI.

We also have the sum constraint (an extension of (6.37) with multiple receive antennas, cf.(8.10)):

$$R_1 + R_2 \leq \mathbb{E}\left[\log\det\left(\mathbf{I}_{n_r} + \frac{1}{N_0}\mathbf{H}\mathbf{K}_x\mathbf{H}^*\right)\right]. \qquad (10.13)$$

Here we have written $\mathbf{H} = [\mathbf{h}_1\mathbf{h}_2]$ and $\mathbf{K}_x = \mathrm{diag}\{P_1, P_2\}$. The capacity region is a pentagon (see Figure 10.7). The two corner points are achieved by the receiver architecture of linear MMSE filters followed by successive cancellation of the decoded user. Appendix B.9.3 provides a formal justification.

Let us focus on the sum capacity in (10.13). This is exactly the capacity of a point-to-point MIMO channel with receiver CSI where the covariance matrix is chosen to be diagonal. The performance gain in the sum capacity over the single receive antenna case (cf. (6.37)) is of the same nature as that of a point-to-point MIMO channel over a point-to-point channel with only a single receive antenna. With a sufficiently random and well-conditioned channel matrix $\mathbf{H}$, the performance gain is significant (cf. our discussion in Section 8.2.2). Since there is a strong likelihood of the users being geographically far apart, the channel matrix is likely to be well-conditioned (recall our discussion in Example 7.4 in Section 7.2.4). In particular, the important observation we can make is that each of the users has one spatial degree of freedom, while with a single receive antenna, the sum capacity itself has one spatial degree of freedom.

Full CSI

We now move to the other scenario, full CSI both at the base-station and at each of the users.[2] We have studied the full CSI case in the uplink for single transmit and receive antennas in Section 6.3 and here we will see the role played by an array of receive antennas.

Now the users can vary their transmit power as a function of the channel realizations; still subject to an average power constraint. If we denote the transmit power of user k at time m by $P_k(\mathbf{h}_1[m], \mathbf{h}_2[m])$, i.e., it is a function of the channel states $\mathbf{h}_1[m], \mathbf{h}_2[m]$ at time m, then the rate pairs (R_1, R_2) at which the users can jointly reliably communicate to the base-station satisfy (analogous to (10.12) and (10.13)):

$$R_k \leq \mathbb{E}\left[\log\left(1 + \frac{\|\mathbf{h}_k\|^2 P_k(\mathbf{h}_1, \mathbf{h}_2)}{N_0}\right)\right], \qquad k = 1, 2, \quad (10.14)$$

$$R_1 + R_2 \leq \mathbb{E}\left[\log\det\left(\mathbf{I}_{n_r} + \frac{1}{N_0}\mathbf{H}\mathbf{K}_x\mathbf{H}^*\right)\right]. \qquad (10.15)$$

Here we have written $\mathbf{K}_x = \mathrm{diag}\{P_1(\mathbf{h}_1, \mathbf{h}_2), P_2(\mathbf{h}_1, \mathbf{h}_2)\}$. By varying the power allocations, the users can communicate at rate pairs in the *union* of the pentagons of the form defined in (10.14) and (10.15). By time sharing between two different power allocation policies, the users can also achieve every rate pair in the *convex hull*[3] of the union of these pentagons; this is the capacity region of the uplink with full CSI. The power allocations are still subject to the average constraint, denoted by P (taken to be the same for each user for notational convenience):

$$\mathbb{E}[P_k(\mathbf{h}_1, \mathbf{h}_2)] \leq P, \qquad k = 1, 2. \qquad (10.16)$$

In the point-to-point channel, we have seen that the power variations are waterfilling over the channel states (cf. Section 5.4.6). To get some insight into how the power variations are done in the uplink with multiple receive antennas, let us focus on the sum capacity

$$C_{\mathrm{sum}} = \max_{P_k(\mathbf{h}_1, \mathbf{h}_2), \; k=1,2} \mathbb{E}\left[\log\det\left(\mathbf{I}_{n_r} + \frac{1}{N_0}\mathbf{H}\mathbf{K}_x\mathbf{H}^*\right)\right], \qquad (10.17)$$

where the power allocations are subject to the average constraint in (10.16). In the uplink with a single receive antenna at the base-station (cf. Section 6.3.3), we have seen that the power allocation that maximizes sum capacity allows only the best user to transmit (a power that is waterfilling over the best user's

---

[2] In an FDD system, the base-station need not feedback all the channel states of all the users to every user. Instead, only the amount of power to be transmitted needs be relayed to the users.

[3] The convex hull of a set is the collection of all points that can be represented as convex combinations of elements of the set.

channel state, cf. (6.47)). Here each user is received as a vector ($\mathbf{h}_k$ for user k) at the base-station and there is no natural ordering of the users to bring this argument forth here. Still, the optimal allocation of powers can be found using the Lagrangian techniques, but the solution is somewhat complicated and is studied in Exercise 10.9.

10.1.6 Multiuser diversity revisited

One of the key insights from the study of the performance of the uplink with full CSI in Chapter 6 was the discovery of multiuser diversity. How do multiple receive antennas affect multiuser diversity? With a single receive antenna and i.i.d. user channel statistics, we have seen (see Section 6.6) that the sum capacity in the uplink can be interpreted as the capacity of the following point-to-point channel with full CSI:

- The power constraint is the sum of the power constraints of the users (equal to KP with equal power constraints for the users $P_i = P$).
- The channel quality is $|h_{k^*}|^2 := \max_{k=1\ldots K} |h_k|^2$, that corresponding to the strongest user k^*.

The corresponding sum capacity is (see (6.49))

$$C_{\text{sum}} = \mathbb{E}\left[\log\left(1 + \frac{P^*(h_{k^*})|h_{k^*}|^2}{N_0}\right)\right], \tag{10.18}$$

where P^* is the waterfilling power allocation (see (5.100) and (6.47)). With multiple receive antennas, the optimal power allocation does not allow a simple characterization. To get some insight, let us first consider (the suboptimal strategy of) transmitting from only one user at a time.

One user at a time policy

In this case, the multiple antennas at the base-station translate into receive beamforming gain for the users. Now we can order the users based on the beamforming power gain due to the multiple receive antennas at the base-station. Thus, as an analogy to the strongest user in the single antenna situation, here we can choose that user which has the largest receive beamforming gain: the user with the largest $\|\mathbf{h}_k\|^2$. Assuming i.i.d. user channel statistics, the sum rate with this policy is

$$\mathbb{E}\left[\log\left(1 + \frac{P_{k^*}^*(\|\mathbf{h}_{k^*}\|)\|\mathbf{h}_{k^*}\|^2}{N_0}\right)\right]. \tag{10.19}$$

Comparing (10.19) with (10.18), we see that the only difference is that the scalar channel gain $|h_k|^2$ is replaced by the receive beamforming gain $\|\mathbf{h}_k\|^2$.

The multiuser diversity gain depends on the probability that the maximum of the users' channel qualities becomes large (the *tail* probability). For example, we have seen (cf. Section 6.7) that the multiuser diversity gain with Rayleigh fading is larger than that in Rician fading (with the same average channel quality). With i.i.d. channels to the receive antenna array (with unit average channel quality), we have by the law of large numbers

$$\frac{\|\mathbf{h}_k\|^2}{n_{\rm r}} \to 1, \qquad n_{\rm r} \to \infty. \tag{10.20}$$

So, the receive beamforming gain can be approximated as $\|\mathbf{h}_k\|^2 \approx n_{\rm r}$ for large enough $n_{\rm r}$. This means that the tail of the receive beamforming gain decays rapidly for large $n_{\rm r}$.

As an illustration, the density of $\|\mathbf{h}_k\|^2$ for i.i.d. Rayleigh fading (i.e., it is a $\chi^2_{2n_{\rm r}}$ random variable) scaled by $n_{\rm r}$ is plotted in Figure 10.8. We see that the larger the $n_{\rm r}$ value is, the more concentrated the density of the scaled random variable $\chi^2_{2n_{\rm r}}$ is around its mean. This illustration is similar in nature to that in Figure 6.23 in Section 6.7 where we have seen the plot of the densities of the channel quality with Rayleigh and Rician fading. Thus, while the array of receive antennas provides a beamforming gain, the multiuser diversity gain is restricted. This effect is illustrated in Figure 10.9 where we see that the sum capacity does not increase much with the number of users, when compared to the corresponding AWGN channel.

Optimal power allocation policy

We have discussed the impact of multiple receive antennas on multiuser diversity under the suboptimal strategy of allowing only one user (the best user) to transmit at any time. Let us now consider how the sum capacity benefits from multiuser diversity; i.e., we have to study the power allocation policy that is optimal for the sum of user rates. In our previous discussions, we have found a simple form for this power allocation policy: for a point-to-point single

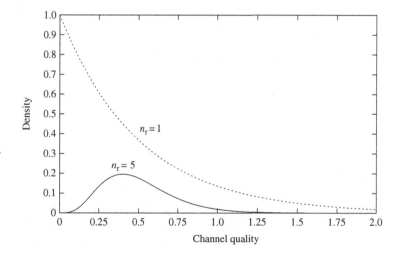

Figure 10.8 Plot of the density of a $\chi^2_{2n_{\rm r}}$ random variable divided by $n_{\rm r}$ for $n_{\rm r} = 1, 5$. The larger the $n_{\rm r}$, the more concentrated the normalized random variable is around its mean of one.

Figure 10.9 Sum capacities of the uplink Rayleigh fading channel with n_r the number of receive antennas, for $n_r = 1, 5$. Here SNR $= 1$ (0 dB) and the Rayleigh fading channel is $\mathbf{h} \sim \mathcal{CN}(0, \mathbf{I}_{n_r})$. Also plotted for comparison is the corresponding performance for the uplink AWGN channel with $n_r = 5$ and SNR $= 5$ (7 dB).

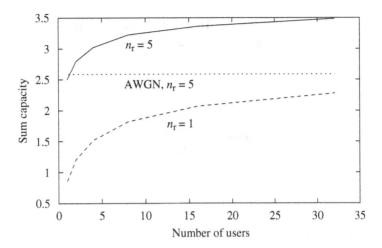

antenna channel, the allocation is waterfilling. For the single antenna uplink, the policy is to allow only the best user to transmit and, further, the power allocated to the best user is waterfilling over its channel quality. In the uplink with multiple receive antennas, there is no such simple expression in general. However, with both n_r and K large and comparable, the following simple policy is very close to the optimal one. (See Exercise 10.10.) *Every* user transmits and the power allocated is waterfilling over its own channel state, i.e.,

$$P_k(\mathbf{H}) = \left(\frac{1}{\lambda} - \frac{I_0}{\|\mathbf{h}_k\|^2} \right)^+, \qquad k = 1, \ldots, K. \qquad (10.21)$$

As usual the water level, λ, is chosen such that the average power constraint is met.

It is instructive to compare the waterfilling allocation in (10.21) with the one in the uplink with a single receive antenna (see (6.47)). The important difference is that when there is only one user transmitting, waterfilling is done over the channel quality with respect to the background noise (of power density N_0). However, here all the users are simultaneously transmitting, using a similar waterfilling power allocation policy. Hence the waterfilling in (10.21) is done over the channel quality (the receive beamforming gain) with respect to the background *interference* plus noise: this is denoted by the term I_0 in (10.21). In particular, at high SNR the waterfilling policy in (10.21) simplifies to the constant power allocation at all times (under the condition that there are more receive antennas than the number of users).

Now the impact on multiuser diversity is clear: it is reduced to the basic opportunistic communication gain by waterfilling in a point-to-point channel. This gain depends solely on how the individual channel qualities of the users fluctuate with time and thus the multiuser nature of the gain is lost. As we have seen earlier (cf. Section 6.6), the gain of opportunistic communication in a point-to-point context is much more limited than that in the multiuser context.

Summary 10.2 Opportunistic communication and multiple receive antennas

Orthogonal multiple access: scheduled user gets a power gain but reduced multiuser diversity gain.

SDMA: multiple users simultaneously transmit.
- Optimal power allocation approximated by *waterfilling* with respect to an intra-cell interference level.
- Multiuser nature of the opportunistic gain is lost.

10.2 MIMO uplink

Figure 10.10 The MIMO uplink with multiple transmit antennas at each user and multiple receive antennas at the base-station.

Now we move to consider the role of multiple transmit antennas (at the mobiles) along with the multiple receive antennas at the base-station (Figure 10.10). Let us denote the number of transmit antennas at user k by $n_{tk}, k = 1, \ldots, K$. We begin with the time-invariant channel; the corresponding model is an extension of (10.1):

$$\mathbf{y}[m] = \sum_{k=1}^{K} \mathbf{H}_k \mathbf{x}_k[m] + \mathbf{w}[m], \qquad (10.22)$$

where $\mathbf{H}_k$ is a fixed n_r by n_{tk} matrix.

10.2.1 SDMA with multiple transmit antennas

There is a natural extension of our SDMA discussion in Section 10.1.2 to multiple transmit antennas. As before, we start with $K = 2$ users.

- **Transmitter architecture** Each user splits its data and encodes them into independent streams of information with user k employing $n_k :=$ $\min(n_{tk}, n_r)$ streams (just as in the point-to-point MIMO channel). Powers $P_{k1}, P_{k2}, \ldots, P_{kn_k}$ are allocated to the n_k data streams, passed through a rotation $\mathbf{U}_k$ and sent over the transmit antenna array at user k. This is analogous to the transmitter structure we have seen in the point-to-point MIMO channel in Chapter 5. In the time-invariant *point-to-point* MIMO channel, the rotation matrix $\mathbf{U}$ was chosen to correspond to the right rotation in the singular value decomposition of the channel and the powers allocated to the data streams correspond to the waterfilling allocations over the squared singular values of the channel matrix (cf. Figure 7.2). The transmitter architecture is illustrated in Figure 10.11.
- **Receiver architecture** The base-station uses the MMSE–SIC receiver to decode the data streams of the users. This is an extension of the receiver

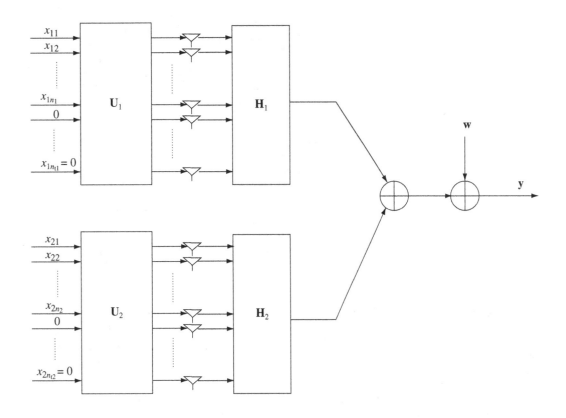

Figure 10.11 The transmitter architecture for the two-user MIMO uplink. Each user splits its data into independent data streams, allocates powers to the data streams and transmits a rotated version over the transmit antenna array.

architecture in Chapter 8 (cf. Figure 8.16). This architecture is illustrated in Figure 10.12.

The rates R_1, R_2 achieved by this transceiver architecture must satisfy the constraints, analogous to (10.2), (10.3) and (10.4):

$$R_k \leq \log \det \left(\mathbf{I}_{n_r} + \frac{1}{N_0} \mathbf{H}_k \mathbf{K}_{xk} \mathbf{H}_k^* \right), \qquad k = 1, 2, \quad (10.23)$$

$$R_1 + R_2 \leq \log \det \left(\mathbf{I}_{n_r} + \frac{1}{N_0} \sum_{k=1}^{2} \mathbf{H}_k \mathbf{K}_{xk} \mathbf{H}_k^* \right). \quad (10.24)$$

Here we have written $\mathbf{K}_{xk} := \mathbf{U}_k \Lambda_k \mathbf{U}_k^*$ and Λ_k to be a diagonal matrix with the n_{tk} diagonal entries equal to the power allocated to the data streams $P_{k1}, \ldots, P_{kn_k}$ (if $n_k < n_{tk}$ then the remaining diagonal entries are equal to zero, see Figure 10.11). The rate region defined by the constraints in (10.23) and (10.24) is a pentagon; this is similar to the one in Figure 10.3 and illustrated in Figure 10.13. The receiver architecture in Figure 10.2, where the data streams of user 1 are decoded first, canceled, and then the data streams of user 2 are decoded, achieves the corner point A in Figure 10.13.

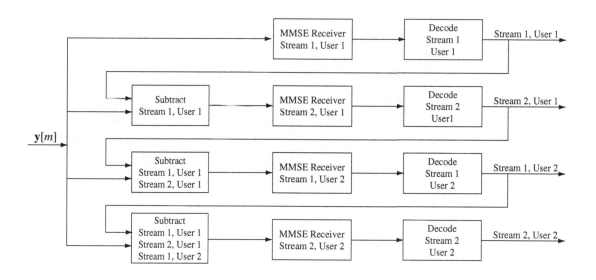

Figure 10.12 Receiver architecture for the two-user MIMO uplink. In this figure, each user has two transmit antennas and splits their data into two data streams each. The base-station decodes the data streams of the users using the linear MMSE filter, successively canceling them as they are decoded.

With a single transmit antenna at each user, the transmitter architecture simplifies considerably: there is only one data stream and the entire power is allocated to it. With multiple transmit antennas, we have a choice of power splits among the data streams and also the choice of the rotation $\mathbf{U}$ before sending the data streams out of the transmit antennas. In general, different choices of power splits and rotations lead to different pentagons (see Figure 10.14), and the capacity region is the convex hull of the union of all these pentagons; thus the capacity region in general is not a pentagon. This is because, unlike the single transmit antenna case, there are no covariance matrices $\mathbf{K}_{x1}, \mathbf{K}_{x2}$ that simultaneously maximize the right hand side of all the three constraints in (10.23) and (10.24). Depending on how one wants to trade off the performance of the two users, one would use different input strategies. This is formulated as a convex programming problem in Exercise 10.12.

Throughout this section, our discussion has been restricted to the two-user uplink. The extension to K users is completely natural. The capacity region is now K dimensional and for fixed transmission filters $\mathbf{K}_{xk}$ modulating the streams of user k (here $k = 1, \ldots, K$) there are $K!$ corner points on the boundary region of the achievable rate region; each corner point is specified by an ordering of the K users and the corresponding rate tuple is achieved by the linear MMSE filter bank followed by successive cancellation of users (and streams within a user's data). The transceiver structure is a K user extension of the pictorial depiction for two users in Figures 10.11 and 10.12.

10.2.2 System implications

Simple engineering insights can be drawn from the capacity results. Consider an uplink channel with K mobiles, each with a single transmit antenna. There

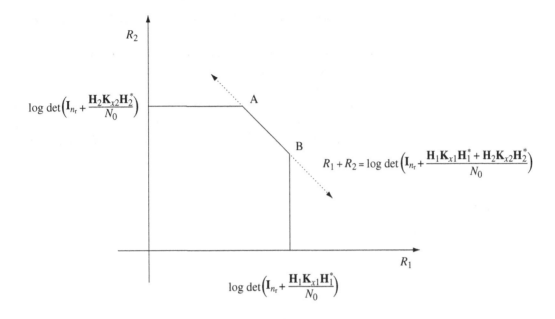

Figure 10.13 The rate region of the two-user MIMO uplink with transmitter strategies (power allocations to the data streams and the choice of rotation before sending over the transmit antenna array) given by the covariance matrices $\mathbf{K}_{x1}$ and $\mathbf{K}_{x2}$.

are n_r receive antennas at the base-station. Suppose the system designer wants to add one more transmit antenna at each mobile. How does this translate to increasing the number of spatial degrees of freedom?

If we look at each user in isolation and think of the uplink channel as a set of isolated SIMO point-to-point links from each user to the base-station, then adding one extra antenna at the mobile increases by one the available spatial degrees of freedom in each such link. However, this is misleading. Due to the sum rate constraint, the *total* number of spatial degrees of freedom is limited by the minimum of K and n_r. Hence, if K is larger than n_r, then the number of spatial degrees of freedom is already limited by the number of receive antennas at the base-station, and increasing the number of transmit antennas at the mobiles will not increase the total number of spatial degrees of freedom further. This example points out the importance of looking at

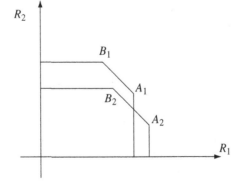

Figure 10.14 The achievable rate region for the two-user MIMO MAC with two specific choices of transmit filter covariances: $\mathbf{K}_{xk}$ for user k, for $k = 1, 2$.

the uplink channel as a whole rather than as a set of isolated point-to-point links.

On the other hand, multiple transmit antennas at each of the users significantly benefit the performance of orthogonal multiple access (which, however, is suboptimal to start with when $n_r > 1$). With a single transmit antenna, the total number of spatial degrees of freedom with orthogonal multiple access is just one. Increasing the number of transmit antennas at the users boosts the number of spatial degrees of freedom; user k has $\min(n_{tk}, n_r)$ spatial degrees of freedom when it is transmitting.

10.2.3 Fast fading

Our channel model is an extension of (10.22):

$$\mathbf{y}[m] = \sum_{k=1}^{K} \mathbf{H}_k[m]\mathbf{x}_k[m] + \mathbf{w}[m]. \tag{10.25}$$

The channel variations $\{\mathbf{H}_k[m]\}_m$ are independent across users k and stationary and ergodic in time m.

Receiver CSI

In the receiver CSI model, the users only have access to the statistical characterization of the channels while the base-station tracks all the users' channel realizations. The users can still follow the SDMA transmitter architecture in Figure 10.11: splitting the data into independent data streams, splitting the total power across the streams and then sending the rotated version of the data streams over the transmit antenna array. However, the power allocations and the choice of rotation can only depend on the channel statistics and not on the explicit realization of the channels at any time m.

In our discussion of the point-to-point MIMO channel with receiver CSI in Section 8.2.1, we have seen some additional structure to the transmit signal. With linear antenna arrays and sufficiently rich scattering so that the channel elements can be modelled as zero mean uncorrelated entries, the capacity achieving transmit signal sends independent data streams over the different *angular windows*; i.e., the covariance matrix is of the form (cf. (8.11)):

$$\mathbf{K}_x = \mathbf{U}_t \mathbf{\Lambda} \mathbf{U}_t^*, \tag{10.26}$$

where $\mathbf{\Lambda}$ is a diagonal matrix with non-negative entries (representing the power transmitted in each of the transmit angular windows). The rotation matrix $\mathbf{U}_t$ represents the transformation of the signal sent over the angular windows to the actual signal sent out of the linear antenna array (cf. (7.68)).

A similar result holds in the uplink MIMO channel as well. When each of the users' MIMO channels (viewed in the angular domain) have zero mean, uncorrelated entries then it suffices to consider covariance matrices of the form in (10.26); i.e., user k has the transmit covariance matrix:

$$\mathbf{K}_{xk} = \mathbf{U}_{tk} \Lambda_k \mathbf{U}_{tk}^*, \tag{10.27}$$

where the diagonal entries of Λ_k represent the powers allocated to the data streams, one in each of the angular windows (so their sum is equal to P_k, the power constraint for user k). (See Exercise 10.13.) With this choice of transmit strategy, the pair of rates (R_1, R_2) at which users can jointly reliably communicate is constrained, as in (10.12) and (10.13), by

$$R_k \leq \mathbb{E}\left[\log\det\left(\mathbf{I}_{n_r} + \frac{1}{N_0}\mathbf{H}_k\mathbf{K}_{xk}\mathbf{H}_k^*\right)\right], \qquad k = 1, 2, \tag{10.28}$$

$$R_1 + R_2 \leq \mathbb{E}\left[\log\det\left(\mathbf{I}_{n_r} + \frac{1}{N_0}\sum_{k=1}^{2}\mathbf{H}_k\mathbf{K}_{xk}\mathbf{H}_k^*\right)\right]. \tag{10.29}$$

This constraint forms a pentagon and the corner points are achieved by the architecture of the linear MMSE filter combined with successive cancellation of data streams (cf. Figure 10.12).

The capacity region is the convex hull of the union of these pentagons, one for each power allocation to the data streams of the users (i.e., the diagonal entries of Λ_1, Λ_2). In the point-to-point MIMO channel, with some additional symmetry (such as in the i.i.d. Rayleigh fading model), we have seen that the capacity achieving power allocation is equal powers to the data streams (cf. (8.12)). An analogous result holds in the MIMO uplink as well. With i.i.d. Rayleigh fading for all the users, the equal power allocation to the data streams, i.e.,

$$\mathbf{K}_{xk} = \frac{P_k}{n_{tk}}\mathbf{I}_{n_{tk}}, \tag{10.30}$$

achieves the entire capacity region; thus in this case the capacity region is simply a pentagon. (See Exercise 10.14.)

The analysis of the capacity region with full CSI is very similar to our previous analysis (cf. Section 10.1.5). Due to the increase in number of parameters to feedback (so that the users can change their transmit strategies as a function of the time-varying channels), this scenario is also somewhat less relevant in engineering practice, at least for FDD systems.

10.3 Downlink with multiple transmit antennas

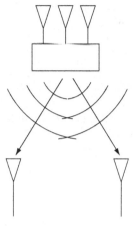

Figure 10.15 The downlink with multiple transmit antennas at the base-station and single receive antenna at each user.

We now turn to the downlink channel, from the base-station to the multiple users. This time the base-station has an array of transmit antennas but each user has a single receive antenna (Figure 10.15). It is often a practically interesting situation since it is easier to put multiple antennas at the base-station than at the mobile users. As in the uplink case we first consider the time-invariant scenario where the channel is fixed. The baseband model of the narrowband downlink with the base-station having n_t antennas and K users with each user having a single receive antenna is

$$y_k[m] = \mathbf{h}_k^* \mathbf{x}[m] + w_k[m], \qquad k = 1, \ldots, K, \qquad (10.31)$$

where $y_k[m]$ is the received vector for user k at time m, $\mathbf{h}_k^*$ is an n_t dimensional row vector representing the channel from the base-station to user k. Geometrically, user k observes the projection of the transmit signal in the spatial direction $\mathbf{h}_k$ in additive Gaussian noise. The noise $w_k[m] \sim \mathcal{CN}(0, N_0)$ and is i.i.d. in time m. An important assumption we are implicitly making here is that the channel's $\mathbf{h}_k$ are known to the base-station as well as to the users.

10.3.1 Degrees of freedom in the downlink

If the users could cooperate, then the resulting MIMO point-to- point channel would have $\min(n_t, K)$ spatial degrees of freedom, assuming that the rank of the matrix $\mathbf{H} = [\mathbf{h}_1, \ldots, \mathbf{h}_K]$ is full. Can we attain this full spatial degrees of freedom even when users cannot cooperate?

Let us look at a special case. Suppose $\mathbf{h}_1, \ldots, \mathbf{h}_K$ are orthogonal (which is only possible if $K \leq n_t$). In this case, we can transmit independent streams of data to each user, such that the stream for the kth user $\{\tilde{x}_k[m]\}$ is along the transmit spatial signature $\mathbf{h}_k$, i.e.,

$$\mathbf{x}[m] = \sum_{k=1}^{K} \tilde{x}_k[m] \mathbf{h}_k. \qquad (10.32)$$

The overall channel decomposes into a set of parallel channels; user k receives

$$y_k[m] = \|\mathbf{h}_k\|^2 \tilde{x}_k[m] + w_k[m]. \qquad (10.33)$$

Hence, one can transmit K parallel non-interfering streams of data to the users, and attain the full number of spatial degrees of freedom in the channel.

What happens in general, when the channels of the users are not orthogonal? Observe that to obtain non-interfering channels for the users in the example above, the key property of the transmit signature $\mathbf{h}_k$ is that $\mathbf{h}_k$ is orthogonal to the spatial direction's $\mathbf{h}_i$ of all the other users. For general channels (but still assuming linear independence among $\mathbf{h}_1, \ldots, \mathbf{h}_K$; thus $K \leq n_t$), we can preserve the same property by replacing the signature $\mathbf{h}_k$ by a vector $\mathbf{u}_k$ that lies in the subspace V_k orthogonal to all the other $\mathbf{h}_i$; the resulting channel for user k is

$$y_k[m] = (\mathbf{h}_k^* \mathbf{u}_k) \tilde{x}_k[m] + w_k[m]. \qquad (10.34)$$

Thus, in the general case too, we can get K spatial degrees of freedom. We can further choose $\mathbf{u}_k \in V_k$ to maximize the SNR of the channel above; geometrically, this is given by the projection of $\mathbf{h}_k$ onto the subspace V_k. This transmit filter is precisely the decorrelating receive filter used in the uplink and also in the point-to-point setting. (See Section 8.3.1 for the geometric derivation of the decorrelator.)

The above discussion is for the case when $K \leq n_t$. When $K \geq n_t$, one can apply the same scheme but transmitting only to n_t users at a time, achieving n_t spatial degrees of freedom. Thus, in all cases, we can achieve a total spatial degrees of freedom of $\min(n_t, K)$, the same as that of the point-to-point link when all the receivers can cooperate.

An important point to observe is that this performance is achieved assuming knowledge of the channels $\mathbf{h}_k$ at the base-station. We required the same channel side information at the base-station when we studied SDMA and showed that it achieves the same spatial degrees of freedom as when the users cooperate. In a TDD system, the base-station can exploit channel reciprocity and measure the uplink channel to infer the downlink channel. In an FDD system, the uplink and downlink channels are in general quite different, and feedback would be required: quite an onerous task especially when the users are highly mobile and the number of transmit antennas is large. Thus the requirement of channel state information at the base-station is quite asymmetric in the uplink and the downlink: it is more onerous in the downlink.

10.3.2 Uplink–downlink duality and transmit beamforming

In the *uplink*, we understand that the decorrelating receiver is the optimal linear filter at high SNR when the interference from other streams dominates over the additive noise. For general SNR, one should use the linear MMSE receiver to balance optimally between interference and noise suppression. This was also called *receive beamforming*. In the previous section, we found a downlink transmission strategy that is the analog of the decorrelating receive strategy. It is natural to look for a downlink transmission strategy analogous to the linear MMSE receiver. In other words, what is "optimal" transmit beamforming?

For a given set of powers, the *uplink* performance of the kth user is a function of only the receive filter $\mathbf{u}_k$. Thus, it is simple to formulate what we mean by an "optimal" linear receiver: the one that maximizes the output SINR. The solution is the MMSE receiver. In the downlink, however, the SINR of each user is a function of *all* of the transmit signatures $\mathbf{u}_1, \ldots, \mathbf{u}_K$ of the users. Thus, the problem is seemingly more complex. However, there is in fact a downlink transmission strategy that is a natural "dual" to the MMSE receive strategy and is optimal in a certain sense. This is in fact a consequence of a more general duality between the uplink and the downlink, which we now explain.

Uplink–downlink duality

Suppose transmit signatures $\mathbf{u}_1, \ldots, \mathbf{u}_K$ are used for the K users. The transmitted signal at the antenna array is

$$\mathbf{x}[m] = \sum_{k=1}^{K} \tilde{x}_k[m] \mathbf{u}_k, \tag{10.35}$$

where $\{\tilde{x}_k[m]\}$ is the data stream of user k. Substituting into (10.31) and focusing on user k, we get

$$y_k[m] = (\mathbf{h}_k^* \mathbf{u}_k) \tilde{x}_k[m] + \sum_{j \neq k} (\mathbf{h}_k^* \mathbf{u}_j) \tilde{x}_j[m] + w_k[m]. \tag{10.36}$$

The SINR for user k is given by

$$\mathsf{SINR}_k := \frac{P_k \, | \mathbf{u}_k^* \mathbf{h}_k |^2}{N_0 + \sum_{j \neq k} P_j \, | \mathbf{u}_j^* \mathbf{h}_k |^2}. \tag{10.37}$$

where P_k is the power allocated to user k.

Denote $\mathbf{a} := (a_1, \ldots, a_K)^t$ where

$$a_k := \frac{\mathsf{SINR}_k}{(1 + \mathsf{SINR}_k) \, | \mathbf{h}_k^* \mathbf{u}_k |^2},$$

and we can rewrite (10.37) in matrix notation as

$$(\mathbf{I}_K - \mathrm{diag}\{a_1, \ldots, a_K\} \mathbf{A}) \mathbf{p} = N_0 \mathbf{a}. \tag{10.38}$$

Here we denoted $\mathbf{p}$ to be the vector of transmitted powers $(P_1, \ldots, P_K)$. We also denoted the $K \times K$ matrix $\mathbf{A}$ to have component (k, j) equal to $| \mathbf{u}_j^* \mathbf{h}_k |^2$.

We now consider an uplink channel that is naturally "dual" to the given downlink channel. Rewrite the downlink channel (10.31) in matrix form:

$$\mathbf{y}_{\mathrm{dl}}[m] = \mathbf{H}^* \mathbf{x}_{\mathrm{dl}}[m] + \mathbf{w}_{\mathrm{dl}}[m], \tag{10.39}$$

where $\mathbf{y}_{\mathrm{dl}}[m] := (y_1[m], \ldots, y_K[m])^t$ is the vector of the received signals at the K users and $\mathbf{H} := [\mathbf{h}_1, \mathbf{h}_2, \ldots, \mathbf{h}_K]$ is an n_t by K matrix. We added the

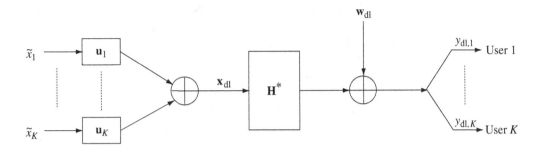

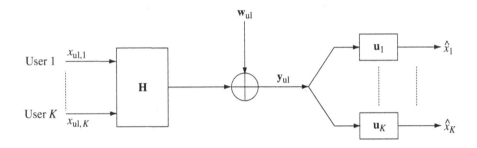

Figure 10.16 The original downlink with linear transmit strategy and its uplink dual with linear reception strategy.

subscript "dl" to emphasize that this is the downlink. The dual uplink channel has K users (each with a single transmit antenna) and n_t receive antennas:

$$\mathbf{y}_{ul}[m] = \mathbf{H}\mathbf{x}_{ul}[m] + \mathbf{w}_{ul}[m], \tag{10.40}$$

where $\mathbf{x}_{ul}[m]$ is the vector of transmitted signals from the K users, $\mathbf{y}_{ul}[m]$ is the vector of received signals at the n_t receive antennas, and $\mathbf{w}_{ul}[m] \sim \mathcal{C}N(0, N_0)$. To demodulate the kth user in this uplink channel, we use the receive filter $\mathbf{u}_k$, which is the transmit filter for user k in the downlink. The two dual systems are shown in Figure 10.16.

In this uplink, the SINR for user k is given by

$$\mathrm{SINR}_k^{ul} := \frac{Q_k \, | \, \mathbf{u}_k^* \mathbf{h}_k \, |^2}{N_0 + \sum_{j \neq k} Q_j \, | \, \mathbf{u}_k^* \mathbf{h}_j \, |^2}, \tag{10.41}$$

where Q_k is the transmit power of user k. Denoting $\mathbf{b} := (b_1, \ldots, b_K)^t$ where

$$b_k := \frac{\mathrm{SINR}_k^{ul}}{(1 + \mathrm{SINR}_k^{ul}) \, | \, \mathbf{u}_k^* \mathbf{h}_k \, |^2},$$

we can rewrite (10.41) in matrix notation as

$$(\mathbf{I}_K - \mathrm{diag}\{b_1, \ldots, b_K\}\mathbf{A}^t)\mathbf{q} = N_0\mathbf{b}. \tag{10.42}$$

Here, $\mathbf{q}$ is the vector of transmit powers of the users and $\mathbf{A}$ is the same as in (10.38).

What is the relationship between the performance of the downlink transmission strategy and its dual uplink reception strategy? We claim that to achieve the same SINR for the users in both the links, the *total transmit power* is the same in the two systems. To see this, we first solve (10.38) and (10.42) for the transmit powers and we get

$$\mathbf{p} = N_0(\mathbf{I}_K - \text{diag}\{a_1, \ldots, a_K\}\mathbf{A})^{-1}\mathbf{a} = N_0(D_a - \mathbf{A})^{-1}\mathbf{1}, \quad (10.43)$$

$$\mathbf{q} = N_0(\mathbf{I}_K - \text{diag}\{b_1, \ldots, b_K\}\mathbf{A}^t)^{-1}\mathbf{b} = N_0(D_b - \mathbf{A}^t)^{-1}\mathbf{1}, \quad (10.44)$$

where $D_a := \text{diag}(1/a_1, \ldots, 1/a_K)$, $D_b := \text{diag}(1/b_1, \ldots, 1/b_K)$ and $\mathbf{1}$ is the vector of all 1's. To achieve the same SINR in the downlink and its dual uplink, $\mathbf{a} = \mathbf{b}$, and we conclude

$$\sum_{k=1}^{K} P_k = N_0\mathbf{1}^t(D_a - \mathbf{A})^{-1}\mathbf{1} = N_0\mathbf{1}^t\left[(D_a - \mathbf{A})^{-1}\right]^t\mathbf{1}$$

$$= N_0\mathbf{1}^t(D_a - \mathbf{A}^t)^{-1}\mathbf{1} = \sum_{k=1}^{K} Q_k. \quad (10.45)$$

It should be emphasized that the *individual* powers P_k and Q_k to achieve the same SINR are not the same in the downlink and the uplink dual; only the *total* power is the same.

Transmit beamforming and optimal power allocation

As observed earlier, the SINR of each user in the downlink depends in general on *all* the transmit signatures of the users. Hence, it is not meaningful to pose the problem of choosing the transmit signatures to maximize each of the SINR separately. A more sensible formulation is to minimize the total transmit power needed to meet a *given* set of SINR requirements. The optimal transmit signatures balance between focusing energy in the direction of the user of interest and minimizing the interference to other users. This transmit strategy can be thought of as performing *transmit beamforming*. Implicit in this problem formulation is also a problem of allocating powers to each of the users.

Armed with the uplink–downlink duality established above, the transmit beamforming problem can be solved by looking at the uplink dual. Since for any choice of transmit signatures, the same SINR can be met in the uplink dual using the transmit signatures as receive filters and the same total transmit power, the downlink problem is solved if we can find receive filters that minimize the total transmit power in the uplink dual. But this problem was already solved in Section 10.1.1. The receive filters are always chosen to be the MMSE filters given the transmit powers of the users; the transmit powers are iteratively updated so that the SINR requirement of each user is just met. (In fact, this algorithm not only minimizes the total transmit power, it minimizes the transmit powers of every user simultaneously.) The MMSE filters at the optimal solution for the uplink dual can now be used as the optimal transmit signatures in the downlink, and the corresponding optimal power allocation **p** for the downlink can be obtained via (10.43).

It should be noted that the MMSE filters are the ones associated with the minimum powers used in the *uplink dual*, not the ones associated with the optimal transmit powers **p** in the *downlink*. At high SNR, each MMSE filter approaches a decorrelator, and since the decorrelator, unlike the MMSE filter, does not depend on the powers of the other interfering users, the same filter is used in the uplink and in the downlink. This is what we have already observed in Section 10.3.1.

Beyond linear strategies

In our discussion of receiver architectures for point-to-point communication in Section 8.3 and the uplink in Section 10.1.1, we boosted the performance of linear receivers by adding successive cancellation. Is there something analogous in the downlink as well?

In the case of the downlink with *single* transmit antenna at the base-station, we have already seen such a strategy in Section 6.2: superposition coding and decoding. If multiple users' signals are superimposed, the user with the strongest channel can decode the signals of the weaker users, strip them off and then decode its own. This is a natural analog to successive cancellation in the uplink. In the multiple transmit antenna case, however, there is no natural ordering of the users. In particular, if a linear superposition of signals is transmitted at the base-station:

$$\mathbf{x}[m] = \sum_{k=1}^{K} \tilde{x}_k[m]\mathbf{u}_k,$$

then each user's signal will be projected differently onto different users, and there is no guarantee that there is a single user who would have sufficient SINR to decode everyone else's data.

In both the uplink and the point-to-point MIMO channel, successive cancellation was possible because there was a single entity (the base-station) that had access to the entire vector of received signals. In the downlink we do not have that luxury since the users cannot cooperate. This was overcome in the special case of single transmit antenna because, from a decodability point of view, it is *as though* a given user has access to the received signals of all the users with weaker channels. In the general multiple transmit antenna case, this property does not hold and a "cancellation" scheme has to be necessarily *at the base-station*, which does indeed have access to the data of all the users. But how does one cancel a signal of a user even before it has been transmitted? We turn to this topic next.

10.3.3 Precoding for interference known at transmitter

Let us consider the precoding problem in a simple point-to-point context:

$$y[m] = x[m] + s[m] + w[m], \qquad (10.46)$$

where $x[m], y[m], w[m]$ are the real transmitted symbol, received symbol and $\mathcal{N}(0, \sigma^2)$ noise at time m respectively. The noise is i.i.d. in time. The interference sequence $\{s[m]\}$ is known in its entirety at the transmitter but not at the receiver. The transmitted signal $\{x[m]\}$ is subject to a power constraint. For simplicity, we have assumed all the signals to be real-valued for now. When applied to the downlink problem, $\{s[m]\}$ is the signal intended for another user, hence known at the transmitter (the base-station) but not necessary at the receiver of the user of interest. This problem also appears in many other scenarios. For example, in *data hiding* applications, $\{s[m]\}$ is the "host" signal in which one wants to hide digital information; typically the encoder has access to the host signal but not the decoder. The power constraint on $\{x[m]\}$ in this case reflects a constraint on how much the host signal can be distorted, and the problem here is to embed as much information as possible given this constraint.[4]

How can the transmitter precode the information onto the sequence $\{x[m]\}$ taking advantage of its knowledge of the interference? How much power penalty must be paid when compared to the case when the interference is also known at the receiver, or equivalently, when the interference does not exist? To get some intuition about the problem, let us first look at symbol-by-symbol precoding schemes.

Symbol-by-symbol precoding: Tomlinson–Harashima

For concreteness, suppose we would like to modulate information using uncoded $2M$-PAM: the constellation points are $\{a(1 + 2i)/2, i = -M, \ldots, M - 1\}$, with a separation of a. We consider only symbol-by-symbol precoding in this subsection, and so to simplify notations below, we drop the index m. Suppose we want to send a symbol u in this constellation. The simplest way to compensate for the interference s is to transmit $x = u - s$ instead of u, so that the received signal is $y = u + w$.[5] However, the price to pay is an increase in the required energy by s^2. This power penalty grows unbounded with s^2. This is depicted in Figure 10.17.

The problem with the naive pre-cancellation scheme is that the PAM symbol may be arbitrarily far away from the interference. Consider the following

---

[4] A good application of data hiding is embedding digital information in analog television broadcast.

[5] This strategy will not work for the downlink channel at all because s contains the message of the other user and cancellation of s at the transmitter means that the other user will get nothing.

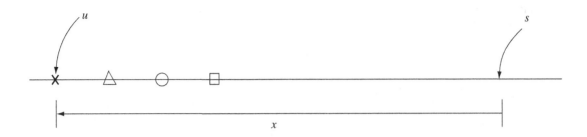

Figure 10.17 The transmitted signal is the difference between the PAM symbol and the interference. The larger the interference, the more the power that is consumed.

precoding scheme which performs better. The idea is to replicate the PAM constellation along the entire length of the real line to get an infinite extended constellation (Figures 10.18 and 10.19). Each of the $2M$ information symbols now corresponds to the equivalence class of points at the same relative position in the replicated constellations. Given the information symbol u, the precoding scheme chooses that representation p in its equivalence class which is closest to the interference s. We then transmit the difference $x = p - s$. Unlike the naive scheme, this difference can be much smaller and does not grow unbounded with s. A visual representation of the precoding scheme is provided in Figure 10.20.

One way to interpret the precoding operation is to think of the equivalence class of any one PAM symbol u as a (uniformly spaced) *quantizer* $q_u(\cdot)$ of the real line. In this context, we can think of the transmitted signal x to be the *quantization error*: the difference between the interference s and the quantized value $p = q_u(s)$, with u being the information symbol to be transmitted.

The received signal is

$$y = (q_u(s) - s) + s + w = q_u(s) + w.$$

The receiver finds the point in the infinite replicated constellation that is closest to y and then decodes to the equivalence class containing that point.

Let us look at the probability of error and the power consumption of this scheme, and how they compare to the corresponding performance when there is no interference. The probability of error is approximately[6]

$$2Q\left(\frac{a}{2\sigma}\right), \tag{10.47}$$

When there is no interference and a $2M$-PAM is used, the error probability of the interior points is the same as (10.47) but for the two exterior points, the error probability is $Q(a/2\sigma)$, smaller by a factor of $1/2$. The probability of error is larger for the exterior points in the precoding case because there is an

---

[6] The reason why this is not exact is because there is a chance that the noise will be so large that the closest point to y just happens to be in the same equivalence class of the information symbol, thus leading to a correct decision. However, the probability of this event is negligible.

Figure 10.18 A four-point PAM constellation.

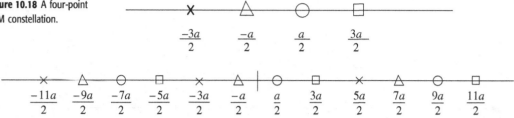

Figure 10.19 The four-point PAM constellation is replicated along the entire real line. Points marked by the same sign correspond to the same information symbol (one of the four points in the original constellation).

additional possibility of confusion *across* replicas. However, the difference is negligible when error probabilities are small.[7]

What about the power consumption of the precoding scheme? The distance between adjacent points in each equivalence class is $2Ma$; thus, unlike in the naive interference pre-cancellation scheme, the quantization error does not grow unbounded with s:

$$|x| \leq Ma.$$

If we assume that s is totally random so that this quantization error is uniform between zero and this value, then the average transmit power is

$$\mathbb{E}[x^2] = \frac{a^2 M^2}{3}. \tag{10.48}$$

In comparison, the average transmit power of the original $2M$-PAM constellation is $a^2 M^2/3 - a^2/12$. Hence, the precoding scheme requires a factor of

$$\frac{4M^2}{4M^2 - 1}$$

Figure 10.20 Depiction of the precoding operation for $M = 2$ and PAM information symbol $u = -3a/2$. The crosses form the equivalence class for this symbol. The difference between s and the closest cross p is transmitted.

more transmit power. Thus, there is still a gap from AWGN detection performance. However, this power penalty is negligible when the constellation size M is large.

Our description is motivated from a similar precoding scheme for the point-to-point frequency-selective (ISI) channel, devised independently by

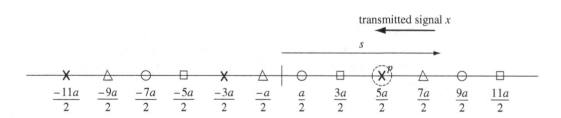

---

[7] This factor of 2 can easily be compensated for by making the symbol separation slightly larger.

Tomlinson [121] and Harashima and Miyakawa [57]. In this context, the interference is inter-symbol interference:

$$s[m] = \sum_{\ell \geq 0} h_\ell x[m - \ell],$$

where h is the impulse response of the channel. Since the previous transmitted symbols are known to the transmitter, the interference is known if the transmitter has knowledge of the channel. In Discussion 8.1 we have alluded to connections between MIMO and frequency-selective channels and precoding is yet another import from one knowledge base to the other. Indeed, Tomlinson–Harashima precoding was devised as an alternative to receiver-based decision-feedback equalization for the frequency-selective channel, the analog to the SIC receiver in MIMO and uplink channels. The precoding approach has the advantage of avoiding the error propagation problem of decision-feedback equalizers, since in the latter the cancellation is based on detected symbols, while the precoding is based on known symbols at the transmitter.

Dirty-paper precoding: achieving AWGN capacity

The precoding scheme in the last section is only for a single-dimensional constellation (such as PAM), while spectrally efficient communication requires coding over multiple dimensions. Moreover, in the low SNR regime, uncoded transmission yields very poor error probability performance and coding is necessary. There has been much work in devising block precoding schemes and it is still a very active research area. A detailed discussion of specific schemes is beyond the scope of this book. Here, we will build on the insights from symbol-by-symbol precoding to give a plausibility argument that *appropriate precoding can in fact completely obliviate the impact of the interference and achieve the capacity of the AWGN channel*. Thus, the power penalty we observed in the symbol-by-symbol precoding scheme can actually be avoided with high-dimensional coding. In the literature, the precoding technique presented here is also called *Costa precoding* or dirty-paper precoding.[8]

A first attempt

Consider communication over a block of length N symbols:

$$\mathbf{y} = \mathbf{x} + \mathbf{s} + \mathbf{w}. \tag{10.49}$$

In the symbol-by-symbol precoding scheme earlier, we started with a basic PAM constellation and replicated it to cover uniformly the entire (one-dimensional) range the interference s spans. For block coding, we would like

---

[8] This latter name comes from the title of Costa's paper: "Writing on dirty-paper" [23]. The writer of the message knows where the dirt is and can adapt his writing to help the reader decipher the message without knowing where the dirt is.

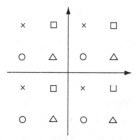

Figure 10.21 A replicated constellation in high dimension. The information specifies an equivalence class of points corresponding to replicas of a codeword (here with the same marking).

to mimic this strategy by starting with a basic AWGN constellation and replicating it to cover the N-dimensional space uniformly. Using a sphere-packing argument, we give an estimate of the maximum rate of reliable communication using this type of scheme.

Consider a domain of volume V in $\Re^N$. The exact size of the domain is not important, as long as we ensure that the domain is large enough for the received signal $\mathbf{y}$ to lie inside. This is the domain on which we replicate the basic codebook. We generate a codebook with M codewords, and replicate each of the codewords K times and place the extended constellation $\mathcal{C}_e$ of MK points on the domain sphere (Figure 10.21). Each codeword then corresponds to an equivalence class of points in $\Re^N$. Equivalently, the given information bits $\mathbf{u}$ define a quantizer $q_u(\cdot)$. The natural generalization of the symbol-by-symbol precoding procedure simply quantizes the known interference $\mathbf{s}$ using this quantizer to a point $\mathbf{p} = q_u(\mathbf{s})$ in $\mathcal{C}_e$ and transmits the quantization error

$$\mathbf{x}_1 = \mathbf{p} - \mathbf{s}. \qquad (10.50)$$

Based on the received signal $\mathbf{y}$, the decoder finds the point in the extended constellation that is closest to $\mathbf{y}$ and decodes to the information bits corresponding to its equivalence class.

Performance

To estimate the maximum rate of reliable communication for a given average power constraint P using this scheme, we make two observations:

- **Sphere-packing** To avoid confusing $\mathbf{x}_1$ with any of the other $K(M-1)$ points in the extended constellation $\mathcal{C}_e$ that belong to other equivalence classes, the noise spheres of radius $\sqrt{N\sigma^2}$ around each of these points should be disjoint. This means that

$$KM < \frac{V}{\text{Vol}[B_N(\sqrt{N\sigma^2})]}, \qquad (10.51)$$

the ratio of the volume of the domain sphere to that of the noise sphere.

- **Sphere-covering** To maintain the average transmit power constraint of P, the quantization error should be no more than $\sqrt{NP}$ for any interference vector $\mathbf{s}$. Thus, the spheres of radius $\sqrt{NP}$ around the K replicas of a codeword should be able to cover the whole domain such that any point is within a distance of $\sqrt{NP}$ from a replica. To ensure that,

$$K > \frac{V}{\text{Vol}[B_N(\sqrt{NP})]}. \qquad (10.52)$$

This in effect imposes a constraint on the *minimal density* of the replication.

Putting the two constraints (10.51) and (10.52) together, we get

$$M < \frac{\text{Vol}[B_N(\sqrt{NP})]}{\text{Vol}[B_N(\sqrt{N\sigma^2})]} = \frac{\left(\sqrt{NP}\right)^N}{\left(\sqrt{N\sigma^2}\right)^N}, \tag{10.53}$$

which implies that the maximum rate of reliable communication is, at most,

$$R := \frac{\log M}{N} = \frac{1}{2} \log \frac{P}{\sigma^2}. \tag{10.54}$$

This yields an upper bound on the rate of reliable communication. Moreover, it can be shown that if the MK constellation points are independently and uniformly distributed on the domain, then with high probability, communication is reliable if condition (10.51) holds and the average power constraint is satisfied if condition (10.52) holds. Thus, the rate (10.54) is also achievable. The proof of this is along the lines of the argument in Appendix B.5.2, where the achievability of the AWGN capacity is shown.

Observe that the rate (10.54) is close to the AWGN capacity $1/2 \log(1 + P/\sigma^2)$ at high SNR. However, the scheme is strictly suboptimal at finite SNR. In fact, it achieves zero rate if the SNR is below $0\,\text{dB}$. How can the performance of this scheme be improved?

Performance enhancement via MMSE estimation

The performance of the above scheme is limited by the two constraints (10.51) and (10.52). To meet the average power constraint, the density of replication cannot be reduced beyond (10.52). On the other hand, constraint (10.51) is a direct consequence of the nearest neighbor decoding rule, and this rule is in fact suboptimal for the problem at hand. To see why, consider the case when the interference vector $\mathbf{s}$ is 0 and the noise variance σ^2 is significantly larger than P. In this case, the transmitted vector $\mathbf{x}_1$ is roughly at a distance $\sqrt{NP}$ from the origin while the received vector $\mathbf{y}$ is at a distance $\sqrt{N(P+\sigma^2)}$, much further away. Blindly decoding to the point in $\mathcal{C}_e$ nearest to $\mathbf{y}$ makes no use of the prior information that the transmitted vector $\mathbf{x}_1$ is of (relatively short) length $\sqrt{NP}$ (Figure 10.22). Without using this prior information, the transmitted vector is thought of by the receiver as anywhere in a large *uncertainty sphere* of radius $\sqrt{N\sigma^2}$ around $\mathbf{y}$ and the extended constellation points have to be spaced that far apart to avoid confusion. By making use of the prior information, the size of the uncertainty sphere can be reduced. In particular, we can consider a linear estimate $\alpha\mathbf{y}$ of $\mathbf{x}_1$. By the law of large numbers, the squared error in the estimate is

$$\|\alpha\mathbf{y} - \mathbf{x}_1\|^2 = \|\alpha\mathbf{w} + (\alpha - 1)\mathbf{x}_1\|^2 \approx N\left[\alpha^2\sigma^2 + (1-\alpha)^2 P\right] \tag{10.55}$$

and by choosing

$$\alpha = \frac{P}{P+\sigma^2}, \tag{10.56}$$

Figure 10.22 MMSE decoding yields a much smaller uncertainty sphere than does nearest neighbor decoding.

Nearest neighbor decoding

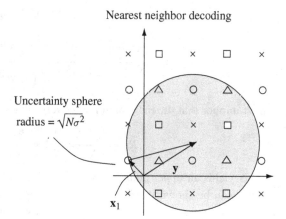

Uncertainty sphere
radius $= \sqrt{N\sigma^2}$

MMSE then nearest neighbor decoding

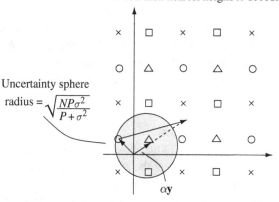

Uncertainty sphere
radius $= \sqrt{\dfrac{NP\sigma^2}{P+\sigma^2}}$

this error is minimized, equalling

$$\frac{NP\sigma^2}{P+\sigma^2}. \tag{10.57}$$

In fact $\alpha \mathbf{y}$ is nothing but the linear MMSE estimate $\hat{\mathbf{x}}_{\text{mmse}}$ of $\mathbf{x}_1$ from $\mathbf{y}$ and $NP\sigma^2/(P+\sigma^2)$ is the MMSE estimation error. If we now use a decoder that decodes to the constellation point nearest to $\alpha \mathbf{y}$ (as opposed to $\mathbf{y}$), then an error occurs only if there is another constellation point closer than this distance to $\alpha \mathbf{y}$. Thus, the uncertainty sphere is now of radius

$$\sqrt{\frac{NP\sigma^2}{P+\sigma^2}}. \tag{10.58}$$

We can now redo the analysis in the above subsection, but with the radius $\sqrt{N\sigma^2}$ of the noise sphere replaced by this radius of the MMSE uncertainty sphere. The maximum achievable rate is now

$$\frac{1}{2}\log\left(1+\frac{P}{\sigma^2}\right), \tag{10.59}$$

thus achieving the AWGN capacity.

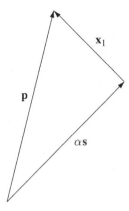

Figure 10.23 The precoding process with the α factor.

In the above, we have simplified the problem by assuming $\mathbf{s} = 0$, to focus on how the decoder has to be modified. For a general interference vector $\mathbf{s}$,

$$\alpha \mathbf{y} = \alpha(\mathbf{x}_1 + \mathbf{s} + \mathbf{w}) = \alpha(\mathbf{x}_1 + \mathbf{w}) + \alpha \mathbf{s} = \hat{\mathbf{x}}_{\text{mmse}} + \alpha \mathbf{s}, \qquad (10.60)$$

i.e., the linear MMSE estimate of $\mathbf{x}_1$ but shifted by $\alpha \mathbf{s}$. Since the receiver does not know $\mathbf{s}$, this shift has to be pre-compensated for at the transmitter. In the earlier scheme, we were using the nearest neighbor rule and we compensated for the effect of $\mathbf{s}$ by pre-subtracting $\mathbf{s}$ from the constellation point $\mathbf{p}$ representing the information, i.e., we sent the error in quantizing $\mathbf{s}$. But now we are using the MMSE rule and hence we should compensate by pre-subtracting $\alpha \mathbf{s}$ instead. Specifically, given the data $\mathbf{u}$, we find within the equivalence class representing $\mathbf{u}$ the point $\mathbf{p}$ that is closest to $\alpha \mathbf{s}$, and transmit $\mathbf{x}_1 = \mathbf{p} - \alpha \mathbf{s}$ (Figure 10.23). Then,

$$\mathbf{p} = \mathbf{x}_1 + \alpha \mathbf{s}$$
$$\alpha \mathbf{y} = \hat{\mathbf{x}}_{\text{mmse}} + \alpha \mathbf{s} = \hat{\mathbf{p}}$$

and

$$\mathbf{p} - \alpha \mathbf{y} = \mathbf{x}_1 - \hat{\mathbf{x}}_{\text{mmse}}. \qquad (10.61)$$

The receiver finds the constellation point nearest to $\alpha \mathbf{y}$ and decodes the information (Figure 10.24). An error occurs only if there is another constellation point closer to $\alpha \mathbf{y}$ than $\mathbf{p}$, i.e., if it lies in the MMSE uncertainty sphere. This is exactly the same situation as in the case of zero interference.

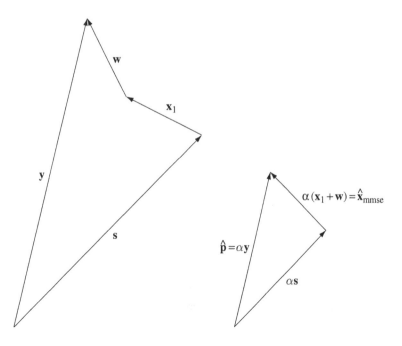

Figure 10.24 The decoding process with the α factor.

Transmitter knowledge of interference is enough

Something quite remarkable has been accomplished: even though the interference is known only at the transmitter and not at the receiver, the performance that can be achieved is as though there were no interference at all. The comparison between the cases with and without interference is depicted in Figure 10.25.

For the plain AWGN channel without interference, the codewords lie in a sphere of radius $\sqrt{NP}$ (x-sphere). When a codeword $\mathbf{x}_1$ is transmitted, the received vector $\mathbf{y}$ lies in the y-sphere, outside the x-sphere. The MMSE rule scales down $\mathbf{y}$ to $\alpha\mathbf{y}$, and the uncertainty sphere of radius $\sqrt{NP\sigma^2/(P+\sigma^2)}$ around $\alpha\mathbf{y}$ lies inside the x-sphere. The maximum reliable rate of communication is given by the number of uncertainty spheres that can be packed into the x-sphere:

$$\frac{1}{N} \log \frac{\mathrm{Vol}[B_N(\sqrt{NP})]}{\mathrm{Vol}[B_N(\sqrt{NP\sigma^2/(P+\sigma^2)})]} = \frac{1}{2} \log \left(1 + \frac{P}{\sigma^2} \right), \qquad (10.62)$$

the capacity of the AWGN channel. In fact, this is how achievability of the AWGN capacity is shown in Appendix B.5.2.

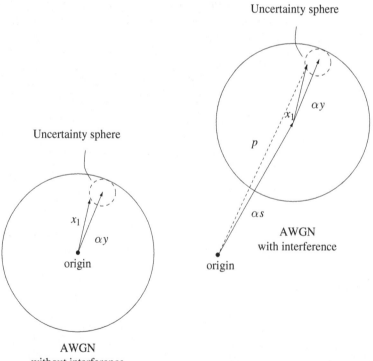

Figure 10.25 Pictorial representation of the cases with and without interference.

With interference, the codewords have to be replicated to cover the entire domain where the interference vector can lie. For any interference vector **s**, consider a sphere of radius $\sqrt{NP}$ around α**s**; this can be thought of as the AWGN x-sphere whose center is shifted to α**s**. A constellation point **p** representing the given information bits lies inside this sphere. The vector $\mathbf{p} - \alpha\mathbf{s}$ is transmitted. By using the MMSE rule, the uncertainty sphere around α**y** again lies inside this shifted x-sphere. Thus, we have the same situation as in the case without interference: the same information rate can be supported.

In the case without interference and where the codewords lie in a sphere of radius $\sqrt{NP}$, both the nearest neighbor rule and the MMSE rule achieve capacity. This is because although **y** lies outside the x-sphere, there are no codewords outside the x-sphere and the nearest neighbor rule will automatically find the codeword in the x-sphere closest to **y**. However, in the precoding problem when there *are* constellation points lying outside the shifted x-sphere, the nearest neighbor rule will lead to confusion with these other points and is therefore strictly suboptimal.

Dirty-paper code design

We have given a plausibility argument of how the AWGN capacity can be achieved without knowledge of the interference at the receiver. It can be shown that randomly chosen codewords can achieve this performance. Construction of practical codes is the subject of current research. One such class of codes is called *nested lattice codes* (Figure 10.26). The design requirements of this nested lattice code are:

- Each sub-lattice should be a good vector quantizer for the scaled interference α**s**, to minimize the transmit power.
- The entire extended constellation should behave as a good AWGN channel code.

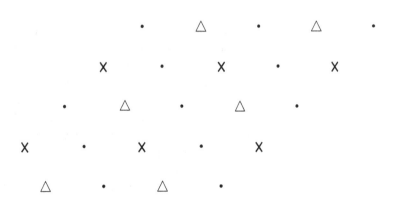

Figure 10.26 A nested lattice code. All the points in each sub-lattice represent the same information bits.

The discussion of such codes is beyond the scope of this book. The design problem, however, simplifies in the low SNR regime. We discuss this below.

Low SNR: opportunistic orthogonal coding

In the infinite bandwidth channel, the SNR per degree of freedom is zero and we can use this as a concrete channel to study the nature of precoding at low SNR. Consider the infinite bandwidth real AWGN channel with additive interference $s(t)$ modelled as real white Gaussian (with power spectral density $N_s/2$) and known non-causally to the transmitter. The interference is independent of both the background real white Gaussian noise and the real transmit signal, which is power constrained, but not bandwidth constrained. Since the interference is known non-causally only to the transmitter, the minimum $\mathcal{E}_b/N_0$ for reliable communication on this channel can be no smaller than that in the plain AWGN channel without interference; thus a lower bound on the minimum $\mathcal{E}_b/N_0$ is -1.59 dB.

We have already seen for the AWGN channel (cf. Section 5.2.2 and Exercises 5.8 and 5.9) that orthogonal codes achieve the capacity in the infinite bandwidth regime. Equivalently, orthogonal codes achieve the minimum $\mathcal{E}_b/N_0$ of -1.59 dB over the AWGN channel. Hence, we start with an orthogonal set of codewords representing M messages. Each of the code-words is replicated K times so that the overall constellation with MK vectors forms an orthogonal set. Each of the M messages corresponds to a set of K orthogonal signals. To convey a specific message, the encoder transmits that signal, among the set of K orthogonal signals corresponding to the message selected, that is closest to the interference $s(t)$, i.e., the one that has the largest correlation with the $s(t)$. This signal is the constellation point to which $s(t)$ is quantized. Note that, in the general scheme, the signal $q_u(\alpha s) - \alpha s$ is transmitted, but since $\alpha \to 0$ in the low SNR regime, we are transmitting $q_u(\alpha s)$ itself.

An equivalent way of seeing this scheme is as *opportunistic pulse position modulation*: classical PPM involves a pulse that conveys information based on the position when it is not zero. Here, every K of the pulse positions corresponds to one message and the encoder opportunistically chooses the position of the pulse among the K possible pulse positions (once the desired message to be conveyed is picked) where the interference is the *largest*.

The decoder first picks the most likely position of the transmit pulse (among the MK possible choices) using the standard largest amplitude detector. Next, it picks the message corresponding to the set in which the most likely pulse occurs. Choosing K large allows the encoder to harness the opportunistic gains afforded by the knowledge of the additive interference. On the other hand, decoding gets harder as K increases since the number of possible pulse positions, MK, grows with K. An appropriate choice of K as a function of the number of messages, M, and the noise and interference powers, N_0 and N_s respectively, trades off the opportunistic gains on the one hand with the increased difficulty in decoding on the other. This tradeoff is evaluated in Exercise 10.16 where we see that the correct choice of K allows the opportunistic orthogonal codes to achieve the infinite bandwidth capacity of the AWGN channel *without* interference. Equivalently, the minimum $\mathcal{E}_b/N_0$ is the *same* as that in the plain AWGN channel and is achieved by opportunistic orthogonal coding.

10.3.4 Precoding for the downlink

We now apply the precoding technique to the downlink channel. We first start with the single transmit antenna case and then discuss the multiple antenna case.

Single transmit antenna

Consider the two-user downlink channel with a single transmit antenna:

$$y_k[m] = h_k x[m] + w_k[m], \qquad k = 1, 2, \tag{10.63}$$

where $w_k[m] \sim \mathcal{CN}(0, N_0)$. Without loss of generality, let us assume that user 1 has the stronger channel: $|h_1|^2 \geq |h_2|^2$. Write $x[m] = x_1[m] + x_2[m]$, where $\{x_k[m]\}$ is the signal intended for user k, $k = 1, 2$. Let P_k be the power allocated to user k. We use a standard i.i.d. Gaussian codebook to encode information for user 2 in $\{x_2[m]\}$. Treating $\{x_2[m]\}$ as interference that is known at the transmitter, we can apply Costa precoding for user 1 to achieve a rate of

$$R_1 = \log\left(1 + \frac{|h_1|^2 P_1}{N_0}\right), \tag{10.64}$$

the capacity of an AWGN channel for user 1 with $\{x_2[m]\}$ completely absent. What about user 2? It can be shown that $\{x_1[m]\}$ can be made to appear like independent Gaussian noise to user 2. (See Exercise 10.17.) Hence, user 2 gets a reliable data rate of

$$R_2 = \log\left(1 + \frac{|h_2|^2 P_2}{|h_2|^2 P_1 + N_0}\right). \tag{10.65}$$

Since we have assumed that user 1 has the stronger channel, these same rates can in fact be achieved by superposition coding and decoding (cf. Section 6.2): we superimpose independent i.i.d. Gaussian codebook for user 1 and 2, with user 2 decoding the signal $\{x_2[m]\}$ treating $\{x_1[m]\}$ as Gaussian noise, and user 1 decoding the information for user 2, canceling it off, and then decoding the information intended for it. Thus, precoding is another approach to achieve rates on the boundary of the capacity region in the single antenna downlink channel.

Superposition coding is a *receiver-centric* scheme: the base-station simply adds the codewords of the users while the stronger user has to do the decoding job of both the users. In contrast, precoding puts a substantial computational burden on the base-station with receivers being regular nearest neighbor decoders (though the user whose signal is being precoded needs to decode the extended constellation, which has more points than the rate would entail). In this sense we can think of precoding as a *transmitter-centric* scheme.

However, there is something curious about this calculation. The precoding strategy described above encodes information for user 1 treating user 2's signal as known interference. But certainly we can reverse the role of user 1 and user 2, and encode information for user 2, treating user 1's signal as interference. This strategy achieves rates

$$R_1' = \log\left(1 + \frac{|h_1|^2 P_1}{|h_1|^2 P_2 + N_0}\right), \qquad R_2' = \log\left(1 + \frac{|h_2|^2 P_2}{N_0}\right). \qquad (10.66)$$

But these rates *cannot* be achieved by superposition coding/decoding under the power allocations P_1, P_2: the weak user cannot remove the signal intended for the strong user. Is this rate tuple then outside the capacity region? It turns out that there is no contradiction and this rate pair is strictly contained inside the capacity region (Exercise 10.19).

In this discussion, we have restricted ourselves to just two users, but the extension to K users is obvious. See Exercise 10.19.

Multiple transmit antennas

We now return to the scenario of real interest, multiple transmit antennas (10.31):

$$y_k[m] = \mathbf{h}_k^* \mathbf{x}[m] + w_k[m], \qquad k = 1, 2, \ldots, K. \qquad (10.67)$$

The precoding technique can be applied to upgrade the performance of the linear beamforming technique described in Section 10.3.2. Recall from (10.35), the transmitted signal is

$$\mathbf{x}[m] = \sum_{k=1}^{K} \tilde{x}_k[m] \mathbf{u}_k, \qquad (10.68)$$

where $\{\tilde{x}_k[m]\}$ is the signal for user k and $\mathbf{u}_k$ is its transmit beamforming vector. The received signal of user k is given by

$$y_k[m] = (\mathbf{h}_k^* \mathbf{u}_k)\tilde{x}_k[m] + \sum_{j \neq k}(\mathbf{h}_k^* \mathbf{u}_j)\tilde{x}_j[m] + w_k[m], \qquad (10.69)$$

$$= (\mathbf{h}_k^* \mathbf{u}_k)\tilde{x}_k[m] + \sum_{j < k}(\mathbf{h}_k^* \mathbf{u}_j)\tilde{x}_j[m]$$

$$+ \sum_{j > k}(\mathbf{h}_k^* \mathbf{u}_j)\tilde{x}_j[m] + w_k[m]. \qquad (10.70)$$

Applying Costa precoding for user k, treating the interference $\sum_{j<k}(\mathbf{h}_k^*\mathbf{u}_j)\tilde{x}_j[m]$ from users $1, \ldots, k-1$ as known and $\sum_{j>k}(\mathbf{h}_k^*\mathbf{u}_j)\tilde{x}_j[m]$ from users $k+1, \ldots K$ as Gaussian noise, the rate that user k gets is

$$R_k = \log(1 + \mathsf{SINR}_k), \qquad (10.71)$$

where SINR_k is the effective signal-to-interference-plus-noise ratio after precoding:

$$\mathsf{SINR}_k = \frac{P_k \mid \mathbf{u}_k^*\mathbf{h}_k \mid^2}{N_0 + \sum_{j>k} P_j \mid \mathbf{u}_j^*\mathbf{h}_k \mid^2}. \qquad (10.72)$$

Here P_j is the power allocated to user j. Observe that unlike the single transmit antenna case, this performance may not be achievable by superposition coding/decoding.

For linear beamforming strategies, an interesting uplink–downlink duality is identified in Section 10.3.2. We can use the downlink transmit signatures (denoted by $\mathbf{u}_1, \ldots, \mathbf{u}_K$) to be the same as the receive filters in the dual uplink channel (10.40) and the same SINR for the users can be achieved in both the uplink and the downlink with appropriate user power allocations such that the sum of these power allocations is the same for both the uplink and the downlink. We now extend this observation to a duality between transmit beamforming with precoding in the downlink and receive beamforming with SIC in the uplink.

Specifically, suppose we use Costa precoding in the downlink and SIC in the uplink, and the transmit signatures of the users in the downlink are the same as the receive filters of the users in the uplink. Then it turns out that the same set SINR of the users can be achieved by appropriate user power allocations in the uplink and the downlink and, further, the sum of these power allocations is the same. This duality holds provided that the order of SIC in the uplink is the *reverse* of the Costa precoding order in the downlink. For example, in the Costa precoding above we employed the order $1, \ldots, K$; i.e., we precoded the user k signal so as to cancel the interference from the signals of users $1, \ldots, k-1$. For this duality to hold, we need to *reverse* this order in the SIC in the uplink; i.e., the users are successively canceled in the order $K, \ldots, 1$ (with user k seeing no interference from the canceled user signals $K, K-1, \ldots, k+1$).

The derivation of this duality follows the same lines as for linear strategies and is done in Exercise 10.20. Note that in this SIC ordering, user 1 sees the least uncanceled interference and user K sees the most. This is exactly the opposite to that under the Costa precoding strategy. Thus, we see that in this duality the ordering of the users is *reversed*. Identifying this duality facilitates the computation of good transmit filters in the downlink. For example, we know that in the uplink the optimal filters for a given set of powers are MMSE filters; the same filters can be used in the downlink transmission.

In Section 10.1.2, we saw that receive beamforming in conjunction with SIC achieves the capacity region of the uplink channel with multiple receive antennas. It has been shown that transmit beamforming in conjunction with Costa precoding achieves the capacity of the downlink channel with multiple transmit antennas.

10.3.5 Fast fading

The time-varying downlink channel is an extension of (10.31):

$$y_k[m] = \mathbf{h}_k^*[m]\mathbf{x}[m] + w_k[m], \qquad k = 1, \dots, K. \tag{10.73}$$

Full CSI

With full CSI, both the base-station and the users track the channel fluctuations and, in this case, the extension of the linear beamforming strategies combined with Costa precoding to the fading channel is natural. Now we can vary the power and transmit signature allocations of the users, and the Costa precoding order as a function of the channel variations. Linear beamforming combined with Costa precoding achieves the capacity of the fast fading downlink channel with full CSI, just as in the time-invariant downlink channel.

It is interesting to compare this sum capacity achieving strategy with that when the base-station has just one transmit antenna (see Section 6.4.2). In this basic downlink channel, we identified the structure of the sum capacity achieving strategy: transmit only to the best user (using a power that is waterfilling over the best user's channel quality, see (6.54)). The linear beamforming strategy proposed here involves in general transmitting to all the users simultaneously and is quite different from the one user at a time policy. This difference is analogous to what we have seen in the uplink with single and multiple receive antennas at the base-station.

Due to the duality, we have a connection between the strategies for the downlink channel and its dual uplink channel. Thus, the impact of multiple transmit antennas at the base-station on multiuser diversity follows the discussion in the uplink context (see Section 10.1.6): focusing on the one user at a time policy, the multiple transmit antennas provide a beamforming power gain; this gain is the same as in the point-to-point context and the multiuser nature of the gain is lost. With the sum capacity achieving strategy, the multiple transmit antennas provide multiple spatial degrees of freedom allowing the users to be transmitted to simultaneously, but the opportunistic gains are of the same form as in the point-to-point case; the multiuser nature of the gain is diminished.

Receiver CSI

So far we have made the full CSI assumption. In practice, it is often very hard for the base-station to have access to the user channel fluctuations and the receiver CSI model is more natural. The major difference here is that now the transmit signatures of the users cannot be allocated as a function of the channel variations. Furthermore, the base-station is not aware of the interference caused by the other users' signals for any specific user k (since the channel to the kth user is unknown) and Costa precoding is ruled out.

Exercise 10.21 discusses how to use the multiple antennas at the base-station without access to the channel fluctuations. One of the important conclusions is that time sharing among the users achieves the capacity region in the symmetric downlink channel with receiver CSI alone. This implies that the total spatial degrees of freedom in the downlink are restricted to one, the same as the degrees of freedom of the channel from the base-station to any individual user. On the other hand, with full CSI at the base-station we have seen (Section 10.3.1) that the spatial degrees of freedom are equal to $\min(n_t, K)$. Thus lack of CSI at the base-station causes a drastic reduction in the degrees of freedom of the channel.

Partial CSI at the base-station: opportunistic beamforming with multiple beams

In many practical systems, there is some form of partial CSI fed back to the base-station from the users. For example, in the IS-856 standard discussed in Chapter 6 each user feeds back the overall SINR of the link to the base-station it is communicating with. Thus, while the base-station does not have exact knowledge of the channel (phase and amplitude) from the transmit antenna array to the users, it does have partial information: the overall quality of the channel (such as $\|\mathbf{h}_k[m]\|^2$ for user k at time m).

In Section 6.7.3 we studied opportunistic beamforming that induces time fluctuations in the channel to increase the multiuser diversity. The multiple transmit antennas were used to induce time fluctuations and the partial CSI was used to schedule the users at appropriate time slots. However, the gain from multiuser diversity is a power gain (boost in the SINR of the user being scheduled) and with just a single user scheduled at any time slot, only one of the spatial degrees of freedom is being used. This basic scheme can be modified, however, allowing multiple users to be scheduled and thus increasing the utilized spatial degrees of freedom.

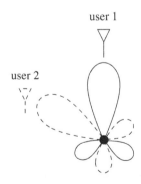

user 1

user 2

Figure 10.27 Opportunistic beamforming with two orthogonal beams. The user "closest" to a beam is scheduled on that beam, resulting in two parallel data streams to two users.

The conceptual idea is to have *multiple beams*, each orthogonal to one another, at the same time (Figure 10.27). Separate pilot symbols are introduced on each of the beams and the users feedback the SINR of *each* beam. Transmissions are scheduled to as many users as there are beams at each time slot. If there are enough users in the system, the user who is beamformed with respect to a specific beam (and orthogonal to the other beams) is scheduled on the specific beam. Let us consider $K \geq n_t$ (if $K < n_t$ then we use only K of the transmit antennas), and at each time m, let $\mathbf{Q}[m] = [\mathbf{q}_1[m], \ldots, \mathbf{q}_{n_t}[m]]$ be an $n_t \times n_t$ unitary matrix, with the columns $\mathbf{q}_1[m], \ldots, \mathbf{q}_{n_t}[m]$ orthonormal. The vector $\mathbf{q}_i[m]$ represents the ith beam at time m.

The vector signal sent out from the antenna array at time m is

$$\sum_{i=1}^{n_t} \tilde{x}_i[m]\mathbf{q}_i[m]. \tag{10.74}$$

Here $\tilde{x}_1, \ldots, \tilde{x}_{n_t}$ are the n_t independent data streams (in the case of coherent downlink reception, these signals include pilot symbols as well). The unitary matrix $\mathbf{Q}[m]$ is varied such that the individual components do not change abruptly in time. Focusing on the kth user, the signal it receives at time m is (substituting (10.74) in (10.73))

$$y_k[m] = \sum_{i=1}^{n_t} \tilde{x}_i[m]\mathbf{h}_k^*[m]\mathbf{q}_i[m] + w_k[m]. \tag{10.75}$$

For simplicity, let us consider the scenario when the channel coefficients are not varying over the time-scale of communication (slow fading), i.e., $\mathbf{h}_k[m] = \mathbf{h}_k$. When the ith beam takes on the value

$$\mathbf{q}_i[m] = \frac{\mathbf{h}_k}{\|\mathbf{h}_k\|}, \tag{10.76}$$

then user k is in beamforming configuration with respect to the ith beam; moreover, it is simultaneously orthogonal to the other beams. The received signal at user k is

$$y_k[m] = \|\mathbf{h}_k\|\tilde{x}_i[m] + w_k[m]. \tag{10.77}$$

If there are enough users in the system, for every beam i some user will be nearly in beamforming configuration with respect to it (and simultaneously nearly orthogonal to the other beams). Thus n_t data streams are transmitted simultaneously in orthogonal spatial directions and the full spatial degrees of freedom are utilized. The limited feedback from the users allows opportunistic scheduling of the user transmissions in the appropriate beams at the appropriate time slots. To achieve close to the beamforming performance and corresponding nulling to all the other beams requires a user population that is larger than in the scenario of Section 6.7.3. In general, depending on the number of the users in the system, the number of spatially orthogonal beams can be designed.

There are extra system requirements to support multiple beams (as compared to just the single time-varying beam introduced in Section 6.7.3). First, multiple pilot symbols have to be inserted (one for each beam) to enable coherent downlink reception; thus the fraction of pilot symbol power increases. Second, the receivers now track n_t separate beams and feedback SINR of each on each of the beams. On a practical note, the receivers could feedback only the *best* SINR and the identification of the beam that yields this SINR; this restriction probably will not degrade the performance by much. Thus, with almost the same amount of feedback as the single beam scheme, the modified opportunistic beamforming scheme utilizes all the spatial degrees of freedom.

10.4 MIMO downlink

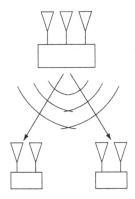

We have seen so far how downlink is affected by the availability of multiple transmit antennas at the base-station. In this section, we study the downlink with multiple receive antennas (at the users) (see Figure 10.28). To focus on the role of multiple receive antennas, we begin with a single transmit antenna at the base-station.

The downlink channel with a single transmit and multiple receive antennas at each user can be written as

$$\mathbf{y}_k[m] = \mathbf{h}_k x[m] + \mathbf{w}_k[m], \qquad k = 1, 2, \tag{10.78}$$

Figure 10.28 The downlink with multiple transmit antennas at the base-station and multiple receive antennas at each user.

where $\mathbf{w}_k[m] \sim \mathcal{CN}(0, N_0 I_{n_r})$ and i.i.d. in time m. The receive spatial signature at user k is denoted by $\mathbf{h}_k$. Let us focus on the time-invariant model first and fix this vector. If there is only one user, then we know from Section 7.2.1 that the user should do receive beamforming: project the received signal in the direction of the vector channel. Let us try this technique here, with both users matched filtering their received signals w.r.t. their channels. This is illustrated in Figure 10.29 and can be shown to be the optimal strategy for both the users (Exercise 10.22). With the matched filter front-end at each user, we have an effective AWGN downlink with a single antenna:

$$\tilde{y}_k[m] := \frac{\mathbf{h}_k^* \mathbf{y}_k[m]}{\|\mathbf{h}_k\|} = \|\mathbf{h}_k\| x[m] + w_k[m], \qquad k = 1, 2. \tag{10.79}$$

Here $w_k[m]$ is $\mathcal{CN}(0, N_0)$ and i.i.d. in time m and the downlink channel in (10.79) is very similar to the basic single antenna downlink channel model of (6.16) in Section 6.2. The only difference is that user k's channel quality $|h_k|^2$ is replaced by $\|\mathbf{h}_k\|^2$.

Thus, to study the downlink with multiple receive antennas, we can now carry over all our discussions from Section 6.2 for the single antenna scenario. In particular, we can order the two users based on their received SNR (suppose $\|\mathbf{h}_1\| \leq \|\mathbf{h}_2\|$) and do superposition coding: the transmit signal is the linear superposition of the signals to the two users. User 1 treats the signal of user 2 as noise and decodes its data from $\tilde{y}_1$. User 2, which has the better SNR, decodes the data of user 1, subtracts the transmit signal of user 1 from $\tilde{y}_2$ and then decodes its data. With a total power constraint of P and splitting this among the two users $P = P_1 + P_2$ we can write the

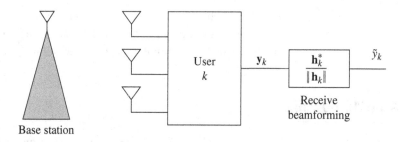

Figure 10.29 Each user with a front-end matched filter converting the SIMO downlink into a SISO downlink.

rate tuple that is achieved with the receiver architecture in Figure 10.29 and superposition coding (cf. (6.22)),

$$R_1 = \log\left(1 + \frac{P_1\|\mathbf{h}_1\|^2}{P_2\|\mathbf{h}_1\|^2 + N_0}\right), \quad R_2 = \log\left(1 + \frac{P_2\|\mathbf{h}_2\|^2}{N_0}\right). \qquad (10.80)$$

Thus we have combined the techniques of Sections 7.2.1 and 6.2, namely receive beamforming and superposition coding into a communication strategy for the single transmit and multiple receive antenna downlink.

The matched filter operation by the users in Figure 10.29 only requires tracking of their channels by the users, i.e., CSI is required at the receivers. Thus, even with fast fading, the architecture in Figure 10.29 allows us to transform the downlink with multiple receive antennas to the basic single antenna downlink channel as long as the users have their channel state information. In particular, analyzing receiver CSI and full CSI for the downlink in (10.78) simplifies to the basic single antenna downlink discussion (in Section 6.4).

In particular, we can ask what impact multiple receive antennas have on multiuser diversity, an important outcome of our discussion in Section 6.4. The only difference here is the distribution of the channel quality: $\|\mathbf{h}_k\|^2$ replacing $|h_k|^2$. This was also the same difference in the uplink when we studied the role of multiple receive antennas in multiuser diversity gain (in Section 10.1.6). We can carry over our main observation: the multiple receive antennas provide a beamforming gain but the tail of $\|\mathbf{h}_k\|^2$ decays more rapidly (Figure 10.8) and the multiuser diversity gain is restricted (Figure 10.9). To summarize, the traditional receive beamforming power gain is balanced by the loss of the benefit of the multiuser diversity gain (which is also a power gain) due to the "hardening" of the effective fading distribution: $\|\mathbf{h}_k\|^2 \approx n_r$ (cf. (10.20)).

With multiple transmit antennas at the base-station and multiple receive antennas at each of the users, we can extend our set of linear strategies from the discussion in Section 10.3.2: now the base-station splits the information for user k into independent data streams, modulates them on different spatial signatures and then transmits them. With full CSI, we can vary these spatial signatures and powers allocated to the users (and the further allocation among the data streams within a user) as a function of the channel fluctuations. We can also embellish the linear strategies with Costa precoding, successively precanceling the data streams. The performance of this scheme (linear beam-forming strategies with and without Costa precoding) can be related to the corresponding performance of a dual MIMO uplink channel (much as in the discussion of Section 10.3.2 with multiple antennas at the base-station alone). This scheme achieves the capacity of the MIMO downlink channel.

10.5 Multiple antennas in cellular networks: a system view

We have discussed the system design implications of multiple antennas in both the uplink and the downlink. These discussions have been in the context of multiple access *within a single cell* and are spread throughout the chapter (Sections 10.1.3, 10.1.6, 10.2.2, 10.3.5 and 10.4). In this section we take stock of these implications and consider the role of multiple antennas in cellular networks with multiple cells. Particular emphasis is on two points:

- the use of multiple antennas in suppressing inter-cell interference;
- how the use of multiple antennas within cells impacts the optimal amount of frequency reuse in the network.

Summary 10.3 System implications of multiple antennas on multiple access

Three ways of using multiple receive antennas in the uplink:
- **Orthogonal multiple access** Each user gets a power gain, but no change in degrees of freedom.
- **Opportunistic communication, one user at a time** Power gain but the multiuser diversity gain is reduced.
- **Space division multiple access** is capacity achieving: users simultaneously transmit and are jointly decoded at the base-station.

Comparison between orthogonal multiple access and SDMA
- Low SNR: performance of orthogonal multiple access comparable to that of SDMA.
- High SNR: SDMA allows up to n_r users to simultaneously transmit with a single degree of freedom each. Performance is significantly better than that with orthogonal multiple access.
- An intermediate access scheme with moderate complexity performs comparably to SDMA at all SNR levels: blocks of approximately n_r users in SDMA mode and orthogonal access for different blocks.

MIMO uplink
- Orthogonal multiple access: each user has multiple degrees of freedom.
- SDMA: the overall degrees of freedom are still restricted by the number of receive antennas.

> Downlink with multiple receive antennas
> Each user gets receive beamforming gain but reduced multiuser diversity gain.
> Downlink with multiple transmit antennas
> - No CSI at the base-station: single spatial degree of freedom.
> - Full CSI: the uplink–downlink duality principle makes this situation analogous to the uplink with multiple receive antennas and now there are up to n_t spatial degrees of freedom.
> - Partial CSI at the base-station: the same spatial degrees of freedom as the full CSI scenario can be achieved by a modification of the opportunistic beamforming scheme: multiple spatially orthogonal beams are sent out and multiple users are simultaneously scheduled on these beams.

10.5.1 Inter-cell interference management

Consider the multiple receive antenna uplink with users operating in SDMA mode. We have seen that successive cancellation is an optimal way to handle interference among the users within the same cell. However, this technique is not suitable to handle interference from neighboring cells: the out-of-cell transmissions are meant to be decoded by their nearest base-stations and the received signal quality is usually too poor to allow decoding at base-stations further away. On the other hand, linear receivers such as the MMSE do not decode the information from the interference and can be used to suppress out-of-cell interference.

The following model captures the essence of out-of-cell interference: the received signal at the antenna array ($\mathbf{y}$) comprises the signal (x) of the user of interest (with the signals of other users in the same cell successfully canceled) and the out-of-cell interference ($\mathbf{z}$):

$$\mathbf{y} = \mathbf{h}x + \mathbf{z}. \tag{10.81}$$

Here $\mathbf{h}$ is the received spatial signature of the user of interest. One model for the random interference $\mathbf{z}$ is as $\mathcal{CN}(0, \mathbf{K}_z)$, i.e., it is *colored* Gaussian noise with covariance matrix $\mathbf{K}_z$. For example, if the interference originates from just one out-of-cell transmission (with transmit power, say, q) and the base-station has an estimate of the received spatial signature of the interfering transmission (say, $\mathbf{g}$), then the covariance matrix is

$$q\mathbf{g}\mathbf{g}^* + N_0\mathbf{I}, \tag{10.82}$$

taking into account the structure of the interference and the background additive Gaussian noise.

Once such a model has been adopted, the multiple receive antennas can be used to suppress interference: we can use the linear MMSE receiver developed in Section 8.3.3 to get the soft estimate (cf. (8.61)):

$$\hat{x} = \mathbf{v}^*_{\text{mmse}}\mathbf{y} = \mathbf{h}^*\mathbf{K}_z^{-1}\mathbf{y}. \qquad (10.83)$$

The expression for the corresponding SINR is in (8.62). This is the best SINR possible with a linear estimate. When the interfering noise is white, the operation is simply traditional receive beamforming. On the other hand, when the interference is very large and not white then the operation reduces to a decorrelator: this corresponds to nulling out the interference. The effect of channel estimation error on interference suppression is explored in Exercise 10.23.

In the uplink, the model for the interference depends on the type of multiple access. In many instances, a natural model for the interference is that it is white. For example, if the out-of-cell interference comes from many geographically spread out users (this situation occurs when there are many users in SDMA mode), then the overall interference is averaged over the multiple users' spatial locations and white noise is a natural model. In this case, the receive antenna array does not explicitly suppress out-of-cell interference. To be able to exploit the interference suppression capability of the antennas, two things must happen:

- The number of simultaneously transmitting users in each cell should be small. For example, in a hybrid SDMA/TDMA strategy, the total number of users in each cell may be large but the number of users simultaneously in SDMA mode is small (equal to or less than the number of receive antennas).
- The out-of-cell interference has to be trackable. In the SDMA/TDMA system, even though the interference at any time comes from a small number of users, the interference depends on the geographic location of the interfering user(s), which changes with the time slot. So either each slot has to be long enough to allow enough time to estimate the color of the interference based only on the pilot signal received in that time slot, or the users are scheduled in a periodic manner and the interference can be tracked across different time slots.

An example of such a system is described in Example 10.1.

On the other hand, interference suppression in the downlink using multiple receive antennas at the mobiles is different. Here the interference comes from a few base-stations of the neighboring cells that reuse the same frequency, i.e., from fixed specific geographic locations. Now, an estimate of the covariance of the interference can be formed and the linear MMSE can be used to manage the inter-cell interference.

We now turn to the role of multiple antennas in deciding the optimal amount of frequency reuse in the cellular network. We consider the effect on both the uplink and the downlink and the role of multiple receive and multiple transmit antennas separately.

10.5.2 Uplink with multiple receive antennas

We begin with a discussion of the impact of multiple antennas at the base-station on the two orthogonal cellular systems studied in Chapter 4 and then move to SDMA.

Orthogonal multiple access

The array of multiple antennas is used to boost the received signal strength from the user within the cell via receive beamforming. One immediate benefit is that each user can lower its transmit power by a factor equal to the beamforming gain (proportional to n_r) to maintain the same signal quality at the base-station. This reduction in transmit power also helps to reduce inter-cell interference, so the effective SINR with the power reduction is in fact more than the SINR achieved in the original setting.

In Example 5.2 we considered a linear array of base-stations and analyzed the tradeoff between reuse and data rates per user for a given cell size and transmit power setting. With an array of antennas at each base-station, the SNR of every user improves by a factor equal to the receive beamforming gain. Much of the insight derived in Example 5.2 on how much to reuse can be naturally extended to the case here with the operating SNR boosted by the receive beamforming gain.

SDMA

If we do not impose the constraint that uplink communication be orthogonal among the users in the cell, we can use the SDMA strategy where many users simultaneously transmit and are jointly decoded at the base-station. We have seen that this scheme significantly betters orthogonal multiple access at high SNR due to the increased spatial degrees of freedom. At low SNR, both orthogonal multiple access and SDMA benefit comparably, with the users getting a receive beamforming gain. Thus, for SDMA to provide significant performance improvement over orthogonal multiple access, we need the operating SNR to be large; in the context of a cellular system, this means less frequency reuse.

Whether the loss in spectral efficiency due to less frequency reuse is fully compensated for by the increase in spatial degrees of freedom depends on the specific physical situation. The frequency reuse ratio ρ represents the loss in spectral efficiency. The corresponding reduction in interference is represented by the fraction f_ρ: this is the fraction of the received power from a user at the edge of the cell that the interference constitutes. For example, in a linear cellular system f_ρ decays roughly as ρ^α, but for a hexagonal cellular system the decay is much slower: f_ρ decays roughly as $\rho^{\alpha/2}$ (cf. Example 5.2).

Suppose all the K users are at the edge of the cell (a worst case scenario) and communicating via SDMA to the base-station with receiver CSI. W is the total bandwidth allotted to the cellular system scaled down by the number of simultaneous SDMA users sharing it within a cell (as with orthogonal multiple access, cf. Example 5.2). With SDMA used in each cell, K users simultaneously transmit over the entire bandwidth $K\rho W$.

The SINR of the user at the edge of the cell is, as in (5.20),

$$\mathsf{SINR} = \frac{\mathsf{SNR}}{\rho K + f_\rho \mathsf{SNR}}, \quad \text{with} \quad \mathsf{SNR} := \frac{P}{N_0 W d^\alpha}. \tag{10.84}$$

The SNR at the edge of the cell is SNR, a function of the transmit power P, the cell size d, and the power decay rate α (cf. (5.21)). The notation for the fraction f_ρ is carried over from Example 5.2. The largest symmetric rate each user gets is, the MIMO extension of (5.22),

$$R_\rho = \rho W \mathbb{E}[\log \det(\mathbf{I}_{n_r} + \mathsf{SINR} \, \mathbf{HH}^*)] \text{ bits/s}. \tag{10.85}$$

Here the columns of $\mathbf{H}$ represent the receive spatial signatures of the users at the base-station and the log det expression is the sum of the rates at which users can simultaneously communicate reliably.

We can now address the engineering question of how much to reuse using the simple formula for the rate in (10.85). At low SNR the situation is analogous to the single receive antenna scenario studied in Example 5.2: the rate is insensitive to the reuse factor and this can be verified directly from (10.85). On the other hand, at large SNR the interference grows as well and the SINR peaks at $1/f_\rho$. The largest rate then is, as in (5.23),

$$\rho W \mathbb{E}\left[\log \det\left(\mathbf{I}_{n_r} + \frac{1}{f_\rho} \mathbf{HH}^*\right)\right] \text{ bits/s}, \tag{10.86}$$

and goes to zero for small values of ρ: thus as in Example 5.2, less reuse does not lead to a favorable situation.

How do multiple receive antennas affect the optimal reuse ratio? Setting $K = n_r$ (a rule of thumb arrived at in Exercise 10.5), we can use the approximation in (8.29) to simplify the expression for the rate in (10.86):

$$R_\rho \approx \rho W n_r c^*\left(\frac{1}{f_\rho}\right). \tag{10.87}$$

The first observation we can make is that since the rate grows linearly in n_r, the optimal reuse ratio does not depend on the number of receive antennas. The optimal reuse ratio thus depends only on how the inter-cell interference f_ρ decays with the reuse parameter ρ, as in the single antenna situation studied in Example 5.2.

Figure 10.30 The symmetric rate for every user (in bps/Hz) with $K = 5$ users in SDMA model in an uplink with $n_r = 5$ receive antennas plotted as a function of the power decay rate α for the linear cellular system. The rates are plotted for reuse ratios 1, 1/2 and 1/3.

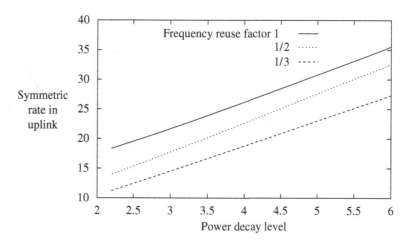

Figure 10.30 The symmetric rate for every user (in bps/Hz) with $K = 5$ users in SDMA model in an uplink with $n_r = 5$ receive antennas plotted as a function of the power decay rate α for the linear cellular system. The rates are plotted for reuse ratios 1, 1/2 and 1/3.

The rates at high SNR with reuse ratios 1, 1/2 and 1/4 are plotted in Figure 10.30 for $n_r = K = 5$ in the linear cellular system. We observe the optimality of universal reuse at all power decay rates: the gain in SINR from less reuse is not worth the loss in spectral reuse. Comparing with the single receive antenna example, the receive antennas provide a performance boost (the rate increases linearly with n_r). We also observe that universal reuse is now preferred. The hexagonal cellular system provides even less improvement in SINR and thus universal reuse is optimal; this is unchanged from the single receive antenna example.

10.5.3 MIMO uplink

An implementation of SDMA corresponds to altering the nature of medium access. For example, there is no simple way of incorporating SDMA in any of the three cellular systems introduced in Chapter 4 without altering the fundamental way resource allocation is done among users. On the other hand, the use of multiple antennas at the base-station to do receive beamforming for each user of interest is a scheme based at the level of a point-to-point communication link and can be implemented regardless of the nature of the medium access. In some contexts where the medium access scheme cannot be altered, a scheme based on improving the quality of individual point-to-point links is preferred. However, an array of multiple antennas at the base-station used to receive beamform provides only a power gain and not an increase in degrees of freedom. If each user has multiple transmit antennas as well, then an increase in the degrees of freedom of each individual point-to-point link can be obtained.

In an orthogonal system, the point-to-point MIMO link provides each user with multiple degrees of freedom and added diversity. With receiver CSI, each user can use its transmit antenna array to harness the spatial degrees of freedom when it is scheduled. The discussion of the role of frequency reuse earlier now carries over to this case. The nature of the tradeoff is similar: there is a loss in spectral degrees of freedom (due to less reuse) but an increase in the spatial degrees of freedom (due to the availability of multiple transmit antennas at the users).

10.5.4 Downlink with multiple receive antennas

In the downlink the interference comes from a few specific locations at fixed transmit powers: the neighboring base-stations that reuse the same frequency. Thus, the interference pattern can be empirically measured at each user and the array of receive antennas used to do linear MMSE (as discussed in Section 10.5.1) and boost the received SINR. For orthogonal systems, the impact on frequency reuse analysis is similar to that in the uplink with the SINR from the MMSE receiver replacing the earlier simpler expression (as in (5.20), for the uplink example).

If the base-station has multiple transmit antennas as well, the interference could be harder to suppress: in the presence of substantial scattering, each of the base-station transmit antennas could have a distinct receive spatial signature at the mobile, and in this case an appropriate model for the interference is white noise. On the other hand, if the scattering is only local (at the base-station and at the mobile) then all the base-station antennas have the same receive spatial signature (cf. Section 7.2.3) and interference suppression via the MMSE receiver is still possible.

10.5.5 Downlink with multiple transmit antennas

With full CSI (i.e., both at the base-station and at the users), the uplink–downlink duality principle (see Section 10.3.2) allows a comparison to the reciprocal uplink with the multiple *receive* antennas and receiver CSI. In particular, there is a one-to-one relationship between linear schemes (with and without successive cancellation) for the uplink and that for the downlink. Thus, many of our inferences in the uplink with multiple receive antennas hold in the downlink as well. However, full CSI may not be so practical in an FDD system: having CSI at the base-station in the downlink requires substantial CSI feedback via the uplink.

> **Example 10.1 SDMA in ArrayComm systems**
> ArrayComm Inc. is one of the early companies implementing SDMA technology. Their products include an SDMA overlay on Japan's PHS cellular system, a fixed wireless local loop system, and a mobile cellular system (iBurst).

An ArrayComm SDMA system exemplifies many of the design features that multiple antennas at the base-station allow. It is TDMA based and is much like the narrowband system we studied in Chapter 4. The main difference is that within each narrowband channel in each time slot, a small number of users are in SDMA mode (as opposed to just a single user in the basic narrowband system of Section 4.2). The array of antennas at the base-station is also used to suppress out-of-cell interference, thus allowing denser frequency reuse than a basic narrowband system. To enable successful SDMA operation and interference suppression in both the uplink and the downlink, the ArrayComm system has several key design features.

- The time slots for TDMA are synchronized across different cells. Further, the time slots are long enough to allow accurate estimation of the interference using the training sequence. The estimate of the color of the interference is then in the same time slot to suppress out-of-cell interference. Channel state information is not kept across slots.
- The small number of SDMA users within each narrowband channel are demodulated using appropriate linear filters: for each user, this operation suppresses both the out-of-cell interference and the in-cell interference from the other users in SDMA mode sharing the same narrowband channel.
- The uplink and the downlink operate in TDD mode with the downlink transmission immediately *following* the uplink transmission and to the *same* set of users. The uplink transmission provides the base-station CSI that is used in the immediately following downlink transmission to perform SDMA and to suppress out-of-cell interference via transmit beamforming and nulling. TDD operation avoids the expensive channel state feedback required for downlink SDMA in FDD systems.

To get a feel for the performance improvement with SDMA over the basic narrowband system, we can consider a specific implementation of the ArrayComm system. There are up to twelve antennas per sector at the base-station with up to four users in SDMA mode over each narrowband channel. This is an improvement of roughly a factor of four over the basic narrowband system, which schedules only a single user over each narrowband channel. Since there are about three antennas per user, substantial out-of-cell interference suppression is possible. This allows us to increase the frequency reuse ratio; this is a further benefit over the basic narrowband system. For example, the SDMA overlay on the PHS system increases the frequency reuse ratio of 1/8 to 1.

In the Flash OFDM example in Chapter 4, we have mentioned that one advantage of orthogonal multiple access systems over CDMA systems is that users can get access to the system without the need to slowly ramp up the power. The interference suppression capability of adaptive antennas provides another way to allow users who are not power controlled to get access to the system quickly without swamping the existing active users. Even in a near–far situation of 40–50 dB, SDMA still works successfully; this means that potentially many users can be kept in the hold state when there are no active transmissions.

These improvements come at an increased cost to certain system design features. For example, while downlink transmissions meant for specific users enjoy a power gain via transmit beamforming, the pilot signal is intended for all users and has to be isotropic, thus requiring a proportionally larger amount of power. This reduces the traditional amortization benefit of the downlink pilot. Another aspect is the forced symmetry between the uplink and the downlink transmissions. To successfully use the uplink measurements (of the channels of the users in SDMA mode and the color of the out-of-cell interference) in the following downlink transmission, the transmission power levels in the uplink and the downlink have to be comparable (see Exercise 10.24). This puts a strong constraint on the system designer since the mobiles operate on batteries and are typically much more power constrained than the base-station, which is powered by an AC supply. Further, the pairing of the uplink or downlink transmissions is ideal when the flow of traffic is symmetric in both directions; this is usually true in the case of voice traffic. On the other hand, data traffic can be asymmetric and leads to wasted uplink (downlink) transmissions if only downlink (uplink) transmissions are desired.

Chapter 10 The main plot

Uplink with multiple receive antennas
Space division multiple access (SDMA) is capacity-achieving: all users simultaneously transmit and are jointly decoded by the base-station.
- Total spatial degrees of freedom limited by number of users and number of receive antennas.
- Rule of thumb is to have a group of n_r users in SDMA mode and different groups in orthogonal access mode.
- *Each* of the n_r user transmissions in a group obtains the full receive diversity gain equal to n_r.

Uplink with multiple transmit and receive antennas
The overall spatial degrees of freedom are still restricted by the number of receive antennas, but the diversity gain is enhanced.

Downlink with multiple transmit antennas

Uplink–downlink duality identifies a correspondence between the down-link and the *reciprocal* uplink.

Precoding is the analogous operation to successive cancelation in the uplink. A precoding scheme that perfectly cancels the intra-cell interference caused to a user was described.

Precoding operation requires full CSI; hard to justify in an FDD system. With only partial CSI at the base-station, an opportunistic beamforming scheme with multiple orthogonal beams utilizes the full spatial degrees of freedom.

Downlink with multiple receive antennas

Each user's link is enhanced by receive beamforming: both a power gain and a diversity gain equal to the number of receive antennas are obtained.

10.6 Bibliographical notes

The precoding technique for communicating on a channel where the transmitter is aware of the channel was first studied in the context of the ISI channel by Tomlinson [121] and Harashima and Miyakawa [57]. More sophisticated precoders for the ISI channel (designed for use in telephone modems) were developed by Eyuboglu and Forney [36] and Laroia *et al.* [71]. A survey on precoding and shaping for ISI channels is contained in an article by Forney and Ungerböck [39].

Information theoretic study of a state-dependent channel where the transmitter has non-causal knowledge of the state was studied, and the capacity characterized, by Gelfand and Pinsker [46]. The calculation of the capacity for the important special case of additive Gaussian noise and an additive Gaussian state was done by Costa [23], who concluded the surprising result that the capacity is the same as that of the channel where the state is known to the receiver also. Practical construction of the binning schemes (involving two steps: a vector quantization step and a channel coding step) is still an ongoing effort and the current progress is surveyed by Zamir *et al.* [154]. The performance of the opportunistic orthogonal signaling scheme, which uses orthogonal signals as both channel codes and vector quantizers, was analyzed by Liu and Viswanath [76].

The Costa precoding scheme was used in the multiple antenna downlink channel by Caire and Shamai [17]. The optimality of these schemes for the sum rate was shown in [17, 135, 138, 153]. Weingarten, *et al.* [141] proved that the Costa precoding scheme achieves the entire capacity region of the multiple antenna downlink.

The reciprocity between the uplink and the downlink was observed in different contexts: linear beamforming (Visotsky and Madhow [134], Farrokhi *et al.* [37]), capacity of the point-to-point MIMO channel (Telatar [119]), and achievable rates of the single antenna Gaussian MAC and BC (Jindal *et al.* [63]). The presentation here is based on a unified understanding of these results (Viswanath and Tse [138]).

10.7 Exercises

Exercise 10.1 Consider the time-invariant uplink with multiple receive antennas (10.1). Suppose user k transmits data at power $P_k, k = 1, \ldots, K$. We would like to employ a bank of linear MMSE receivers at the base-station to decode the data of the users:

$$\hat{x}_k[m] = \mathbf{c}_k^* \mathbf{y}[m], \tag{10.88}$$

is the estimate of the data symbol $x_k[m]$.

1. Find an explicit expression for the linear MMSE filter $\mathbf{c}_k$ (for user k). *Hint*: Recall the analogy between the uplink here with independent data streams being transmitted on a point-to-point MIMO channel and see (8.66) in Section 8.3.3.
2. Explicitly calculate the SINR of user k using the linear MMSE filter. *Hint*: See (8.67).

Exercise 10.2 Consider the bank of linear MMSE receivers at the base-station decoding the user signals in the uplink (as in Exercise 10.1). We would like to tune the transmit powers of the users $P_1, \ldots, P_K$ such that the SINR of each user (calculated in Exercise 10.1(2)) is at least equal to a target level β. Show that, if it is possible to find a set of power levels that meet this requirement, then there exists a component-wise minimum power setting that meets the SINR target level. This result is on similar lines to the one in Exercise 4.5 and is proved in [128].

Exercise 10.3 In this problem, a sequel to Exercise 10.2, we will see an adaptive algorithm that updates the transmit powers and linear MMSE receivers for each user in a greedy fashion. This algorithm is closely related to the one we studied in Exercise 4.8 and is adapted from [128].

Users begin (at time 1) with an arbitrary power setting $p_1^{(1)}, \ldots, p_K^{(1)}$. The bank of linear MMSE receivers $(\mathbf{c}_1^{(1)}, \ldots, \mathbf{c}_K^{(1)})$ at the base-station is tuned to these transmit powers. At time $m + 1$, each user updates its transmit power and its MMSE filter as a function of the power levels of the other users at time m so that its SINR is exactly equal to β. Show that if there exists a set of powers such that the SINR requirement can be met, then this synchronous update algorithm will converge to the component-wise minimal power setting identified in Exercise 10.2.

In this exercise, the update of the user powers (and corresponding MMSE filters) is synchronous among the users. An asynchronous algorithm, analogous to the one in Exercise 4.9, works as well.

Exercise 10.4 Consider the two-user uplink with multiple receive antennas (10.1):

$$\mathbf{y}[m] = \sum_{k=1}^{2} \mathbf{h}_k x_k[m] + \mathbf{w}[m]. \tag{10.89}$$

Suppose user k has an average power constraint $P_k, k = 1, 2$.

1. Consider orthogonal multiple access: with α the fraction of the degrees of freedom allocated to user 1 (and $1 - \alpha$ the fraction to user 2), the reliable communication rates of the two users are given in Eq. (10.7). Calculate the fraction α that yields the largest sum rate achievable by orthogonal multiple access and the corresponding sum rate. *Hint*: Recall the result for the uplink with a single receive antenna in Section 6.1.3 that the largest sum rate with orthogonal multiple access is *equal* to the sum capacity of the uplink, cf. Figure 6.4.

2. Consider the difference between the sum capacity of the uplink with multiple receive antennas (see (10.4)) with the largest sum rate of this uplink with orthogonal multiple access.

   (a) Show that this difference is zero *exactly* when $\mathbf{h}_1 = c\mathbf{h}_2$ for some (complex) constant c.

   (b) Suppose $\mathbf{h}_1$ and $\mathbf{h}_2$ are not scalar complex multiples of each other. Show that at high SNR (N_0 goes to zero) the difference between the two sum rates becomes arbitrarily large. With $P_1 = P_2 = P$, calculate the rate of growth of this difference with SNR (P/N_0). We conclude that at high SNR (large values of P_1, P_2 as compared to N_0) orthogonal multiple access is very suboptimal in terms of the sum of the rates of the users .

Exercise 10.5 Consider the K-user uplink and focus on the *sum* and *symmetric* capacities. The base-station has an array of n_r receive antennas. With receiver CSI and fast fading, we have the following expression: the symmetric capacity is

$$C_{\text{sym}} = \frac{1}{K} \mathbb{E}[\log_2 \det(\mathbf{I}_{n_r} + \text{SNR}\,\mathbf{H}\mathbf{H}^*)] \text{ bits/s/Hz.} \qquad (10.90)$$

and the sum capacity C_{sum} is KC_{sym}. Here the columns of $\mathbf{H}$ represent the receive spatial signatures of the users and are modeled as i.i.d. $\mathcal{CN}(0, 1)$. Each user has an identical transmit power constraint P, and the common SNR is equal to P/N_0.

1. Show that the sum capacity increases monotonically with the number of users.

2. Show that the symmetric capacity, on the other hand, goes to zero as the number of users K grows large, for every fixed SNR value and n_r. *Hint*: You can use Jensen's inequality to get a bound.

3. Show that the sum capacity increases linearly in K at low SNR. Thus the symmetric capacity is independent of K at low SNR values.

4. Argue that at high SNR the sum capacity only grows logarithmically in K as K increases beyond n_r.

5. Plot C_{sum} and C_{sym} as a function of K for sample SNR values (from 0 dB to 30 dB) and sample n_r values (3 through 6). Can you conclude some general trends from your plots? In particular, focus on the following issues.

   (a) How does the value of K at which the sum capacity starts to grow slowly depend on n_r?

   (b) How does the value of K beyond which the symmetric capacity starts to decay rapidly depend on n_r?

   (c) How does the answer to the previous two questions change with the operating SNR value?

You should be able to arrive at the following rule of thumb: $K = n_r$ is a good operating point at most SNR values in the sense that increasing K beyond it does not increase the sum capacity by much, and in fact reduces the symmetric capacity by quite a bit.

Exercise 10.6 Consider the K-user uplink with n_r multiple antennas at the base-station as in Exercise 10.5. The expression for the symmetric capacity is in (10.90). Argue that the symmetric capacity at low SNR is comparable to the symmetric rate with orthogonal multiple access. *Hint*: Recall the discussion on the low SNR MIMO performance gain in Section 8.2.2.

Exercise 10.7 In a slow fading uplink, the multiple receive antennas can be used to improve the reliability of reception (diversity gain), improve the rate of communication at a fixed reliability level (multiplexing gain), and also spatially separate the signals of the users (multiple access gain). A reading exercise is to study [86] and [125] which derive the fundamental tradeoff between these gains.

Exercise 10.8 In this exercise, we further study the comparison between orthogonal multiple access and SDMA with multiple receive antennas at the base-station. While orthogonal multiple access is simple to implement, SDMA is the capacity achieving scheme and outperforms orthogonal multiple access in certain scenarios (cf. Exercise 10.4) but requires complex joint decoding of the users at the base-station.

Consider the following access mechanism, which is a cross between purely orthogonal multiple access (where all the users' signals are orthogonal) and purely SDMA (where all the K users share the bandwidth and time simultaneously). Divide the K users into groups of approximately n_r users each. We provide orthogonal resource allocation (time, frequency or a combination) to each of the groups but within each group the users (approximately n_r of them) operate in an SDMA mode.

We would like to compare this intermediate scheme with orthogonal multiple access and SDMA. Let us use the largest symmetric rate achievable with each scheme as the performance criterion. The uplink model (same as the one in Exercise 10.5) is the following: receiver CSI with i.i.d. Rayleigh fast fading. Each user has the same average transmit power constraint P, and SNR denotes the ratio of P to the background complex Gaussian noise power N_0.

1. Write an expression for the symmetric rate with the intermediate access scheme (the expression for the symmetric rate with SDMA is in (10.90)).
2. Show that the intermediate access scheme has performance comparable to both orthogonal multiple access and SDMA at low SNR, in the sense that the ratio of the performances goes to 1 as SNR $\rightarrow$ 0.
3. Show that the intermediate access scheme has performance comparable to SDMA at high SNR, in the sense that the ratio of the performances goes to 1 as SNR $\rightarrow \infty$.
4. Fix the number of users K (to, say, 30) and the number of receive antennas n_r (to, say, 5). Plot the symmetric rate with SDMA, orthogonal multiple access and the intermediate access scheme as a function of SNR (0 dB to 30 dB). How does the intermediate access scheme compare with SDMA and orthogonal multiple access for the intermediate SNR values?

Exercise 10.9 Consider the K-user uplink with multiple receive antennas (10.1):

$$\mathbf{y}[m] = \sum_{k=1}^{K} \mathbf{h}_k x_k[m] + \mathbf{w}[m]. \tag{10.91}$$

Consider the sum capacity with full CSI (10.17):

$$C_{\text{sum}} = \max_{P_k(\mathbf{H}), k=1,\dots,K} \mathbb{E}\left[\log\det\left(\mathbf{I}_{n_r} + \sum_{k=1}^{K} P_k(\mathbf{H})\mathbf{h}_k\mathbf{h}_k^*\right)\right], \tag{10.92}$$

where we have assumed the noise variance $N_0 = 1$ and have written $\mathbf{H} = [\mathbf{h}_1, \dots, \mathbf{h}_K]$. User k has an average power constraint P; due to the ergodicity in the channel fluctuations, the average power is equal to the ensemble average of the power transmitted at each fading state ($P_k(\mathbf{H})$ when the channel state is $\mathbf{H}$). So the average power constraint can be written as

$$\mathbb{E}[P_k(\mathbf{H})] \leq P. \tag{10.93}$$

We would like to understand what power allocations maximize the sum capacity in (10.92).

1. Consider the map from a set of powers to the corresponding sum rate in the uplink:

$$(P_1, \dots, P_K) \mapsto \log\det\left(\mathbf{I}_{n_r} + \sum_{k=1}^{K} P_k\mathbf{h}_k\mathbf{h}_k^*\right). \tag{10.94}$$

   Show that this map is jointly concave in the set of powers. *Hint*: You will find useful the following generalization (to higher dimensions) of the elementary observation that the map $x \mapsto \log x$ is concave for positive real x:

$$\mathbf{A} \mapsto \log\det(\mathbf{A}) \tag{10.95}$$

   is concave in the set of positive definite matrices $\mathbf{A}$.

2. Due to the concavity property, we can characterize the optimal power allocation policy using the Lagrangian:

$$\mathcal{L}(P_1(\mathbf{H}), \dots, P_K(\mathbf{H})) := \mathbb{E}\left[\log\det\left(\mathbf{I}_{n_r} + \sum_{k=1}^{K} P_k(\mathbf{H})\mathbf{h}_k\mathbf{h}_k^*\right)\right]$$
$$- \sum_{k=1}^{K} \lambda_k \mathbb{E}[P_k(\mathbf{H})]. \tag{10.96}$$

   The optimal power allocation policy $P_k^*(\mathbf{H})$ satisfies the Kuhn–Tucker equations:

$$\frac{\partial L}{\partial P_k(\mathbf{H})} \begin{cases} = 0 & \text{if } P_k^*(\mathbf{H}) > 0, \\ \leq 0 & \text{if } P_k^*(\mathbf{H}) = 0. \end{cases} \tag{10.97}$$

   Calculate the partial derivative explicitly to arrive at:

$$\mathbf{h}_k^*\left(\mathbf{I}_{n_r} + \sum_{j=1}^{K} P_j^*(\mathbf{H})\mathbf{h}_j\mathbf{h}_j^*\right)^{-1}\mathbf{h}_k \begin{cases} = \lambda_k & \text{if } P_k^*(\mathbf{H}) > 0, \\ \leq \lambda_k & \text{if } P_k^*(\mathbf{H}) = 0. \end{cases} \tag{10.98}$$

   Here $\lambda_1, \dots, \lambda_K$ are constants such that the average power constraint in (10.93) is met. With i.i.d. channel fading statistics (i.e., $\mathbf{h}_1, \dots, \mathbf{h}_K$ are i.i.d. random vectors), these constants can be taken to be equal.

3. The optimal power allocation $P_k^*(\mathbf{H})$, $k = 1, \ldots, K$ satisfying (10.98) is also the solution to the following optimization problem:

$$\max_{P_1, \ldots, P_K \geq 0} \log \det \left(\mathbf{I}_{n_r} + \sum_{k=1}^{K} P_k \mathbf{h}_k \mathbf{h}_k^* \right) - \sum_{k=1}^{K} \lambda_k P_k. \tag{10.99}$$

In general, no closed form solution to this problem is known. However, efficient algorithms yielding numerical solutions have been designed; see [15]. Solve numerically an instance of the optimization problem in (10.99) with $n_r = 2$, $K = 3$,

$$\mathbf{h}_1 = \begin{bmatrix} 1 \\ 0 \end{bmatrix}, \qquad \mathbf{h}_2 = \begin{bmatrix} 0 \\ 1 \end{bmatrix}, \qquad \mathbf{h}_3 = \begin{bmatrix} 1 \\ 1 \end{bmatrix}, \tag{10.100}$$

and $\lambda_1 = \lambda_2 = \lambda_3 = 0.1$. You might find the software package [82] useful.

4. To get a feel for the optimization problem in (10.99) let us consider a few illustrative examples.

    (a) Consider the uplink with a single receive antenna, i.e., $n_r = 1$. Further suppose that each of the $|h_k|^2/\lambda_k$, $k = 1, \ldots, K$ are distinct. Show that an optimal solution to the problem in (10.99) is to allocate positive power to at most one user:

    $$P_k^* = \begin{cases} \left(\frac{1}{\lambda_k} - \frac{1}{|h_k|^2} \right)^+ & \text{if } \frac{|h_k|^2}{\lambda_k} = \max_{j=1 \ldots K} \frac{|h_j|^2}{\lambda_j}, \\ 0 & \text{else.} \end{cases} \tag{10.101}$$

    This calculation is a reprise of that in Section 6.3.3.

    (b) Now suppose there are three users in the uplink with two receive antennas, i.e., $K = 3$ and $n_r = 2$. Suppose $\lambda_k = \lambda$, $k = 1, 2, 3$ and

    $$\mathbf{h}_1 = \begin{bmatrix} 1 \\ 1 \end{bmatrix}, \quad \mathbf{h}_2 = \begin{bmatrix} 1 \\ \exp(j2\pi/3) \end{bmatrix}, \quad \mathbf{h}_3 = \begin{bmatrix} 1 \\ \exp(j4\pi/3) \end{bmatrix}. \tag{10.102}$$

    Show that the optimal solution to (10.99) is

    $$P_k^* = \frac{2}{9} \left(\frac{3}{\lambda} - 1 \right)^+, \qquad k = 1, 2, 3. \tag{10.103}$$

    Thus for $n_r > 1$ the optimal solution in general allocates positive power to more than one user. *Hint*: First show that for any set of powers P_1, P_2, P_3 with their sum constrained (to say P), it is always optimal to choose them all equal (to $P/3$).

Exercise 10.10 In this exercise, we look for an approximation to the optimal power allocation policy derived in Exercise 10.9. To simplify our calculations, we take i.i.d. fading statistics of the users so that $\lambda_1, \ldots, \lambda_K$ can all be taken equal (and denoted by λ).

1. Show that

$$\mathbf{h}_k^* \left(\mathbf{I}_{n_r} + \sum_{j=1}^{K} P_j \mathbf{h}_j \mathbf{h}_j^* \right)^{-1} \mathbf{h}_k = \frac{\mathbf{h}_k^* \left(\mathbf{I}_{n_r} + \sum_{j \neq k} P_j \mathbf{h}_j \mathbf{h}_j^* \right)^{-1} \mathbf{h}_k}{1 + \mathbf{h}_k^* \left(\mathbf{I}_{n_r} + \sum_{j \neq k} P_j \mathbf{h}_j \mathbf{h}_j^* \right)^{-1} \mathbf{h}_k P_k}. \tag{10.104}$$

Hint: You will find the matrix inversion lemma (8.124) useful.

2. Starting from (10.98), use (10.104) to show that the optimal power allocation policy can be rewritten as

$$P_k^*(\mathbf{H}) = \left(\frac{1}{\lambda} - \frac{1}{\mathbf{h}_k^*(\mathbf{I}_{n_r} + \sum_{j \neq k} P_j^*(\mathbf{H})\mathbf{h}_j\mathbf{h}_j^*)^{-1}\mathbf{h}_k} \right)^+ . \tag{10.105}$$

3. The quantity

$$\mathsf{SINR}_k := \mathbf{h}_k^* \left(\mathbf{I}_{n_r} + \sum_{j \neq k} P_j^*(\mathbf{H})\mathbf{h}_j\mathbf{h}_j^* \right)^{-1} \mathbf{h}_k P_k^*(\mathbf{H}) \tag{10.106}$$

can be interpreted as the SINR at the output of an MMSE filter used to demodulate user k's data (cf. (8.67)). If we define

$$I_0 := \frac{P_k^*(\mathbf{H})\|\mathbf{h}_k\|^2}{\mathsf{SINR}_k}, \tag{10.107}$$

then I_0 can be interpreted as the *interference* plus noise seen by user k. Substituting (10.107) in (10.105) we see that the optimal power allocation policy can be written as

$$P_k(\mathbf{H}) = \left(\frac{1}{\lambda} - \frac{I_0}{\|\mathbf{h}_k\|^2} \right)^+ . \tag{10.108}$$

While this power allocation appears to be the same as that of waterfilling, we have to be careful since I_0 itself is a function of the power allocations of the other users (which themselves depend on the power allocated to user k, cf. (10.105)). However, in a large system with K and n_r large enough (but the ratio of K and n_r being fixed) I_0 converges to a constant in probability (with i.i.d. zero mean entries of $\mathbf{H}$, the constant it converges to depends only on the variance of the entries of $\mathbf{H}$, the ratio between K and n_r and the background noise density N_0). This convergence result is essentially an application of a general convergence result that is of the same nature as the singular values of a large random matrix (discussed in Section 8.2.2). This justifies (10.21) and the details of this result can be found in [136].

Exercise 10.11 Consider the two-user MIMO uplink (see Section 10.2.1) with input covariances $\mathbf{K}_{x1}, \mathbf{K}_{x2}$.

1. Consider the corner point A in Figure 10.13, which depicts the achievable rate region using this input strategy. Show (as an extension of (10.5)) that at the point A the rates of the two users are

$$R_2 = \log \det(\mathbf{I}_{n_r} + \frac{1}{N_0}\mathbf{H}_2\mathbf{K}_{x2}\mathbf{H}_2^*), \tag{10.109}$$

$$R_1 = \log \det(\mathbf{I}_{n_r} + \mathbf{H}_1\mathbf{K}_{x1}\mathbf{H}_1^*(N_0\mathbf{I}_{n_r} + \mathbf{H}_2\mathbf{K}_{x2}\mathbf{H}_2^*)^{-1}). \tag{10.110}$$

2. Analogously, calculate the rate pair represented by the point B.

Exercise 10.12 Consider the capacity region of the two-user MIMO uplink (the convex hull of the union of the pentagon in Figure 10.13 for all possible input strategies parameterized by $\mathbf{K}_{x1}$ and $\mathbf{K}_{x2}$). Let us fix positive weights $a_1 \leq a_2$ and consider maximizing $a_1 R_1 + a_2 R_2$ over all rate pairs (R_1, R_2) in the capacity region.

1. Fix an input strategy $(\mathbf{K}_{xk}, k = 1, 2)$ and consider the value of $a_1 R_1 + a_2 R_2$ at the two corner points A and B of the corresponding pentagon (evaluated in Exercise 10.12). Show that the value of the linear functional is always no less at the vertex A than at the vertex B. You can use the expression for the rate pairs at the two corner points A and B derived in Exercise 10.11. This result is analogous to the polymatroid property derived in Exercise 6.9 for the capacity region of the single antenna uplink.

2. Now we would like to optimize $a_1 R_1 + a_2 R_2$ over all possible input strategies. Since the linear functional will always be optimized at one of the two vertices A or B in one of the pentagons, we only need to evaluate $a_1 R_1 + a_2 R_2$ at the corner point A (cf. (10.110) and (10.109)) and then maximize over the different input strategies:

$$\max_{\mathbf{K}_{xk}, \mathrm{Tr}\mathbf{K}_{xk} \leq P_k, k=1,2} a_1 \log \det(\mathbf{I}_{n_r} + \mathbf{H}_1 \mathbf{K}_{x1} \mathbf{H}_1^* (N_0 \mathbf{I}_{n_r} + \mathbf{H}_2 \mathbf{K}_{x2} \mathbf{H}_2^*)^{-1})$$
$$+ a_2 \log \det(\mathbf{I}_{n_r} + \frac{1}{N_0} \mathbf{H}_2 \mathbf{K}_{x2} \mathbf{H}_2^*). \tag{10.111}$$

Show that the function being maximized above is jointly concave in the input $\mathbf{K}_{x1}, \mathbf{K}_{x2}$. *Hint*: Show that $a_1 R_1 + a_2 R_2$ evaluated at the point A can also be written as

$$a_1 \log \det(\mathbf{I}_{n_r} + \frac{1}{N_0} \mathbf{H}_1 \mathbf{K}_{x1} \mathbf{H}_1^* + \frac{1}{N_0} \mathbf{H}_2 \mathbf{K}_{x2} \mathbf{H}_2^*) + (a_2 - a_1) \log \det(\mathbf{I}_{n_r} + \frac{1}{N_0} \mathbf{H}_2 \mathbf{K}_{x2} \mathbf{H}_2^*). \tag{10.112}$$

Now use the concavity property in (10.95) to arrive at the desired result.

3. In general there is no closed-form solution to the optimization problem in (10.111). However, the concavity property of the function being maximized has been used to design efficient algorithms that arrive at numerical solutions to this problem, [15].

Exercise 10.13 Consider the two-user fast fading MIMO uplink (see (10.25)). In the angular domain representation (see (7.70))

$$\mathbf{H}_k^a[m] = \mathbf{U}_r^* \mathbf{H}_k[m] \mathbf{U}_t, \qquad k = 1, 2, \tag{10.113}$$

suppose that the stationary distribution of $\mathbf{H}_k^a[m]$ has entries that are zero mean and uncorrelated (and further independent across the two users). Now consider maximizing the linear functional $a_1 R_1 + a_2 R_2$ (with $a_1 \leq a_2$) over all rate pairs (R_1, R_2) in the capacity region.

1. As in Exercise 10.12, show that the maximal value of the linear functional is attained at the vertex A in Figure 10.7 for some input covariances. Thus conclude that, analogous to (10.112), the maximal value of the linear functional over the capacity region can be written as

$$\max_{\mathbf{K}_{xk}, \mathrm{Tr}\mathbf{K}_{xk} \leq P_k, k=1,2} a_1 \mathbb{E}[\log \det(\mathbf{I}_{n_r} + \frac{1}{N_0} \mathbf{H}_1 \mathbf{K}_{x1} \mathbf{H}_1^* + \frac{1}{N_0} \mathbf{H}_2 \mathbf{K}_{x2} \mathbf{H}_2^*)]$$
$$+ (a_2 - a_1) \mathbb{E}[\log \det(\mathbf{I}_{n_r} + \frac{1}{N_0} \mathbf{H}_2 \mathbf{K}_{x2} \mathbf{H}_2^*)]. \tag{10.114}$$

2. Analogous to Exercise 8.3 show that the input covariances of the form in (10.27) achieve the maximum above in (10.114).

Exercise 10.14 Consider the two-user fast fading MIMO uplink under i.i.d. Rayleigh fading. Show that the input covariance in (10.30) achieves the maximal value of every linear functional $a_1 R_1 + a_2 R_2$ over the capacity region. Thus the capacity region in this case is simply a pentagon. *Hint*: Show that the input covariance in (10.30) *simultaneously* maximizes each of the constraints (10.28) and (10.29).

Exercise 10.15 Consider the (primal) point-to-point MIMO channel

$$\mathbf{y}[m] = \mathbf{H}\mathbf{x}[m] + \mathbf{w}[m], \tag{10.115}$$

and its reciprocal

$$\mathbf{y}_{\text{rec}}[m] = \mathbf{H}^*\mathbf{x}_{\text{rec}}[m] + \mathbf{w}_{\text{rec}}[m]. \tag{10.116}$$

The MIMO channel $\mathbf{H}$ has n_t transmit antennas and n_r receive antennas (so the reciprocal channel $\mathbf{H}^*$ is n_t times n_r). Here $\mathbf{w}[m]$ is i.i.d. $\mathcal{CN}(0, N_0\mathbf{I}_{n_r})$ and $\mathbf{w}_{\text{rec}}[m]$ is i.i.d. $\mathcal{CN}(0, N_0\mathbf{I}_{n_t})$. Consider sending $n_{\min}$ independent data streams on both these channels. The data streams are transmitted on the channels after passing through linear transmit filters (represented by unit norm vectors): $\mathbf{v}_1, \ldots, \mathbf{v}_{n_{\min}}$ for the primal channel and $\mathbf{u}_1, \ldots, \mathbf{u}_{n_{\min}}$ for the reciprocal channel. The data streams are then recovered from the received signal after passing through linear receive filters: $\mathbf{u}_1, \ldots, \mathbf{u}_{n_{\min}}$ for the primal channel and $\mathbf{v}_1, \ldots, \mathbf{v}_{n_{\min}}$ for the reciprocal channel. This process is illustrated in Figure 10.31.

1. Suppose powers $Q_1, \ldots, Q_{n_{\min}}$ are allocated to the data streams on the primal channel and powers $P_1, \ldots, P_{n_{\min}}$ are allocated to the data streams on the reciprocal channel. Show that the SINR for data stream k on the primal channel is

$$\text{SINR}_k = \frac{Q_k \mathbf{u}_k^* \mathbf{H} \mathbf{v}_k}{N_0 + \sum_{j \neq k} Q_j \mathbf{u}_k^* \mathbf{H} \mathbf{v}_j}, \tag{10.117}$$

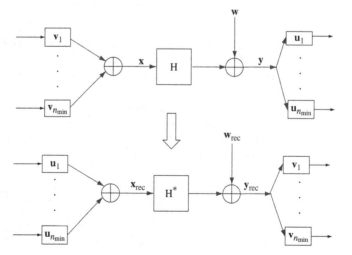

Figure 10.31 The data streams transmitted and received via linear filters on the primal (top) and reciprocal (bottom) channels.

and that on the reciprocal channel is

$$\text{SINR}_k^{\text{rec}} = \frac{P_k \mathbf{v}_k^* \mathbf{H}^* \mathbf{u}_k}{N_0 + \sum_{j \neq k} P_j \mathbf{v}_k^* \mathbf{H}^* \mathbf{u}_j}. \tag{10.118}$$

2. Suppose we fix the linear transmit and receive filters and want to allocate powers to meet a target SINR for each data stream (in both the primal and reciprocal channels). Find an expression analogous to (10.43) for the component-wise minimal set of power allocations.

3. Show that to meet the same SINR requirement for a given data stream on both the primal and reciprocal channels, the sum of the minimal set of powers is the same in both the primal and reciprocal channels. This is a generalization of (10.45).

4. We can use this general result to see earlier results in a unified way.

   (a) With the filters $\mathbf{v}_k = [0, \ldots, 0, 1, 0, \ldots, 0]^t$ (with the single 1 in the kth position), show that we capture the uplink–downlink duality result in (10.45).

   (b) Suppose $\mathbf{H} = \mathbf{U}\boldsymbol{\Lambda}\mathbf{V}^*$ is the singular value decomposition. With the filters $\mathbf{u}_k$ equal to the first $n_{\min}$ rows of $\mathbf{U}$ and the filters $\mathbf{v}_k$ equal to the first $n_{\min}$ columns of $\mathbf{V}$, show that this transceiver architecture achieves the capacity of the point-to-point MIMO primal and reciprocal channels with the same overall transmit power constraint, cf. Figure 7.2. Thus conclude that this result captures the reciprocity property discussed in Exercise 8.1.

Exercise 10.16 [76] Consider the opportunistic orthogonal signaling scheme described in Section 10.3.3. Each of the M messages corresponds to K (real) orthogonal signals. The encoder transmits the signal that has the largest correlation (among the K possible choices corresponding to the message to be conveyed) with the interference (real white Gaussian process with power spectral density $N_s/2$). The decoder decides the most likely transmit signal (among the MK possible choices) and then decides on the message corresponding to the most likely transmit signal. Fix the number of messages, M, and the number of signals for each message, K. Suppose that message 1 is to be conveyed.

1. Derive a good upper bound on the error probability of opportunistic orthogonal signaling. Here you can use the technique developed in the upper bound on the error probability of regular orthogonal signaling in Exercise 5.9. What is the appropriate choice of the threshold, γ, as a function of M, K and the power spectral densities $N_s/2, N_0/2$?

2. By an appropriate choice of K as a function of M, N_s, N_0 show that the upper bound you have derived converges to zero as M goes to infinity as long as $\mathcal{E}_b/N_0$ is larger than $-1.59\,\text{dB}$.

3. Can you explain why opportunistic orthogonal signaling achieves the capacity of the infinite bandwidth AWGN channel with no interference by interpreting the correct choice of K?

4. We have worked with the assumption that the interference $s(t)$ is white Gaussian. Suppose $s(t)$ is still white but not Gaussian. Can you think of a simple way to modify the opportunistic orthogonal signaling scheme presented in the text so that we still achieve the same minimal $\mathcal{E}_b/N_0$ of $-1.59\,\text{dB}$?

Exercise 10.17 Consider a real random variable x_1 that is restricted to the range $[0,1]$ and x_2 is another random variable that is jointly distributed with x_1. Suppose u is a uniform random variable on [0,1] and is jointly independent of x_1 and x_2. Consider the new random variable

$$\tilde{x}_1 = \begin{cases} x_1 + u & \text{if } x_1 + u \le 1, \\ x_1 + u - 1 & \text{if } x_1 + u > 1. \end{cases} \qquad (10.119)$$

The random variable $\tilde{x}_1$ can be thought of as the right cyclic addition of x_1 and u.
1. Show that $\tilde{x}_1$ is uniformly distributed on [0,1].
2. Show that $\tilde{x}_1$ and (x_1, x_2) are independent.

Now suppose x_1 is the Costa-precoded signal containing the message to user 1 in a two-user single antenna downlink based on x_2, the signal of user 2 (cf. Section 10.3.4). If the realization of the random variable u is known to user 1 also, then $\tilde{x}_1$ and x_1 contain the same information (since the operation in (10.119) is invertible). Thus we could transmit $\tilde{x}_1$ in place of x_1 without any change in the performance of user 1. But the important change is that the transmit signal $\tilde{x}_1$ is now independent of x_2.

The common random variable u, shared between the base-station and user 1, is called the *dither*. Here we have focused on a single time symbol and made $\tilde{x}_1$ uniform. With a large block length, this basic argument can be extended to make the transmit vector $\tilde{\mathbf{x}}_1$ appear Gaussian and independent of $\mathbf{x}_2$; this dithering idea is used to justify (10.65).

Exercise 10.18 Consider the two-user single antenna downlink (cf. (10.63)) with $|h_1| > |h_2|$. Consider the rate tuple (R'_1, R'_2) achieved via Costa precoding in (10.66). In this exercise we show that this rate pair is strictly inside the capacity region of the downlink. Suppose we allocate powers Q_1, Q_2 to the two users and do superposition encoding and decoding (cf. Figures 6.7 and 6.8) and aim to achieve the same rates as the pair in (10.66).
1. Calculate Q_1, Q_2 such that

$$R'_1 = \log\left(1 + \frac{|h_1|^2 Q_1}{N_0}\right), \qquad R'_2 = \log\left(1 + \frac{|h_2|^2 Q_2}{N_0 + |h_2|^2 Q_1}\right), \qquad (10.120)$$

where R'_1 and R'_2 are the rate pair in (10.66).
2. Using the fact that user 1 has a stronger channel than user 2 (i.e., $|h_1| > |h_2|$) show that the total power used in the superposition strategy to achieve the same rate pair (i.e., $Q_1 + Q_2$ from the previous part) is strictly smaller than $P_1 + P_2$, the transmit power in the Costa precoding strategy.
3. Observe that an increase in transmit power strictly increases the capacity region of the downlink. Hence conclude that the rate pair in (10.66) achieved by the Costa precoding strategy is strictly within the capacity region of the downlink.

Exercise 10.19 Consider the K-user downlink channel with a single antenna (an extension of the two-user channel in (10.63)):

$$y_k[m] = h_k x[m] + w_k[m], \qquad k = 1, \dots, K. \qquad (10.121)$$

Show that the following rates are achievable using Costa precoding, extending the argument in Section 10.3.4:

$$R_k = \log\left(1 + \frac{|h_k|^2 P_k}{\sum_{j=k+1}^{K} |h_j|^2 P_j + N_0}\right), \qquad k = 1, \ldots, K. \qquad (10.122)$$

Here $P_1, \ldots, P_K$ are some non-negative numbers that sum to P, the transmit power constraint at the base-station. You should not need to assume any specific ordering of the channels qualities $|h_1|, |h_2|, \ldots, |h_K|$ in arriving at your result. On the other hand, if we have

$$|h_1| \leq |h_2| \leq \cdots \leq |h_K|, \qquad (10.123)$$

then the superposition coding approach, discussed in Section 6.2, achieves the rates in (10.122).

Exercise 10.20 Consider the reciprocal uplink channel in (10.40) with the receive filters $\mathbf{u}_1, \ldots, \mathbf{u}_K$ as in Figure 10.16. This time we embellish the receiver with successive cancellation, canceling users in the order K through 1 (i.e., user k does not see any interference from users $K, K-1, \ldots, k+1$). With powers $Q_1, \ldots, Q_K$ allocated to the users, show that the SINR for user k can be written as

$$\mathsf{SINR}_k^{\mathsf{ul}} = \frac{Q_k \,|\, \mathbf{u}_k^* \mathbf{h}_k \,|^2}{N_0 + \sum_{j<k} Q_j \,|\, \mathbf{u}_k^* \mathbf{h}_j \,|^2}. \qquad (10.124)$$

To meet the same SINR requirement as in the downlink with Costa precoding in the reverse order (the expression for the corresponding SINR is in (10.72)) show that the sum of the minimal powers required is the same for the uplink and the downlink. This is an extension of the conservation of sum-of-powers property seen without cancellation in (10.45).

Exercise 10.21 Consider the fast fading multiple transmit antenna downlink (cf. (10.73)) where the channels from antenna i to user k are modeled as i.i.d. $\mathcal{CN}(0, 1)$ random variables (for each antenna $i = 1, \ldots, n_t$ and for each user $k = 1, \ldots, K$). Each user has a single receive antenna. Further suppose that the channel fluctuations are i.i.d. over time as well. Each user has access to the realization of its channel fluctuations, while the base-station only has knowledge of the statistics of the channel fluctuations (the receiver CSI model). There is an overall power constraint P on the transmit power.

1. With just one user in the downlink, we have a MIMO channel with receiver only CSI. Show that the capacity of this channel is equal to

$$\mathbb{E}\left[\log\left(1 + \frac{\mathsf{SNR}\|\mathbf{h}\|^2}{n_t}\right)\right], \qquad (10.125)$$

   where $\mathbf{h} \sim \mathcal{CN}(0, \mathbf{I}_{n_t})$ and $\mathsf{SNR} = P/N_0$. *Hint*: Recall (8.15) and Exercise 8.4.

2. Since the statistics of the user channels are identical, argue that if user k can decode its data reliably, then all the other users can also successfully decode user k's data (as we did in Section 6.4.1 for the single antenna downlink). Conclude that the

sum of the rates at which the users are being simultaneously reliably transmitted to is bounded as

$$\sum_{k=1}^{K} R_k \leq \mathbb{E}\left[\log\left(1 + \frac{\mathrm{SNR}\|\mathbf{h}\|^2}{n_t}\right)\right], \tag{10.126}$$

analogous to (6.52).

Exercise 10.22 Consider the downlink with multiple receive antennas (cf. (10.78)). Show that the random variables $x[m]$ and $\mathbf{y}_k[m]$ are independent conditioned on $\tilde{y}_k[m]$. Hence conclude that

$$I(x; \mathbf{y}_k) = I(x; \tilde{y}_k), \qquad k = 1, 2. \tag{10.127}$$

Thus there is no loss in information by having a matched filter front end at each of the users converting the SIMO downlink into a single antenna channel to each user.

Exercise 10.23 Consider the two-user uplink fading channel with multiple antennas at the base-station:

$$\mathbf{y}[m] = \mathbf{h}_1[m]x_1[m] + \mathbf{h}_2[m]x_2[m] + \mathbf{w}[m]. \tag{10.128}$$

Here the user channels $\{\mathbf{h}_1[m]\}, \{\mathbf{h}_2[m]\}$ are statistically independent. Suppose that $\mathbf{h}_1[m]$ and $\mathbf{h}_2[m]$ are $\mathcal{CN}(0, N_0\mathbf{I}_{n_r})$. We operate the uplink in SDMA mode with the users having the same power P. The background noise $\mathbf{w}[m]$ is i.i.d. $\mathcal{CN}(0, N_0\mathbf{I}_{n_r})$. An SIC receiver decodes user 1 first, removes its contribution from $\{\mathbf{y}[m]\}$ and then decodes user 2. We would like to assess the effect of channel estimation error of $\mathbf{h}_2$ on the performance of user 1.

1. Suppose the users send training symbols using orthogonal multiple access and they spend 20% of their power on sending the training signal, repeated every T_c seconds, which is the channel coherence time of the users. What is the mean square estimation error of $\mathbf{h}_1$ and $\mathbf{h}_2$?

2. The first step of the SIC receiver is to decode user 1's information suppressing the user 2's signal. Using the linear MMSE filter to suppress the interference, numerically evaluate the average output SINR of the filter due to the channel estimation error, as compared to that with perfect channel estimation (cf. (8.62)). Plot the degradation (ratio of the SINR with imperfect and perfect channel estimates) as a function of the SNR, P/N_0, with $T_c = 10\,\mathrm{ms}$.

3. Argue using the previous calculation that better channel estimates are required to fully harness the gains of interference suppression. This means that the pilots in the uplink with SDMA have to be stronger than in the uplink with a single receive antenna.

Exercise 10.24 In this exercise, we explore the effect of channel measurement error on the reciprocity relationship between the uplink and the downlink. To isolate the situation of interest, consider just a single user in the uplink and the downlink (this is the natural model whenever the multiple access is orthogonal) with only the base-station having an array of antennas. The uplink channel is (cf. (10.40))

$$\mathbf{y}_{\mathrm{ul}}[m] = \mathbf{h}x_{\mathrm{ul}}[m] + \mathbf{w}_{\mathrm{ul}}[m], \tag{10.129}$$

with a power constraint of P_{ul} on the uplink transmit symbol x_{ul}. The downlink channel is (cf. (10.39))

$$y_{dl}[m] = \mathbf{h}^* \mathbf{x}_{dl}[m] + w_{dl}[m], \qquad (10.130)$$

with a power constraint of P_{dl} on the downlink transmit vector $\mathbf{x}_{dl}$.

1. Suppose a training symbol is sent with the full power P_{ul} over one symbol time in the uplink to estimate the channel $\mathbf{h}$ at the base-station. What is the mean square error in the best estimate $\hat{\mathbf{h}}$ of the channel $\mathbf{h}$?

2. Now suppose the channel estimate $\hat{\mathbf{h}}$ from the previous part is used to beamform in the downlink, i.e., the transmit signal is

$$\mathbf{x}_{dl} = \frac{\hat{\mathbf{h}}}{\|\hat{\mathbf{h}}\|} x_{dl},$$

with the power in the data symbol x_{dl} equal to P_{dl}. What is the average received SNR in the downlink? The degradation in SNR is measured by the ratio of the average received SNR with imperfect and perfect channel estimates. For a fixed uplink SNR, P_{ul}/N_0, plot the average degradation for different values of the downlink SNR, P_{dl}/N_0.

3. Argue using your calculations that using the reciprocal channel estimate in the downlink is most beneficial when the uplink power P_{ul} is larger than or of the same order as the downlink power P_{dl}. Further, there is a huge degradation in performance when P_{dl} is much larger than P_{ul}.

Appendix A Detection and estimation in additive Gaussian noise

A.1 Gaussian random variables

A.1.1 Scalar real Gaussian random variables

A *standard Gaussian* random variable w takes values over the real line and has the probability density function

$$f(w) = \frac{1}{\sqrt{2\pi}} \exp\left(-\frac{w^2}{2}\right), \qquad w \in \Re. \tag{A.1}$$

The mean of w is zero and the variance is 1. A (general) Gaussian random variable x is of the form

$$x = \sigma w + \mu. \tag{A.2}$$

The mean of x is μ and the variance is equal to σ^2. The random variable x is a one-to-one function of w and thus the probability density function follows from (A.1) as

$$f(x) = \frac{1}{\sqrt{2\pi\sigma^2}} \exp\left(-\frac{(x-\mu)^2}{2\sigma^2}\right), \qquad x \in \Re. \tag{A.3}$$

Since the random variable is completely characterized by its mean and variance, we denote x by $\mathcal{N}(\mu, \sigma^2)$. In particular, the standard Gaussian random variable is denoted by $\mathcal{N}(0, 1)$. The *tail* of the Gaussian random variable w

$$Q(a) := \mathbb{P}\{w > a\} \tag{A.4}$$

is plotted in Figure A.1. The plot and the computations $Q(1) = 0.159$ and $Q(3) = 0.00135$ give a sense of how rapidly the tail decays. The tail decays *exponentially* fast as evident by the following upper and lower bounds:

$$\frac{1}{\sqrt{2\pi}a}\left(1 - \frac{1}{a^2}\right)e^{-a^2/2} < Q(a) < e^{-a^2/2}, \qquad a > 1. \tag{A.5}$$

Figure A.1 The Q
function.

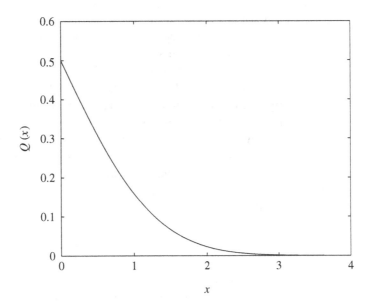

An important property of Gaussianity is that it is preserved by linear transformations: linear combinations of independent Gaussian random variables are still Gaussian. If $x_1, \ldots, x_n$ are independent and $x_i \sim \mathcal{N}(\mu_i, \sigma_i^2)$ (where the $\sim$ notation represents the phrase "is distributed as"), then

$$\sum_{i=1}^{n} c_i x_i \sim \mathcal{N}\left(\sum_{i=1}^{n} c_i \mu_i, \sum_{i=1}^{n} c_i^2 \sigma_i^2\right). \tag{A.6}$$

A.1.2 Real Gaussian random vectors

A *standard Gaussian random vector* $\mathbf{w}$ is a collection of n independent and identically distributed (i.i.d.) standard Gaussian random variables $w_1, \ldots, w_n$. The vector $\mathbf{w} = (w_1, \ldots, w_n)^t$ takes values in the vector space $\Re^n$. The probability density function of $\mathbf{w}$ follows from (A.1):

$$f(\mathbf{w}) = \frac{1}{\left(\sqrt{2\pi}\right)^n} \exp\left(-\frac{\|\mathbf{w}\|^2}{2}\right), \qquad \mathbf{w} \in \Re^n. \tag{A.7}$$

Here $\|\mathbf{w}\| := \sqrt{\sum_{i=1}^{n} w_i^2}$, is the Euclidean distance from the origin to $\mathbf{w} := (w_1, \ldots, w_n)^t$. Note that the density depends only on the *magnitude* of the argument. Since an orthogonal transformation $\mathbf{O}$ (i.e., $\mathbf{O}^t\mathbf{O} = \mathbf{O}\mathbf{O}^t = \mathbf{I}$) preserves the magnitude of a vector, we can immediately conclude:

If $\mathbf{w}$ is standard Gaussian, then $\mathbf{O}\mathbf{w}$ is also standard Gaussian.

(A.8)

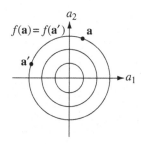

Figure A.2 The isobars, i.e., level sets for the density $f(\mathbf{w})$ of the standard Gaussian random vector, are circles for $n = 2$.

What this result says is that $\mathbf{w}$ has the same distribution in any orthonormal basis. Geometrically, the distribution of $\mathbf{w}$ is invariant to rotations and reflections and hence $\mathbf{w}$ does not prefer any specific direction. Figure A.2 illustrates this *isotropic* behavior of the density of the standard Gaussian random vector $\mathbf{w}$. Another conclusion from (A.8) comes from observing that the rows of matrix $\mathbf{O}$ are orthonormal: the projections of the standard Gaussian random vector in orthogonal directions are independent.

How is the squared magnitude $\|\mathbf{w}\|^2$ distributed? The squared magnitude is equal to the sum of the square of n i.i.d. zero-mean Gaussian random variables. In the literature this sum is called a χ-squared random variable with n degrees of freedom and denoted by χ_n^2. With $n = 2$, the squared magnitude has density

$$f(a) = \frac{1}{2} \exp\left(-\frac{a}{2}\right), \qquad a \geq 0, \tag{A.9}$$

and is said to be *exponentially* distributed. The density of the χ_n^2 random variable for general n is derived in Exercise A.1.

Gaussian random vectors are defined as linear transformations of a standard Gaussian random vector plus a constant vector, a natural generalization of the scalar case (cf. (A.2)):

$$\mathbf{x} = \mathbf{A}\mathbf{w} + \boldsymbol{\mu}. \tag{A.10}$$

Here $\mathbf{A}$ is a matrix representing a linear transformation from $\Re^n$ to $\Re^n$ and $\boldsymbol{\mu}$ is a fixed vector in $\Re^n$. Several implications follow:

1. A standard Gaussian random vector is also Gaussian (with $\mathbf{A} = \mathbf{I}$ and $\boldsymbol{\mu} = \mathbf{0}$).
2. For any $\mathbf{c}$, a vector in $\Re^n$, the random variable

$$\mathbf{c}^t\mathbf{x} \sim \mathcal{N}(\mathbf{c}^t\boldsymbol{\mu}, \mathbf{c}^t\mathbf{A}\mathbf{A}^t\mathbf{c}); \tag{A.11}$$

this follows directly from (A.6). Thus any linear combination of the elements of a Gaussian random vector is a Gaussian random variable.[1] More generally, any linear transformation of a Gaussian random vector is also Gaussian.
3. If $\mathbf{A}$ is invertible, then the probability density function of $\mathbf{x}$ follows directly from (A.7) and (A.10):

$$f(\mathbf{x}) = \frac{1}{\left(\sqrt{2\pi}\right)^n \sqrt{\det(\mathbf{A}\mathbf{A}^t)}} \exp\left(-\frac{1}{2}(\mathbf{x} - \boldsymbol{\mu})^t(\mathbf{A}\mathbf{A}^t)^{-1}(\mathbf{x} - \boldsymbol{\mu})\right), \quad \mathbf{x} \in \Re^n. \tag{A.12}$$

---

[1] This property can be used to define a Gaussian random vector; it is equivalent to our definition in (A.10).

Figure A.3 The isobars of a general Gaussian random vector are ellipses. They corresponds to level sets $\{\mathbf{x} : \|\mathbf{A}^{-1}(\mathbf{x} - \boldsymbol{\mu})\|^2 = c\}$ for constants c.

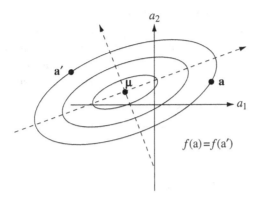

The isobars of this density are ellipses; the circles of the standard Gaussian vectors being rotated and scaled by $\mathbf{A}$ (Figure A.3). The matrix $\mathbf{AA}^t$ replaces σ^2 in the scalar Gaussian random variable (cf. (A.3)) and is equal to the *covariance matrix* of $\mathbf{x}$:

$$\mathbf{K} := \mathbb{E}[(\mathbf{x} - \boldsymbol{\mu})(\mathbf{x} - \boldsymbol{\mu})^t] = \mathbf{AA}^t. \qquad (A.13)$$

For invertible $\mathbf{A}$, the Gaussian random vector is completely characterized by its mean vector $\boldsymbol{\mu}$ and its covariance matrix $\mathbf{K} = \mathbf{AA}^t$, which is a symmetric and non-negative definite matrix. We make a few inferences from this observation:

(a) Even though the Gaussian random vector is defined via the matrix $\mathbf{A}$, only the covariance matrix $\mathbf{K} = \mathbf{AA}^t$ is used to characterize the density of $\mathbf{x}$. Is this surprising? Consider two matrices $\mathbf{A}$ and $\mathbf{AO}$ used to define two Gaussian random vectors as in (A.10). When $\mathbf{O}$ is orthogonal, the covariance matrices of both these random vectors are the same, equal to $\mathbf{AA}^t$; so the two random vectors must be distributed identically. We can see this directly using our earlier observation (see (A.8)) that $\mathbf{Ow}$ has the same distribution as $\mathbf{w}$ and thus $\mathbf{AOw}$ has the same distribution as $\mathbf{Aw}$.

(b) A Gaussian random vector is composed of independent Gaussian random variables exactly when the covariance matrix $\mathbf{K}$ is diagonal, i.e., the component random variables are *uncorrelated*. Such a random vector is also called a *white* Gaussian random vector.

(c) When the covariance matrix $\mathbf{K}$ is equal to identity, i.e., the component random variables are uncorrelated and have the same unit variance, then the Gaussian random vector reduces to the standard Gaussian random vector.

4. Now suppose that $\mathbf{A}$ is not invertible. Then $\mathbf{Aw}$ maps the standard Gaussian random vector $\mathbf{w}$ into a subspace of dimension less than n, and the density of $\mathbf{Aw}$ is equal to zero outside that subspace and impulsive inside. This means that some components of $\mathbf{Aw}$ can be expressed as linear

combinations of the others. To avoid messy notation, we can focus only on those components of $\mathbf{Aw}$ that are linearly independent and represent them as a lower dimensional vector $\tilde{\mathbf{x}}$, and represent the other components of $\mathbf{Aw}$ as (deterministic) linear combinations of the components of $\tilde{\mathbf{x}}$. By this strategem, we can always take the covariance $\mathbf{K}$ to be invertible.

In general, a Gaussian random vector is completely characterized by its mean $\boldsymbol{\mu}$ and by the covariance matrix $\mathbf{K}$; we denote the random vector by $\mathcal{N}(\boldsymbol{\mu}, \mathbf{K})$.

A.1.3 Complex Gaussian random vectors

So far we have considered real random vectors. In this book, we are primarily interested in *complex* random vectors; these are of the form $\mathbf{x} = \mathbf{x}_R + j\mathbf{x}_I$ where $\mathbf{x}_R, \mathbf{x}_I$ are real random vectors. *Complex Gaussian* random vectors are ones in which $[\mathbf{x}_R, \mathbf{x}_I]^t$ is a real Gaussian random vector. The distribution is completely specified by the mean and covariance matrix of the real vector $[\mathbf{x}_R, \mathbf{x}_I]^t$. Exercise A.3 shows that the same information is contained in the mean $\boldsymbol{\mu}$, the covariance matrix $\mathbf{K}$, and the *pseudo-covariance* matrix $\mathbf{J}$ of the complex vector $\mathbf{x}$, where

$$\boldsymbol{\mu} := \mathbb{E}[\mathbf{x}], \tag{A.14}$$

$$\mathbf{K} := \mathbb{E}[(\mathbf{x} - \boldsymbol{\mu})(\mathbf{x} - \boldsymbol{\mu})^*], \tag{A.15}$$

$$\mathbf{J} := \mathbb{E}[(\mathbf{x} - \boldsymbol{\mu})(\mathbf{x} - \boldsymbol{\mu})^t]. \tag{A.16}$$

Here, $\mathbf{A}^*$ is the transpose of the matrix $\mathbf{A}$ with each element replaced by its complex conjugate, and $\mathbf{A}^t$ is just the transpose of $\mathbf{A}$. Note that in general the covariance matrix $\mathbf{K}$ of the complex random vector $\mathbf{x}$ by itself is not enough to specify the full second-order statistics of $\mathbf{x}$. Indeed, since $\mathbf{K}$ is Hermitian, i.e., $\mathbf{K} = \mathbf{K}^*$, the diagonal elements are real and the elements in the lower and upper triangles are complex conjugates of each other. Hence it is specified by n^2 real parameters, where n is the (complex) dimension of $\mathbf{x}$. On the other hand, the full second-order statistics of $\mathbf{x}$ are specified by the $n(2n + 1)$ real parameters in the symmetric $2n \times 2n$ covariance matrix of $[\mathbf{x}_R, \mathbf{x}_I]^t$.

For reasons explained in Chapter 2, in wireless communication we are almost exclusively interested in complex random vectors that have the *circular symmetry* property:

> $\mathbf{x}$ is circular symmetric if $e^{j\theta}\mathbf{x}$ has the same distribution of $\mathbf{x}$ for any θ.

$$\tag{A.17}$$

For a circular symmetric complex random vector $\mathbf{x}$,

$$\mathbb{E}[\mathbf{x}] = \mathbb{E}[e^{j\theta}\mathbf{x}] = e^{j\theta}\mathbb{E}[\mathbf{x}] \tag{A.18}$$

for any θ; hence the mean $\boldsymbol{\mu} = \mathbf{0}$. Moreover

$$\mathbb{E}[\mathbf{x}\mathbf{x}^t] = \mathbb{E}[e^{j\theta}\mathbf{x}(e^{j\theta}\mathbf{x})^t] = e^{j2\theta}\mathbb{E}[\mathbf{x}\mathbf{x}^t] \tag{A.19}$$

for any θ; hence the pseudo-covariance matrix $\mathbf{J}$ is also zero. Thus, the covariance matrix $\mathbf{K}$ fully specifies the first- and second-order statistics of a circular symmetric random vector. And if the complex random vector is also Gaussian, $\mathbf{K}$ in fact specifies its entire statistics. A circular symmetric Gaussian random vector with covariance matrix $\mathbf{K}$ is denoted as $\mathcal{CN}(0,\mathbf{K})$.

Some special cases:

1. A complex Gaussian random variable $w = w_R + jw_I$ with i.i.d. zero-mean Gaussian real and imaginary components is circular symmetric. The circular symmetry of w is in fact a restatement of the rotational invariance of the real Gaussian random vector $[w_R, w_I]^t$ already observed (cf. (A.8)). In fact, a circular symmetric Gaussian random variable *must* have i.i.d. zero-mean real and imaginary components (Exercise A.5). The statistics are fully specified by the variance $\sigma^2 := \mathbb{E}[|w|^2]$, and the complex random variable is denoted as $\mathcal{CN}(0, \sigma^2)$. (Note that, in contrast, the statistics of a general complex Gaussian random variable are specified by five real parameters: the means and the variances of the real and imaginary components and their correlation.) The phase of w is *uniform* over the range $[0, 2\pi]$ and independent of the magnitude $\|w\|$, which has a density given by

$$f(r) = \frac{2r}{\sigma^2} \exp\left\{\frac{-r^2}{\sigma^2}\right\}, \qquad r \geq 0 \tag{A.20}$$

and is known as a *Rayleigh* random variable. The square of the magnitude, i.e., $w_1^2 + w_2^2$, is χ_2^2, i.e., exponentially distributed, cf. (A.9). A random variable distributed as $\mathcal{CN}(0, 1)$ is said to be *standard*, with the real and imaginary parts each having variance 1/2.

2. A collection of n i.i.d. $\mathcal{CN}(0, 1)$ random variables forms a standard circular symmetric Gaussian random vector $\mathbf{w}$ and is denoted by $\mathcal{CN}(0, \mathbf{I})$. The density function of $\mathbf{w}$ can be explicitly written as, following from (A.7),

$$f(\mathbf{w}) = \frac{1}{\pi^n} \exp(-\|\mathbf{w}\|^2), \qquad \mathbf{w} \in \mathcal{C}^n. \tag{A.21}$$

As in the case of a real Gaussian random vector $\mathcal{N}(0, \mathbf{I})$ (cf. (A.8)), we have the property that

$$\boxed{\mathbf{U}\mathbf{w} \text{ has the same distribution as } \mathbf{w},} \tag{A.22}$$

for any complex orthogonal matrix $\mathbf{U}$ (such a matrix is called a *unitary* matrix and is characterized by the property $\mathbf{U}^*\mathbf{U} = \mathbf{I}$). The property (A.22) is the complex extension of the isotropic property of the real standard Gaussian random vector (cf. (A.8)). Note the distinction between the *circular* *symmetry* (A.17) and the *isotropic* (A.22) properties: the latter is in general much stronger than the former except that they coincide when $\mathbf{w}$ is scalar.

The square of the magnitude of $\mathbf{w}$, as in the real case, is a χ^2_{2n} random variable.

3. If $\mathbf{w}$ is $\mathcal{CN}(0, \mathbf{I})$ and $\mathbf{A}$ is a complex matrix, then $\mathbf{x} = \mathbf{Aw}$ is also circular symmetric Gaussian, with covariance matrix $\mathbf{K} = \mathbf{AA}^*$, i.e., $\mathcal{CN}(0, \mathbf{K})$. Conversely, any circular symmetric Gaussian random vector with covariance matrix $\mathbf{K}$ can be written as a linearly transformed version of a standard circular symmetric random vector. If $\mathbf{A}$ is invertible, the density function of $\mathbf{x}$ can be explicitly calculated via (A.21), as in (A.12),

$$f(\mathbf{x}) = \frac{1}{\pi^n \det \mathbf{K}} \exp\left(-\mathbf{x}^* \mathbf{K}^{-1} \mathbf{x}\right), \qquad \mathbf{x} \in \mathcal{C}^n. \qquad (A.23)$$

When $\mathbf{A}$ is not invertible, the earlier discussion for real random vectors applies here as well: we focus only on the linearly independent components of $\mathbf{x}$, and treat the other components as deterministic linear combinations of these. This allows us to work with a compact notation.

Summary A.1 Complex Gaussian random vectors

- An n-dimensional complex Gaussian random vector $\mathbf{x}$ has real and imaginary components which form a $2n$-dimensional real Gaussian random vector.

- $\mathbf{x}$ is *circular symmetric* if for any θ,

$$e^{j\theta}\mathbf{x} \sim \mathbf{x}. \qquad (A.24)$$

- A circular symmetric Gaussian $\mathbf{x}$ has zero mean and its statistics are fully specified by the covariance matrix $\mathbf{K} := \mathbb{E}[\mathbf{xx}^*]$. It is denoted by $\mathcal{CN}(0, \mathbf{K})$.

- The scalar complex random variable $w \sim \mathcal{CN}(0, 1)$ has i.i.d. real and imaginary components each distributed as $\mathcal{N}(0, 1/2)$. The phase of w is uniformly distributed in $[0, 2\pi]$ and independent of its magnitude $|w|$, which is Rayleigh distributed:

$$f(r) = r \exp\left(-\frac{r^2}{2}\right), \qquad r \geq 0. \qquad (A.25)$$

$|w|^2$ is exponentially distributed.

- If the random vector $\mathbf{w} \sim \mathcal{CN}(0, \mathbf{I})$, then its real and imaginary components are all i.i.d., and $\mathbf{w}$ is *isotropic*, i.e., for any unitary matrix $\mathbf{U}$,

$$\mathbf{Uw} \sim \mathbf{w}. \qquad (A.26)$$

Equivalently, the projections of $\mathbf{w}$ onto orthogonal directions are i.i.d. $\mathcal{CN}(0, 1)$. The squared magnitude $\|\mathbf{w}\|^2$ is distributed as χ^2_{2n} with mean n.

- If $\mathbf{x} \sim \mathcal{CN}(0, \mathbf{K})$ and $\mathbf{K}$ is invertible, then the density of $\mathbf{x}$ is

$$f(\mathbf{x}) = \frac{1}{\pi^n \det \mathbf{K}} \exp(-\mathbf{x}^* \mathbf{K}^{-1} \mathbf{x}), \qquad \mathbf{x} \in \mathcal{C}^n. \qquad (A.27)$$

A.2 Detection in Gaussian noise

A.2.1 Scalar detection

Consider the real additive Gaussian noise channel:

$$y = u + w, \qquad (A.28)$$

where the transmit symbol u is equally likely to be u_A or u_B and $w \sim \mathcal{N}(0, N_0/2)$ is real Gaussian noise. The *detection* problem involves making a decision on whether u_A or u_B was transmitted based on the observation y. The optimal detector, with the least probability of making an erroneous decision, chooses the symbol that is most likely to have been transmitted given the received signal y, i.e., u_A is chosen if

$$\mathbb{P}\{u = u_A | y\} \geq \mathbb{P}\{u = u_B | y\}. \qquad (A.29)$$

Since the two symbols u_A, u_B are equally likely to have been transmitted, Bayes' rule lets us simplify this to the *maximum likelihood* (ML) receiver, which chooses the transmit symbol that makes the observation y most likely. Conditioned on $u = u_i$, the received signal $y \sim \mathcal{N}(u_i, N_0/2)$, $i = A, B$, and the decision rule is to choose u_A if

$$\frac{1}{\sqrt{\pi N_0}} \exp\left(-\frac{(y - u_A)^2}{N_0}\right) \geq \frac{1}{\sqrt{\pi N_0}} \exp\left(-\frac{(y - u_B)^2}{N_0}\right), \qquad (A.30)$$

and u_B otherwise. The ML rule in (A.30) further simplifies: choose u_A when

$$|y - u_A| < |y - u_B|. \qquad (A.31)$$

The rule is illustrated in Figure A.4 and can be interpreted as corresponding to choosing the *nearest neighboring* transmit symbol. The probability of making an error, the same whether the symbol u_A or u_B was transmitted, is equal to

$$\mathbb{P}\left\{y < \frac{u_A + u_B}{2} \Big| u = u_B\right\} = \mathbb{P}\left\{w > \frac{|u_A - u_B|}{2}\right\} = Q\left(\frac{|u_A - u_B|}{2\sqrt{N_0/2}}\right). \quad (A.32)$$

Figure A.4 The ML rule is to choose the symbol that is *closest* to the received symbol.

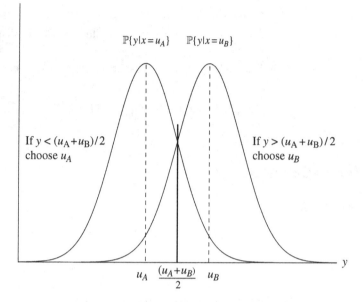

$\mathbb{P}\{y|x=u_A\}$ $\mathbb{P}\{y|x=u_B\}$

If $y < (u_A + u_B)/2$
choose u_A

If $y > (u_A + u_B)/2$
choose u_B

u_A $\dfrac{(u_A + u_B)}{2}$ u_B

Thus, the error probability only depends on the distance between the two transmit symbols u_A, u_B.

A.2.2 Detection in a vector space

Now consider detecting the transmit *vector* $\mathbf{u}$ equally likely to be $\mathbf{u}_A$ or $\mathbf{u}_B$ (both elements of $\Re^n$). The received vector is

$$\mathbf{y} = \mathbf{u} + \mathbf{w}, \qquad (A.33)$$

and $\mathbf{w} \sim \mathcal{N}(0, (N_0/2)\mathbf{I})$. Analogous to (A.30), the ML decision rule is to choose $\mathbf{u}_A$ if

$$\frac{1}{(\pi N_0)^{n/2}} \exp\left(-\frac{\|\mathbf{y} - \mathbf{u}_A\|^2}{N_0} \right) \geq \frac{1}{(\pi N_0)^{n/2}} \exp\left(-\frac{\|\mathbf{y} - \mathbf{u}_B\|^2}{N_0} \right), \qquad (A.34)$$

which simplifies to, analogous to (A.31),

$$\|\mathbf{y} - \mathbf{u}_A\| < \|\mathbf{y} - \mathbf{u}_B\|, \qquad (A.35)$$

the same *nearest neighbor* rule. By the isotropic property of the Gaussian noise, we expect the error probability to be the same for both the transmit symbols $\mathbf{u}_A, \mathbf{u}_B$. Suppose $\mathbf{u}_A$ is transmitted, so $\mathbf{y} = \mathbf{u}_A + \mathbf{w}$. Then an error occurs when the event in (A.35) does not occur, i.e., $\|\mathbf{w}\| > \|\mathbf{w} + \mathbf{u}_A - \mathbf{u}_B\|$. So, the error probability is equal to

$$\mathbb{P}\{\|\mathbf{w}\|^2 > \|\mathbf{w} + \mathbf{u}_A - \mathbf{u}_B\|^2\} = \mathbb{P}\left\{ (\mathbf{u}_A - \mathbf{u}_B)^t \mathbf{w} < -\frac{\|\mathbf{u}_A - \mathbf{u}_B\|^2}{2} \right\}. \qquad (A.36)$$

Geometrically, this says that the decision regions are the two sides of the hyperplane perpendicular to the vector $\mathbf{u}_B - \mathbf{u}_A$, and an error occurs when the received vector lies on the side of the hyperplane opposite to the transmit vector (Figure A.5). We know from (A.11) that $(\mathbf{u}_A - \mathbf{u}_B)^t\mathbf{w} \sim \mathcal{N}(0, \|\mathbf{u}_A - \mathbf{u}_B\|^2 N_0/2)$. Thus the error probability in (A.36) can be written in compact notation as

$$Q\left(\frac{\|\mathbf{u}_A - \mathbf{u}_B\|}{2\sqrt{N_0/2}}\right). \qquad (A.37)$$

The quantity $\|\mathbf{u}_A - \mathbf{u}_B\|/2$ is the distance from each of the vectors $\mathbf{u}_A, \mathbf{u}_B$ to the decision boundary. Comparing the error probability in (A.37) with that in the scalar case (cf. (A.32)), we see that the the error probability depends only on the Euclidean distance between $\mathbf{u}_A$ and $\mathbf{u}_B$ and not on the specific orientations and magnitudes of $\mathbf{u}_A$ and $\mathbf{u}_B$.

An alternative view

To see how we could have reduced the vector detection problem to the scalar one, consider a small change in the way we think of the transmit vector $\mathbf{u} \in \{\mathbf{u}_A, \mathbf{u}_B\}$. We can write the transmit vector $\mathbf{u}$ as

$$\mathbf{u} = x(\mathbf{u}_A - \mathbf{u}_B) + \frac{1}{2}(\mathbf{u}_A + \mathbf{u}_B), \qquad (A.38)$$

where the information is in the *scalar x*, which is equally likely to be $\pm 1/2$. Substituting (A.38) in (A.33), we can subtract the constant vector $(\mathbf{u}_A + \mathbf{u}_B)/2$ from the received signal $\mathbf{y}$ to arrive at

$$\mathbf{y} - \frac{1}{2}(\mathbf{u}_A + \mathbf{u}_B) = x(\mathbf{u}_A - \mathbf{u}_B) + \mathbf{w}. \qquad (A.39)$$

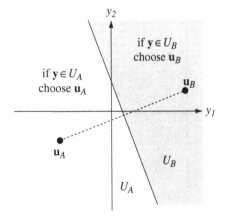

Figure A.5 The decision region for the nearest neighbor rule is partitioned by the hyperplane perpendicular to $\mathbf{u}_B - \mathbf{u}_A$ and halfway between $\mathbf{u}_A$ and $\mathbf{u}_B$.

We observe that the transmit symbol (a scalar x) is only in a specific direction:

$$\mathbf{v} := (\mathbf{u}_A - \mathbf{u}_B)/\|\mathbf{u}_A - \mathbf{u}_B\|. \tag{A.40}$$

The components of the received vector $\mathbf{y}$ in the directions orthogonal to $\mathbf{v}$ contain purely noise, and, due to the isotropic property of $\mathbf{w}$, the noise in these directions is also independent of the noise in the signal direction. This means that the components of the received vector in these directions are *irrelevant* for detection. Therefore projecting the received vector along the signal direction $\mathbf{v}$ provides all the necessary information for detection:

$$\tilde{y} := \mathbf{v}^t \left(\mathbf{y} - \frac{1}{2}(\mathbf{u}_A + \mathbf{u}_B) \right). \tag{A.41}$$

We have thus reduced the vector detection problem to the scalar one. Figure A.6 summarizes the situation.

More formally, we are viewing the received vector in a different orthonormal basis: the first direction is that given by $\mathbf{v}$, and the other directions are orthogonal to each other and to the first one. In other words, we form an orthogonal matrix $\mathbf{O}$ whose first row is $\mathbf{v}$, and the other rows are orthogonal to each other and to the first one and have unit norm. Then

$$\mathbf{O} \left(\mathbf{y} - \frac{1}{2}(\mathbf{u}_A + \mathbf{u}_B) \right) = \begin{bmatrix} x\|\mathbf{u}_A - \mathbf{u}_B\| \\ 0 \\ \vdots \\ 0 \end{bmatrix} + \mathbf{O}\mathbf{w}. \tag{A.42}$$

Since $\mathbf{O}\mathbf{w} \sim \mathcal{N}(0, (N_0/2)\mathbf{I})$ (cf. (A.8)), this means that all but the first component of the vector $\mathbf{O}(\mathbf{y} - \frac{1}{2}(\mathbf{u}_A + \mathbf{u}_B))$ are independent of the transmit symbol x and the noise in the first component. Thus it suffices to make a decision on the transmit symbol x, using only the first component, which is precisely (A.41).

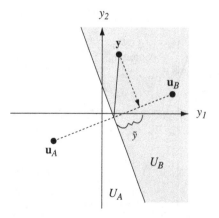

Figure A.6 Projecting the received vector **y** onto the signal direction **v** reduces the vector detection problem to the scalar one.

This important observation can be summarized:

1. In technical jargon, the scalar $\tilde{y}$ in (A.41) is called a *sufficient statistic* of the received vector $\mathbf{y}$ to detect the transmit symbol u.
2. The sufficient statistic $\tilde{y}$ is a projection of the received signal in the signal direction $\mathbf{v}$: in the literature on communication theory, this operation is called a *matched filter*; the linear filter at the receiver is "matched" to the direction of the transmit signal.
3. This argument explains why the error probability depends on $\mathbf{u}_A$ and $\mathbf{u}_B$ only through the distance between them: the noise is isotropic and the entire detection problem is rotationally invariant.

We now arrive at a scalar detection problem:

$$\tilde{y} = x\|\mathbf{u}_A - \mathbf{u}_B\| + w, \tag{A.43}$$

where w, the first component of $\mathbf{O}\mathbf{w}$ is $\mathcal{N}(0, N_0/2)$ and independent of the transmit symbol u. The effective distance between the two constellation points is $\|\mathbf{u}_A - \mathbf{u}_B\|$. The error probability is, from (A.32),

$$Q\left(\frac{\|\mathbf{u}_A - \mathbf{u}_B\|}{2\sqrt{N_0/2}}\right), \tag{A.44}$$

the same as that arrived at in (A.37), via a direct calculation.

The above argument for binary detection generalizes naturally to the case when the transmit vector can be one of M vectors $\mathbf{u}_1, \ldots, \mathbf{u}_M$. The projection of $\mathbf{y}$ onto the subspace spanned by $\mathbf{u}_1, \ldots, \mathbf{u}_M$ is a sufficient statistic for the detection problem. In the special case when the vectors $\mathbf{u}_1, \ldots, \mathbf{u}_M$ are collinear, i.e., $\mathbf{u}_i = \mathbf{h}x_i$ for some vector $\mathbf{h}$ (for example, when we are transmitting from a PAM constellation), then a projection onto the direction $\mathbf{h}$ provides a sufficient statistic.

A.2.3 Detection in a complex vector space

Consider detecting the transmit symbol $\mathbf{u}$, equally likely to be one of two complex vectors $\mathbf{u}_A, \mathbf{u}_B$ in additive standard complex Gaussian noise. The received complex vector is

$$\mathbf{y} = \mathbf{u} + \mathbf{w}, \tag{A.45}$$

where $\mathbf{w} \sim \mathcal{CN}(0, N_0\mathbf{I})$. We can proceed as in the real case. Write

$$\mathbf{u} = x(\mathbf{u}_A - \mathbf{u}_B) + \frac{1}{2}(\mathbf{u}_A + \mathbf{u}_B). \tag{A.46}$$

The signal is in the direction

$$\mathbf{v} := (\mathbf{u}_A - \mathbf{u}_B)/\|\mathbf{u}_A - \mathbf{u}_B\|. \tag{A.47}$$

Projection of the received vector $\mathbf{y}$ onto $\mathbf{v}$ provides a (complex) scalar sufficient statistic:

$$\tilde{y} := \mathbf{v}^* \left(\mathbf{y} - \frac{1}{2}(\mathbf{u}_A + \mathbf{u}_B) \right) = x\|\mathbf{u}_A - \mathbf{u}_B\| + w, \tag{A.48}$$

where $w \sim \mathcal{CN}(0, N_0)$. Note that since x is real ($\pm 1/2$), we can further extract a sufficient statistic by looking only at the real component of $\tilde{y}$:

$$\Re[\tilde{y}] = x\|\mathbf{u}_A - \mathbf{u}_B\| + \Re[w], \tag{A.49}$$

where $\Re[w] \sim N(0, N_0/2)$. The error probability is exactly as in (A.44):

$$Q\left(\frac{\|\mathbf{u}_A - \mathbf{u}_B\|}{2\sqrt{N_0/2}} \right), \tag{A.50}$$

Note that although $\mathbf{u}_A$ and $\mathbf{u}_B$ are *complex* vectors, the transmit vectors

$$x(\mathbf{u}_A - \mathbf{u}_B) + \frac{1}{2}(\mathbf{u}_A + \mathbf{u}_B), \qquad x = \pm 1, \tag{A.51}$$

lie in a subspace of one *real* dimension and hence we can extract a *real* sufficient statistic. If there are more than two possible transmit vectors and they are of the form $\mathbf{h}x_i$, where x_i is *complex* valued, $\mathbf{h}^*\mathbf{y}$ is still a sufficient statistic but $\Re[\mathbf{h}^*\mathbf{y}]$ is sufficient only if x is real (for example, when we are transmitting a PAM constellation).

The main results of our discussion are summarized below.

Summary A.2 Vector detection in complex Gaussian noise

Binary signals
The transmit vector $\mathbf{u}$ is either $\mathbf{u}_A$ or $\mathbf{u}_B$ and we wish to detect $\mathbf{u}$ from received vector

$$\mathbf{y} = \mathbf{u} + \mathbf{w}, \tag{A.52}$$

where $\mathbf{w} \sim \mathcal{CN}(0, N_0\mathbf{I})$. The ML detector picks the transmit vector closest to $\mathbf{y}$ and the error probability is

$$Q\left(\frac{\|\mathbf{u}_A - \mathbf{u}_B\|}{2\sqrt{N_0/2}} \right). \tag{A.53}$$

Collinear signals

The transmit symbol x is equally likely to take one of a finite set of values in $\mathcal{C}$ (the *constellation* points) and the received vector is

$$\mathbf{y} = \mathbf{h}x + \mathbf{w}, \tag{A.54}$$

where $\mathbf{h}$ is a fixed vector.

Projecting $\mathbf{y}$ onto the unit vector $\mathbf{v} := \mathbf{h}/\|\mathbf{h}\|$ yields a scalar sufficient statistic:

$$\mathbf{v}^*\mathbf{y} = \|\mathbf{h}\|x + w. \tag{A.55}$$

Here $w \sim \mathcal{CN}(0, N_0)$.

If further the constellation is real-valued, then

$$\Re[\mathbf{v}^*\mathbf{y}] = \|\mathbf{h}\|x + \Re[w] \tag{A.56}$$

is sufficient. Here $\Re[w] \sim \mathcal{N}(0, N_0/2)$.

With antipodal signalling, $x = \pm a$, the ML error probability is simply

$$Q\left(\frac{a\|\mathbf{h}\|}{\sqrt{N_0/2}}\right). \tag{A.57}$$

Via a translation, the binary signal detection problem in the first part of the summary can be reduced to this antipodal signalling scenario.

A.3 Estimation in Gaussian noise

A.3.1 Scalar estimation

Consider a zero-mean real signal x embedded in independent additive real Gaussian noise ($w \sim \mathcal{N}(0, N_0/2)$):

$$y = x + w. \tag{A.58}$$

Suppose we wish to come up with an estimate $\hat{x}$ of x and we use the mean squared error (MSE) to evaluate the performance:

$$\text{MSE} := \mathbb{E}[(x - \hat{x})^2], \tag{A.59}$$

where the averaging is over the randomness of both the signal x and the noise w. This problem is quite different from the detection problem studied in Section A.2. The estimate that yields the smallest mean squared error is the classical *conditional mean*:

$$\hat{x} = \mathbb{E}[x|y], \tag{A.60}$$

which has the important *orthogonality* property: the error is independent of the observation. In particular, this implies that

$$\mathbb{E}[(\hat{x} - x)y] = 0. \tag{A.61}$$

The *orthogonality principle* is a classical result and all standard textbooks dealing with probability theory and random variables treat this material.

In general, the conditional mean $\mathbb{E}[x|y]$ is some complicated non-linear function of y. To simplify the analysis, one studies the restricted class of linear estimates that minimize the MSE. This restriction is without loss of generality in the important case when x is a Gaussian random variable because, in this case, the conditional mean operator is actually *linear*.

Since x is zero mean, linear estimates are of the form $\hat{x} = cy$ for some real number c. What is the best coefficient c? This can be derived directly or via using the orthogonality principle (cf. (A.61)):

$$c = \frac{\mathbb{E}[x^2]}{\mathbb{E}[x^2] + N_0/2}. \tag{A.62}$$

Intuitively, we are weighting the received signal y by the transmitted signal energy as a fraction of the received signal energy. The corresponding minimum mean squared error (MMSE) is

$$\text{MMSE} = \frac{\mathbb{E}[x^2]N_0/2}{\mathbb{E}[x^2] + N_0/2}. \tag{A.63}$$

A.3.2 Estimation in a vector space

Now consider estimating x in a vector space:

$$\mathbf{y} = \mathbf{h}x + \mathbf{w}. \tag{A.64}$$

Here x and $\mathbf{w} \sim \mathcal{N}(0, (N_0/2)\mathbf{I})$ are independent and $\mathbf{h}$ is a fixed vector in $\Re^n$. We have seen that the projection of $\mathbf{y}$ in the direction of $\mathbf{h}$,

$$\tilde{y} = \frac{\mathbf{h}^t\mathbf{y}}{\|\mathbf{h}\|^2} = x + w, \tag{A.65}$$

is a sufficient statistic: the projections of $\mathbf{y}$ in directions orthogonal to $\mathbf{h}$ are independent of both the signal x and w, the noise in the direction of $\mathbf{h}$. Thus we can convert this problem to a scalar one: estimate x from $\tilde{y}$, with $w \sim \mathcal{N}(0, N_0/(2\|\mathbf{h}\|^2))$. Now this problem is identical to the scalar estimation problem in (A.58) with the energy of the noise w suppressed by a factor of $\|\mathbf{h}\|^2$. The best linear estimate of x is thus, as in (A.62),

$$\frac{\mathbb{E}[x^2]\|\mathbf{h}\|^2}{\mathbb{E}[x^2]\|\mathbf{h}\|^2 + N_0/2}\, \tilde{y}. \tag{A.66}$$

We can combine the sufficient statistic calculation in (A.65) and the scalar linear estimate in (A.66) to arrive at the best linear estimate $\hat{x} = \mathbf{c}'\mathbf{y}$ of x from $\mathbf{y}$:

$$\mathbf{c} = \frac{\mathbb{E}[x^2]}{\mathbb{E}[x^2]\|\mathbf{h}\|^2 + N_0/2}\, \mathbf{h}. \tag{A.67}$$

The corresponding minimum mean squared error is

$$\mathrm{MMSE} = \frac{\mathbb{E}[x^2]N_0/2}{\mathbb{E}[x^2]\|\mathbf{h}\|^2 + N_0/2}. \tag{A.68}$$

An alternative performance measure to evaluate linear estimators is the *signal-to-noise ratio* (SNR) defined as the ratio of the signal energy in the estimate to the noise energy:

$$\mathrm{SNR} := \frac{(\mathbf{c}'\mathbf{h})^2\mathbb{E}[x^2]}{\|\mathbf{c}\|^2 N_0/2}. \tag{A.69}$$

That the matched filter ($\mathbf{c} = \mathbf{h}$) yields the maximal SNR at the output of any linear filter is a classical result in communication theory (and is studied in all standard textbooks on the topic). It follows directly from the Cauchy–Schwartz inequality:

$$(\mathbf{c}'\mathbf{h})^2 \leq \|\mathbf{c}\|^2\, \|\mathbf{h}\|^2, \tag{A.70}$$

with equality exactly when $\mathbf{c} = \mathbf{h}$. The fact that the matched filter maximizes the SNR and when appropriately scaled yields the MMSE is not coincidental; this is studied in greater detail in Exercise A.8.

A.3.3 Estimation in a complex vector space

The extension of our discussion to the complex field is natural. Let us first consider scalar complex estimation, an extension of the basic real setup in (A.58):

$$y = x + w, \tag{A.71}$$

where $w \sim \mathcal{CN}(0, N_0)$ is independent of the complex zero-mean transmitted signal x. We are interested in a linear estimate $\hat{x} = c^*y$, for some complex constant c. The performance metric is

$$\text{MSE} = \mathbb{E}[|x - \hat{x}|^2]. \tag{A.72}$$

The best linear estimate $\hat{x} = c^*y$ can be directly calculated to be, as an extension of (A.62),

$$c = \frac{\mathbb{E}[|x|^2]}{\mathbb{E}[|x|^2] + N_0}. \tag{A.73}$$

The corresponding minimum MSE is

$$\text{MMSE} = \frac{\mathbb{E}[|x|^2]N_0}{\mathbb{E}[|x|^2] + N_0}. \tag{A.74}$$

The orthogonality principle (cf. (A.61)) for the complex case is extended to:

$$\mathbb{E}[(\hat{x} - x)y^*] = 0. \tag{A.75}$$

The linear estimate in (A.73) is easily seen to satisfy (A.75).

Now let us consider estimating the scalar complex zero mean x in a complex vector space:

$$\mathbf{y} = \mathbf{h}x + \mathbf{w}, \tag{A.76}$$

with $\mathbf{w} \sim \mathcal{CN}(0, N_0\mathbf{I})$ independent of x and $\mathbf{h}$ a fixed vector in $\mathcal{C}^n$. The projection of $\mathbf{y}$ in the direction of $\mathbf{h}$ is a sufficient statistic and we can reduce the vector estimation problem to a scalar one: estimate x from

$$\tilde{y} = \frac{\mathbf{h}^*\mathbf{y}}{\|\mathbf{h}\|^2} = x + w, \tag{A.77}$$

where $w \sim \mathcal{CN}(0, N_0/\|\mathbf{h}\|^2)$.

Thus the best linear estimator is, as an extension of (A.67),

$$\mathbf{c} = \frac{\mathbb{E}[|x|^2]}{\mathbb{E}[|x|^2]\|\mathbf{h}\|^2 + N_0}\mathbf{h}. \tag{A.78}$$

The corresponding minimum MSE is, as an extension of (A.68),

$$\text{MMSE} = \frac{\mathbb{E}[x^2]N_0}{\mathbb{E}[x^2]\|\mathbf{h}\|^2 + N_0}. \tag{A.79}$$

Summary A.3 Mean square estimation in a complex vector space

The linear estimate with the smallest mean squared error of x from

$$y = x + w, \tag{A.80}$$

with $w \sim \mathcal{CN}(0, N_0)$, is

$$\hat{x} = \frac{\mathbb{E}[|x|^2]}{\mathbb{E}[|x|^2] + N_0} y. \tag{A.81}$$

To estimate x from

$$\mathbf{y} = \mathbf{h}x + \mathbf{w}, \tag{A.82}$$

where $\mathbf{w} \sim \mathcal{CN}(0, N_0 \mathbf{I})$,

$$\mathbf{h}^* \mathbf{y} \tag{A.83}$$

is a sufficient statistic, reducing the vector estimation problem to the scalar one.

The best *linear* estimator is

$$\hat{x} = \frac{\mathbb{E}[|x|^2]}{\mathbb{E}[|x|^2]\|\mathbf{h}\|^2 + N_0} \mathbf{h}^* \mathbf{y}. \tag{A.84}$$

The corresponding minimum mean squared error (MMSE) is:

$$\mathrm{MMSE} = \frac{\mathbb{E}[|x|^2] N_0}{\mathbb{E}[|x|^2]\|\mathbf{h}\|^2 + N_0}. \tag{A.85}$$

In the special case when $x \sim \mathcal{CN}(\mu, \sigma^2)$, this estimator yields the minimum mean squared error among *all* estimators, linear or non-linear.

A.4 Exercises

Exercise A.1 Consider the n-dimensional standard Gaussian random vector $\mathbf{w} \sim \mathcal{N}(0, \mathbf{I}_n)$ and its squared magnitude $\|\mathbf{w}\|^2$.

1. With $n = 1$, show that the density of $\|\mathbf{w}\|^2$ is

$$f_1(a) = \frac{1}{\sqrt{2\pi a}} \exp\left(-\frac{a}{2}\right), \qquad a \geq 0. \tag{A.86}$$

2. For any n, show that the density of $\|\mathbf{w}\|^2$ (denoted by $f_n(\cdot)$) satisfies the recursive relation:

$$f_{n+2}(a) = \frac{a}{n} f_n(a), \qquad a \geq 0. \qquad (A.87)$$

3. Using the formulas for the densities for $n = 1$ and 2 ((A.86) and (A.9), respectively) and the recursive relation in (A.87) determine the density of $\|\mathbf{w}\|^2$ for $n \geq 3$.

Exercise A.2 Let $\{w(t)\}$ be white Gaussian noise with power spectral density $N_0/2$. Let $\mathbf{s}_1, \ldots, \mathbf{s}_M$ be a set of finite orthonormal waveforms (i.e., orthogonal and unit energy), and define $z_i = \int_{-\infty}^{\infty} w(t) s_i(t) \mathrm{d}t$. Find the joint distribution of $\mathbf{z}$. *Hint*: Recall the isotropic property of the normalized Gaussian random vector (see (A.8)).

Exercise A.3 Consider a complex random vector $\mathbf{x}$.
1. Verify that the second-order statistics of $\mathbf{x}$ (i.e., the covariance matrix of the real representation $[\Re[\mathbf{x}], \Im[\mathbf{x}]]^t$) can be completely specified by the covariance and pseudo-covariance matrices of $\mathbf{x}$, defined in (A.15) and (A.16) respectively.
2. In the case where $\mathbf{x}$ is circular symmetric, express the covariance matrix $[\Re[\mathbf{x}], \Im[\mathbf{x}]]^t$ in terms of the covariance matrix of the complex vector $\mathbf{x}$ only.

Exercise A.4 Consider a complex Gaussian random vector $\mathbf{x}$.
1. Show that a necessary and sufficient condition for $\mathbf{x}$ to be circular symmetric is that the mean $\boldsymbol{\mu}$ and the pseudo-covariance matrix $\mathbf{J}$ are zero.
2. Now suppose the relationship between the covariance matrix of $[\Re[\mathbf{x}], \Im[\mathbf{x}]]^t$ and the covariance matrix of $\mathbf{x}$ in part (2) of Exercise A.3 holds. Can we conclude that $\mathbf{x}$ is circular symmetric?

Exercise A.5 Show that a circular symmetric complex Gaussian random variable must have i.i.d. real and imaginary components.

Exercise A.6 Let $\mathbf{x}$ be an n-dimensional i.i.d. complex Gaussian random vector, with the real and imaginary parts distributed as $\mathcal{N}(0, \mathbf{K}_x)$ where $\mathbf{K}_x$ is a 2×2 covariance matrix. Suppose $\mathbf{U}$ is a unitary matrix (i.e., $\mathbf{U}^*\mathbf{U} = \mathbf{I}$). Identify the conditions on $\mathbf{K}_x$ under which $\mathbf{U}\mathbf{x}$ has the same distribution as $\mathbf{x}$.

Exercise A.7 Let $\mathbf{z}$ be an n-dimensional i.i.d. complex Gaussian random vector, with the real and imaginary parts distributed as $\mathcal{N}(0, \mathbf{K}_x)$ where $\mathbf{K}_x$ is a 2×2 covariance matrix. We wish to detect a scalar x, equally likely to be ± 1 from

$$\mathbf{y} = \mathbf{h}x + \mathbf{z}, \qquad (A.88)$$

where x and $\mathbf{z}$ are independent and $\mathbf{h}$ is a fixed vector in $\mathcal{C}^n$. Identify the conditions on $\mathbf{K}_x$ under which the scalar $\mathbf{h}^*\mathbf{y}$ is a sufficient statistic to detect x from $\mathbf{y}$.

Exercise A.8 Consider estimating the real zero-mean scalar x from:

$$\mathbf{y} = \mathbf{h}x + \mathbf{w}, \qquad (A.89)$$

where $\mathbf{w} \sim \mathcal{N}(0, N_0/2\mathbf{I})$ is uncorrelated with x and $\mathbf{h}$ is a fixed vector in $\Re^n$.

1. Consider the scaled linear estimate $\mathbf{c}^t\mathbf{y}$ (with the normalization $\|\mathbf{c}\| = 1$):

$$\hat{x} := a\mathbf{c}^t\mathbf{y} = (a\mathbf{c}^t\mathbf{h})\,x + a\mathbf{c}^t\mathbf{z}. \qquad (A.90)$$

Show that the constant a that minimizes the mean square error ($\mathbb{E}[(x - \hat{x})^2]$) is equal to

$$\frac{\mathbb{E}[x^2]\mathbf{c}^t\mathbf{h}}{\mathbb{E}[x^2]|\mathbf{c}^t\mathbf{h}|^2 + N_0/2}. \qquad (A.91)$$

2. Calculate the minimal mean square error (denoted by MMSE) of the linear estimate in (A.90) (by using the value of a in (A.91)). Show that

$$\frac{\mathbb{E}[x^2]}{\text{MMSE}} = 1 + \text{SNR} := 1 + \frac{\mathbb{E}[x^2]|\mathbf{c}^t\mathbf{h}|^2}{N_0/2}. \qquad (A.92)$$

For every fixed linear estimator $\mathbf{c}$, this shows the relationship between the corresponding SNR and MMSE (of an appropriately scaled estimate). In particular, this relation holds when we optimize over all $\mathbf{c}$ leading to the best linear estimator.

Appendix B Information theory from first principles

This appendix discusses the information theory behind the capacity expressions used in the book. Section 8.3.4 is the only part of the book that supposes an understanding of the material in this appendix. More in-depth and broader expositions of information theory can be found in standard texts such as [26] and [43].

B.1 Discrete memoryless channels

Although the transmitted and received signals are continuous-valued in most of the channels we considered in this book, the heart of the communication problem is *discrete* in nature: the transmitter sends one out of a finite number of codewords and the receiver would like to figure out which codeword is transmitted. Thus, to focus on the essence of the problem, we first consider channels with discrete input and output, so-called *discrete memoryless channels* (DMCs).

Both the input $x[m]$ and the output $y[m]$ of a DMC lie in finite sets $\mathcal{X}$ and $\mathcal{Y}$ respectively. (These sets are called the input and output alphabets of the channel respectively.) The statistics of the channel are described by conditional probabilities $\{p(j|i)\}_{i \in \mathcal{X}, j \in \mathcal{Y}}$. These are also called *transition probabilities*. Given an input sequence $\mathbf{x} = (x[1], \ldots, x[N])$, the probability of observing an output sequence $\mathbf{y} = (y[1], \ldots, y[N])$ is given by[1]

$$p(\mathbf{y}|\mathbf{x}) = \prod_{m=1}^{N} p(y[m]|x[m]). \tag{B.1}$$

The interpretation is that the channel noise corrupts the input symbols independently (hence the term *memoryless*).

---

[1] This formula is only valid when there is no feedback from the receiver to the transmitter, i.e., the input is not a function of past outputs. This we assume throughout.

Example B.1 Binary symmetric channel

The binary symmetric channel has binary input and binary output ($\mathcal{X} = \mathcal{Y} = \{0, 1\}$). The transition probabilities are $p(0|1) = p(1|0) = \epsilon$, $p(0|0) = p(1|1) = 1 - \epsilon$. A 0 and a 1 are both flipped with probability ϵ. See Figure B.1(a).

Example B.2 Binary erasure channel

The binary erasure channel has binary input and ternary output ($\mathcal{X} = \{0, 1\}$, $\mathcal{Y} = \{0, 1, e\}$). The transition probabilities are $p(0|0) = p(1|1) = 1 - \epsilon$, $p(e|0) = p(e|1) = \epsilon$. Here, symbols cannot be flipped but can be erased. See Figure B.1(b).

An abstraction of the communication system is shown in Figure B.2. The sender has one out of several equally likely messages it wants to transmit to the receiver. To convey the information, it uses a codebook $\mathcal{C}$ of block length N and size $|\mathcal{C}|$, where $\mathcal{C} = \{\mathbf{x}_1, \ldots, \mathbf{x}_{|\mathcal{C}|}\}$ and $\mathbf{x}_i$ are the codewords. To transmit the ith message, the codeword $\mathbf{x}_i$ is sent across the noisy channel. Based on the received vector $\mathbf{y}$, the decoder generates an estimate $\hat{i}$ of the correct message. The error probability is $p_e = \mathbb{P}\{\hat{i} \neq i\}$. We will assume that the maximum likelihood (ML) decoder is used, since it minimizes the error probability for a given code. Since we are transmitting one of $|\mathcal{C}|$ messages, the number of bits conveyed is $\log|\mathcal{C}|$. Since the block length of the code is N, the rate of the code is $R = \frac{1}{N}\log|\mathcal{C}|$ bits per unit time. The data rate R and the ML error probability p_e are the two key performance measures of a code.

$$R = \frac{1}{N}\log|\mathcal{C}|. \tag{B.2}$$

$$p_e = \mathbb{P}\{\hat{i} \neq i\}. \tag{B.3}$$

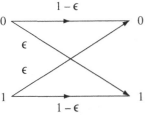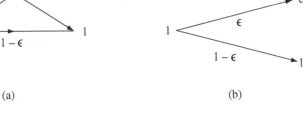

Figure B.1 Examples of discrete memoryless channels: (a) binary symmetric channel; (b) binary erasure channel.

(a)   (b)

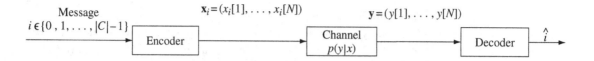

Message
$i \in \{0, 1, \ldots, |C|-1\}$

$\mathbf{x}_i = (x_i[1], \ldots, x_i[N])$

$\mathbf{y} = (y[1], \ldots, y[N])$

Encoder

Channel
$p(y|x)$

Decoder

$\hat{i}$

Figure B.2 Abstraction of a communication system à la Shannon.

Information is said to be *communicated reliably* at rate R if for every $\delta > 0$, one can find a code of rate R and block length N such that the error probability $p_e < \delta$. The capacity C of the channel is the maximum rate for which reliable communication is possible.

Note that the key feature of this definition is that one is allowed to code over arbitrarily large block length N. Since there is noise in the channel, it is clear that the error probability cannot be made arbitrarily small if the block length is fixed a priori. (Recall the AWGN example in Section 5.1.) Only when the code is over long block length is there hope that one can rely on some kind of law of large numbers to average out the random effect of the noise. Still, it is not clear a priori whether a non-zero reliable information rate can be achieved in general.

Shannon showed not only that $C > 0$ for most channels of interest but also gave a simple way to compute C as a function of $\{p(y|x)\}$. To explain this we have to first define a few statistical measures.

B.2 Entropy, conditional entropy and mutual information

Let x be a discrete random variable taking on values in $\mathcal{X}$ and with a probability mass function p_x. Define the *entropy* of x to be[2]

$$H(x) := \sum_{i \in \mathcal{X}} p_x(i) \log(1/p_x(i)). \tag{B.4}$$

This can be interpreted as a measure of the amount of uncertainty associated with the random variable x. The entropy $H(x)$ is always non-negative and equal to zero if and only if x is deterministic. If x can take on K values, then it can be shown that the entropy is maximized when x is uniformly distributed on these K values, in which case $H(x) = \log K$ (see Exercise B.1).

Example B.3 Binary entropy
The entropy of a binary-valued random variable x which takes on the values with probabilities p and $1-p$ is

$$H(p) := -p \log p - (1-p) \log(1-p). \tag{B.5}$$

---

[2] In this book, all logarithms are taken to the base 2 unless specified otherwise.

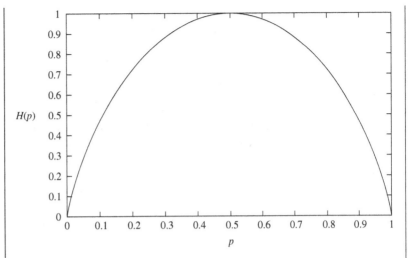

Figure B.3 The binary entropy function.

The function $H(\cdot)$ is called the *binary entropy function*, and is plotted in Figure B.3. It attains its maximum value of 1 at $p = 1/2$, and is zero when $p = 0$ or $p = 1$. Note that we never mentioned the actual *values* x takes on; the amount of uncertainty depends only on the probabilities.

Let us now consider two random variables x and y. The joint entropy of x and y is defined to be

$$H(x, y) := \sum_{i \in \mathcal{X}, j \in \mathcal{Y}} p_{x,y}(i, j) \log(1/p_{x,y}(i, j)). \tag{B.6}$$

The entropy of x conditional on $y = j$ is naturally defined to be

$$H(x|y = j) := \sum_{i \in \mathcal{X}} p_{x|y}(i|j) \log(1/p_{x|y}(i|j)). \tag{B.7}$$

This can be interpreted as the amount of uncertainty left in x after observing that $y = j$. The conditional entropy of x given y is the expectation of this quantity, averaged over all possible values of y:

$$H(x|y) := \sum_{j \in \mathcal{Y}} p_y(j) H(x|y = j) = \sum_{i \in \mathcal{X}, j \in \mathcal{Y}} p_{x,y}(i, j) \log(1/p_{x|y}(i|j)). \tag{B.8}$$

The quantity $H(x|y)$ can be interpreted as the average amount of uncertainty left in x after observing y. Note that

$$H(x, y) = H(x) + H(y|x) = H(y) + H(x|y). \qquad (B.9)$$

This has a natural interpretation: the total uncertainty in x and y is the sum of the uncertainty in x plus the uncertainty in y conditional on x. This is called the *chain rule for entropies*. In particular, if x and y are independent, $H(x|y) = H(x)$ and hence $H(x, y) = H(x) + H(y)$. One would expect that conditioning reduces uncertainty, and in fact it can be shown that

$$H(x|y) \leq H(x), \qquad (B.10)$$

with equality if and only if x and y are independent. (See Exercise B.2.) Hence,

$$H(x, y) = H(x) + H(y|x) \leq H(x) + H(y), \qquad (B.11)$$

with equality if and only if x and y are independent.

The quantity $H(x) - H(x|y)$ is of special significance to the communication problem at hand. Since $H(x)$ is the amount of uncertainty in x before observing y, this quantity can be interpreted as the *reduction* in uncertainty of x from the observation of y, i.e., the amount of information in y about x. Similarly, $H(y) - H(y|x)$ can be interpreted as the reduction in uncertainty of y from the observation of x. Note that

$$H(y) - H(y|x) = H(y) + H(x) - H(x, y) = H(x) - H(x|y). \qquad (B.12)$$

So if one defines

$$I(x; y) := H(y) - H(y|x) = H(x) - H(x|y), \qquad (B.13)$$

then this quantity is symmetric in the random variables x and y. $I(x; y)$ is called the *mutual information* between x and y. A consequence of (B.10) is that the mutual information $I(x; y)$ is a non-negative quantity, and equal to zero if and only if x and y are independent.

We have defined the mutual information between scalar random variables, but the definition extends naturally to random vectors. For example, $I(x_1, x_2; y)$ should be interpreted as the mutual information between the random vector (x_1, x_2) and y, i.e., $I(x_1, x_2; y) = H(x_1, x_2) - H(x_1, x_2|y)$. One can also define a notion of conditional mutual information:

$$I(x; y|z) := H(x|z) - H(x|y, z). \qquad (B.14)$$

Note that since

$$H(x|z) = \sum_k p_z(k) H(x|z = k), \qquad (B.15)$$

and

$$H(x|y, z) = \sum_k p_z(k) H(x|y, z = k), \tag{B.16}$$

it follows that

$$I(x; y|z) = \sum_k p_z(k) I(x; y|z = k). \tag{B.17}$$

Given three random variables x_1, x_2 and y, observe that

$$
\begin{aligned}
I(x_1, x_2; y) &= H(x_1, x_2) - H(x_1, x_2|y) \\
&= H(x_1) + H(x_2|x_1) - [H(x_1|y) + H(x_2|x_1, y)] \\
&= I(x_1; y) + I(x_2; y|x_1).
\end{aligned}
$$

This is the *chain rule for mutual information*:

$$\boxed{I(x_1, x_2; y) = I(x_1; y) + I(x_2; y|x_1).} \tag{B.18}$$

In words: the information that x_1 and x_2 jointly provide about y is equal to the sum of the information x_1 provides about y plus the additional information x_2 provides about y after observing x_1. This fact is very useful in Chapters 7 to 10.

B.3 Noisy channel coding theorem

Let us now go back to the communication problem shown in Figure B.2. We convey one of $|\mathcal{C}|$ equally likely messages by mapping it to its N-length codeword in the code $\mathcal{C} = \{\mathbf{x}_1, \ldots, \mathbf{x}_{|\mathcal{C}|}\}$. The input to the channel is then an N-dimensional random vector $\mathbf{x}$, uniformly distributed on the codewords of $\mathcal{C}$. The output of the channel is another N-dimensional vector $\mathbf{y}$.

B.3.1 Reliable communication and conditional entropy

To decode the transmitted message correctly with high probability, it is clear that the conditional entropy $H(\mathbf{x}|\mathbf{y})$ has to be close to zero[3]. Otherwise, there is too much uncertainty in the input, given the output, to figure out what the right message is. Now,

$$H(\mathbf{x}|\mathbf{y}) = H(\mathbf{x}) - I(\mathbf{x}; \mathbf{y}), \tag{B.19}$$

---

[3] This statement can be made precise in the regime of large block lengths using Fano's inequality.

i.e., the uncertainty in $\mathbf{x}$ subtracting the reduction in uncertainty in $\mathbf{x}$ by observing $\mathbf{y}$. The entropy $H(\mathbf{x})$ is equal to $\log |\mathcal{C}| = NR$, where R is the data rate. For reliable communication, $H(\mathbf{x}|\mathbf{y}) \approx 0$, which implies

$$R \approx \frac{1}{N} I(\mathbf{x}; \mathbf{y}). \tag{B.20}$$

Intuitively: for reliable communication, the *rate of flow* of mutual information across the channel should match the rate at which information is generated. Now, the mutual information depends on the distribution of the random input $\mathbf{x}$, and this distribution is in turn a function of the code $\mathcal{C}$. By optimizing over all codes, we get an upper bound on the reliable rate of communication:

$$\max_{\mathcal{C}} \frac{1}{N} I(\mathbf{x}; \mathbf{y}). \tag{B.21}$$

B.3.2 A simple upper bound

The optimization problem (B.21) is a high-dimensional combinatorial one and is difficult to solve. Observe that since the input vector $\mathbf{x}$ is uniformly distributed on the codewords of $\mathcal{C}$, the optimization in (B.21) is over only a subset of possible input distributions. We can derive a further upper bound by relaxing the feasible set and allowing the optimization to be over *all* input distributions:

$$\bar{C} := \max_{p_{\mathbf{x}}} \frac{1}{N} I(\mathbf{x}; \mathbf{y}), \tag{B.22}$$

Now,

$$I(\mathbf{x}; \mathbf{y}) = H(\mathbf{y}) - H(\mathbf{y}|\mathbf{x}) \tag{B.23}$$

$$\leq \sum_{m=1}^{N} H(y[m]) - H(\mathbf{y}|\mathbf{x}) \tag{B.24}$$

$$= \sum_{m=1}^{N} H(y[m]) - \sum_{m=1}^{N} H(y[m]|x[m]) \tag{B.25}$$

$$= \sum_{m=1}^{N} I(x[m]; y[m]). \tag{B.26}$$

The inequality in (B.24) follows from (B.11) and the equality in (B.25) comes from the memoryless property of the channel. Equality in (B.24) is attained if the output symbols are independent over time, and one way to achieve this is to make the inputs independent over time. Hence,

$$\bar{C} = \frac{1}{N} \sum_{m=1}^{N} \max_{p_{x[m]}} I(x[m]; y[m]) = \max_{p_{x[1]}} I(x[1]; y[1]). \tag{B.27}$$

Thus, the optimizing problem over input distributions on the N-length block reduces to an optimization problem over input distributions on single symbols.

B.3.3 Achieving the upper bound

To achieve this upper bound $\bar{C}$, one has to find a code whose mutual information $I(\mathbf{x}; \mathbf{y})/N$ per symbol is close to $\bar{C}$ *and* such that (B.20) is satisfied. A priori it is unclear if such a code exists at all. The cornerstone result of information theory, due to Shannon, is that indeed such codes exist *if the block length N is chosen sufficiently large.*

Theorem B.1 *(Noisy channel coding theorem [109]) Consider a discrete memoryless channel with input symbol x and output symbol y. The capacity of the channel is*

$$C = \max_{p_x} I(x; y). \tag{B.28}$$

Shannon's proof of the existence of optimal codes is through a randomization argument. Given any symbol input distribution p_x, we can randomly generate a code $\mathcal{C}$ with rate R by choosing each symbol in each codeword independently according to p_x. The main result is that with the rate as in (B.20), the code with large block length N satisfies, with high probability,

$$\frac{1}{N}I(\mathbf{x}; \mathbf{y}) \approx I(x; y). \tag{B.29}$$

In other words, reliable communication is possible at the rate of $I(x; y)$. In particular, by choosing codewords according to the distribution p_x^* that maximizes $I(x; y)$, the maximum reliable rate is achieved. The smaller the desired error probability, the larger the block length N has to be for the law of large numbers to average out the effect of the random noise in the channel as well as the effect of the random choice of the code. We will not go into the details of the derivation of the noisy channel coding theorem in this book, although the sphere-packing argument for the AWGN channel in Section B.5 suggests that this result is plausible. More details can be found in standard information theory texts such as [26].

The maximization in (B.28) is over all distributions of the input random variable x. Note that the input distribution together with the channel transition probabilities specifies a joint distribution on x and y. This determines the

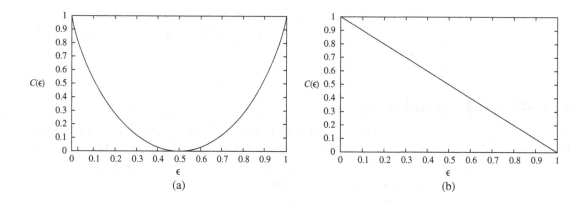

Figure B.4 The capacity of (a) the binary symmetric channel and (b) the binary erasure channel.

value of $I(x; y)$. The maximization is over all possible input distribution It can be shown that the mutual information $I(x; y)$ is a concave function of the input probabilities and hence the input maximization is a convex optimization problem, which can be solved very efficiently. Sometimes one can even appeal to symmetry to obtain the optimal distribution in closed form.

Example B.4 Binary symmetric channel
The capacity of the binary symmetric channel with crossover probability ϵ is

$$C = \max_{p_x} H(y) - H(y|x)$$

$$= \max_{p_x} H(y) - H(\epsilon)$$

$$= 1 - H(\epsilon) \text{ bits per channel use} \tag{B.30}$$

where $H(\epsilon)$ is the binary entropy function (B.5). The maximum is achieved by choosing x to be uniform so that the output y is also uniform. The capacity is plotted in Figure B.4. It is 1 when $\epsilon = 0$ or 1, and 0 when $\epsilon = 1/2$.

Note that since a fraction ϵ of the symbols are flipped in the long run, one may think that the capacity of the channel is $1 - \epsilon$ bits per channel use, the fraction of symbols that get through unflipped. However, this is too naive since the receiver does not know which symbols are flipped and which are correct. Indeed, when $\epsilon = 1/2$, the input and output are independent and there is no way we can get any information across the channel. The expression (B.30) gives the correct answer.

> **Example B.5 Binary erasure channel**
> The optimal input distribution for the binary symmetric channel is uniform because of the symmetry in the channel. Similar symmetry exists in the binary erasure channel and the optimal input distribution is uniform too. The capacity of the channel with erasure probability ϵ can be calculated to be
>
> $$C = 1 - \epsilon \text{ bits per channel use.} \qquad (B.31)$$
>
> In the binary symmetric channel, the receiver does not know which symbols are flipped. In the erasure channel, on the other hand, the receiver knows exactly which symbols are erased. If the *transmitter* also knows that information, then it can send bits only when the channel is not erased and a long-term throughput of $1 - \epsilon$ bits per channel use is achieved. What the capacity result says is that no such feedback information is necessary; (forward) coding is sufficient to get this rate reliably.

B.3.4 Operational interpretation

There is a common misconception that needs to be pointed out. In solving the input distribution optimization problem (B.22) for the capacity C, it was remarked that, at the optimal solution, the outputs $y[m]$ should be independent, and one way to achieve this is for the inputs $x[m]$ to be independent. Does that imply no coding is needed to achieve capacity? For example, in the binary symmetric channel, the optimal input yields i.i.d. equally likely symbols; does it mean then that we can send equally likely information bits raw across the channel and still achieve capacity?

Of course not: to get very small error probability one needs to code over many symbols. The fallacy of the above argument is that reliable communication *cannot* be achieved at *exactly* the rate C and when the outputs are *exactly* independent. Indeed, when the outputs and inputs are i.i.d.,

$$H(\mathbf{x}|\mathbf{y}) = \sum_{m=1}^{N} H(x[m]|y[m]) = NH(x[m]|y[m]), \qquad (B.32)$$

and there is a lot of uncertainty in the input given the output: the communication is hardly reliable. But once one shoots for a rate *strictly* less than C, no matter how close, the coding theorem guarantees that reliable communication is possible. The mutual information $I(\mathbf{x}; \mathbf{y})/N$ per symbol is close to C, the outputs $y[m]$ are *almost* independent, but now the conditional entropy $H(\mathbf{x}|\mathbf{y})$ is reduced abruptly to (close to) zero since reliable decoding is possible. But to achieve this performance, *coding* is crucial; indeed the entropy per input symbol is close to $I(\mathbf{x}; \mathbf{y})/N$, less than $H(x[m])$ under uncoded transmission.

For the binary symmetric channel, the entropy per coded symbol is $1 - H(\epsilon)$, rather than 1 for uncoded symbols.

The bottom line is that while the *value* of the input optimization problem (B.22) has operational meaning as the maximum rate of reliable communication, it is incorrect to interpret the i.i.d. input distribution which attains that value as the statistics of the input symbols which achieve reliable communication. Coding is *always* needed to achieve capacity. What *is* true, however, is that if we randomly pick the codewords according to the i.i.d. input distribution, the resulting code is very likely to be good. But this is totally different from sending uncoded symbols.

B.4 Formal derivation of AWGN capacity

We can now apply the methodology developed in the previous sections to formally derive the capacity of the AWGN channel.

B.4.1 Analog memoryless channels

So far we have focused on channels with discrete-valued input and output symbols. To derive the capacity of the AWGN channel, we need to extend the framework to analog channels with continuous-valued input and output. There is no conceptual difficulty in this extension. In particular, Theorem B.1 can be generalized to such analog channels.[4] The definitions of entropy and conditional entropy, however, have to be modified appropriately.

For a continuous random variable x with pdf f_x, define the *differential entropy* of x as

$$h(x) := \int_{-\infty}^{\infty} f_x(u) \log(1/f_x(u)) \mathrm{d}u. \tag{B.33}$$

Similarly, the conditional differential entropy of x given y is defined as

$$h(x|y) := \int_{-\infty}^{\infty} f_{x,y}(u, v) \log(1/f_{x|y}(u|v)) \mathrm{d}u \mathrm{d}v. \tag{B.34}$$

The mutual information is again defined as

$$I(x; y) := h(x) - h(x|y). \tag{B.35}$$

---

[4] Although the underlying channel is analog, the communication process is still digital. This means that discrete symbols will still be used in the encoding. By formulating the communication problem directly in terms of the underlying analog channel, this means we are not constraining ourselves to using a particular symbol constellation (for example, 2-PAM or QPSK) a priori.

Observe that the chain rules for entropy and for mutual information extend readily to the continuous-valued case. The capacity of the continuous-valued channel can be shown to be

$$C = \max_{f_x} I(x; y).$$ (B.36)

This result can be proved by discretizing the continuous-valued input and output of the channel, approximating it by discrete memoryless channels with increasing alphabet sizes, and taking limits appropriately.

For many channels, it is common to have a *cost constraint* on the transmitted codewords. Given a cost function $c : \mathcal{X} \to \Re$ defined on the input symbols, a cost constraint on the codewords can be defined: we require that every codeword $\mathbf{x}_n$ in the codebook must satisfy

$$\frac{1}{N} \sum_{m=1}^{N} c(x_n[m]) \le A.$$ (B.37)

One can then ask: what is the maximum rate of reliable communication subject to this constraint on the codewords? The answer turns out to be

$$C = \max_{f_x : E[c(x)] \le A} I(x; y).$$ (B.38)

B.4.2 Derivation of AWGN capacity

We can now apply this result to derive the capacity of the power-constrained (real) AWGN channel:

$$y = x + w,$$ (B.39)

The cost function is $c(x) = x^2$. The differential entropy of a $\mathcal{N}(\mu, \sigma^2)$ random variable w can be calculated to be

$$h(w) = \frac{1}{2} \log(2\pi e \sigma^2).$$ (B.40)

Not surprisingly, $h(w)$ does not depend on the mean μ of W: differential entropies are invariant to translations of the pdf. Thus, conditional on the input x of the Gaussian channel, the differential entropy $h(y|x)$ of the output y is just $(1/2) \log(2\pi e \sigma^2)$. The mutual information for the Gaussian channel is, therefore,

$$I(x; y) = h(y) - h(y|x) = h(y) - \frac{1}{2} \log(2\pi e \sigma^2).$$ (B.41)

The computation of the capacity

$$C = \max_{f_x : E[x^2] \le P} I(x; y)$$ (B.42)

is now reduced to finding the input distribution on x to maximize $h(y)$ subject to a second moment constraint on x. To solve this problem, we use a key fact about Gaussian random variables: they are differential entropy maximizers. More precisely, given a constraint $E[u^2] \leq A$ on a random variable u, the distribution u is $\mathcal{N}(0, A)$ maximizes the differential entropy $h(u)$. (See Exercise B.6 for a proof of this fact.) Applying this to our problem, we see that the second moment constraint of P on x translates into a second moment constraint of $P + \sigma^2$ on y. Thus, $h(y)$ is maximized when y is $\mathcal{N}(0, P + \sigma^2)$, which is achieved by choosing x to be $\mathcal{N}(0, P)$. Thus, the capacity of the Gaussian channel is

$$C = \frac{1}{2}\log(2\pi e(P + \sigma^2)) - \frac{1}{2}\log(2\pi e\sigma^2) = \frac{1}{2}\log\left(1 + \frac{P}{\sigma^2}\right), \quad \text{(B.43)}$$

agreeing with the result obtained via the heuristic sphere-packing derivation in Section 5.1. A capacity-achieving code can be obtained by choosing each component of each codeword i.i.d. $\mathcal{N}(0, P)$. Each codeword is therefore isotropically distributed, and, by the law of large numbers, with high probability lies near the surface of the sphere of radius $\sqrt{NP}$. Since in high dimensions most of the volume of a sphere is near its surface, this is effectively the same as picking each codeword uniformly from the sphere.

Now consider a *complex* baseband AWGN channel:

$$y = x + w \quad \text{(B.44)}$$

where w is $\mathcal{CN}(0, N_0)$. There is an average power constraint of P per (complex) symbol. One way to derive the capacity of this channel is to think of each use of the complex channel as two uses of a real AWGN channel, with $\mathsf{SNR} = (P/2)/(N_0/2) = P/N_0$. Hence, the capacity of the channel is

$$\frac{1}{2}\log\left(1 + \frac{P}{N_0}\right) \text{ bits per real dimension,} \quad \text{(B.45)}$$

or

$$\log\left(1 + \frac{P}{N_0}\right) \text{ bits per complex dimension.} \quad \text{(B.46)}$$

Alternatively we may just as well work directly with the complex channel and the associated complex random variables. This will be useful when we deal with other more complicated wireless channel models later on. To this end, one can think of the differential entropy of a complex random variable x as that of a real random vector $(\Re(x), \Im(x))$. Hence, if w is $\mathcal{CN}(0, N_0)$, $h(w) = h(\Re(w)) + h(\Im(w)) = \log(\pi e N_0)$. The mutual information $I(x; y)$ of the complex AWGN channel $y = x + w$ is then

$$I(x; y) = h(y) - \log(\pi e N_0). \quad \text{(B.47)}$$

With a power constraint $E[|x|^2] \leq P$ on the complex input x, y is constrained to satisfy $E[|y|^2] \leq P + N_0$. Here, we use an important fact: among all complex random variables, the *circular symmetric* Gaussian random variable maximizes the differential entropy for a given second moment constraint. (See Exercise B.7.) Hence, the capacity of the complex Gaussian channel is

$$C = \log(\pi e(P + N_0)) - \log(\pi e N_0) = \log\left(1 + \frac{P}{N_0}\right), \qquad \text{(B.48)}$$

which is the same as Eq. (5.11).

B.5 Sphere-packing interpretation

In this section we consider a more precise version of the heuristic sphere-packing argument in Section 5.1 for the capacity of the real AWGN channel. Furthermore, we outline how the capacity as predicted by the sphere-packing argument can be achieved. The material here is particularly useful when we discuss precoding in Chapter 10.

B.5.1 Upper bound

Consider transmissions over a block of N symbols, where N is large. Suppose we use a code $\mathcal{C}$ consisting of $|\mathcal{C}|$ equally likely codewords $\{\mathbf{x}_1, \ldots, \mathbf{x}_{|\mathcal{C}|}\}$. By the law of large numbers, the N-dimensional received vector $\mathbf{y} = \mathbf{x} + \mathbf{w}$ will with high probability lie approximately[5] within a y-sphere of radius $\sqrt{N(P + \sigma^2)}$, so without loss of generality we need only to focus on what happens inside this y-sphere. Let $\mathcal{D}_i$ be the part of the maximum-likelihood decision region for $\mathbf{x}_i$ within the y-sphere. The sum of the volumes of the $\mathcal{D}_i$ is equal to V_y, the volume of the y-sphere. Given this total volume, it can be shown, using the spherical symmetry of the Gaussian noise distribution, that the error probability is lower bounded by the (hypothetical) case when the $\mathcal{D}_i$ are all perfect spheres of equal volume $V_y/|\mathcal{C}|$. But by the law of large numbers, the received vector $\mathbf{y}$ lies near the surface of a *noise sphere* of radius $\sqrt{N\sigma^2}$ around the transmitted codeword. Thus, for reliable communication, $V_y/|\mathcal{C}|$ should be no smaller than the volume V_w of this noise sphere, otherwise even in the ideal case when the decision regions are all spheres of equal volume, the error probability will still be very large. Hence, the number of

---

[5] To make this and other statements in this section completely rigorous, appropriate δ and ε have to be added.

codewords is at most equal to the ratio of the volume of the y-sphere to that of a noise sphere:

$$\frac{V_y}{V_w} = \frac{\left[\sqrt{N(P+\sigma^2)}\right]^N}{\left[\sqrt{N\sigma^2}\right]^N}.$$

(See Exercise B.10(3) for an explicit expression of the volume of an N-dimensional sphere of a given radius.) Hence, the number of bits per symbol time that can be reliably communicated is at most

$$\frac{1}{N}\log\left(\frac{\left[\sqrt{N(P+\sigma^2)}\right]^N}{\left[\sqrt{N\sigma^2}\right]^N}\right) = \frac{1}{2}\log\left(1+\frac{P}{\sigma^2}\right). \tag{B.49}$$

The geometric picture is in Figure B.5.

B.5.2 Achievability

The above argument only gives an upper bound on the rate of reliable communication. The question is: can we design codes that can perform this well?

Let us use a codebook $\mathcal{C} = \{\mathbf{x}_1,\ldots,\mathbf{x}_{|\mathcal{C}|}\}$ such that the N-dimensional codewords lie in the sphere of radius $\sqrt{NP}$ (the "x-sphere") and thus satisfy the power constraint. The optimal detector is the maximum likelihood nearest neighbor rule. For reasons that will be apparent shortly, we instead consider the following suboptimal detector: given the received vector $\mathbf{y}$, decode to the codeword $\mathbf{x}_i$ nearest to $\alpha\mathbf{y}$, where $\alpha := P/(P+\sigma^2)$.

It is not easy to design a specific code that yields good performance, but suppose we just randomly and independently choose each codeword to be

Figure B.5 The number of noise spheres that can be packed into the y-sphere yields the maximum number of codewords that can be reliably distinguished.

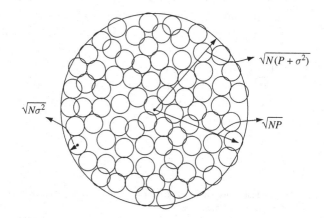

uniformly distributed in the sphere[6]. In high dimensions, most of the volume of the sphere lies near its surface, so in fact the codewords will with high probability lie near the surface of the x-sphere.

What is the performance of this random code? Suppose the transmitted codeword is $\mathbf{x}_1$. By the law of large numbers again,

$$
\begin{aligned}
\|\alpha\mathbf{y} - \mathbf{x}_1\|^2 &= \|\alpha\mathbf{w} + (\alpha - 1)\mathbf{x}_1\|^2, \\
&\approx \alpha^2 N\sigma^2 + (\alpha - 1)^2 NP, \\
&= N\frac{P\sigma^2}{P + \sigma^2},
\end{aligned}
$$

i.e., the transmitted codeword lies inside an *uncertainty* sphere of radius $\sqrt{NP\sigma^2/(P+\sigma^2)}$ around the vector $\alpha\mathbf{y}$. Thus, as long as all the other codewords lie outside this uncertainty sphere, then the receiver will be able to decode correctly (Figure B.6). The probability that the random codeword $\mathbf{x}_i$ ($i \neq 1$) lies inside the uncertainty sphere is equal to the ratio of the volume of the uncertainty sphere to that of the x-sphere:

$$
p = \frac{\left(\sqrt{NP\sigma^2/(P+\sigma^2)}\right)^N}{(\sqrt{NP})^N} = \left(\frac{\sigma^2}{P + \sigma^2}\right)^{\frac{N}{2}}. \tag{B.50}
$$

By the union bound, the probability that *any* of the codewords $(\mathbf{x}_2, \ldots, \mathbf{x}_{|\mathcal{C}|})$ lie inside the uncertainty sphere is bounded by $(|\mathcal{C}| - 1)p$. Thus, as long as the number of codewords is much smaller than $1/p$, then the probability of error is small (in particular, we can take the number of codewords $|\mathcal{C}|$ to be

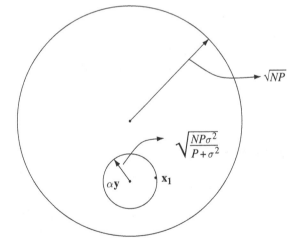

Figure B.6 The ratio of the volume of the uncertainty sphere to that of the x-sphere yields the probability that a given random codeword lies inside the uncertainty sphere. The inverse of this probability yields a lower bound on the number of codewords that can be reliably distinguished.

---

[6]   Randomly and independently choosing each codeword to have i.i.d. $\mathcal{N}(0, P)$ components would work too but the argument is more complex.

$1/pN$). In terms of the data rate R bits per symbol time, this means that as long as

$$R = \frac{\log |\mathcal{C}|}{N} = \frac{\log 1/p}{N} - \frac{\log N}{N} < \frac{1}{2} \log \left(1 + \frac{P}{\sigma^2}\right),$$

then reliable communication is possible.

Both the upper bound and the achievability arguments are based on calculating the ratio of volumes of spheres. The ratio is the same in both cases, but the spheres involved are different. The sphere-packing picture in Figure B.5 corresponds to the following decomposition of the capacity expression:

$$\frac{1}{2} \log \left(1 + \frac{P}{\sigma^2}\right) = I(x; y) = h(y) - h(y|x), \tag{B.51}$$

with the volume of the y-sphere proportional to $2^{Nh(y)}$ and the volume of the noise sphere proportional to $2^{Nh(y|x)}$. The picture in Figure B.6, on the other hand, corresponds to the decomposition:

$$\frac{1}{2} \log \left(1 + \frac{P}{\sigma^2}\right) = I(x; y) = h(x) - h(x|y), \tag{B.52}$$

with the volume of the x-sphere proportional to $2^{Nh(x)}$. Conditional on y, x is $N(\alpha y, \sigma_{\text{mmse}}^2)$, where $\alpha = P/(P+\sigma^2)$ is the coefficient of the MMSE estimator of x given y, and

$$\sigma_{\text{mmse}}^2 = \frac{P\sigma^2}{P+\sigma^2},$$

is the MMSE estimation error. The radius of the uncertainty sphere considered above is $\sqrt{N\sigma_{\text{mmse}}^2}$ and its volume is proportional to $2^{Nh(x|y)}$. In fact the proposed receiver, which finds the nearest codeword to $\alpha\mathbf{y}$, is motivated precisely by this decomposition. In this picture, then, the AWGN capacity formula is being interpreted in terms of the number of MMSE error spheres that can be packed inside the x-sphere.

B.6 Time-invariant parallel channel

Consider the parallel channel (cf. (5.33)):

$$\tilde{y}_n[i] = \tilde{h}_n \tilde{d}_n[i] + \tilde{w}_n[i] \qquad n = 0, 1, \ldots, N_c - 1, \tag{B.53}$$

subject to an average power per sub-carrier constraint of P (cf. (5.37)):

$$E[|\tilde{\mathbf{d}}[i]|^2] \leq N_c P. \tag{B.54}$$

The capacity in bits per symbol is

$$C_{N_c} = \max_{\mathbb{E}[\|\tilde{\mathbf{d}}\|^2] \leq N_c P} I(\tilde{\mathbf{d}}; \tilde{\mathbf{y}}). \tag{B.55}$$

Now

$$I(\tilde{\mathbf{d}}; \tilde{\mathbf{y}}) = h(\tilde{\mathbf{y}}) - h(\tilde{\mathbf{y}}|\tilde{\mathbf{d}}) \tag{B.56}$$

$$\leq \sum_{n=0}^{N_c-1} \left(h(\tilde{y}_n) - h(\tilde{y}_n|\tilde{d}_n) \right) \tag{B.57}$$

$$\leq \sum_{n=0}^{N_c-1} \log \left(1 + \frac{P_n |\tilde{h}_n|^2}{N_0} \right). \tag{B.58}$$

The inequality in (B.57) is from (B.11) and P_n denotes the variance of $\tilde{d}_n$ in (B.58). Equality in (B.57) is achieved when $\tilde{d}_n, n = 0, \ldots, N_c - 1$, are independent. Equality is achieved in (B.58) when $\tilde{d}_n$ is $\mathcal{CN}(0, P_n), n = 0, \ldots, N_c - 1$. Thus, computing the capacity in (B.55) is reduced to a power allocation problem (by identifying the variance of $\tilde{d}_n$ with the power allocated to the nth sub-carrier):

$$C_{N_c} = \max_{P_0, \ldots, P_{N_c-1}} \sum_{n=0}^{N_c-1} \log \left(1 + \frac{P_n |\tilde{h}_n|^2}{N_0} \right), \tag{B.59}$$

subject to

$$\frac{1}{N_c} \sum_{n=0}^{N_c-1} P_n = P, \qquad P_n \geq 0, \quad n = 0, \ldots, N_c - 1. \tag{B.60}$$

The solution to this optimization problem is waterfilling and is described in Section 5.3.3.

B.7 Capacity of the fast fading channel

B.7.1 Scalar fast fading channel

Ideal interleaving

The fast fading channel with ideal interleaving is modeled as follows:

$$y[m] = h[m]x[m] + w[m], \tag{B.61}$$

where the channel coefficients $h[m]$ are i.i.d. in time and independent of the i.i.d. $\mathcal{CN}(0, N_0)$ additive noise $w[m]$. We are interested in the situation when the receiver tracks the fading channel, but the transmitter only has access to the statistical characterization; the receiver CSI scenario. The capacity of the power-constrained fast fading channel with receiver CSI can be written as, by viewing the receiver CSI as part of the output of the channel,

$$C = \max_{p_x : \mathbb{E}[x^2] \leq P} I(x; y, h). \qquad (B.62)$$

Since the fading channel h is independent of the input, $I(x; h) = 0$. Thus, by the chain rule of mutual information (see (B.18)),

$$I(x; y, h) = I(x; h) + I(x; y|h) = I(x; y|h). \qquad (B.63)$$

Conditioned on the fading coefficient h, the channel is simply an AWGN one, with SNR equal to $P|h|^2/N_0$, where we have denoted the transmit power constraint by P. The optimal input distribution for a power constrained AWGN channel is $\mathcal{CN}$, *regardless* of the operating SNR. Thus, the maximizing input distribution in (B.62) is $\mathcal{CN}(0, P)$. With this input distribution,

$$I(x; y|h = \mathrm{h}) = \log\left(1 + \frac{P|\mathrm{h}|^2}{N_0}\right),$$

and thus the capacity of the fast fading channel with receiver CSI is

$$C = \mathbb{E}_h\left[\log\left(1 + \frac{P|h|^2}{N_0}\right)\right], \qquad (B.64)$$

where the average is over the stationary distribution of the fading channel.

Stationary ergodic fading

The above derivation hinges on the i.i.d. assumption on the fading process $\{h[m]\}$. Yet in fact (B.64) holds as long as $\{h[m]\}$ is stationary and ergodic. The alternative derivation below is more insightful and valid for this more general setting.

We first fix a realization of the fading process $\{h[m]\}$. Recall from (B.20) that the rate of reliable communication is given by the average rate of flow of mutual information:

$$\frac{1}{N} I(\mathbf{x}; \mathbf{y}) = \frac{1}{N} \sum_{m=1}^{N} \log(1 + |h[m]|^2 \mathrm{SNR}). \qquad (B.65)$$

For large N, due to the ergodicity of the fading process,

$$\frac{1}{N} \sum_{m=1}^{N} \log(1 + |h[m]|^2 \mathrm{SNR}) \to \mathbb{E}[\log(1 + |h|^2 \mathrm{SNR})], \qquad (B.66)$$

for almost all realizations of the fading process $\{h[m]\}$. This yields the same expression of capacity as in (B.64).

B.7.2 Fast fading MIMO channel

We have only considered the scalar fast fading channel so far; the extension of the ideas to the MIMO case is very natural. The fast fading MIMO channel with ideal interleaving is (cf. (8.7))

$$\mathbf{y}[m] = \mathbf{H}[m]\mathbf{x}[m] + \mathbf{w}[m], \qquad m = 1, 2, \ldots, \tag{B.67}$$

where the channel $\mathbf{H}$ is i.i.d. in time and independent of the i.i.d. additive noise, which is $\mathcal{CN}(0, N_0\mathbf{I}_{n_r})$. There is an average total power constraint of P on the transmit signal. The capacity of the fast fading channel with receiver CSI is, as in (B.62),

$$C = \max_{p_\mathbf{x}:\mathbb{E}[|\mathbf{x}|^2] \leq P} I(\mathbf{x}; \mathbf{y}, \mathbf{H}). \tag{B.68}$$

The observation in (B.63) holds here as well, so the capacity calculation is based on the conditional mutual information $I(\mathbf{x}; \mathbf{y}|\mathbf{H})$. If we fix the MIMO channel at a specific realization, we have

$$
\begin{aligned}
I(\mathbf{x}; \mathbf{y}|\mathbf{H} = \mathtt{H}) &= h(\mathbf{y}) - h(\mathbf{y}|\mathbf{x}) \\
&= h(\mathbf{y}) - h(\mathbf{w}) \tag{B.69} \\
&= h(\mathbf{y}) - n_\mathrm{r}\log(\pi e N_0). \tag{B.70}
\end{aligned}
$$

To proceed, we use the following fact about Gaussian random vectors: they are entropy maximizers. Specifically, among all n-dimensional complex random vectors with a given covariance matrix $\mathbf{K}$, the one that maximizes the differential entropy is complex circular-symmetric jointly Gaussian $\mathcal{CN}(0, \mathbf{K})$ (Exercise B.8). This is the *vector* extension of the result that Gaussian random variables are entropy maximizers for a fixed variance constraint. The corresponding maximum value is given by

$$\log(\det(\pi e \mathbf{K})). \tag{B.71}$$

If the covariance of $\mathbf{x}$ is $\mathbf{K}_x$ and the channel is $\mathbf{H} = \mathtt{H}$, then the covariance of $\mathbf{y}$ is

$$N_0\mathbf{I}_{n_r} + \mathtt{H}\mathbf{K}_x\mathtt{H}^*. \tag{B.72}$$

Calculating the corresponding maximal entropy of $\mathbf{y}$ (cf. (B.71)) and substituting in (B.70), we see that

$$
\begin{aligned}
I(\mathbf{x}; \mathbf{y}|\mathbf{H} = \mathtt{H}) &\leq \log((\pi e)^{n_r}\det(N_0\mathbf{I}_{n_r} + \mathtt{H}\mathbf{K}_x\mathtt{H}^*)) - n_\mathrm{r}\log(\pi e N_0) \\
&= \log\det\left(\mathbf{I}_{n_r} + \frac{1}{N_0}\mathtt{H}\mathbf{K}_x\mathtt{H}^*\right), \tag{B.73}
\end{aligned}
$$

with equality if $\mathbf{x}$ is $\mathcal{CN}(0, \mathbf{K}_x)$. This means that even if the transmitter does not know the channel, there is no loss of optimality in choosing the input to be $\mathcal{CN}$.

Finally, the capacity of the fast fading MIMO channel is found by averaging (B.73) with respect to the stationary distribution of $\mathbf{H}$ and choosing the appropriate covariance matrix subject to the power constraint:

$$C = \max_{\mathbf{K}_x : \mathrm{Tr}[\mathbf{K}_x] \leq P} \mathbb{E}_{\mathbf{H}} \left[\log \det \left(\mathbf{I}_{n_r} + \frac{1}{N_0} \mathbf{H} \mathbf{K}_x \mathbf{H}^* \right) \right]. \qquad (B.74)$$

Just as in the scalar case, this result can be generalized to any stationary and ergodic fading process $\{\mathbf{H}[m]\}$.

B.8 Outage formulation

Consider the slow fading MIMO channel (cf. (8.79))

$$\mathbf{y}[m] = \mathbf{H}\mathbf{x}[m] + \mathbf{w}[m]. \qquad (B.75)$$

Here the MIMO channel, represented by $\mathbf{H}$ (an $n_r \times n_t$ matrix with complex entries), is random but not varying with time. The additive noise is i.i.d. $\mathcal{CN}(0, N_0)$ and independent of $\mathbf{H}$.

If there is a positive probability, however small, that the entries of $\mathbf{H}$ are small, then the capacity of the channel is zero. In particular, the capacity of the i.i.d. Rayleigh slow fading MIMO channel is zero. So we focus on characterizing the ϵ-outage capacity: the largest rate of reliable communication such that the error probability is no more than ϵ. We are aided in this study by viewing the slow fading channel in (B.75) as a *compound channel*.

The basic compound channel consists of a collection of DMCs $p_\theta(y|x)$, $\theta \in \Theta$ with the same input alphabet $\mathcal{X}$ and the same output alphabet $\mathcal{Y}$ and parameterized by θ. Operationally, the communication between the transmitter and the receiver is carried out over one specific channel based on the (arbitrary) choice of the parameter θ from the set Θ. The transmitter does not know the value of θ but the receiver does. The capacity is the largest rate at which a single coding strategy can achieve reliable communication regardless of which θ is chosen. The corresponding capacity achieving strategy is said to be *universal* over the class of channels parameterized by $\theta \in \Theta$. An important result in information theory is the characterization of the capacity of the compound channel:

$$\boxed{C = \max_{p_x} \inf_{\theta \in \Theta} I_\theta(x; y).} \qquad (B.76)$$

Here, the mutual information $I_\theta(x; y)$ signifies that the conditional distribution of the output symbol y given the input symbol x is given by the channel $p_\theta(y|x)$. The characterization of the capacity in (B.76) offers a natural interpretation: there exists a coding strategy, parameterized by the input distribution p_x, that achieves reliable communication at a rate that is the minimum mutual information among all the allowed channels. We have considered only discrete input and output alphabets, but the generalization to continuous input and output alphabets and, further, to cost constraints on the input follows much the same line as our discussion in Section B.4.1. The tutorial article [69] provides a more comprehensive introduction to compound channels.

We can view the slow fading channel in (B.75) as a compound channel parameterized by $\mathbf{H}$. In this case, we can simplify the parameterization of coding strategies by the input distribution p_x: for any fixed $\mathbf{H}$ and channel input distribution p_x with covariance matrix $\mathbf{K}_x$, the corresponding mutual information

$$I(\mathbf{x}; \mathbf{y}) \leq \log \det \left(\mathbf{I}_{n_r} + \frac{1}{N_0} \mathbf{H} \mathbf{K}_x \mathbf{H}^* \right). \tag{B.77}$$

Equality holds when p_x is $\mathcal{CN}(0, \mathbf{K}_x)$ (see Exercise B.8). Thus we can reparameterize a coding strategy by its corresponding covariance matrix (the input distribution is chosen to be $\mathcal{CN}$ with zero mean and the corresponding covariance). For every fixed covariance matrix $\mathbf{K}_x$ that satisfies the power constraint on the input, we can reword the compound channel result in (B.76) as follows. Over the slow fading MIMO channel in (B.75), there exists a universal coding strategy at a rate R bits/s/Hz that achieves reliable communication over all channels $\mathbf{H}$ which satisfy the property

$$\log \det \left(\mathbf{I}_{n_r} + \frac{1}{N_0} \mathbf{H} \mathbf{K}_x \mathbf{H}^* \right) > R. \tag{B.78}$$

Furthermore, no reliable communication using the coding strategy parameterized by $\mathbf{K}_x$ is possible over channels that are in *outage*: that is, they do not satisfy the condition in (B.78). We can now choose the covariance matrix, subject to the input power constraints, such that we minimize the probability of outage. With a total power constraint of P on the transmit signal, the outage probability when communicating at rate R bits/s/Hz is

$$p_{\text{out}}^{\text{mimo}} := \min_{\mathbf{K}_x : \text{Tr}[\mathbf{K}_x] \leq P} \mathbb{P} \left\{ \log \det \left(\mathbf{I}_{n_r} + \frac{1}{N_0} \mathbf{H} \mathbf{K}_x \mathbf{H}^* \right) < R \right\}. \tag{B.79}$$

The ϵ-outage capacity is now the largest rate R such that $p_{\text{out}}^{\text{mimo}} \leq \epsilon$.

By restricting the number of receive antennas n_r to be 1, this discussion also characterizes the outage probability of the MISO fading channel. Further, restricting the MIMO channel $\mathbf{H}$ to be diagonal we have also characterized the outage probability of the parallel fading channel.

B.9 Multiple access channel

B.9.1 Capacity region

The uplink channel (with potentially multiple antenna elements) is a special case of the multiple access channel. Information theory gives a formula for computing the capacity region of the multiple access channel in terms of mutual information, from which the corresponding region for the uplink channel can be derived as a special case.

The capacity of a memoryless point-to-point channel with input x and output y is given by

$$C = \max_{p_x} I(x; y),$$

where the maximization is over the input distributions subject to the average cost constraint. There is an analogous theorem for multiple access channels. Consider a two-user channel, with inputs x_k from user k, $k = 1, 2$ and output y. For given input distributions p_{x_1} and p_{x_2} and *independent* across the two users, define the pentagon $\mathcal{C}(p_{x_1}, p_{x_2})$ as the set of all rate pairs satisfying:

$$R_1 < I(x_1; y|x_2), \tag{B.80}$$

$$R_2 < I(x_2; y|x_1), \tag{B.81}$$

$$R_1 + R_2 < I(x_1, x_2; y). \tag{B.82}$$

The capacity region of the multiple access channel is the convex hull of the union of these pentagons over all possible independent input distributions subject to the appropriate individual average cost constraints, i.e.,

$$\mathcal{C} = \text{convex hull of} (\cup_{p_{x_1}, p_{x_2}} \mathcal{C}(p_{x_1}, p_{x_2})). \tag{B.83}$$

The convex hull operation means that we not only include points in $\cup \mathcal{C}(p_{x_1}, p_{x_2})$ in $\mathcal{C}$, but also all their convex combinations. This is natural since the convex combinations can be achieved by time-sharing.

The capacity region of the uplink channel with single antenna elements can be arrived at by specializing this result to the scalar Gaussian multiple access channel. With average power constraints on the two users, we observe that Gaussian inputs for user 1 and 2 *simultaneously* maximize $I(x_1; y|x_2)$, $I(x_2; y|x_1)$ and $I(x_1; x_2; y)$. Hence, the pentagon from this input distribution is a superset of all other pentagons, and the capacity region itself is this pentagon. The same observation holds for the time-invariant uplink channel with single transmit antennas at each user and multiple receive antennas at the base-station. The expressions for the capacity regions of the uplink with a single receive antenna are provided in (6.4), (6.5) and (6.6). The capacity region of the uplink with multiple receive antennas is expressed in (10.6).

Figure B.7 The achievable rate
regions (pentagons)
corresponding to two different
input distributions may not
fully overlap with respect to
one another.

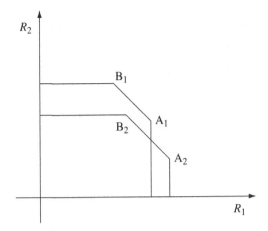

In the uplink with single transmit antennas, there was a unique set of input distributions that simultaneously maximized the different constraints ((B.80), (B.81) and (B.82)). In general, no single pentagon may dominate over the other pentagons, and in that case the overall capacity region may not be a pentagon (see Figure B.7). An example of this situation is provided by the uplink with multiple transmit antennas at the users. In this situation, zero mean circularly symmetric complex Gaussian random vectors still simultaneously maximize all the constraints, but with different covariance matrices. Thus we can restrict the user input distributions to be zero mean $\mathcal{CN}$, but leave the covariance matrices of the users as parameters to be chosen. Consider the two-user uplink with multiple transmit and receive antennas. Fixing the kth user input distribution to be $\mathcal{CN}(0, \mathbf{K}_k)$ for $k = 1, 2$, the corresponding pentagon is expressed in (10.23) and (10.24). In general, there is no single choice of covariance matrices that simultaneously maximize the constraints: the capacity region is the convex hull of the union of the pentagons created by all the possible covariance matrices (subject to the power constraints on the users).

B.9.2 Corner points of the capacity region

Consider the pentagon $\mathcal{C}(p_{x_1}, p_{x_2})$ parameterized by fixed independent input distributions on the two users and illustrated in Figure B.8. The two corner points A and B have an important significance: if we have coding schemes that achieve reliable communication to the users at the rates advertised by these two points, then the rates at every other point in the pentagon can be achieved by appropriate time-sharing between the two strategies that achieved the points A and B. Below, we try to get some insight into the nature of the two corner points and properties of the receiver design that achieves them.

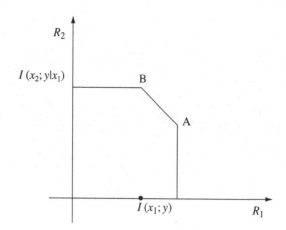

Consider the corner point B. At this point, user 1 gets the rate $I(x_1; y)$. Using the chain rule for mutual information we can write

$$I(x_1, x_2; y) = I(x_1; y) + I(x_2; y|x_1).$$

Since the sum rate constraint is tight at the corner point B, user 2 achieves its highest rate $I(x_2; y|x_1)$. This rate pair can be achieved by a successive interference cancellation (SIC) receiver: decode user 1 first, treating the signal from user 2 as part of the noise. Next, decode user 2 conditioned on the already decoded information from user 1. In the uplink with a single antenna, the second stage of the successive cancellation receiver is very explicit: given the decoded information from user 1, the receiver simply subtracts the decoded transmit signal of user 1 from the received signal. With multiple receive antennas, the successive cancellation is done in conjunction with the MMSE receiver. The MMSE receiver is information lossless (this aspect is explored in Section 8.3.4) and we can conclude the following intuitive statement: the MMSE–SIC receiver is optimal because it "implements" the chain rule for mutual information.

B.9.3 Fast fading uplink

Consider the canonical two-user fast fading MIMO uplink channel:

$$\mathbf{y}[m] = \mathbf{H}_1[m]\mathbf{x}_1[m] + \mathbf{H}_2[m]\mathbf{x}_2[m] + \mathbf{w}[m], \tag{B.84}$$

where the MIMO channels $\mathbf{H}_1$ and $\mathbf{H}_2$ are independent and i.i.d. over time. As argued in Section B.7.1, interleaving allows us to convert stationary channels with memory to this canonical form. We are interested in the receiver CSI situation: the receiver tracks both the users' channels perfectly. For fixed independent input distributions p_{x_1} and p_{x_2}, the achievable rate region consists of tuples (R_1, R_2) constrained by

$$R_1 < I(\mathbf{x}_1; \mathbf{y}, \mathbf{H}_1, \mathbf{H}_2 | \mathbf{x}_2), \tag{B.85}$$

$$R_2 < I(\mathbf{x}_2; \mathbf{y}, \mathbf{H}_1, \mathbf{H}_2 | \mathbf{x}_1), \tag{B.86}$$

$$R_1 + R_2 < I(\mathbf{x}_1, \mathbf{x}_2; \mathbf{y}, \mathbf{H}_1, \mathbf{H}_2). \tag{B.87}$$

Here we have modeled receiver CSI as the MIMO channels being part of the output of the multiple access channel. Since the channels are independent of the user inputs, we can use the chain rule of mutual information, as in (B.63), to rewrite the constraints on the rate tuples as

$$R_1 < I(\mathbf{x}_1; \mathbf{y} | \mathbf{H}_1, \mathbf{H}_2, \mathbf{x}_2), \tag{B.88}$$

$$R_2 < I(\mathbf{x}_2; \mathbf{y} | \mathbf{H}_1, \mathbf{H}_2, \mathbf{x}_1), \tag{B.89}$$

$$R_1 + R_2 < I(\mathbf{x}_1, \mathbf{x}_2; \mathbf{y} | \mathbf{H}_1, \mathbf{H}_2). \tag{B.90}$$

Fixing the realization of the MIMO channels of the users, we see again (as in the time-invariant MIMO uplink) that the input distributions can be restricted to be zero mean $\mathcal{CN}$ but leave their covariance matrices as parameters to be chosen later. The corresponding rate region is a pentagon expressed by (10.23) and (10.24). The conditional mutual information is now the average over the stationary distributions of the MIMO channels: an expression for this pentagon is provided in (10.28) and (10.29).

B.10 Exercises

Exercise B.1 Suppose x is a discrete random variable taking on K values, each with probability $p_1, \ldots, p_K$. Show that

$$\max_{p_1, \ldots, p_K} H(x) = \log K,$$

and further that this is achieved only when $p_i = 1/K$, $i = 1, \ldots, K$, i.e., x is uniformly distributed.

Exercise B.2 In this exercise, we will study when conditioning does not reduce entropy.
1. A concave function f is defined in the text by the condition $f''(x) \le 0$ for x in the domain. Give an alternative geometric definition that does not use calculus.
2. Jensen's inequality for a random variable x states that for any concave function f

$$\mathbb{E}[f(x)] \le f(\mathbb{E}[x]). \tag{B.91}$$

Prove this statement. *Hint*: You might find it useful to draw a picture and visualize the proof geometrically. The geometric definition of a concave function might come in handy here.

3. Show that $H(x|y) \leq H(x)$ with equality if and only if x and y are independent. Give an example in which $H(x|y = k) > H(x)$. Why is there no contradiction between these two statements?

Exercise B.3 Under what condition on x_1, x_2, y does it hold that

$$I(x_1, x_2; y) = I(x_1; y) + I(x_2; y)? \tag{B.92}$$

Exercise B.4 Consider a continuous real random variable x with density $f_x(\cdot)$ non-zero on the entire real line. Suppose the second moment of x is fixed to be P. Show that among all random variables with the constraints as those on x, the Gaussian random variable has the maximum differential entropy. *Hint*: The differential entropy is a concave function of the density function and fixing the second moment corresponds to a linear constraint on the density function. So, you can use the classical Lagrangian techniques to solve this problem.

Exercise B.5 Suppose x is now a non-negative random variable with density non-zero for all non-negative real numbers. Further suppose that the mean of x is fixed. Show that among all random variables of this form, the exponential random variable has the maximum differential entropy.

Exercise B.6 In this exercise, we generalize the results in Exercises B.4 and B.5. Consider a continuous real random variable x with density $f_x(\cdot)$ on a support set S (i.e., $f_x(u) = 0, u \notin S$). In this problem we will study the structure of the random variable x with maximal differential entropy that satisfies the following *moment* conditions:

$$\int_S r_i(u) f_x(u) \, du = A_i, \qquad i = 1, \ldots, m. \tag{B.93}$$

Show that x with density

$$f_x(u) = \exp\left(\lambda_0 - 1 + \sum_{i=1}^m \lambda_i r_i(u)\right), u \in S, \tag{B.94}$$

has the maximal differential entropy subject to the moment conditions (B.93). Here $\lambda_0, \lambda_1, \ldots, \lambda_m$ are chosen such that the moment conditions (B.93) are met and that $f_x(\cdot)$ is a density function (i.e., it integrates to unity).

Exercise B.7 In this problem, we will consider the differential entropy of a vector of continuous random variables with moment conditions.

1. Consider the class of continuous real random vectors $\mathbf{x}$ with the covariance condition: $\mathbb{E}[\mathbf{x}\mathbf{x}^t] = \mathbf{K}$. Show that the jointly Gaussian random vector with covariance $\mathbf{K}$ has the maximal differential entropy among this set of covariance constrained random variables.

2. Now consider a complex random variable x. Show that among the class of continuous complex random variables x with the second moment condition $\mathbb{E}[|x|^2] \leq P$,

the *circularly symmetric* Gaussian complex random variable has the maximal differential entropy. *Hint*: View **x** as a length 2 vector of real random variables and use the previous part of this question.

Exercise B.8 Consider a zero mean complex random vector **x** with fixed covariance $\mathbb{E}[\mathbf{xx}^*] = \mathbf{K}$. Show the following upper bound on the differential entropy:

$$h(\mathbf{x}) \leq \log \det(\pi e \mathbf{K}),\qquad\qquad (B.95)$$

with equality when **x** is $\mathcal{CN}(0, \mathbf{K})$. *Hint*: This is a generalization of Exercise B.7(2).

Exercise B.9 Show that the structure of the input distribution in (5.28) optimizes the mutual information in the MISO channel. *Hint*: Write the second moment of **y** as a function of the covariance of **x** and see which covariance of **x** maximizes the second moment of y. Now use Exercise B.8 to reach the desired conclusion.

Exercise B.10 Consider the real random vector **x** with i.i.d. $\mathcal{N}(0, P)$ components. In this exercise, we consider properties of the scaled vector $\tilde{\mathbf{x}} := (1/\sqrt{N})\mathbf{x}$. (The material here is drawn from the discussion in Chapter 5.5 in [148].)

1. Show that $(\mathbb{E}[\|\mathbf{x}\|^2])/N = P$, so the scaling ensured that the mean length of $\|\tilde{\mathbf{x}}\|^2$ is P, independent of N.
2. Calculate the variance of $\|\tilde{\mathbf{x}}\|^2$ and show that $\|\tilde{\mathbf{x}}\|^2$ converges to P in probability. Thus, the scaled vector is *concentrated* around its mean.
3. Consider the event that $\tilde{\mathbf{x}}$ lies in the *shell* between two concentric spheres of radius $\rho - \delta$ and ρ. (See Figure B.9.) Calculate the volume of this shell to be

$$B_N \left(\rho^N - (\rho - \delta)^N \right), \quad \text{where } B_N = \begin{cases} \pi^{N/2}/(\frac{N}{2})! & N \text{ even} \\ (2^N \pi^{(N-1)/2})[(N-1)/2]!/N! & N \text{ odd.} \end{cases}$$
$$(B.96)$$

4. Show that we can approximate the volume of the shell by

$$N B_N \rho^{N-1} \delta, \qquad \text{for } \delta/\rho \ll 1. \qquad\qquad (B.97)$$

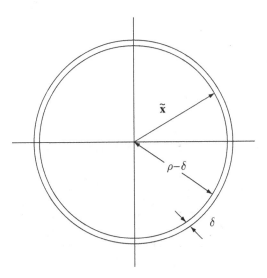

Figure B.9 The shell between two concentric spheres of radius $\rho - \delta$ and ρ.

Figure B.10 Behavior of $\mathbb{P}(\rho - \delta \leq \|\tilde{\mathbf{x}}\| < \rho)$ as a function of ρ.

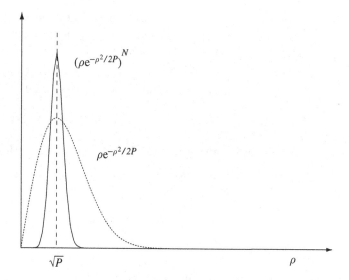

5. Let us approximate the density of $\tilde{\mathbf{x}}$ inside this shell to be

$$f_{\tilde{\mathbf{x}}}(\mathbf{a}) \approx \left(\frac{N}{2\pi P}\right)^{N/2} \exp\left(-\frac{N\rho^2}{2P}\right), \qquad r - \delta < \|\mathbf{a}\| \leq \rho. \tag{B.98}$$

Combining (B.98) and (B.97), show that for $\delta/\rho = $ a constant $\ll 1$,

$$\mathbb{P}(\rho - \delta \leq \|\tilde{\mathbf{x}}\| < \rho) \approx \left[\rho \exp\left(-\frac{\rho^2}{2P}\right)\right]^N. \tag{B.99}$$

6. Show that the right hand side of (B.99) has a single maximum at $\rho^2 = P$ (see Figure B.10).

7. Conclude that as N becomes large, the consequence is that only values of $\|\tilde{\mathbf{x}}\|^2$ in the vicinity of P have significant probability. This phenomenon is called *sphere hardening*.

Exercise B.11 Calculate the mutual information achieved by the isotropic input distribution $\mathbf{x}$ is $\mathcal{CN}(0, P/L \cdot \mathbf{I}_L)$ in the MISO channel (cf. (5.27)) with given channel gains $h_1, \ldots, h_L$.

Exercise B.12 In this exercise, we will study the capacity of the L-tap frequency-selective channel directly (without recourse to the cyclic prefix idea). Consider a length N_c vector input $\mathbf{x}$ on to the channel in (5.32) and denote the vector output (of length $N_c + L - 1$) by $\mathbf{y}$. The input and output are linearly related as

$$\mathbf{y} = \mathbf{G}\mathbf{x} + \mathbf{w}, \tag{B.100}$$

where $\mathbf{G}$ is a matrix whose entries depend on the channel coefficients $h_0, \ldots, h_{L-1}$ as follows: $G[i, j] = h_{i-j}$ for $i \geq j$ and zero everywhere else. The channel in (B.100) is a *vector* version of the basic AWGN channel and we consider the rate of reliable communication $I(\mathbf{x}; \mathbf{y})/N_c$.

1. Show that the optimal input distribution is $\mathbf{x}$ is $\mathcal{CN}(0, \mathbf{K}_x)$, for some covariance matrix $\mathbf{K}_x$ meeting the power constraint. (*Hint*: You will find Exercise B.8 useful.)

2. Show that it suffices to consider only those covariances $\mathbf{K}_x$ that have the same set of eigenvectors as $\mathbf{G}^*\mathbf{G}$. (*Hint*: Use Exercise B.8 to explicitly write the reliable rate of communiation in the vector AWGN channel of (B.100).)

3. Show that

$$(\mathbf{G}^*\mathbf{G})_{ij} = r_{i-j}, \tag{B.101}$$

where

$$r_n := \sum_{\ell=0}^{L-l-1} (h_\ell)^* h[\ell + n], \qquad n \geq 0, \tag{B.102}$$

$$r_n := r_{-n}^*, \qquad n \leq 0. \tag{B.103}$$

Such a matrix $\mathbf{G}^*\mathbf{G}$ is said to be *Toeplitz*.

4. An important result about the Hermitian Toeplitz matrix $\mathbf{G}\mathbf{G}^*$ is that the empirical distribution of its eigenvalues converges (weakly) to the discrete-time Fourier transform of the sequence $\{r_l\}$. How is the discrete-time Fourier transform of the sequence $\{r_l\}$ related to the discrete-time Fourier transform $H(f)$ of the sequence $h_0, \ldots, h_{L-1}$?

5. Use the result of the previous part and the nature of the optimal $\mathbf{K}_x^*$ (discussed in part (2)) to show that the rate of reliable communication is equal to

$$\int_0^W \log\left(1 + \frac{P^*(f)|H(f)|^2}{N_0}\right) df. \tag{B.104}$$

Here the waterfilling power allocation $P^*(f)$ is as defined in (5.47). This answer is, of course, the same as that derived in the text (cf. (5.49)). The cyclic prefix converted the frequency-selective channel into a parallel channel, reliable communication over which is easier to understand. With a direct approach we had to use analytical results about Toeplitz forms; more can be learnt about these techniques from [53].

References

[1] I. C. Abou-Faycal, M. D. Trott and S. Shamai, "The capacity of discrete-time memoryless Rayleigh-fading channels", *IEEE Transactions on Information Theory*, **47**(4), 2001, 1290–1301.

[2] R. Ahlswede, "Multi-way communication channels", *IEEE International Symposium on Information Theory*, Tsahkadsor USSR, 1971, pp. 103–135.

[3] S. M. Alamouti, "A simple transmitter diversity scheme for wireless communication", *IEEE Journal on Selected Areas in Communication*, **16**, 1998, 1451–1458.

[4] J. Barry, E. Lee and D. G. Messerschmitt, *Digital Communication*, Third Edition, Kluwer, 2003.

[5] J.-C. Belfiore, G. Rekaya and E. Viterbo, "The Golden Code: a 2 × 2 fullrate space-time code with non-vanishing determinants", *Proceedings of the IEEE International Symposium on Information Theory*, Chicago June 2004 p. 308.

[6] P. Bender, P. Black, M. Grob, R. Padovani, N. T. Sindhushayana and A. J. Viterbi, "CDMA/HDR: A bandwidth-efficient high-speed wireless data service for nomadic users", *IEEE Communications Magazine*, July 2000.

[7] C. Berge, *Hypergraphs*, Amsterdam, North-Holland, 1989.

[8] P. P. Bergmans, "A simple converse for broadcast channels with additive white Gaussian noise", *IEEE Transactions on Information Theory*, **20**, 1974, 279–280.

[9] E. Biglieri, J. Proakis and S. Shamai, "Fading channels: information theoretic and communications aspects", *IEEE Transactions on Information Theory*, **44**(6), 1998, 2619–2692.

[10] D. Blackwell, L. Breiman and A. J. Thomasian, "The capacity of a class of channels", *Annals of Mathematical Statistics*, **30**, 1959, 1229–1241.

[11] H. Boche and E. Jorswieck, "Outage probability of multiple antenna systems: optimal transmission and impact of correlation", *International Zurich Seminar on Communications*, February 2004.

[12] S. C. Borst and P. A. Whiting, "Dynamic rate control algorithms for HDR throughput optimization", *IEEE Proceedings of Infocom*, **2**, 2001, 976–985.

[13] J. Boutros and E. Viterbo, "Signal space diversity: A power and bandwidth-efficient diversity technique for the Rayleigh fading channel", *IEEE Transactions on Information Theory*, **44**, 1998, 1453–1467.

[14] S. Boyd, "Multitone signals with low crest factor", *IEEE Transactions on Circuits and Systems*, **33**, 1986, 1018–1022.

[15] S. Boyd and L. Vandenberge, *Convex Optimization*, Cambridge University Press, 2004.

[16] R. Brualdi, *Introductory Combinatorics*, New York, North Holland, Second Edition, 1992.

[17] G. Caire and S. Shamai, "On the achievable throughput in multiple antenna Gaussian broadcast channel", *IEEE Transactions on Information Theory*, **49**(7), 2003, 1691–1706.

[18] R. W. Chang, "Synthesis of band-limited orthogonal signals for multichannel data transmission", *Bell System Technical Journal*, **45**, 1966, 1775–1796.

[19] E. F. Chaponniere, P. Black, J. M. Holtzman and D. Tse, *Transmitter directed, multiple receiver system using path diversity to equitably maximize throughput*, U.S. Patent No. 6449490, September 10, 2002.

[20] R. S. Cheng and S. Verdú, "Gaussian multiaccess channels with ISI: Capacity region and multiuser water-filling", *IEEE Transactions on Information Theory*, **39**, 1993, 773–785.

[21] C. Chuah, D. Tse, J. Kahn and R. Valenzuela, "Capacity scaling in MIMO wireless systems under correlated fading", *IEEE Transactions on Information Theory*, **48**(3), 2002, 637–650.

[22] R. H. Clarke, "A statistical theory of mobile-radio reception", *Bell System Technical Journal*, **47**, 1968, 957–1000.

[23] M. H. M. Costa, "Writing on dirty-paper", *IEEE Transactions on Information Theory*, **29**, 1983, 439–441.

[24] T. Cover, "Comments on broadcast channels", *IEEE Transactions on Information Theory*, **44**(6), 1998, 2524–2530.

[25] T. Cover, "Broadcast channels", *IEEE Transactions on Information Theory*, **18**(1), 1972, 2–14.

[26] T. Cover and J. Thomas, *Elements of Information Theory*, John Wiley and Sons, 1991.

[27] R. Jean-Merc Cramer, *An Evaluation of Ultra-Wideband Propagation Channels*, Ph.D. Thesis, University of Southern California, December 2000.

[28] H. A. David, *Order Statistics*, Wiley, First Edition, 1970.

[29] P. Dayal and M. Varanasi, "An optimal two transmit antenna space-time code and its stacked extensions", *Proceedings of Asilomar Conference on Signals, Systems and Computers*, CA, November 2003.

[30] D. Divsalar and M. K. Simon, "The Design of trellis-coded MPSK for fading channels: Performance criteria", *IEEE Transactions on Communications*, **36**(9), 1988, 1004–1012.

[31] R. L. Dobrushin, "Optimum information transmission through a channel with unknown parameters", *Radio Engineering and Electronics*, **4**(12), 1959, 1–8.

[32] A. Edelman, *Eigenvalues and Condition Numbers of Random Matrices*, Ph.D. Dissertation, MIT, 1989.

[33] A. El Gamal, "Capacity of the product and sum of two unmatched broadcast channels", *Problemi Peredachi Informatsii*, **16**(1), 1974, 3–23.

[34] H. El Gamal, G. Caire and M. O. Damen, "Lattice coding and decoding achieves the optimal diversity–multiplexing tradeoff of MIMO channels", *IEEE Transactions on Information Theory*, **50**, 2004, 968–985.

[35] P. Elia, K. R. Kumar, S. A. Pawar, P. V. Kumar and Hsiao-feng Lu, "Explicit construction of space-time block codes achieving the diversity–multiplexing gain tradeoff", ISIT, Adelaide 2005.

[36] M. V. Eyuboglu and G. D. Forney, Jr., "Trellis precoding: Combined coding, precoding and shaping for intersymbol interference channels", *IEEE Transactions on Information Theory*, **38**, 1992, 301–314.

[37] F. R. Farrokhi, K. J. R. Liu and L. Tassiulas, "Transmit beamforming and power control in wireless networks with fading channels", *IEEE Journal on Selected Areas in Communications*, **16**(8), 1998, 1437–1450.

[38] *Flash-OFDM, OFDM Based All-IP Wireless Technology*, IEEE C802.20-03/16, www.flarion.com.

[39] G. D. Forney and G. Ungerböck, "Modulation and coding for linear Gaussian channels", *IEEE Transactions on Information Theory*, **44**(6), 1998, 2384–2415.

[40] G. J. Foschini, "Layered space-time architecture for wireless communication in a fading environment when using multi-element antennas", *Bell Labs Technical Journal*, **1**(2), 1996, 41–59.

[41] G. J. Foschini and M. J. Gans, "On limits of wireless communication in a fading environment when using multiple antennas", *Wireless Personal Communications*, **6**(3), 1998, 311–335.

[42] M. Franceschetti, J. Bruck and M. Cook, "A random walk model of wave propagation", *IEEE Transactions on Antenna Propagation*, **52**(5), 2004, 1304–1317.

[43] R. G. Gallager, *Information Theory and Reliable Communication*, John Wiley and Sons, 1968.

[44] R. G. Gallager, "An inequality on the capacity region of multiple access multipath channels", in *Communications and Cryptography: Two Sides of One Tapestry*, 1994, Boston, Kluwer, pp. 129–139

[45] R. G. Gallager, "A perspective on multiaccess channels", *IEEE Transactions on Information Theory*, **31**, 1985, 124–142.

[46] S. Gelfand and M. Pinsker, "Coding for channel with random parameters", *Problems of Control and Information Theory*, **9**, 1980, 19–31.

[47] D. Gesbert, H. Bölcskei, D. A. Gore and A. J. Paulraj, "Outdoor MIMO wireless channels: Models and performance prediction", *IEEE Transactions on Communications*, **50**, 2002, 1926–1934.

[48] M. J. E. Golay, "Multislit spectrometry", *Journal of the Optical Society of America*, **39**, 1949, 437–444.

[49] M. J. E. Golay, "Static multislit spectrometry and its application to the panoramic display of infrared spectra", *Journal of the Optical Society of America*, **41**, 1951, 468–472.

[50] M. J. E. Golay, "Complementary sequences", *IEEE Transactions on Information Theory*, **7**, 1961, 82–87.

[51] A. Goldsmith and P. Varaiya, "Capacity of fading channel with channel side information", *IEEE Transactions on Information Theory*, **43**, 1995, 1986–1992.

[52] S. W. Golomb, *Shift Register Sequences*, Revised Edition, Aegean Park Press, 1982.

[53] U. Grenander and G. Szego, *Toeplitz Forms and Their Applications*, Second Edition, New York, Chelsea, 1984.

[54] L. Grokop and D. Tse, "Diversity–multiplexing tradeoff of the ISI channel", *Proceedings of the International Symposium on Information Theory*, Chicago, 2004.

[55] Jiann-Ching Guey, M. P. Fitz, M. R. Bell and Wen-Yi Kuo, "Signal design for transmitter diversity wireless communication systems over Rayleigh fading channels", *IEEE Transactions on Communications*, **47**, 1999, 527–537.

[56] S. V. Hanly, "An algorithm for combined cell-site selection and power control to maximize cellular spread-spectrum capacity", *IEEE Journal on Selected Areas in Communications*, **13**(7), 1995, 1332–1340.

[57] H. Harashima and H. Miyakawa, "Matched-transmission technique for channels with intersymbol interference", *IEEE Transactions on Communications*, **20**, 1972, 774–780.

[58] R. Heddergott and P. Truffer, *Statistical Characteristics of Indoor Radio Propagation in NLOS Scenarios*, Technical Report: COST 259 TD(00) 024, January 2000.

[59] J. Y. N. Hui, "Throughput analysis of the code division multiple accessing of the spread-spectrum channel", *IEEE Journal on Selected Areas in Communications*, **2**, 1984, 482–486.

[60] IS-136 Standard (TIA/EIA), Telecommunications Industry Association.

[61] IS-95 Standard (TIA/EIA), Telecommunications Industry Association.

[62] W. C. Jakes, *Microwave Mobile Communications*, Wiley, 1974.

[63] N. Jindal, S. Vishwanath and A. Goldsmith, "On the duality between multiple access and broadcast channels", *Annual Allerton Conference*, 2001.

[64] A. E. Jones, T. A. Wilkinson, "Combined coding error control and increased robustness to system non-linearities in OFDM", *IEEE Vehicular Technology Conference*, April 1996, pp. 904–908.

[65] R. Knopp, and P. Humblet, "Information capacity and power control in single cell multiuser communications", *IEEE International Communications Conference*, Seattle, June 1995.

[66] R. Knopp and P. Humblet, "Multiuser diversity", unpublished manuscript.

[67] C. Kose and R. D. Wesel, "Universal space-time trellis codes," *IEEE Transactions on Information Theory*, **40**(10), 2003, 2717–2727.

[68] A. Lapidoth and S. Moser, "Capacity bounds via duality with applications to multiple-antenna systems on flat fading channels", *IEEE Transactions on Information Theory*, **49**(10), 2003, 2426–2467.

[69] A. Lapidoth and P. Narayan, "Reliable communication under channel uncertainty", *IEEE Transactions on Information Theory*, **44**(6), 1998, 2148–2177.

[70] R. Laroia, T. Richardson and R. Urbanke, "Reduced peak power requirements in ofdm and related systems", unpublished manuscript, available at http://lthcwww.epfl.ch/papers/LRU.ps.

[71] R. Laroia, S. Tretter and N. Farvardin, "A simple and effective precoding scheme for noise whitening on ISI channels", *IEEE Transactions on Communication*, **41**, 1993, 1460–1463.

[72] E. G. Larsson, P. Stoica and G. Ganesan, *Space-Time Block Coding for Wireless Communication*, Cambridge University Press, 2003.

[73] H. Liao, "A coding theorem for multiple access communications", *International Symposium on Information Theory*, Asilomar, CA, 1972.

[74] L. Li and A. Goldsmith, "Capacity and optimal resource allocation for fading broadcast channels: Part I: Ergodic capacity", *IEEE Transactions on Information Theory*, **47**(3), 2001, 1082–1102.

[75] K. Liu, R. Vasanthan and A. M. Sayeed, "Capacity scaling and spectral efficiency in wideband correlated MIMO channels", *IEEE Transactions on Information Theory*, **49**(10), 2003, 2504–2526.

[76] T. Liu and P. Viswanath, "Opportunistic orthogonal writing on dirty-paper", submitted to *IEEE Transactions on Information Theory*, 2005.

[77] R. Lupas and S. Verdú, "Linear multiuser detectors for synchronous code-division multiple-access channels", *IEEE Transactions on Information Theory*, **35**(1), 1989, 123–136.

[78] V. A. Marčenko and L. A. Pastur, "Distribution of eigenvalues for some sets of random matrices", *Math USSR Sbornik*, **1**, 1967, 457–483.

[79] U. Madhow and M. L. Honig, "MMSE interference suppression for direct-sequence spread-spectrum CDMA", *IEEE Transactions on Communications*, **42**(12), 1994, 3178–3188.

[80] A. W. Marshall and I. Olkin, *Inequalities: Theory of Majorization and Its Applications*, Academic Press, 1979.

[81] K. Marton, "A coding theorem for the discrete memoryless broadcast channel", *IEEE Transactions on Information Theory*, **25**, 1979, 306–311.

[82] *MAXDET: A Software for Determinant Maximization Problems*, available at http://www.stanford.edu/~boyd/MAXDET.html.

[83] T. Marzetta and B. Hochwald, "Capacity of a mobile multiple-antenna communication link in rayleigh flat fading", *IEEE Transactions on Information Theory*, **45**(1), 1999, 139–157.

[84] R. J. McEliece and K. N. Sivarajan, "Performance limits for channelized cellular telephone systems", *IEEE Transactions on Information Theory*, **40**(1), 1994, 21–34.

[85] M. Médard and R. G. Gallager, "Bandwidth scaling for fading multipath channels", *IEEE Transactions on Information Theory*, **48**(4), 2002, 840–852.

[86] N. Prasad and M. K. Varanasi, "Outage analysis and optimization for multiaccess/V-BLAST architecture over MIMO Rayleigh fading channels", *Forty-First Annual Allerton Conference on Communication, Control, and Computing*, Monticello, IL, October 2003.

[87] A. Oppenheim and R. Schafer, *Discrete-Time Signal Processing*, Englewood Cliffs, NJ, Prentice-Hall, 1989.

[88] L. Ozarow, S. Shamai and A. D. Wyner, "Information-theoretic considerations for cellular mobile radio", *IEEE Transactions on Vehicular Technology*, **43**(2), 1994, 359–378.

[89] A. Paulraj, D. Gore and R. Nabar, *Introduction to Space-Time Wireless Communication*, Cambridge University Press, 2003.

[90] A. Poon, R. Brodersen and D. Tse, "Degrees of freedom in multiple-antenna channels: a signal space approach", *IEEE Transactions on Information Theory*, **51**, 2005, 523–536.

[91] A. Poon and M. Ho, "Indoor multiple-antenna channel characterization from 2 to 8 GHz", *Proceedings of the IEEE International Conference on Communications*, May 2003, pp. 3519–23.

[92] A. Poon, D. Tse and R. Brodersen, "Impact of scattering on the capacity, diversity, and propagation range of multiple-antenna channels", submitted to *IEEE Transactions on Information Theory*.

[93] B. M. Popović, "Synthesis of power efficient multitone signals with flat amplitude spectrum", *IEEE Transactions on Communication*, **39**, 1991, 1031–1033.

[94] G. Pottie and R. Calderbank, "Channel coding strategies for cellular mobile radio", *IEEE Transactions on Vehicular Technology*, **44**(3), 1995, 763–769.

[95] R. Price and P. Green, "A communication technique for multipath channels", *Proceedings of the IRE*, **46**, 1958, 555–570.

[96] J. Proakis, *Digital Communications*, Fourth Edition, McGraw Hill, 2000.

[97] G. G. Raleigh and J. M. Cioffi, "Spatio-temporal coding for wireless communication", *IEEE Transactions on Communications*, **46**, 1998, 357–366.

[98] T. S. Rappaport, *Wireless Communication: Principle and Practice*, Second Edition, Prentice Hall, 2002.

[99] S. Redl, M. Weber M. W. Oliphant, *GSM and Personal Communications Handbook*, Artech House, 1998.

[100] T. J. Richardson and R. Urbanke, *Modern Coding Theory*, to be published.

[101] B. Rimoldi and R. Urbanke, "A rate-splitting approach to the Gaussian multiple-access channel", *IEEE Transactions on Information Theory*, **42**(2), 1996, 364–375.

[102] N. Robertson, D. P. Sanders, P. D. Seymour and R. Thomas, "The four colour theorem", *Journal of Combinatorial Theory, Series B.* **70**, 1997, 2–44.

[103] W. L. Root and P. P. Varaiya, "Capacity of classes of Gaussian channels", *SIAM Journal of Applied Mathematics*, **16**(6), 1968, 1350–1393.

[104] B. R. Saltzberg, "Performance of an efficient parallel data transmission system", *IEEE Transactions on Communications*, **15**, 1967, 805–811.

[105] A. M. Sayeed, "Deconstructing multi-antenna fading channels", *IEEE Transactions on Signal Processing*, **50**, 2002, 2563–2579.

[106] E. Seneta, *Non-negative Matrices*, New York, Springer, 1981.

[107] N. Seshadri and J. H. Winters, "Two signaling schemes for improving the error performance of frequency-division duplex (FDD) transmission systems using transmitter antenna diversity", *International Journal on Wireless Information Networks*, **1**(1), 1994, 49–60.

[108] S. Shamai and A. D. Wyner, "Information theoretic considerations for symmetric, cellular, multiple-access fading channels: Part I", *IEEE Transactions on Information Theory*, **43**(6), 1997, 1877–1894.

[109] C. E. Shannon, "A mathematical theory of communication", *Bell System Technical Journal*, **27**, 1948, 379–423 and 623–656.

[110] C. E. Shannon, "Communication in the presence of noise", *Proceedings of the IRE*, **37**, 1949, 10–21.

[111] D. S. Shiu, G. J. Foschini, M. J. Gans and J. M. Kahn, "Fading correlation and its effect on the capacity of multielement antenna systems", *IEEE Transactions on Communications*, **48**, 2000, 502–513.

[112] Q. H. Spencer *et al.*, "Modeling the statistical time and angle of arrival characteristics of an indoor multipath channel", *IEEE Journal on Selected Areas in Communication*, **18**, 2000, 347–360.

[113] V. G. Subramanian and B. E. Hajek, "Broadband fading channels: signal burstiness and capacity", *IEEE Transactions on Information Theory*, **48**(4), 2002, 809–827.

[114] G. Taricco and M. Elia, "Capacity of fading channels with no side information", *Electronics Letters*, **33**, 1997, 1368–1370.

[115] V. Tarokh, N. Seshadri and A. R. Calderbank, "Space-time codes for high data rate wireless communication: performance, criterion and code construction", *IEEE Transactions on Information Theory*, **44**(2), 1998, 744–765.

[116] V. Tarokh and H. Jafarkhani, "On the computation and reduction of the peak-to-average power ratio in multicarrier communications", *IEEE Transactions on Communication*, **48**(1), 2000, 37–44.

[117] V. Tarokh, H. Jafarkhani and A. R. Calderbank, "Space-time block codes from orthogonal designs", *IEEE Transactions on Information Theory*, **48**(5), 1999, 1456–1467.

[118] S. R. Tavildar and P. Viswanath, "Approximately universal codes over slow fading channels", submitted to *IEEE Transactions on Information Theory*, 2005.

[119] E. Telatar, "Capacity of the multiple antenna Gaussian channel", *European Transactions on Telecommunications*, **10**(6), 1999, 585–595.

[120] E. Telatar and D. Tse, "Capacity and mutual information of wideband multipath fading channels", *IEEE Transactions on Information Theory*, **46**(4), 2000, 1384–1400.

[121] M. Tomlinson, "New automatic equaliser employing modulo arithmetic", *IEE Electronics Letters*, **7**(5/6), 1971, 138–139.

[122] D. Tse and S. Hanly, "Multi-access fading channels: Part I: Polymatroidal structure, optimal resource allocation and throughput capacities", *IEEE Transactions on Information Theory*, **44**(7), 1998, 2796–2815.

[123] D. Tse and S. Hanly, "Linear Multiuser Receivers: Effective Interference, Effective Bandwidth and User Capacity", *IEEE Transactions on Information Theory*, **45**(2), 1999, 641–657.

[124] D. Tse, "Optimal power allocation over parallel Gaussian broadcast channels", *IEEE International Symposium on Information Theory*, Ulm Germany, June 1997, p. 27.

[125] D. Tse, P. Viswanath and L. Zheng, "Diversity–multiplexing tradeoff in multiple access channels", *IEEE Transactions on Information Theory*, **50**(9), 2004, 1859–1874.

[126] A. M. Tulino, A. Lozano and S. Verdú, "Capacity-achieving input covariance for correlated multi-antenna channels", *Forty-first Annual Allerton Conference on Communication, Control and Computing*, Monticello IL, October 2003.

[127] A. M. Tulino and S. Verdú, "Random matrices and wireless communication", *Foundations and Trends in Communications and Information Theory*, **1**(1), 2004.

[128] S. Ulukus and R. D. Yates, "Adaptive power control and MMSE interference suppression", *ACM Wireless Networks*, **4**(6), 1998, 489–496.

[129] M. K. Varanasi and T. Guess, "Optimum decision feedback multiuser equalization and successive decoding achieves the total capacity of the Gaussian multiple-access channel", *Proceedings of the Asilomar Conference on Signals, Systems and Computers*, 1997.

[130] V. V. Veeravalli, Y. Liang and A. M. Sayeed, "Correlated MIMO Rayleigh fading channels: capacity, optimal signaling, and scaling laws", *IEEE Transactions on Information Theory*, 2005, in press.

[131] S. Verdú, *Multiuser Detection*, Cambridge University Press, 1998.

[132] S. Verdú and S. Shamai, "Spectral efficiency of CDMA with random spreading", *IEEE Transactions on Information Theory*, **45**(2), 1999, 622–640.

[133] H. Vikalo and B. Hassibi, *Sphere Decoding Algorithms for Communications*, Cambridge University Press, 2004.

[134] E. Visotsky and U. Madhow, "Optimal beamforming using tranmit antenna arrays", *Proceedings of Vehicular Technology Conference*, 1999.

[135] S. Vishwanath, N. Jindal and A. Goldsmith, "On the capacity of multiple input multiple output broadcast channels", *IEEE Transactions on Information Theory*, **49**(10), 2003, 2658–2668.

[136] P. Viswanath, D. Tse and V. Anantharam, "Asymptotically optimal waterfilling in vector multiple access channels", *IEEE Transactions on Information Theory*, **47**(1), 2001, 241–267.

[137] P. Viswanath, D. Tse and R. Laroia, "Opportunistic beamforming using dumb antennas", *IEEE Transactions on Information Theory*, **48**(6), 2002, 1277–1294.

[138] P. Viswanath and D. Tse, "Sum capacity of the multiple antenna broadcast channel and uplink-downlink duality", *IEEE Transactions on Information Theory*, **49**(8), 2003, 1912–1921.

[139] A. J. Viterbi, "Error bounds for convolution codes and an asymptotically optimal decoding algorithm", *IEEE Transactions on Information Theory*, **13**, 1967, 260–269.

[140] A. J. Viterbi, *CDMA: Principles of Spread-Spectrum Communication*, Addison-Wesley Wireless Communication, 1995.

[141] H. Weingarten, Y. Steinberg and S. Shamai, "The capacity region of the Gaussian MIMO broadcast channel", submitted to *IEEE Transactions on Information Theory*, 2005.

[142] R. D. Wesel, "Trellis Code Design for Correlated Fading and Achievable Rates for Tomlinson–Harashima Precoding", PhD Dissertation, Stanford University, August 1996.

[143] R. D. Wesel, and J. Cioffi, "Fundamentals of Coding for Broadcast OFDM", in *Twenty-Ninth Asilomar Conference on Signals, Systems, and Computers*, October 30, 1995.

[144] S. G. Wilson and Y. S. Leung, "Trellis-coded modulation on Rayleigh faded channels", *International Conference on Communications*, Seattle, June 1987.

[145] J. H. Winters, J. Salz and R. D. Gitlin, "The impact of antenna diversity on the capacity of wireless communication systems", *IEEE Transactions on Communication*, **42**(2–4), Part 3, 1994, 1740–1751.

[146] J. Wolfowitz, "Simultaneous channels", *Archive for Rational Mechanics and Analysis*, **4**, 1960, 471–386.

[147] P. W. Wolniansky, G. J. Foschini, G. D. Golden and R. A. Valenzuela, "V-BLAST: an architecture for realizing very high data rates over the rich-scattering wireless channel", *Proceedings of the URSI International Symposium on Signals, Systems, and Electronics Conference*, New York, 1998, pp. 295–300.

[148] J. M. Wozencraft and I. M. Jacobs, *Principles of Communication Engineering*, John Wiley and Sons, 1965, Reprinted by Waveland Press.

[149] Q. Wu and E. Esteves, "The cdma2000 high rate packet data system", in *Advances in 3G Enhanced Technologies for Wireless Communication*, Editors J. Wang and T.-S. Ng, Chapter 4, Artech House, 2002.

[150] A. D. Wyner, *Multi-tone Multiple Access for Cellular Systems*, AT&T Bell Labs Technical Memorandum, BL011217-920812- 12TM, 1992.

[151] R. Yates, "A framework for uplink power control in cellular radio systems", *IEEE Journal on Selected Areas in Communication*, **13**(7), 1995, 1341–1347.

[152] H. Yao and G. Wornell, "Achieving the full MIMO diversity–multiplexing frontier with rotation-based space-time codes", *Annual Allerton Conference on Communication, Control and Computing*, Monticello IL, October 2003.

[153] W. Yu and J. Cioffi, "Sum capacity of Gaussian vector broadcast channels", *IEEE Transactions on Information Theory*, **50**(9), 2004, 1875–1892.

[154] R. Zamir, S. Shamai and U. Erez, "Nested linear/lattice codes for structured multiterminal binning", *IEEE Transactions on Information Theory*, **48**, 2002, 1250–1276.

[155] L. Zheng and D. Tse, "Communicating on the Grassmann manifold: a geometric approach to the non-coherent multiple antenna channel", *IEEE Transactions on Information Theory*, **48**(2), 2002, 359–383.

[156] L. Zheng and D. Tse, "Diversity and multiplexing: a fundamental tradeoff in multiple antenna channels", *IEEE Transactions on Information Theory*, **48**(2), 2002, 359–383.

Index